3rd International Conference on Sustainable Agriculture and Food Security 2018

Innovation and Technology

IOP Conference Series: Earth and Environmental Science
Volume 334

Bandung, Indonesia
29 - 30 August 2018

ISBN: 978-1-7138-0134-4
ISSN: 1755-1307

Printed from e-media with permission by:

Curran Associates, Inc.
57 Morehouse Lane
Red Hook, NY 12571

Some format issues inherent in the e-media version may also appear in this print version.

Institute of Physics
Dirac House, Temple Back
Bristol BS1 6BE UK

Phone: 44 1 17 929 7481
Fax: 44 1 17 920 0979

techtracking@iop.org

Additional copies of this publication are available from:

Curran Associates, Inc.
57 Morehouse Lane
Red Hook, NY 12571 USA
Phone: 845-758-0400
Fax: 845-758-2633
Email: curran@proceedings.com
Web: www.proceedings.com

TABLE OF CONTENTS

BIOREMEDIATION OF AGRICULTURAL

CONSERVATION OF WATER AND SOIL RESOURCES

CULTIVATION TECHNOLOGY

ENVIRONMENT AND ENERGY

NUTRITION AND PHYSIOLOGY

PEST AND DISEASES CONTROL

POST-HARVEST AND FOOD TECHNOLOGY

SOCIAL-ECONOMY

SUSTAINABLE AGRICULTURE AND FOOD SYSTEMS

IOP Conf. Series: Earth and Environmental Science 334 (2019) 011001 doi:10.1088/1755-1315/334/1/011001

PREFACE

United Nation reported that the world population reached 7.5 billion last year and it is projected to become 9.1 billion in 2050 and so there will be about 21% more population than there were last year. In this situation, agricultural sector has a huge task to increase production to meet increasing demand for food, feed, fiber, and fuel.

However, agricultural sectors have currently faced a lot of obstacles. Climate change and massive uses of agricultural lands for housing and industries are few examples of those challenges. Public perceptions in considering health and safety aspects of food production for both consumption and the environment should also be taken into account. Here, innovation and technology are greatly important to be involved in dealing comprehensively with the issues of sustainable agriculture and food security along with addressing the issues across sectors including agriculture, biodiversity, conservation, and the policies.

In this situation, Universitas Padjadjaran had taken an opportunity to take a part in solving issues by conducting the first and the second International Conference on Sustainable Agriculture and Food Security which were successfully held in 2011 and 2015, respectively. It is a right time now to conduct the third conference on the same theme with a special emphasis on the use of innovation and technology.

On behalf of the committee, I greatly appreciate and warmly welcome all speakers and participants in particular to our respected overseas guests who have made their long journeys meaningful. In this special opportunity, I would like to express my sincere appreciation to the Rector of Universitas Padjadjaran, the Vice Rectors, Deans of Faculty of Agriculture, Faculty of Agricultural Industry and Technology, and Faculty of Fisheries and Marine Sciences, for their sincere attention and collaboration as well as to all sponsors for their cooperation.

Sincerely yours

Prof. Husmy Yurmiati

IOP Conf. Series: Earth and Environmental Science **334** (2019) 011001 doi:10.1088/1755-1315/334/1/011001

CONFERENCE PHOTOGRAPHS

SPONSOR OR FUNDING ACKNOWLEDGEMENTS

The Third ICSAFS was initiated by four faculties in cluster agro-complex in Universitas Padjadjaran. We were very glad that we got a lot of support from all steering committee and directorate research, community service and innovation.

We would like to thank to all sponsor Petrokimia Gresik, BRI, AGRO, Biotropia and Heryaki powder for all your support to our conference.

IOP Conf. Series: Earth and Environmental Science **334** (2019) 011001 doi:10.1088/1755-1315/334/1/011001

LISTS OF COMMITTEES

STEERING COMMITTEE

1. Vice Rector for Cooperation Universitas Padjadjaran
2. Dean of Faculty of Agriculture Universitas Padjadjaran
3. Dean of Faculty of Animal Husbandry Universitas Padjadjaran
4. Dean of Faculty of Fisheries and Marine Science Universitas Padjadjaran
5. Dean of Faculty of Agricultural Industrial Technology Universitas Padjadjaran
6. Director of Research, Community Service and Innovation Universitas Padjadjaran

ORGANIZING COMMITTEE

Chairperson : Prof. Dr. Ir. Husmy Yurmiati, M.S.
Vice Chairperson : Novi Mayasari, S.Pt., M.Sc., Ph.D

Secretary :
- Syariful Mubarok, S.P., M.Sc., Ph.D.
- Diky Ramdani,S.Pt.,M.Anim.St., Ph.D

Finance :
- Dr. Eulis Tanti Marlina, S. Pt., M.P.
- Vira Kusumadewi, S.P., M.Sc., Ph.D.
- Apong Sandrawati, S.P., M.Si.

Organizer Committee (OC) :
- Dr. Emma Rochima, S.Pi., M.Si.
- Dr. Iman Hernaman, M.Si
- Wahyu K. Sugandi, STP., M.Si.
- Ichsan Nurul Bari, SP., M.Si., Ph.D.
- Lizah Khairani, S.Pt., M.T. M.Agr.

Academic Committee (AC) :
- Dr. rer.pol Ernah, SP., M.Si
- Dr. Souvia Rahimah, STP., M.Sc.
- Dr. Agr. Mochamad Arief Soleh, S.P., M.Agr.Sc.
- Dr. Jajang Gumilar, S.Pt., M.M.
- Rusky Intan Pratama, STP., M.Si.
- Yuniar Mulyani, S.P., M.Si.

Logistic Committee (LC) :
- Dr. Heni Indrijani, S.Pt., M.Si.
- Ibnu Bangkit, S.Pi., M.Si
- Dr. Dwi Rustam Kendarto. S.Si., M.T.
- Ir. Evi Liviawaty, M.P.
- M. Rudyansyah Ismail, S.Pi., M.Si.

Information and Communication Committee (ICC) :
- Dr. Sophia Dwiratna N.P., STP., MT.
- Dr. Gemilang Lara Utama S., S.Pt., M.I.L.
- Ahmad Thoriq, STP., M.Si.
- Sulistyodewi Nur Wiyono, SP., MP.

IOP Conf. Series: Earth and Environmental Science **334** (2019) 011001 doi:10.1088/1755-1315/334/1/011001

Secretariat :

- Andry Pratama, S.Pt., MP.
- Raden Febrianto Christi, S.Pt., M.S.
- Fauzi Reza Umara
- Urip Rosani, S.Pt., M.Si.
- Nurhaeni, S.Sos., M.M.
- Bety Nubaniah, S.IP.

EDITORIAL

- Syariful Mubarok, S.P., M.Sc., Ph.D.
- Novi Mayasari, S.Pt., M.Sc., Ph.D.
- Dicky Ramdani, S.Pt., M.Sc., Ph.D.
- Vira Kusumadewi, S.P., M.Si., Ph.D.
- Dr. Souvia Rahimah, STP., M.Sc.

SCIENTIFIC COMMITTEE

1. Novi Mayasari, PhD (Universitas Padjadjaran, Indonesia)
2. Diky Ramdani, PhD (Universitas Padjadjaran, Indonesia)
3. Dr. Iman Hernaman (Universitas Padjadjaran, Indonesia)
4. Dr. Eulis Tanti Marlina (Universitas Padjadjaran, Indonesia)
5. Prof. Lovita Adriani (Universitas Padjadjaran, Indonesia)
6. Dr. Marina Sulystiawati (Universitas Padjadjaran, Indonesia)
7. Dr. Jajang Gumilar (Universitas Padjadjaran, Indonesia)
8. Dr. Unang Yunasaf (Universitas Padjadjaran, Indonesia)
9. Prof. Koesmajadi Suradi (Universitas Padjadjaran, Indonesia)
10. Dr. Hasan Hadiana (Universitas Padjadjaran, Indonesia)
11. Prof. Tuti Widjastuti (Universitas Padjadjaran, Indonesia)
12. Dr. Siti Darodjah (Universitas Padjadjaran, Indonesia)
13. Dr. Heni Indrijani (Universitas Padjadjaran, Indonesia)
14. Syariful Mubarok, S.P., M.Sc., Ph.D. (Universitas Padjadjaran, Indonesia)
15. Dr. rer.pol Ernah, SP., M.Si (Universitas Padjadjaran, Indonesia)
16. Dr. Agr. Mochamad Arief Soleh, S.P., M.Agr.Sc. (Universitas Padjadjaran, Indonesia)
17. M. Kadapi, S.P., M.Sc., Ph.D. (Universitas Padjadjaran, Indonesia)
18. Dr. Mira Ariyanti, S.P., M.P (Universitas Padjadjaran, Indonesia)
19. Nadia N. Kamaluddin, S.P., M.Sc., Ph.D. (Universitas Padjadjaran, Indonesia)
20. Anne Nurabity, Ph.D. (Universitas Padjadjaran, Indonesia)
21. Eliana Wulandari, Ph.D. (Universitas Padjadjaran, Indonesia)
22. Prof. Tarkus Suganda, Ph.D. (Universitas Padjadjaran, Indonesia)
23. Nono Carsono, Ph.D. (Universitas Padjadjaran, Indonesia)
24. Bambang Nurhadi, Ph.D. (Universitas Padjadjaran, Indonesia)
25. Adi Nugraha, Ph.D. (Universitas Padjadjaran, Indonesia)
26. Yusup Hidayat, Ph.D. (Universitas Padjadjaran, Indonesia)
27. Dr. rer. nat. Yusuf Effendi. (Universitas Al Azhar, Indonesia)
28. Yonathan Asikin, Ph.D. (University of Tsukuba, Japan)
29. Deden Derajat Matra, Ph.D. (Bogor Agriculture University, Indonesia)

PRODUCTION TEAM

1. Raden Febrianto Christi, M.S.

IOP Conf. Series: Earth and Environmental Science **334** (2019) 011001 doi:10.1088/1755-1315/334/1/011001

2. Apong Sandrawati, M.P.
3. Muhamad Abdillah Hasan Qanit, S.P.
4. Ully Ngesti Pratiwi
5. Fajar Edy Maretno Sitanggang
6. Laila Tien Nurjanah
7. Siti Azzahra A. Noor
8. Aisyah Ananda Nugraha
9. Mutia Nurulita
10. Erna Ainal Mardiah

Peer review statement

All papers published in this volume of *IOP Conference Series: Earth and Environmental Science* have been peer reviewed through processes administered by the proceedings Editors. Reviews were conducted by expert referees to the professional and scientific standards expected of a proceedings journal published by IOP Publishing.

Test Performance and Economic Analysis of Straw Chopper Machine

W Sugandi[1], T Herwanto[1], Handarto[1] and Juliya[1]

[1] Department of Agriculture and Biosystem Engineering Faculty of Agricultural Industrial Technology, Universitas Padjadjaran, Sumedang, Indonesia

E-mail: sugandiwahyu@gmail.com

Abstract. The usage of rice straw as an alternative basic compost fertilizer is needed. Rice straw processing into compost fertilizer through stage chopper. Chopper is the size reducing stage that it is easily decomposed. The chopper processing can be done by using straw chop machine. This research aims for performance test and economic analysis. Research use descriptive analysis. The results of performance test of straw chop machine have actual capacity is 103.32±12.068 kg/h, engine efficiency is 79.59%, power requirement with load and without load are 1.1±0.067 HP and 0.698±0.109 HP, specific load 7.94 W/kg, chopper yield is 94.33±0.97%, the percentage of chopper chunks is 83.55±0.04%, engine noise level 76.6±0.5 dB, and vibration level on the framework, hopper and chopper cylinder are 18±4.9 mm/s^2, 53.94±1.3 mm/s^2 and 21.6±4.6 mm/s^2. Based on the result of economic analysis of rice straw chop machine have the basic cost is Rp.46,159,621.- with gross income Rp.55,792,800.-/year. The breakeven point has been reached when the machine has chopped 41,078 kg of rice straw. This rice straw chopper machine has met the economic feasibility criteria of NPV>0 that is Rp.52,821,891. BC Ratio>1 with value is 1.17. IRR value> MARR interest rate (9%) with IRR 102%.

1. Introduction

Rice straw is a waste of agricultural products that can easily be encountered. Most farmers only consider rice straw as a waste that has no economic value. Only a small portion of straw is used by farmers while the rest is left or even burned on the ground directly [1].

Straw has a variety of benefits that can improve the welfare of farmers. In many countries (including China, Korea, and Japan), these wastes are used for compost, animal feed, mulch for vegetables and fruits, household fuels, craft materials, rooftops, and place of mushroom growing [1].

According to [2], national rice production in 2015 reached 75.40 million tons. While in West Java, in the same year rice production reached 11.39 million tons. Based on rice production data, the amount of straw that can be produced in Indonesia is estimated to reach 113.1 million tons, while for West Java area is estimated to reach 17.1 million tons.

The production of agricultural waste, especially rice straw in Indonesia is quite large as previously described. Therefore, it is very important for the community, especially farmers to develop and utilize the existence of such waste in order to provide greater benefits. People, especially farmers, must open their horizons to the utilization of the agricultural wastes they produce. Utilization of waste has been widely applied by the community to household wastes and other organic wastes. However, there are still few farmers who utilize rice straw waste to the maximum. Straw has many benefits that can be developed, among others as fuel, animal feed, cage or barn floor, packing of agricultural materials (eg eggs), building materials (roofs, walls, floors), mulch, and handicrafts.

Most of the farmers do the burning of rice straw produced after harvest without seeing the great benefits of the straw itself. The burning of this straw aims to accelerate the preparation or processing of the soil for the next planting period. However, the burning of straw can result in some nutrients being lost, especially volatile nutrients and other nutrients that are not available on the soil.

To overcome this, one way of utilizing the right rice straw for farmers is to use it as a raw material for composting. In the process of making straw into compost, of course, the process of enumeration of straw into smaller sizes. It is necessary to expand the contact surface of the material with microorganisms. The smaller the size of the raw materials used, the process of material decomposition will be faster. The process of enumeration of straw is of course not be done manually so it requires a simple mechanical tool capable of producing rice straw reproduce.

Agricultural Machinery and Equipment Laboratory, Department of Agricultural Engineering and Biosystem of Padjadjaran University has developed rice straw machine. However, the machine has not been specifically tested in terms of performance or economic analysis. Performance testing and economic analysis of the engine is necessary to know the quality, quantity, and efficiency of the machine work made. In addition, this test can also provide information on the deficiencies and advantages of the machine. Therefore, research on performance test and economic analysis of rice straw machine has been made.

The purpose of this research is to test the performance of rice straw enactment machine which include: theoretical capacity, actual capacity, machine efficiency, actual power requirement, specific power, reedmen enumeration, percentage of length of enumeration, noise, vibration, and physical work load.

This research was also conducted to determine the economic feasibility analysis on rice straw enumerator which includes calculation of fixed cost, variable cost, Break Even Point, Pay Back Period, Net Present Value, Benefit Cost Ratio, and Internal Rate of Return.

2. Methodology

This research was conducted in September 2017 until January 2018. This research took place in Agricultural Machinery and Equipment Workshop, Faculty of Agricultural Industrial Technology, Padjadjaran University. The research method used is descriptive analysis method. Descriptive analysis is doing measurement, observation and calculation of structural components and performance / functional machine [3]. The data obtained then analyzed so as to get an idea of the performance of straw chopper machine which in the end can provide an image of the feasibility of the machine. The research can be seen in Figure 1.

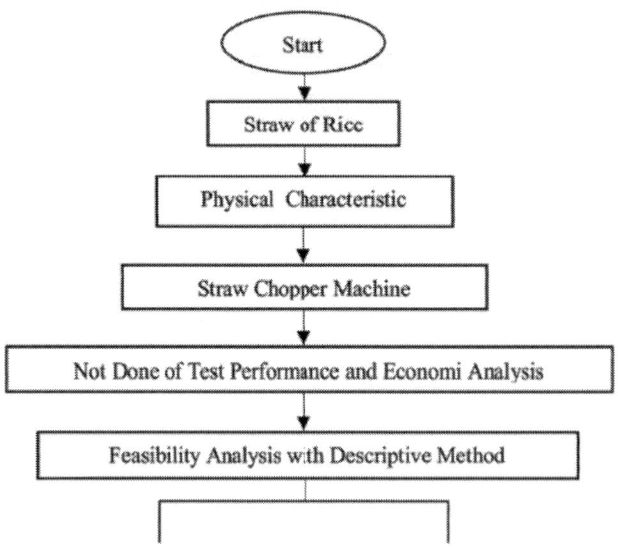

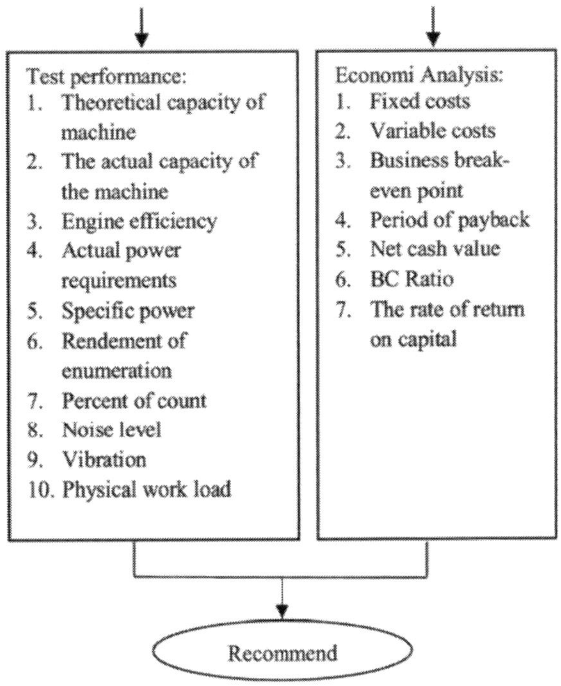

Figure 1. Research Approach

3. Tools and materials

The tools and materials used for this research are:

3.1. Tools.

Tools used are clamp meter, tachometer, vibration meter, soundlevel meter, laptop, meter, stopwatch, scales.

3.2. Material

The material used in this research is rice straw as much as 250 kg.

4. Research stages

The research stages explanation as follows:

4.1. Performance Test

Engine performance tests are performed to evaluate engine capabilities while operating. The performance test parameters to be tested are: a. theoretical capacity; b. actual capacity; c. engine efficiency; d. actual power requirements; e. specific energy; f. rendemen enumeration; g. percentage of length of count; h. noise levels, and vibrations. Performance test is conducted based on [4] standard test of Chopper Machine of Organic Fertilizer Raw Material - Quality Requirement and Test Method. Testing is done with 5 repetitions for each repetition requiring ± 50 kg of rice straw.

4.2. Economic Analysis

To find out the economic feasibility of straw chopper machine is done calculation of economic analysis. Economic analysis is done is to calculate the cost component of the series of rice straw enumeration process. The stage of economic analysis of straw chopper machine includes: a. calculation of basic costs; b. the amount of acceptance; c. determination of business break-even point; and D. economic feasibility analysis by using Net Cash Value Method (NPV), Method of Benefit and Cost Ratio (BCR), and Method of Return on Capital (IRR).

IOP Conf. Series: Earth and Environmental Science **334** (2019) 012001 doi:10.1088/1755-1315/334/1/012001

4.3. Data Calculation

The data calculation is done by using equation of formula as follows:

4.3.1. Theoretical Capacity [5]..

$$Kt = (\rho j \times Lt \times At \times \lambda k \times Nc) / (6 \times 103) \quad (1)$$

Kt = Theoretical capacity of machine (kg/h)
ρj = Density of straw cage (kg/m3)
Nc = Rotating speed of enumerator cylinder (rpm)
At = Area of enumeration area (m2)
Lt = The length of the theoretical slice (m)
λk = Number of blades

4.3.2. Actual Capacity.

$$K_a = (M_{bc} / t_p) \times 3600 \quad (2)$$

Ka = Actual capacity of enumeration (kg/h)
Mbc = The total mass of material that comes out within a certain time (kg)
Tp = Time specified to accommodate material (s)

4.3.3. Machine Efficiency.

$$\eta = Ka / Kt \quad (3)$$

η = Machine efficiency
Ka = Actual capacity (kg/h)
Kt = Theoretical capacity (kg/h)

4.3.4. Theoretical Power Requirement.

$$Pt = (2\pi \times Mt \times Nm) / 60 \quad (4)$$

Pt = The required power of the electric motor (W)
Mt = torque moment (Nm)
Nm = Round on motor fuel shaft (rpm)

4.3.5. Specific Power.

$$Psp = Pa / Ka \quad (5)$$

Psp = Specific power of enumeration (kW/kg)
Pa = Actual power of enumeration (kW)
Ka = Actual capacity of machine (kg)

4.3.6. Rendemen Enumeration.

$$Rp = Mbo / Mbi \quad (6)$$

Rp = Rendemen Enumeration
Mbo = Mass of paddy rice straw out (kg)
Mbi = Mass of incoming rice straw (kg)

4.3.7. Percentage Length of Haystack.

$$Ppk = (Mc1 / (Mci + Mc2)) \times 100\% \quad (7)$$

Ppk = Percentage of length of haystack (%)
Mc1 = Straw pulp mass less than 50 mm (kg)
Mc2 = Straw pulp mass more than 50 mm (kg)

4.3.8. Fixed Cost.

$$BT = D + H + M + I + L \quad (8)$$

BT = Fixed cost (Rp/year)
D = Depreciation Cost (Rp)

H = Place or building cost (Rp)
M = Cost of repair and maintenance (Rp)
I = Tax expense (Rp)
L = Capital interest (Rp)

4.3.9. Biaya Variabel.

$$BV = Bop + Bl \quad (9)$$

BV = Variable cost (Rp/h)
Bop = Operator fee (Rp/h)
Bl = Electricity cost (Rp/h)

4.3.10. Break Even Point (BEP).

$$BEP = BT / (HP - BV) \quad (10)$$

BEP = Business break-even point (kg)
HP = Product price (Rp)

4.3.11. Payback Period (PBP).

$$PBP = \text{Total Investasi} / \text{Keuntungan} \quad (11)$$

PBP = Payback Period (year)

4.3.12. Net Present Value (NPV).

$$NPV = \sum PV\ \text{income} - \sum PV\ \text{spending} \quad (12)$$

NPV = Net Present Value

4.3.13. Benefit Cost Ratio (BCR).

$$BCR = (\sum PV\ \text{income} / \sum PV\ \text{spending}) \geq 1 \quad (13)$$

BCR = Benefit cost ratio

4.3.14. Rate of Return on Capital (IRR).

$$IRR = i1 - (NPV1 \times ((i2 - i1) / (NPV2 - NPV1))) \quad (14)$$

IRR = rate of return on capital
I1 = 1st interest rate
NPV1 = Net Present Value at 1st interest rate
I2 = 2nd interest rate
NPV2 = Net Present Value at the 2nd interest rate

5. Result and discussion

5.1. Physical Characteristics of Rice Straw

Characteristic testing aims to determine the physical characteristics of paddy straw to be chopped. Characteristics of the materials used affects the results of rice straw enumeration. Testing of material characteristics include measurement of water content, bulk density, length and diameter of rice straw. Based on the result of the test, it was found that the water content of wet straw base was 74.28 ± 1.068%, straw density of 160.57 ± 2.04 kg / m3, average straw length 689 ± 59.3 mm. diameter in rice straw 1.2 ± 0.3 mm, and outer diameter of 3.2 ± 0.6 mm. Measurement of moisture content of the substance before the enumeration is to determine the suitability of the water content of the material to be used for standard enumeration. Bulk density measurements are used to calculate the capacity of the chamber, theoretical capacity, and theoretical driving power requirements.

5.2. Performance Test Engine

Performance test is to find out how big the work of machine in quantity and quality. The engine performance test covers the actual engine capacity aspect, engine efficiency, rendemen enumeration, actual power requirements, specific power, percentage of length of churn, physical workload, vibration level, and noise level. Testing is done based on [4] about Choper Machine Organic Fertilizer Raw Material - Quality and Test Requirements. Prior to the test, the calculation of power and engine capacity theoretically. The theoretical capacity of the straw chop machine is 129.8 kg/h can be seen in Appendix 4. Based on the calculation data in Appendix 7, the magnitude of the theoretical power without load is 0.7 horse power [HP] while the theoretical power when there is a load that is 2.03 HP.

Based on the results of performance test, the rice straw machine has an actual capacity of 103.32 ± 1.068 kg/h, the machine efficiency is 79.59%, the engine power requirement when there is load and no load 1.1 ± 0.067 HP and 0.698 ± 0.109 HP , specific power of 7.94 W/kg, yield of 94.33 ± 0.97% counting, percentage of length of length 83.55 ± 0.04%, machine noise level at 76.6 ± 0.5 dB, and vibration level at frame, hopper and process i.e. 18 ± 4.9 mm /s2, 53.94 ± 1.3 mm /s2 and 21.86 ± 4.6 mm/s2.

The specification of the performance test results can be determined by comparing the test standard with the results obtained through measurement and calculation. If the result meets, it will be known how well this machine enumerates. Parameters tested to find out the specification of straw chopper performance test can be seen in Table 1.

Table 1. Specification of Straw Counter Test Results

Parameter	Test Standard	Test Result	Terms	Information
Actual Capacity	Class A with capacity <500 kg/h	103.32 kg/h	Test results <Terms of test	conform
Round knife blade	1200-1300 rpm	1243 rpm	-	conform
Engine efficiency	80%	79.59%	Test results >Terms of test	un conform
Rendemen Enumeration	65%	94.33%	Test results >Terms of test	conform
Percentage length of count	80%	83.55%	Test results >Terms of test	conform
Noise level	90 dB	76.6 dB	Test results <Terms of test	conform
Level of vibration	≤ 4.5 mm/s	47.48 mm/s	Test results <Terms of test	un conform

5.3. Economic Analysis

Economic analysis is conducted to determine the level of investment profit of rice straw enumerator machine when the machine is used for enumeration. Economic analysis includes the calculation of basic cost (fixed cost and variable cost), break even analysis, and economic feasibility analysis including Net Present Value, Internal Rate of Return, Benefit and Cost (Benefit Cost Ratio Analysis), as well as the Method of Payback Period (Pay Back Period Analysis).

Based on economic analysis of rice straw machine is feasible to be operated with the application in the field in accordance with the calculation data. The amount of basic cost Rp. 46,159,621,- with annual gross revenue of Rp. 55,792,800,-. The break point of the straw enumeration was reached when the machine had chopped 41,078 kg of rice straw. This paddy straw machine has met the economic feasibility criteria of NPV> 0 that is Rp. 52,821,891,-. The magnitude of BC Ratio> 1 with value 1.17. IRR value> MARR interest rate (9%) with IRR 102%.

Economic feasibility analysis is used to determine whether the investment of rice straw enumerator machine is profitable or not. This can be seen from the calculation of NPV, BC Ratio, IRR, and payback period. Criteria of economic feasibility include if NPV> 0, BC Ratio> 1, IRR> MARR interest rate [6].

IOP Conf. Series: Earth and Environmental Science **334** (2019) 012001 doi:10.1088/1755-1315/334/1/012001

Of the three criteria are compared with the calculation results of economic feasibility analysis of the machine Table 2.

Table 2. Calculation Result of Economic Feasibility Analysis of Enumerator Machine

Parameter	Requirement	The Calculation Results	Information
NPV	NPV > 0	52,821,891	Feasible
BC Ratio	BC Ratio > 1	1.17	Feasible
IRR	IRR > interest rate MARR	102%	Feasible
Pay Back Period	Fast is better	0.93 year	Feasible

6. Conclusion

Based on the results of performance test, the rice straw machine has an actual capacity of 103.32 ± 1.068 kg / hour, the machine efficiency is 79.59%, the engine power requirement when there is load and no load 1.1 ± 0.067 HP and 0.698 ± 0.109 HP , specific power of 7.94 W / kg, yield of 94.33 ± 0.97% counting, percentage of length of length 83.55 ± 0.04%, machine noise level 76.6 ± 0.5 dB, and vibration level at frame, hopper and process ie 18 ± 4.9 mm/s2, 53.94 ± 1.3 mm/s2 and 21.86 ± 4.6 mm/s2. Overall performance of this jrami counter machine is good enough.

Based on economic analysis of rice straw machine is feasible to be operated with the application in the field in accordance with the calculation data. The amount of basic cost Rp. 46,159,621,- with annual gross revenue of Rp. 55,792,800,-. The break point of the straw enumeration was reached when the machine had chopped 41,078 kg of rice straw. This paddy straw machine has met the economic feasibility criteria of NPV> 0 that is Rp. 52,821,891, -. The magnitude of BC Ratio> 1 with value 1.17. IRR value> MARR interest rate (9%) with IRR 102%.

References

[1] Makarim A K, Sumarno, Suyamto 2007 Jerami Padi Pengelolaan dan Pemanfaatan *Pusat Penelitian dan Pengembangan Tanaman* Badan Penelitian dan Pengembangan Pertanian Bogor.

[2] Badan Pusat Statistik 2016 Produksi Tanaman Padi *Pusat Biro Statistik Indonesia* Jakarta.

[3] Kuncoro 2003 Metode Riset untuk Ekonomi dan Bisnis *Erlangga* Jakarta.

[4] SNI 7580 2010 Mesin Pencacah (Choper) Bahan Baku Pupuk Organik-Syarat Mutu dan Cara Uji *Badan Standarisasi Nasional*.

[5] Srivastava 1993 Engineering Principle of Agricultural Machine *Society of Agricultural Engineering* USA.

[6] Kastaman R 2004 Ekonomi Teknik untuk Pengembangan Kewirausahaan *Pustaka Giratuna dan ELOC-UNPAD* Bandung.

IOP Conf. Series: Earth and Environmental Science **334** (2019) 012002 doi:10.1088/1755-1315/334/1/012002

Actual and Prediction Body Weight Performance from Birth Until of 18 Months of Age in Dairy Cattle Friesian Holstein

D S Tasripin[1], H Indrijani[1], A Anang[1], H Purnamasari[1] and E D Nanda[1]

[1]Faculty of Animal Husbandry, Padjadjaran University, Sumedang Bandung

E-mail : dstasripin@gmail.com

Abstract. The aim of this research is to get the body weight performance and growth pattern of Friesian Holstein dairy cows from birth to 18 months. This study used monthly weight data of dairy Friesian Holstein cows from birth to 18 months (508 head). Obtained standard equation of body weight by using logistic regression model ŷ=363.53⁄=1+ 〖6.57e〗 ^(-0.21 (x))) (Se = 14.78 ; r = 0.989), inflection point occurs at age 10 months (body weight 232.4 kg). The actual body weight of calves at birth until the age of 2 months is lower than the weight of the alleged body. The actual body weight of the calf aged 3 to 6 months is higher than the body weight of the estimation and the actual body weight of the calf above the estimation weight standard. The actual body weight of 7-9 months is still higher than the weight but at 10-12 months the actual body weight is lower than the expected weight, (puberty phase) that will affect growth. Month at 13-16 actual body weight is lower than the weight of the estimation body, as well as the previous month that is the first period of lust and puberty.

1. Introduction

Dairy cattle diversity plays an important role in the development of dairy cattle breeding. Currently some dairy farms have been managed in the form of commercial dairy farms and some are still smallholder farms, unstructured populations and have not used a targeted breeding system. The development of dairy cattle breeding has considerable potential in reducing the dependence on imports of dairy products and the importation of dairy cattle. Therefore, the government is obliged to foster and create a business climate that supports the business of dairy cattle breeding so that it can produce livestock seeds to meet the needs of quantity and quality according to the standard, in addition to providing facilities for increasing the added value of seed products.

Balai Besar Pembibitan Ternak Unggul (BBPTU) Dairy Cattle Baturraden is a government-owned hall that plays a role of quality dairy cows and get dairy cows with optimal milk production performance for the benefit of national dairy farmers. This hall has a recording system for the purpose of selecting superior seeds and improving milk production performance that can serve as the basis of information to conduct selection on livestock.

Efforts to select quality seeds to obtain dairy cows with high milk production continue to be pursued. Standards that can be used as the basis for selection are body weight, age of first child and milk production performance. Quality of seed and milk production performance is influenced by external factors and internal factors. These factors include environmental, physiological conditions of livestock, body weight achievement in accordance with the age of livestock, genetics, feed (quality and quantity), and maintenance management. The management of cattle raising from birth to adult livestock is important. It is very influential on the reproduction and production of milk.

Achievement of body weight from birth to age 18 months and ideal age of the first child will enable the birth of calves with a good birth weight and produce milk with optimal production. The weight of the livestock body is usually assumed with survival and growth performance with milk productivity during lactation, so evaluation of the growth pattern, the age of the first child and the milk production performance becomes very significant in the management of dairy farming. Purpose and objective this

IOP Conf. Series: Earth and Environmental Science **334** (2019) 012002 doi:10.1088/1755-1315/334/1/012002

research is to get the performance of body weight and growth pattern of Fries Holland dairy cattle from birth to 18 months old.

2. Materials and Methods

This study used monthly weight record of dairy cattle at the BBPTU-HPT Baturraden. The data used are of monthly weight records from 2014-2017 as many as 8,696 records derived from 213 head. Data on monthly weight of dairy Friesian Holstein cattle at birth to 18 months the observed variables included Body Weight (kg) from birth weight to 18 months, cattle number, age of cattle, time of cattle birth and daily gain body weight

2.1 Statistical Analysis

Standard body weight and growth curve Body weight data obtained was analyzed by logistic regression equation model with the following formula:

$$\hat{y} = \frac{a}{1+be^{-cx}} \qquad (1)$$

Information :
ŷ = Estimation Body weight (Kg)
x = cattle age (month)
a, b, c = coefficients
e = natural number (2,718)

The accuracy of the model is determined by the standard error (Se), the correlation coefficient (r) between the estimated value with the actual value, the minimum body weight standard, and the estimated weight gain of the body to find the inflection point and the inflection weight [1] with the following formula:

a. Standard error (Se) = √KTr es , the smaller the value of Se, the better
b. The correlation coefficient (r) = √(R^2)
   R2 is obtained from the formula R2 = (JK reg)/(JK total), the greater the value of r, the more accurate the estimator.
c. Standard Minimum Body Weight
   The minimum body weight standard is calculated by reducing the average body weight estimate (ŷ) by the standard error (Se) = ŷ – Se
d. Daily Gain Body Weight Estimation
   The weight gain of the estimation body is useful for obtaining the infectious point of age and body weight by using quadratic equation with the following formula:

$$y = a + bx + cx^2 \text{ and } x = \frac{-b}{2c} \qquad (2)$$

Information :
y = Added Body Weight Estimation (kg)
x = Inflection Point (age)
a, b, c = coefficients

3. Result and Discussions

3.1 Result

Table 1. Standard Weight Body (actual, estimation and minimum standards) and Added Body Weight Estimation Friesian Holstein cow

Month	Body Weight Actual (kg)	Body Weight Estimation (kg)	Standard Minimum (kg)	Average daily Gain Estimation (kg)
0	39.51	55.88	47.81	-
1	54.68	65.44	57.37	0.32
2	69.98	76.26	68.20	0.36
3	88.62	88.39	80.33	0.40
4	108.97	101.82	93.76	0.45
5	127.35	116.51	108.45	0.49
6	144.06	132.35	124.29	0.53
7	157.55	149.17	141.11	0.56
8	170.57	166.75	158.69	0.59
9	185.48	184.82	176.76	0.60
10	199.98	203.07	195.01	0.61
11	216.94	221.19	213.13	0.60
12	232.24	238.87	230.81	0.59
13	249.54	255.83	247.77	0.57
14	267.17	271.85	263.79	0.53
15	283.44	286.74	278.67	0.50
16	299.17	300.39	292.32	0.45
17	317.29	312.73	304.67	0.41
18	331.52	323.77	315.71	0.37

3.2 Discussion

Based on the calculation results obtained equation logistic regression model $\hat{y} = \frac{363{,}53}{1+6{,}57e^{-0{,}21\,(x)}}$ and has standard value error (Se) 14.78 and correlation (r) between the estimated weight of the body with a very high actual body weight value is 0.99 close to 1 or 100%, which means that the greater the value of r then the body weight of the guesswork is more appropriate.

Table 1 shows that the actual body weight of calves at birth to age 2 months is lower than the Estimation body weight, but the results of this study is higher than the research conducted [2, 3], namely birth weight to age 2 months in the tropics (36 kg, 45 kg and 62 kg [4, 1, 5] that the normal birth weight in Indonesia is 35-40 kg, so the birth weight until the age of 2 months is still within the normal range. factors that cause low birth weight are genetics of the male and the parent, the age and condition of the cow's body when the conception, quality and maturity of the ovum when fertilized, the number of calves born, the nutrients of growth, the presence of infectious diseases and stress level from the parent [2, 6]. The preparation of rations for dairy cattle should be balanced, in accordance with the number and proportion of dairy cattle nutrient needs in 24 hours worthy condition [7, 8,4]. The actual body weight of the calf aged 3 to 6 months is higher than the body weight of the Estimation and the actual body weight of the calf above the Estimation weight standard. The greater the actual body weight indicates the maintenance and feeding management in BBPTU-SP Baturraden done well.

The actual body weight of 7-9 months is still higher than the Estimation weight but at 10-12 months the actual body weight is lower than the expected weight, this is because the cow begins to enter the puberty phase so that it will affect growth. At puberty marked estrus and ovulation, the cow becomes more active as a result of reduced nutritional intake so as not to meet the nutritional adequacy for growth. which states that puberty or adult sex occurs before the adult body is reached, so that the cow will be more active at the time of preparation of the first mate so that the consumption of feed becomes lower than the cattle that have grown the body. Growth after puberty is slow, parallel to the increase in adult body weight [2, 9, 10, 11].

Month 13-16 actual body weight is lower than the weight of the Estimation body, as well as the previous month that is the first period of lust and puberty. Based on calculation result obtained by quadratic equation that is y = 0,2264 + 0,0739x + 0,037x^2 and x= (-(0,0739))/(2(0,0037)), so got infection point at age 10 months with a body weight of 232.4 kg.

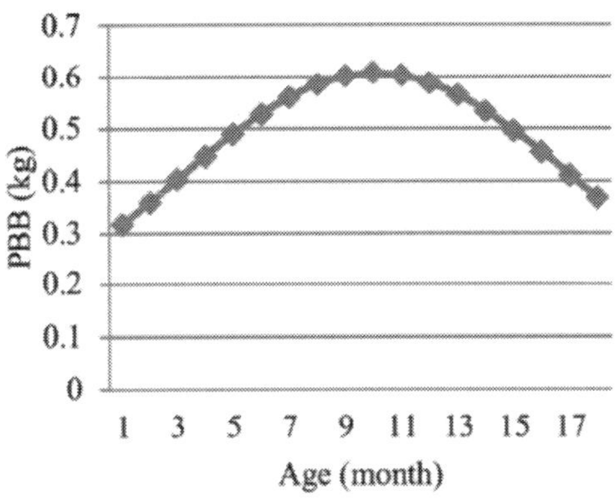

Figure 1. Growth Curve Body Weight Estimation and Inflection Point Dairy Cow Friesian Holstein

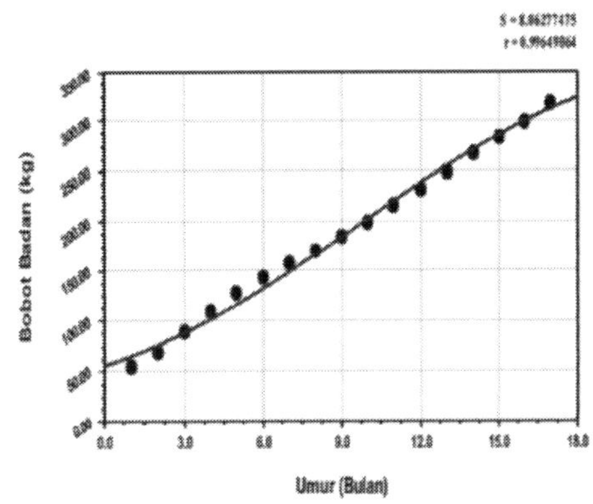

Figure 2. Friesian Holstein Dairy Cattle Growth Curve

Figure 1. shows the weight gain of the Estimation body and the inflection point seen at the age of 10 months with an infectious weight of 232.4 kg. Age 10-12 months is the period of puberty, where before the growth curve will take place rapidly and after puberty growth will slow again [1, 12]. the logistic regression model growth curve has advantages in accuracy and has good biological interpretation in explaining inflection points and inflection weights [1, 11, 13, 5].

Figure 2. shows that the growth curve of Friesian Holstein female cow in BBPTU-SP Baturraden is shaped sigmoid curve according to logistic model. The pattern of actual body weight growth and Estimation body weight does not very much, it is shown by mutual its coincided actual body weight line and weight of the estimation body on each month. Friesian Holstein cattle growth curve from conception to adult with a sigmoid or "S" curve has accelerated to the inflection point and undergoes a slowing phase until it reaches adulthood. The rate of growth is accelerated to the point of inflection, usually associated with puberty is the body weight of 10-12 months [10, 5] which is the most decisive weight for further livestock productivity [6, 13], because after puberty the growth will slow back to reach adulthood. The inflection point occurs in the virgin after a rapid phase growth indicating that the livestock is experiencing puberty. The age of puberty in cows under normal feeding conditions is 9 months, but may range from 5 to 15 months [4, 12, 5].

4. Conclusion

Based on the results of the analysis and discussion can be drawn conclusion that:
The standard weight equation $\hat{y} = 363{,}53/1 + 6{,}57e^{-0{,}21\,(x)}$ with Se = 14,78 and r = 0,989, follows the sigmoid curve according to the logistic regression model. Based on quadratic equation that is $y = 0{,}2264 + 0{,}0739x + 0{,}037x^2$ and $x = \frac{-(0{,}0739)}{2(0{,}0037)}$, got inflection point that is 10 months age with body weight 232. 4 kg

Acknowledgement
The authors would like to say thank you to Academic Leadership Grant Padjadjaran University that fund this research and BBPTU HPT Baturraden for provided the data.

References

[1] Brown J E, Fitzhugh H A, Cartwright T C 1976 A comparison of non linear models for describing weight-age relationship of cattle *J. Anim. Sci.* **42** 810-811.
[2] Bath D L, Dickinson F N, Tucker H A, Appleman R D 1985 Dairy Cattle ; Principles, Practice, Problem and Profits *3rd Edition Lea ang Febiger* Philadelphia.
[3] McDowell 1989 Dairying In Tropical Environments *International Seminar On Frisian Holstein* Bandung Indonesia.
[4] Makin M 1990 Study of the Nature of Growth, Reproduction and Production of Milk Dairy Milk Sahiwal *Sahiwal X Fries Holland Bogor Agricultural Institute* Bogor
[5] Van Amburgh M E, Galton D M, Bauman D E, Everett R W, Fox D G, Chase L E, Erb H N 1998 Effects of three prepubertal body growth rates on performance of Holstein heifers during first lactation *J. Dairy Sci.* **81** 527–538.
[6] Lawrence T L J, Fowler V R 2002 Growth of Farm Animals 2nd edition *CABI Publishing* CABI International Wallingford Oxon Ox 10 8de UK.
[7] Eckles C H, Comb W B, Macy H 1980 Milk and Milk Product 5th edition *The McMillan Company* Newyork.
[8] Ensminger M E, Tyler H D 2006 Dairy Cattle Science 4th edition *Pearson Education Inc.* Upper Saddle River New Jersey.
[9] Ettema J F, Santos J E P 2004 Impact of Age at Calving on Lactation, Reproduction, Health, and Income on First-Parity Holsteins on Commercial Farms *J. Dairy Sci.* **87** 2730–2742
[10] Heinrichs A J 1993 Raising dairy replacements to meet the needs of the 21st century *J. Dairy Sci.* **76** 3179–3187
[11] Heinrichs A J, Vazquez-Anon M 1993 Changes in first lactation dairy herd improvement records *J. Dairy Sci.* **76** 671–675.
[12] Soeharsono 1980 Growth *Physiology* Department Faculty of Animal Husbandry University Padjadjaran.
[13] Mazzini A R A, Muniz J A, Aquino L H D, Silva F F E 2003 Growth curve analysis for Hereford cattle males *Cienc. Agrotec. Lavras.* **27** 1105-1112.

IOP Conf. Series: Earth and Environmental Science 334 (2019) 012003 doi:10.1088/1755-1315/334/1/012003

Potential Use of Banana Peels Waste at Different Ripening Stages for Sheep Feeding on Chemical, Tannin, and *In Vitro* Assessments

D Ramdani[1], I Hernaman[2], A A Nurmeidiansyah[1], D Heryadi[1] and S Nurachma[1]

[1]Department of Animal Production, The Faculty of Animal Husbandry, Universitas Padjadjaran, Jatinangor Campus, Sumedang 45363, Indonesia
[2]Department of Nutrition and Feed Technology, The Faculty of Animal Husbandry, Universitas Padjadjaran, Jatinangor Campus, Sumedang 45363, Indonesia

Email: diky.ramdani@unpad.ac.id

Abstract. This research aimed to investigate the potential use of banana peels waste for sheep feeding. Completely randomized design was used to compare the chemical, phenol, and tannin compositions between raw and ripened banana peels from 4 different varieties (Ambon, Muli, Nangka, Kapas). Moreover, a 2x4 factorial design was used to test the main effects of ripening stages (raw, ripened) and doses (10, 20, 30, 40%) of Ambon banana peels on *in vitro* dry matter and organic matter digestibility (DMD and OMD, %), ammonia (NH_3, mM), volatile fatty acids (VFA, mM), pH, and total gas production (TGP, ml). Raw banana peels contained less ($P<0.05$) dry matter (DM, %) and total digestible nutrients (TDN, %) but it contained higher ($P<0.05$) crude fiber (CF, %) and gross energy (GE, kcal/kg) compared with ripened banana peels. Raw and ripened banana peels contained considerable amount of total phenols (TP, %) and total tannins (TT, %) although being not different ($P>0.05$). Based on *in vitro* assessments, raw Ambon banana peels had lower ($P<0.001$) DMD, OMD, and VFA but higher NH_3 ($P<0,001$) than ripened ones. Adding Ambon banana peels from 10 to 40% replacing roughage in the diet increased ($P<0.001$) DMD, OMD, and VFA but decreased ($P<0,001$) NH_3. Both raw and ripened Banana peels have the potential for sheep feeding based on their chemical and *in vitro* assessments.

1. Introduction

Banana is a tropical fruit and easily grown in tropical countries such as Indonesia. The use of banana either for direct consumption or processed products i.e. banana chips creates waste products including banana peels. This waste can be used as an alternative feed or roughages for sheep. So far, one of the challenges that Indonesian sheep farmers have faced is lack of high quality roughages since many pastures have massively changed into crops, housing, and industries, especially in Java Island. Conversely, the market of lambs in this highly populated island is high, not only for daily consumption but also for Muslim-related-celebrations called Aqeqah and Iedul Adha. Wina [1] stated that banana peels waste has not been optimally used yet for ruminant feeding although it contains organic biomass including protein, fiber, minerals, and other nutrients which are needed by ruminants. Inside the rumen, banana peel waste will be fermented in anaerobic condition by rumen microbes resulting in more available nutrients to be absorbed in the small intestine and distributed in the blood for the needs of growth and production such as meat and milk [2,3].

As a tropical fruit, banana should be able to adapt to tropical climate characterized by high temperature, rain, and humidity with so many varieties of insect predators, bacteria, and fungi so that most tropical plants contain plant bioactive compounds to protect them from predators but they are not directly involved in a biochemical process of plant metabolite systems [4]. Banana peels are supposed to contain considerable high in plant bioactive compounds to protect the inside of banana from predators' attack. Tannin is one of the bioactive compounds that is possibly contained in banana peels so that the waste of banana peels can be used as a source of natural feed additive to improve the productivity of ruminant animals. This research was done to (1) analyze proximate, CF, nitrogen-free extract (NFE, %), TDN, GE, mineral, TP, and TT in different local varieties of banana peel (Ambon, Muli, Nangka, and Kapas) at two ripening stages (raw, ripened) and (2) tested the effect of raw and ripened Ambon banana peels at different doses on *in vitro* digestibility, fermentation profiles, and TGP.

2. Materials and methods

2.1. *Banana peels' sampling preparation*

The sample of each variety of banana peels, either Ambon, Muli, Nangka, or Kapas, was obtained from whole bananas bought in 3 different markets: Pasar Induk Caringin Bandung, Pasar Induk Gedebage Bandung, and Pasar Tanjungsari Sumedang. Pasar Induk means a region main wet market while Pasar is a common wet market. At our laboratory, each of banana was peeled at raw and ripened stages and each peel was then sun dried, ground, and filtered at 20 mesh. After becoming dried powder, each sample was put into coded plastic and ready for the next analyses.

2.2. *Nutrients, phenols, and tannins analyses*

Proximate (DM, OM, CP, ash, fats), CF, NFE, TDN, GE, Ca, and P analyses were done at Laboratory of Feed Chemical and Nutrition, Faculty of Animal Husbandry, Universitas Padjadjaran using the method of AOAC [5]. Meanwhile, TP and TT analyses have followed the Folin-Ciocalteu procedure utilizing a spectrophotometer as explained in Makkar [6] using a tannic acid as a standard equivalent. Most unit measurements of analyses were calculated in % DM except DM (% fresh weight) and GE (kcal/kg).

2.3. *Experimental diets*

The experimental diets were formulated following a 2x4 factorial arrangement to test the effect of 2 ripening stages (raw and ripened) and 4 doses (10, 20, 30, 40%) of Ambon banana peels to replace grass in the diets on in vitro degradability and fermentation profiles. The experimental diets were iso-protein (10.7% ± 0.163) and iso-TDN (61.5% ± 0.503) and they consisted of raw or ripened Ambon banana peels, field grass, and a mixed concentrate. The nutrient contents of raw and ripened Ambon banana peels can be seen in the results of this study while field grass contained 21.9 % DM, 9.33% ash, 9.10% CP, 28.8% CF, 4.72% Fats, 48.1% NFE, 60.6% TDN, 2994 kcal.kg GE, 0.33% Ca, and 0.18% P. Concentrate contained 92,7% DM, 14.2% ash, 13.8% CP, 18.8% CF, 9.4% Fats, 43.9% NFE, 65.2% TDN, 3499 kcal.kg GE, 0.87% Ca, and 0.61% P.

2.4. *DMD and OMD Measurements*

In vitro DMD and OMD were measured using the procedure of Tilley and Terry [7]. Briefly, about 1 gram of each diet sample was put inside a fermenter tube and 40 ml of artificial McDougall saliva and 10 ml fresh rumen fluid added. CO_2 was flown into the fermenter tube to keep an anaerobic condition and the tube was closed with a ventilated rubber and incubated in a water bath at about 39 °C for 48 hours. During incubation, each tube was gently shaken every 3 hours (phase 1). After 48 hours incubation, 0.2 ml $HgCl_2$ was added into each tube and it was centrifuged at 4000 rpm for 15 minutes to separate supernatant and its residue. About 50 ml Pepsin-HCL 10% solution (pepsin activity 1:10,000) was added into the tube containing the residue (without supernatant) and it was incubated again at 39 °C for 48 hours in an aerobic condition (phase 2). After that, the residue was filtered using Whatman paper No. 41 and rinsed with distilled water to make sure all the residue went down into the filter paper. Filtered residue was then dried in an oven at 105°C for 24 hours, cooled in a desiccator, and weighed for calculating DMD. Dried residue was finally heated at 550 °C in a furnace, cooled in a desiccator, and weighed for calculating OMD.

2.5. VFA and NH3 measurements
A similar in vitro method of Tilley dan Terry [7] was used in VFA and NH3 measurements. After 24 hours incubation, about 0.2 ml $HgCl_2$ was added into each of fermenter tube and the tube was then centrifuged at 4000 rpm for 15 minutes to separate supernatant and the residue. The supernatant was then taken off and stored in a bottle in a refrigerator for the next analysis of VFA and NH3, respectively. About 5 mL of each supernatant was placed in a steam heated distillation tube and 1 mL H_2SO_4 15% added and closed the tube tightly. Heated steam forced VFA in the cooling tube to be condensed and stored in an Erlenmeyer tube containing 5 mL NaOH 0,5 N until the volume reached 200-300 mL. After that, 2 drops of phenolphthalein (pp) indicator was added into the tube and titrated with HCl 0,5 N. Titration was ended at the beginning of color change from red to no color. While for NH_3 analysis, 1 mL supernatant was placed in the left side of Conway cup and 1 mL Na_2CO_3 was placed in the right side. Small cup in the middle was filled with 1 mL boric acid with red methyl and green brome cresol indicators. Conway cup was tightly closed with vasline-lubricated cap and shaken so that the supernatant could be well mixed with Na_2CO_3. Then, it was kept at room temperature for 24 hours to let the ammonia gas bounded with boric acid. After 24 hours, it was titrated with H_2SO_4 0.006 N until the color changed into red.

2.6. pH and TGP measurements
TGP was measured following the procedure of Menke and Steingass [8] using measurable syringes at 24 hours of in vitro incubation. After incubation, the in vitro solution was measured for pH using a calibrated pH meter.

2.7. Data tabulation and statistical analysis
Each sample of banana feels was analyzed for nutrients, TP, and TT in duplicate while the data were presented in average. The different of chemical compositions between raw and ripened banana peels was analyzed using one-way ANOVA with different varieties of banana peels as replicates (n=4). Meanwhile, the data of a 2x4 factorial design with 5 replicates in each experimental unit, to test the main effects of ripening stages (raw, ripened) and doses (10, 20, 30, 40%) of Ambon banana peels on in vitro digestibility and fermentation profiles, were analyzed using two-way ANOVA. Here, Minitab 16 Statistical Software was used for all the statistical analyses.

3. Results and discussion
Table 1. describes nutrients, TP, and TT contents (% DM except % fresh weight for DM and kcal/kg for GE) in different varieties of banana peels at raw and ripened stages. At raw stage, there were averagely not different among varieties of banana peels (Ambon, Muli, Nangka, and Kapas) in DM, OM, ash, GE, and Ca. However, Kapas banana peels had averagely the highest CP than others while Nangka banana peels had averagely the highest fats, Ca, and P than others. Muli banana peels had averagely the highest NFE but the lowest CP compared to others whilst Ambon banana peels had averagely the highest TP and TT but the lowest fats than others. At ripened stage, the average OM, ash, TDN, and GE contents among varieties of banana peels seemed the same. However, Ambon banana peels had averagely the highest DM, TP, and TT but the lowest Ca and P in comparison with others while Kapas banana peels had averagely the highest CP, Ca, and P but the lowest NFE, TP, and TT.

Table 2. shows the different of nutrients, TP, and TT compositions (% DM except % fresh weight for DM and kcal/kg for GE) between raw and ripened banana peels. Raw banana peels had significantly ($P<0.01$) less DM and TDN but higher CF ($P<0.01$) and GE ($P<0.05$) in comparison with ripened banana peels. Raw banana peels had averagely higher fats but less TP than ripened banana peels although they were not significant ($P = 0.073$ and 0.085, respectively).

Table 3. outlines the means *in vitro* digestibility, fermentation profiles, and TGP, for only the main effects of raw and ripened stages of Ambon banana peels and their doses (10, 20, 30, 40%) of inclusion as these were mostly significant (P<0.001) but not their interactions. Diets containing ripened Ambon banana peels had significantly higher DMD, OMD, and VFA but lower NH_3 and TGP compared with those containing raw Ambon banana peels. Meanwhile, adding Ambon banana peels into a sheep diet up to 40% could significantly increase DMD, OMD, and VFA but reduce NH_3. Interestingly, the highest TGP was reached by adding Ambon banana peels up 30% but not 40%. There was no significant different between ripening stages and doses on pH.

It can be seen that banana peels has considerable DM, OM, ash, CP, CF, fats, NFE, TDN, GE, Ca, P, TP, and TT that can be beneficial for sheep feeding but their contents in the peels might be different depend upon varieties and ripening stages. Mohapatra et al. [9] added that banana peels also rich in polyunsaturated fatty acids such as linoleic and α-linolenic, essential amino acids such as leucine, valine, phenylalanine, threonine, starch and pectine containing glucose, galactose, arrabinose, rhamnose, xylose, celulose and hemicelulose. Good nutrient qualities of Ambon banana peels might be the reason why adding the peels into a sheep diet up to 40% in replacing grass could significantly increase DMD, OMD, and VFA. Ambon banana peels have also lower CF but higher GE compared with grass which was used in the current study.

Meanwhile, decreased NH_3 production due to Ambon banana peels addition is supposedly caused by considerable tannin content in the peels. Generally, tannins can reduce the solubility and rumen degradability of most leaf proteins due to their potential binding with proteins. Consequently, they can reduce rumen NH_3 production and increase the availability of by-pass protein and non-ammonia nitrogen (N) supply to be absorbed in the small intestine [10,11,12]. Although NH_3 is an important source of N for rumen microbes, its over or fast production may exceed the ability of microbes to utilize it. This can lead to an excessive NH_3 supply that, after absorption through the rumen wall, can enter the blood stream, liver, and eventually excreted in urine as an N waste [13,14]. Except increasing potential by-pass protein as previously explained, tannins can also lower CH_4 production by slowing the inter-species transfer of H_2 into methanogenic bacteria and thus depress their growth [12,15], improve animal health through their antioxidant properties to prevent bloat and break protein-rich cells of nematodes [16,17], and increase the rumenic acid and polyunsaturated fatty acids (PUFA), and decrease saturated fatty acids (SFA) in ruminant products such as meat and milk through altered bio-hydrogenation by changing the microbial population in the rumen [18,19,20].

Table 1. Nutrients, TP, and TT compositions (% DM or otherwise stated) in different varieties of banana peels at raw and ripened stages.

Nutrients (% DM)	Raw banana peels				Ripened banana peels			
	Ambon	Muli	Nangka	Kapas	Ambon	Muli	Nangka	Kapas
DM (%)	37.9	38.1	38.3	34.5	67.3	50.7	58.8	54.0
OM	88.8	89.5	87.7	86.8	88.2	88.8	87.1	84.9
Ash	11,2	10.5	12.3	13.3	11.8	11.3	12.9	15.1
CP	7.82	6.36	7.81	10.8	8.54	8.92	8.98	11.7
CF	20.4	18.9	18.7	20.7	16.2	16.0	13.7	15.2
Fats	1.87	2.13	2.51	2.02	1.85	2.00	1.62	16.3
NFE	58.7	62.1	58.7	53.2	61.6	61.9	62.8	56.3
TDN	56.7	58.7	58.4	56.0	61.1	62.1	63.0	60.4
GE (kcal/kg)	3472	3369	3494	3445	3386	3329	3368	3321
Ca	0.63	0.65	0.78	0.67	0.57	0.62	0.69	0.73
P	0.19	0.23	0.31	0.29	0.18	0.23	0.21	0.28
TP	5.76	4.80	4.83	4.57	6.49	5.88	5.74	5.08
TT	5.32	4.34	4.47	4.19	5.85	5.30	5.11	4.61

DM, dry matter; OM, organic matter; CP, crude protein; CF, crude fibre; NFE, nitrogen free extract; TDN, total digestible nutrients; GE, gross energy; TP. Total phenols; TT, total tannins.

Table 2. The different of nutrients, TP, and TT contents (% DM or otherwise stated) between raw and ripened banana peels.

Banana Peels	DM %	OM	Abu	CP	CF	Fats	NFE	TDN	GE kkal/kg	Ca	P	TP	TT
Raw	38.1	88.2	11.8	8.20	19.7	2.13	58.2	57.4	3445	0.68	0.26	4.99	4.58
Ripened	63.0	87.2	12.8	9.54	15.3	1.78	60.6	61.7	3351	0.65	0.23	5.80	5.22
SEM	2.62	0.75	0.750	0.84	0.54	0.12	1.66	0.60	2.21	0.04	0.02	0.20	0.26
P value[a]	0.001	0.398	0.398	0,303	0.001	0.073	0.337	0.003	0.024	0.562	0.420	0.085	0.127

[a]Mean values were significantly different at $P<0.05$. DM, dry matter; OM, organic matter; CP, crude protein; CF, crude fibre; NFE, nitrogen free extract; TDN, total digestible nutrients; GE, gross energy; TP. Total phenols; TT, total tannins; SEM, standard error of mean.

Table 3. Means for the main effects of raw and ripened stages of Ambon banana peels and their doses (10, 20, 30, 40%) of inclusion into sheep diets on *in vitro* digestibility, fermentation profiles, and TGP.

In vitro tests	Ripe stages (n=20)		Doses (%) (n=10)				SEM and Significances		
	Raw	Ripened	10	20	30	40	Stages	Doses	Interaction
DMD (%)	62.5	66.8	60.7^D	63.0^C	65.5^B	69.3^A	0.13^{***}	0.18^{***}	0.25^{NS}
OMD (%)	51.4	55.1	48.2^D	51.6^C	54.8^B	58.4^A	0.12^{***}	0.17^{***}	0.24^{**}
NH_3 (mM)	6.33	5.38	6.71^A	6.12^B	5.55^C	5.05^D	0.06^{***}	0.09^{***}	0.13^{NS}
VFA (mM)	135	142	119^D	136^C	144^B	156^A	0.81^{***}	1.15^{***}	1.62^{**}
pH	6.54	6.49	6.51	6.57	6.48	6.50	0.04^{NS}	0.05^{NS}	0.07^{NS}
TGP (ml)	244	193	179^C	245^A	244^A	207^B	4.60^{***}	6.50^{***}	9.20^{*}

Here *, ** and *** represent significant differences between means at P<0.05 or P<0.01 or P<0.001 respectively; SEM, standard error of mean; n, number of replicates; DMD, *in vitro* dry matter digestibility; OMD, *in vitro* organic matter digestibility; NH_3, ammonia; VFA, volatile fatty acids; TGP, total gas production.

4. Conclusion

Banana peels are considerably rich in nutrients, phenols and tannins and their compositions may be different depending upon banana varieties and the levels of ripe. Based on *in vitro* assessments, Ambon banana peels can be included in a diet of sheep up to 40% to increase degradability, VFA production, and potential by-pass protein, but *in vivo* experiments to further prove their potential need to be done.

Acknowledgement

Authors would like to thank Directorate General of Higher Education, Ministry of Research, Technology, and Higher Education and Directorate of Research and Services, Universitas Padjadjaran for funding this research via PUPT Scheme 2015/2016.

References

[1] Wina E 2001 Banana plant as ruminant feed *Wartazoa* **11**: 20-27.
[2] Dijkstra J 1994 Production and absorption of volatile fatty acids in the rumen *Livest. Prod. Sci.* **39**: 61-69.
[3] McDonald P, Edwards R A, Greenhalgh J F D, Morgan, C A, Sinclair L A, Wilkinson R G 2011 Animal Nutrition 7th edition England UK *Pearson Education Limited.*
[4] Patra A K, Saxena J 2009 Dietary phytochemicals as rumen midifiers: a review of the effects on microbial populations *Antoine van Leeuwenhoek* **96**: 363-375.
[5] AOAC 2005 Animal Feed (Chapter 4) Official methods of analysis of AOAC international ed. by Horwitz W, Latimer G W, Thiex N J W (Chapter Ed.) *AOAC International*, Suite 500 481 North Frederick Avenue Gaithersburg Maryland 20877-2417 USA.
[6] Makkar H P S 2003 Quantification of tannins in tree and shrub foliage: a laboratory manual *Kluwer Academic Publishers* Dordrecht The Netherlands.
[7] Tilley J M A, Terry R A 1963 A two-stage technique for the in vitro digestion of forage crops *J. British Grassland Society* **18**: 104-111.
[8] Menke K H, Steingass H 1988 Estimation of the energetic feed value obtained from chemical analysis and in vitro gas production using rumen fluid *Anim. Res. Develop.* **28**: 7-55.
[9] Mohapatra D, Mishra S, Sutar N 2010 Banana and its by-product utilisation: an overview *J. Sci. Ind. Res.* **69**: 323-329.
[10] McSweeney C S, Palmer B, McNeill D M, and Krause D O 2001 Microbial interactions with tannins: nutritional consequences for ruminants *Anim. Feed Sci. Technol.* **91**: 83-93.
[11] Min B R, Barry T N, Attwood G T, McNabb W C 2003 The effect of condensed tannins on the nutrition and health of ruminants fed fresh temperate forages: a review *Anim. Feed Sci.Technol.* **106**: 3-19.
[12] Mueller-Harvey I 2006 Review unravelling the conundrum of tannins in animal nutrition and

health *J. Sci. Food Agric.* **86**: 2010-2037.
[13] Attwood G T, Klieve A V, Ouwerkerk D, Patel B K C 1998 Ammonia-Hyperproducing Bacteria from New Zealand Ruminants *Appl. Environ. Microbiol.* **64**: 1796-1804.
[14] Szumacher-Strabel M, Cieslak A 2010 Potential of phytofactors to mitigate rumen ammonia and methane production *J. Anim. Feed Sci.* **19**: 319-337.
[15] Boadi D, Benchar C, Chiquette J, Masse D 2004 Mitigation strategies to reduce enteric methane emissions from dairy cows: Update review *Can. J. Anim. Sci.* **84**: 319-335.
[16] Ishihara N, Akachi S 1997 Chemistry and applications of green tea ed. by Yamamoto T, Juneja L R, Chu C D, Kim M Boca Raton Florida *CRC Press* pp. 137-144.
[17] Ishihara N, Chu D C, Akachi S, Juneja L R 2001 Improvement of intestinal microflora balance and prevention of digestive and respiratory organ diseases in calves by green tea extracts *Livest. Prod. Sci.* **68**: 217-229.
[18] Vasta V, Mere M, Serra A, Scerra M, Luciano G, Lanza M, Priolo A 2009 Metabolic fate of fatty acids involved in ruminal biohydrogenation in sheep fed concentrate or herbage with or wihout tannins *J. Anim. Sci.* **87**: 2674-2684.
[19] Vasta V, Yanez-Ruiz D R, Mere M, Serra A, Luciano G, Lanza M, Biondi L, Priolo A 2010 Bacterial and protozoal communities and fatty acid profile in the rumen of sheep fed a diet containing added tannins *Appl. Environ. Microbiol.* **76**: 2549-2010.
[20] Wood T A, Ramos-Morales E, McKain N, Shen X, Atasoglu C, Wallace R J 2010 *Chrysanthemum coronarium* as a modulator of fatty acid biohydrogenation in the rumen *Anim. Feed Sci. Technol.* **161**: 28-37.

IOP Conf. Series: Earth and Environmental Science **334** (2019) 012004 doi:10.1088/1755-1315/334/1/012004

Comparison of Five Different Lactation Curve Models to Estimate Milk Yield of Friesian Holstein Cows at BBPTU HPT Baturraden

E D Nanda[1], L B Salman[2], H Indrijani[2], D S Tasripin[2] and A Anang[2]

[1] Student of Faculty of Animal Husbandry, Padjadjaran University, Indonesia
[2] Faculty of Animal Husbandry, Padjadjaran University, Indonesia

E-mail : erinnedwi@gmail.com

Abstract. The objective of this study was to compare the accuracy of five different lactation curve models to estimate 305 days milk yields in Friesian Holstein (FH) dairy cow at BBPTU HPT Baturraden. Appropriate lactation curve could provide useful information for genetic breeding programs, herd nutritional management, decision taking on the culling cows and milk production simulation systems. The analysis was carried out on a total of 8.167 milk yield records based on test day (TD) from first to fourth lactation from 2014 to 2017. The models compared were Wood, Yadav, Ali-Schaeffer, Wilmink, and Guo and Swalve models. The results showed that the Ali-Schaeffer's model was the best model to predict milk production. It is known from the coefficient of determination at first, second, third, and fourth lactation respectively were 0.99156, 0.99074, 0.98985, 0.96706, correlation between predicted and observed values (r) were 0.99577, 0.99536, 0.99491, 0.98339, and Se were 0.30, 0.37, 0.38, 0.63.

1. Introduction

Lactation curve refers to a graphic representation of the ratios between milk production and lactation time starting at calving and can be studied using mathematical models [1,2]. Mathematical models is a tools that can mimic fundamental processes of milk production in different scenarios [3]. The curves of animal lactation facilitate the prediction of the total yield from partial yield, which aids in early animal disposal and breeding evaluation when the lactation records of the progeny are incomplete. The lactation curve models used the function parameters to estimate lactation curve characteristics with economic significance, including early yield, peak yield, time from onset to lactation peak, persistency and total lactation yield [2]. Appropriate curve models provide useful information for genetic breeding programs, herd nutritional management, decision taking on the culling cows and milk production simulation systems [1,4,5,6,7].

The first mathematical model of the lactation curve in dairy cattle was developed by Brody *et al.* [8], who proposed a decreasing exponential function to fit the declining phase of the lactation pattern in dairy cattle [3]. Brody model could not represent an increase in production in the beginning of lactation and underestimating in the peak yield [1]. Brody model was useful to outline only in declining phase of the curve [9]. Therefore, this model was not suitable to describe the entire lactations of cows that were highly specialized in milk production, but the model was adequately fitted for low milk production crossbred animals reared in tropical conditions [1]. Since than, studies on lactation curve modelling have experienced a rather waving pattern and numerous mathematical models have been developed by specific needs of the dairy industry [3,10,11].

The use of lactation models become more popular after Wood's model [12]. The model was proposed to use incomplete gamma function to describe the curve and it obtained a better fit at ascending and descending phases of the curve, than the models used until that time [9]. Another important feature of Wood's model was its simplicity and a reduced number of parameters [12], but limitations of the Wood model are also well known [3]. Overestimation of daily milk yield in the first part of the curve, underestimation around and after the peak have been reported [13].

After Wood's model, there have been numerous studies to develop linier and nonlinear lactation curve models. In the linier, models proposed by Ali and Schaeffer [14] and Guo and Swalve [15], while in the nonlinear, models proposed by Yadav [16] and Wilmink [17] have been developed. The models are considered worthy to be investigated, because they have been proven to have a better fit than Wood in several studies. But these model need to be tested before it used, because not all models are fit to be used in Indonesia. Using the different set data and number of lactation with previous study, the objective of this study is to compare the accuracy of five different lactation curve models on test days milk yields in Friesian Holstein dairy cows at BBPTU HPT Baturraden.

2. Materials and Methods

The data were obtained from nucleus breeding centre of dairy cattle, BBPTU HTP Baturraden in Central Java, Indonesia from 2014 to 2017. The study used 2.354, 2.711, 1.903, and 1.199 records for first, second, third, and fourth lactation, respectively. Each lactation consisted of 11 point of measurement from day 8 to 305 with interval time of 30 days. The models evaluated are presented in Table 1.

Table 1. Five Different Model were Evaluated

Researcher	Model
Wood (1967)	$y_t = at^b\,2{,}7183^{-ct}$
Yadav (1977)	$y_t = \frac{t}{a + bt + ct^2}$
Ali-Schaeffer (1987)	$y_t = a+b\left(\frac{t}{305}\right)+c\left(\frac{t}{305}\right)+d\,ln\left(\frac{305}{t}\right)+f\,ln^2\left(\frac{305}{t}\right)$
Wilmink (1987)	$y_t = a + b.\,2{,}7183^{-0{,}05t} + ct$
Guo and Swalve (1995)	$y_t = a + b\sqrt{t} + cln(t)$

In all functions, t was DIM, a, b, c, d, and f were parameters to be estimated, and γ was milk yield at DIM t. The parameters were estimated with the help of proc-nonlin within SAS 9.0. The best fit was compared with the value of coefficient of determination, correlation between observed and predicted data (r), and standard error (Se). Models resulting in higher (R^2) and (r) and smaller (Se) were considered to be superior.

3. Result and Discussions

The results shows that the estimated parameters value is differs for each lactation even though it analized with the same model. Parameter 'a' affected by parity and production level of animals; parameter 'b' affected by farm, season of calving, and lactation number; parameter 'c' affected by farm and season of calving, these parameters are crucial for estimating lactation features, including early yield, milk yield at lactation peak, time between the onset of lactation and lactation peak and persistency [2]. Parameter 'a' was explained as a beginning of milk yield. The lowest value found in Ali-Schaeffer's model, while the highest value was found in Wilmink's model. Parameter 'b' was explained as a slope until accession of the value of highest milk yield. The values was found as the highest in Ali-Schaeffer's model, while the lowest value found in Guo and Swalve's model. Parameter 'c' was explained as slope after the value of highest milk yield. This value was found as the highest in Wilmink's model, and the lowest value was found in Ali-Schaffer's model for all lactation. For parameter b value, the results in this study have similarities with the research done by Cilek and Keskin [18], but the value of other parameters are different. The estimated parameters value for all models from first to fourth lactation are shown in Table 2.

Table 2. Parameters Estimated of Five Different Lactation Curve Models

Lactation	Parameters	Wood	Yadav	Ali-Schaeffer	Wilmink	Guo and Swalve
	A	9.5417	0.2908	-40.7939	18.3908	5.9354
	B	0.1754	0.0431	68.3256	0.0280	-1.5793
	C	0.0032	0.000183	-16.5073	8.2357	5.5023
	D	-	-	38.8568	-	-
1	E	2.7183	-	-	2.7183	-
	F	-	-	-6.8334	-	-
	R^2	0.70342	0.88275	0.99156	0.74435	0.75283
	R	0.83870	0.93955	0.99577	0.86276	0.86766
	Se	1.54	0.97	0.30	1.43	1.40
	A	9.7339	0.2797	-16.1965	19.6199	6.4085
	B	0.1963	0.0384	32.87	0.0360	-1.8868
	C	0.00406	0.000223	-7.0602	8.3912	6.1570
	D	-	-	24.4301	-	-
2	E	2.7183	-	-	2.7183	-
	F	-	-	-4.5599	-	-
	R^2	0.89577	0.98277	0.99074	0.89827	0.90952
	r	0.94645	0.99135	0.99536	0.94777	0.95369
	Se	1.09	0.44	0.37	1.07	1.01
	a	9.7840	0.2558	-20.8679	19.0331	6.8356
	b	0.1871	0.0411	44.2568	0.0353	-1.8125
	c	0.00401	0.000223	-14.7461	7.5186	5.7969
	d	-	-	25.5256	-	-
3	e	2.7183	-	-	2.7183	-
	f	-	-	-4.5204	-	-
	R^2	0.94047	0.98200	0.98985	0.94140	0.95215
	r	0.96978	0.99096	0.99491	0.97026	0.97578
	Se	0.79	0.43	0.38	0.79	0.71
	a	7.3413	0.3347	-26.8922	17.9893	3.3450
	b	0.2448	0.0444	59.1551	0.0335	-1.8795
	c	0.00436	0.000227	-25.0056	9.0893	6.4148
	d	-	-	26.8399	-	-
4	e	2.7183	-	-	2.7183	-
	f	-	-	-4.5971	-	-
	R^2	0.92756	0.91397	0.96706	0.93822	0.94191
	r	0.96310	0.95602	0.98339	0.96862	0.97052
	Se	0.81	0.89	0.63	0.75	0.72

Despite the different of parameters value, all five models can properly describe the lactation curve. In the beginning of the lactation period the milk yield is low and then it will rise rapidly (first phase) until it reaches the peak yield, and then gradually descend (second phase) [3]. Based on the results of the analysis, Ali-Schaeffer's and Yadav's model reach peak yield in the second month of recording or at DIM 35 in all lactation. Meanwhile, Wood's model reached peak yield at DIM 35 in the 2nd and 3rd lactation, and reached peak yield at DIM 65 in 1st and 4th lactation. The Wilmink's model reached peak yield at DIM 35 in 3rd lactation, and reached peak yield at DIM 65 on 1st, 2nd, and 4th lactation. In the Guo and Swalve's model, peak yield at DIM 35 occurs in 2nd, 3rd, and 4th lactation, whereas in 1st lactation peak yield was reached at DIM 65. The difference in peak yield are because of these functions differ chiefly in the type of regression, the number of parameters, and the degree of associations with the main characteristics of a typical lactation pattern: i.e., peak yield, time at peak, and persistency [19].

3.1. Coefficient of Determination (R^2)

Table 2 shows value of the coefficient of determination (R^2 value) for each model in first, second, third, and fourth lactation. It was observed that Ali-Schaeffer's model has the highest coefficient of determination (R^2 value) compared to the other model in all lactation, while Wood's model has the lowest coefficient of determination (R^2 value) in almost all lactation. It means that Ali-Schaeffer's model indicated a small diversity, while Wood's model indicated a wide diversity between the observed and predicted yield, starting from the initial phase until the end of lactation [6]. The R^2 values obtained in this study were higher than the value reported by Koçak and Ekiz [4] and Biswal et al. [20]. However, the result showed in this study has similarity with those reported by Koçak and Ekiz [4] and Biswal et al. [20] that the Ali and Schaeffer Model has the highest R^2 values and gave the best fit to model the lactation curve.

3.2. Correlation Values (r)

The correlation for each model in first, second, third, and fourth lactation are presented in Table 2. The results showed that Ali-Schaeffer's model has the highest correlation value (r) compared to the other model in all lactation. The correlation value (r) of Ali-Schaeffer's model for first to fourth lactation were 0.99577, 0.99536, 0.99491, 0.98339, respectively, which means that Ali-Schaeffer's model has a high accuracy in estimating actual milk yield at BBPTU HPT Baturraden. According to Surrakhman [21] correlation value (r) range between 0.9 and 1 has a very high level of accuracy. Although the other four models also have correlation values (r) at 0.9 in almost all lactation periods, Ali-Schaeffer's model has a correlation value (r) that is closer to one compared to other models. The study by Nugroho et al. [22] showed similar result with this current study. In his study Nugroho et al. [22] compare eleven models of lactation curves in first and second lactation, and the results showed that Ali-Schaeffer's model was the best lactation curve to estimate milk yield at BBPTU HPT Baturraden.

3.3. Standard Error (SE)

SE has been used as a standard statistical metric to measure model performance [23]. Se indicate the deviation of the lactation curve in estimating actual milk yield. Table 2 also shows the value of Se for each model in first, second, third, and fourth lactation. The result showed that Ali-Schaeffer's model has the lowest Se compared to the other four models in all observed lactation periods. It means the deviation of Ali-Schaeffer's model in estimating milk production is smaller than other models. The study by Nugroho et al. [22] also shows that Ali-Schaeffer's model has the lowest Se compared to the other models. Although Se in this current study was lower than the study done by Nugroho et al. [22] in 1st and 2nd lactation period. Therefore the Ali-Schaeffer's model was proven to be the best model to predict milk yields at BBPTU HPT Baturraden. The shape of Ali-Schaeffer lactation curve model compare with the actual milk yield are presented in Figure 1.

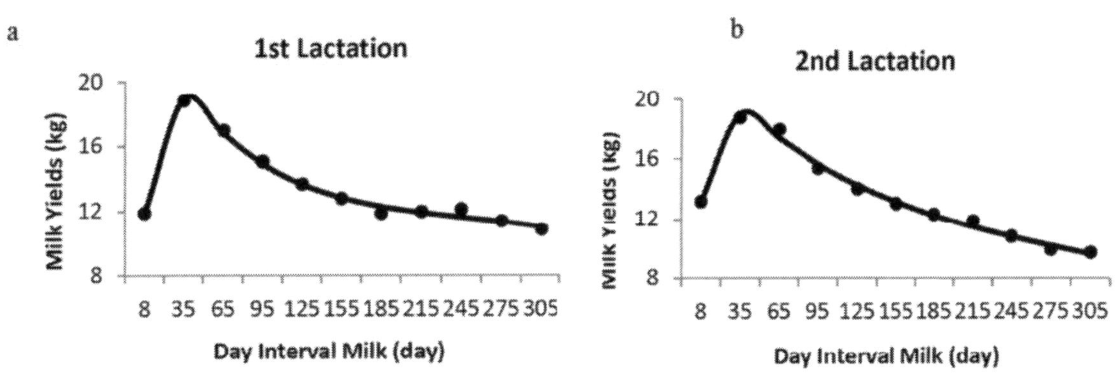

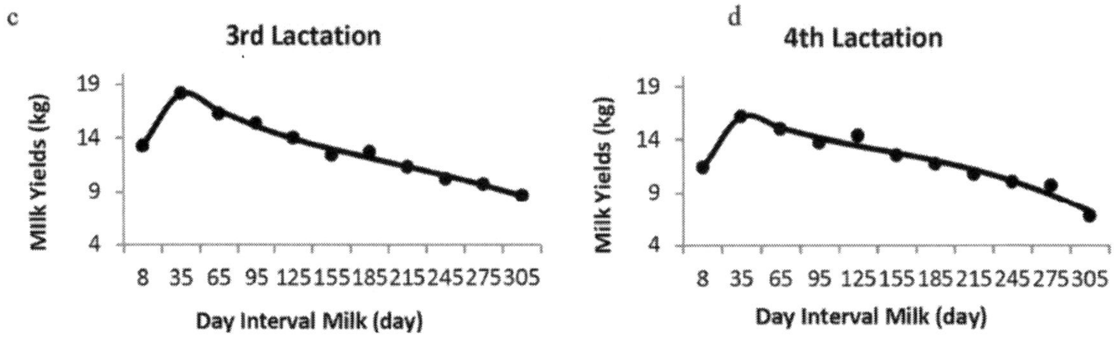

Figure 1. Shape of Ali-schaeffer Lactation Curve Model Compare with Actual Milk Yield in (a) 1st, (b) 2nd, (c) 3rd, and (d) 4th Lactation, respectively. The actual milk yield (—) ; The Ali-Scheffer Model (•)

4. Conclusion

Five mathematical functions for modeling the lactation curve in friesian holstein dairy cows were compared for accuracy of predicting milk yields from test day records. The results showed that Ali-Schaeffer's model was the best model to predict milk yields at BBPTU HPT Baturraden. It is known from the coefficient of determination were 0.99156, 0.99074, 0.98985, 0.96706, correlation between predicted and observed values (r) were 0.99577, 0.99536, 0.99491, 0.98339, and Se were 0.30, 0.37, 0.38, 0.63 at 1st, 2nd, 3rd, and 4th lactation respectively.

Acknowledgement
We would like thanks Academic Leadership Grant that fund this research and BBPTU HPT Baturraden for the research permit.

References

[1] Cunha D N F V, José C P, Fabyano F S, Oriel F C, José L B, Janaina A M 2010 Selection of models of lactation curves to use in milk production simulation systems *R. Bras. Zootec* **39**: 891-902
[2] Glória J R, José A G B, Célia R Q, José R M R, Jonas C C P, Ronaldo B R, Sandra G C, Martinho A S 2012 Environmental and genetic effects on the lactation curves of four genetic groups of crossbred Holstein-Zebu cows *R. Bras. Zootec* **41**: 2309-15
[3] Macciotta N P P, Corrado D, Salvatore P G R, Roberto S, Giuseppe P 2011 The mathematical description of lactation curves in dairy cattle *Ital. J. Anim. Sci.* **10**: 213-23
[4] Koçak Ö, Bülent E 2008 Comparison of different lactation curve models in holstein cows raised on a farm in the south-eastern anatolia region *Arch. Tierz. Dummerstorf* **51**: 329-37
[5] Torshizi M E, Aslamenejad A A, Nassiri M R, Farhangfar H 2011 Comparison and evaluation of mathematical lactation curve functions of Iranian primiparous Holsteins *S. Afr. J. Anim. Sci.* **41**: 104-15
[6] Dongre V, Gandhi R S, Singh A 2012 Comparison of different lactation curve models in Sahiwal cows *Turk. J. Vet. Anim. Sci.* **36**: 723-26
[7] Otwinowska-Mindur A, Ptak E, Jagusiak W, Satola A 2014 Modeling lactation curves of polish holstein-friesian cows part II: Prediction of 305-d lactation milk, fat and protein yields *J. Anim. Feed. Sci.* **23**: 29-36
[8] Brody S, Arthur C R, Turner C W 1923 The rate of decline of milk secretion with the advance of the period of lactation *J. Gen. Physiol* **5**: 442-44
[9] Ferreira A G T, Douglass S H, Ricardo A M V, Emilyn M M, Altair A V 2015 Fitting

mathematical models to lactation curves from holstein cows in the southwestern region of the state of parana, brazil *An. Acad. Bras. Cienc.* **87**: 503-17
[10] Macciotta N P P, Dimauro C, Steri R, Cappio-Borlino A 2008 Dairy goats feeding and nutrition UK: *CAB International* p 31-46
[11] Gonzalez-Pena D, Acosta J, Guerra D, Gonzalez N, Acosta M, Sosa D, Torres-Hernandez G 2012 Modeling of individual lactation curves for milk production in a population of Alpine goats in Cuba *Livestock Science* **150**: 42–50
[12] Wood P D P 1967 Algebratic model of the lactation curve in cattle *Nature* **216**: 164–5
[13] Dijkstra J, Lopez S, Bannink A, Dhanoa M S, Kebreab E, Odongo N E, Fathi N M H, Behera U K, Hernandez-Ferrer D, France J 2010 Evaluation of a mechanistic lactation model using cow, goat and sheep data *J. Agric. Sci.* **148**: 249–62
[14] Ali T E, Schaeffer L R 1987 Accounting For Covariances Among Test Day Milk Yield In Dairy Cows *Can. J. Anim. Sci.* **67**: 637-44
[15] Guo Z, Swalve H H 1995 Proc. Open Session Interbull Annual Meetingg Prague Czech Republic *Interbull Bulletin* **11**: 4–7
[16] Yadav M P, Katpatal B G, Kaushik S N 1977 Components of inverse polynomial function of lactation curve, and factors affecting them in harina and its friesian crosses *Indian J. Anim. Sci.* **47**: 777–81
[17] Wilmink J B M 1987 Adjustment of test-day milk, fat and protein yield for age, season and stage of lactation *Livest. Prod. Scie.* **16**: 335–48
[18] Cilek S, Keskin I 2008 Comparisson of six different mathematical models to the lactation curve of simmental cows reared in kazova state farm *J. Anim. Vet. Adv.* **7**
[19] Otwinowska-Mindur A, Ptak E, Jagusiak W, Satola A 2013 Modeling lactation curves of polish holstein-friesian cows part I: the accuracy of five lactation curve models *J. Anim. Feed. Sci.* **22**: 19-25
[20] Biswal S, Thirunavukkarasu M, Selvam S, Venkataramanan R, Pandian A S S 2017 Modeling lactation curve in jersey crossbred cows *J. Entomology and zoology studies* **5**: 1282-85
[21] Surakhman W 1998 Pengantar penelitian ilmiah Bandung: *Tarsito*
[22] Nugroho K, Anang A, Indrijani H 2015 Perbandingan model kurva produksi susu pada periode laktasi 1 dan 2 friesian holstein berdasarkan catatan harian *Jurnal Ilmu Ternak* **15**: 30-35
[23] Chai T, Draxler R R 2014 Root mean square error (RMSE) or mean absolute error (MAE) *Geosci. Model Dev.* **7**: 1247–50

IOP Conf. Series: Earth and Environmental Science **334** (2019) 012005 doi:10.1088/1755-1315/334/1/012005

Milk Production Curve on Various Test Day Patterns (Case in BBPTU-HPT Baturraden)

H Indrijani[1], **A Anang**[1], **D Tasripin**[2] **and L B Salman**[2]

[1]Breeding and Biometric Laboratory, Faculty of Animal Husbandry Padjadjaran University, Jl. Raya Bandung-Sumedang Km 21, West Java, Indonesia
[2]Dairy Cow Laboratory, Faculty of Animal Husbandry Padjadjaran University, Jl. Raya Bandung-Sumedang Km 21, West Java, Indonesia

E-mail : heni.indrijani@unpad.ac.id

Abstract. Information about milk production in dairy cows is obtained most accurately by daily recording, but with consideration, such as time and cost. This study aimed to obtain accurate and efficient recording patterns to estimate milk production in dairy cows. This research used recording of milk production in BBPTU-HPT Baturraden. The data used is the record of milk production of Test Day (TD) from the year 2014-2017 as many as 8696 records of TD that derived from 213 head of cows on first lactation. The record of TD milk production is the recording of milk production performed only on certain days during a lactation period. The milk production expectant curve used is the Ali-Schaeffer. The result of this research that Ali-Schaeffer equation with correlation between estimated and actual value (r) 0.991, standard error (se) 0.64, and standard deviation (stdv) 0,01. Recording of milk production Pattern 5, namely the recording of milk production in TD 1, 2, 5, 8, and 11, showed the highest correlation value (r=0.998). It is recommended to record the milk production in the early months of lactation and combined with the recording of TD milk production periodically every three months, from mid to late production.

1. Introduction

Information on milk production in dairy cattle, obtained most accurately with daily recordings, but with a variety of considerations, including time and cost, then now generally record milk production done periodically during lactation. Cumulative milk production recording can be done weekly, biweekly, monthly, bimonthly, or trimonthly. Another recording system that is now widely use in the world is Test Day (TD), which is the recording of milk production only on certain days in one lactation period. TD recording system is different from the system of recording cumulative milk production. In cumulative recording, for estimating total milk production, daily production is multiplied by the number of recording time intervals, whereas in TD recording, to estimate total milk production, it is not necessary to multiply the milking day interval, but simply by using curves.

TD recording can be done in two ways, recorded on the same date or on the same production day. That TD can be made into various recording combinations, so that from various combinations of TD recording patterns will be obtained a simple TD record pattern, but quite accurate to prediction of milk production. The recent TD record attracts many parties, as it is more flexible in handling records derived from different recording patterns, and is cheaper when compared to cumulative records, especially for large companies demanding more efficiency.

Milk production in dairy cattle forms a certain curve that has the same tendency, which is increased at the beginning of production, then reaches its peak and then decreases until the end of production. An accurate dairy milk dairy production curve can be used to determine the total of milk production at certain time based on production records. In Indonesia, until now, there are only few farmers who record milk production, even if the recording is generally incomplete. Incomplete records will make it difficult to predict milk production during a lactation period, so it is worth looking for the right milk production curve to predict it.

Ali and Schaeffer [1] have studied three milk production curves in first lactation and the results of the correlation are as follows: Gamma 0.88, IQP 0.87, and Ali-Schaeffer regression 0.95. Pallawarukka (1989) used gamma curves and non-linear curves to estimate the production of FH in Wisconsin from various recording systems. Jamrozik [2] obtained the correlation results for the Ali-Schaeffer curve of 0.975, Wood of 0.951, and Wilmink of 0.953. The results of the study of Dimauro [3] showed that from 13,925 recorded of milk production, which had a determination above 80%, 49.8% for the Wood curve; 48.8% for the Wilmink; 62.8% for the Ali-Schaeffer; and 60.9% for the Legendre Orthogonal Polynomial. So based on the results of the study, it can be conclude that the milk production curve from Ali-Schaeffer is more accurate in estimating milk production in dairy cattle.

Estimating milk production curves requires simultaneous and complete milk production records. One of the agencies that can meet these data needs is *Balai Besar Pembibitan Ternak Unggul dan Hijauan Pakan Ternak* (BBPTU-HPT) Baturraden. The BBPTU also functions as a place for assessment and dissemination of livestock technology and as a technical research center for livestock and has a good recording system, so that it can support the research that will be carried out. Based on this information, it is necessary to do research on estimation of milk production curve in various patterns of recording of Test Day (TD). The results of this study are expected to obtain accurate and efficient recording patterns to estimate milk production in dairy cattle.

2. Materials and Methods

2.1. Materials

This study used milk production record of dairy cattle at the BBPTU-HPT Baturraden. The data used are TD milk production records from 2014-2017 as many as 8,696 records of TD derived from 213 female cows on first lactation. The complete data required as follows:

1. Livestock identity data : cattle number, age of cattle, time of calf birth
2. Records of TD milk production based on certain milking days.

2.2. Method

The research used case study at BBPTU-HPT Baturraden. Data analysis using regression and correlation. Determination of livestock data retrieval using purposive sampling with consideration of the availability of data on milk production at the first lactation and has the required data completeness.

Records of daily milk production (test day/TD) are recorded twice a day, ie in morning milking and afternoon, so that daily milk production is total milk production as a result of the sum of milk production in morning and afternoon milking. In this study milk production record required on certain days, that is as follows: TD 1 is milk production recorded on milking day 5th, TD 2 day 35th, TD 3 day 65th, TD 4 day 95th, TD 5 day 125th, TD 6 day 155th, TD 7 day 185th, TD 8 day 215th, TD 9 day 245th, TD 10 day 275th, and TD 11 day 305th.

Daily record data collection was carried out starting on day 5th with the assumption that in the previous days, colostrum milk produced by dam was given to calves and it was expected that production would be relatively stable at that time. Recording is done with an interval of 30 days, this is done to simplify data processing and make it easier for farmers to apply the results of this study.

TD milk production is made into several patterns, some are simultaneous and some are periodic as follows:

1. Pattern 1: Complete TD records (TD 1, 2, 3, 4, 5, 6, 7, 8, 9, 10, 11)
2. Pattern 2: Odd TD records (TD 1, 3, 5, 7, 9, 11)
3. Pattern 3: Even TD records (TD 2, 4, 6, 8, 10)
4. Pattern 4: TD records 1, 2, 3, 5, 7, 9, 11
5. Pattern 5: TD records 1, 2, 5, 8, 11
6. Pattern 6: TD records 1, 2, 3, 7, 11

7. Pattern 7: TD 1, 2, 6, 10

2.3. Test of Milk Production Accuracy with Ali-Schaeffer Regression Equation

Non-linear regression analysis using SAS software 9. Accuracy is seen from the correlation between the estimated values and the observations (r) and standard error (SE). The estimator equation of milk production curve used is Ali-Schaeffer regression as follows:

$$y = a + b\left(\frac{t}{305}\right) + c\left(\frac{t}{305}\right)^2 + d\left(\ln\frac{305}{t}\right) + e\left(\ln\frac{305}{t}\right)^2$$

y = Milk yield predicted
t = DIM (*Day Interval Milk*) = 5, 35, 65, ..., 305
a,b,c,d,e = regression coefficient

3. Result and Discussions

3.1. Data Description

TD milk production is the amount of milk production taken on certain days. This milk production is the result of the addition of milk production in the morning and afternoon milking that is done on the same day. The results can be seen in Table 1.

Table 1. Data Description Test Day (TD) Milk Production on First Lactation

Test Day (TD)	Milking day (day)	Amount of data (n)	Actual Milk Production (kg)
TD 1	5	1174	17.47
TD 2	35	2645	20.71
TD 3	65	2304	21.08
TD 4	95	1264	18.08
TD 5	125	776	17.12
TD 6	155	259	14.16
TD 7	185	126	14.06
TD 8	215	96	12.44
TD 9	245	20	11.86
TD 10	275	17	11.19
TD 11	305	15	10.65

In general, the level of milk production of dairy cows will increase at the beginning of lactation to the peak of production, and then will decrease slowly to low production. Based on the description of milk production data in Table 1, the average milk production increased from TD 1 to TD 3 (milking day 65th), and then continued to decrease until the end of lactation at TD 11 (milking day 305th). Generally dairy cow will be dried up (no longer milking) at the end of lactation or about two months before to the next partus.

The result of this research get the average of TD milk production on first lactation period at BBPTU-HPT Baturraden equal to 15.35 kg. Milk production average is higher when compared with the average production of TD milk lactation in the same place in the year 1997-2005 as much as 12.24 kg [4], 2006-2011 as much as 12.91 kg [5], and 2006-2013 as much as 12.99 kg. Increase in production from year to year may be possible due to the selection process of dairy cows conducted by BBPTU.

3.2. Estimating Milk Production Curves

Based on the calculations using of SAS 9 software, we obtain the coefficients needed to complete the equation of Ali-Schaeffer milk production estimator, which will be tested for accuracy with actual TD milk production. The formula as follows:

$$y = 43.22 - 56.20\left(\frac{t}{305}\right) + 23.88\left(\frac{t}{305}\right)^2 - 8.92\left(\ln\frac{305}{t}\right) + 0.70\left(\ln\frac{305}{t}\right)^2$$

The next statistical analysis is find the correlation value. The curve equation accuracy will be based on the correlation between the expected value with the actual result (r) and the standard error (se). The results showed that this equation can be used to estimate milk production, because it has a correlation between the estimated value with the actual value (r) of 0.991 with the standard error (se) 0.64. The standard deviation (stdv) between actual milk production and the allegation of 0.01 as can be seen in Table 2.

Table 2. Actual milk production, predicted, and deviation

Test Day (TD)	Milking Day (day)	Actual Milk Production (kg)	Predicted Milk Production (kg)	Deviation (kg)
TD 1	5	17.47	17.45	0.02
TD 2	35	20.71	21.06	-0.35
TD 3	65	21.08	20.22	0.87
TD 4	95	18.08	18.58	-0.50
TD 5	125	17.12	16.80	0.32
TD 6	155	14.16	15.11	-0.95
TD 7	185	14.06	13.64	0.43
TD 8	215	12.44	12.44	0.00
TD 9	245	11.86	11.57	0.29
TD 10	275	11.19	11.05	0.14
TD 11	305	10.65	10.90	-0.25

Observing the increase or decrease in milk production based on the TD record, it would be easier if the available data is made into curve form, as can be seen in figure 1. Generally milk production follows a regular pattern on each lactation. Milk production will rise for 45th to 60th days after the cow are partus until it reaches the peak of production and then slowly decreases until the end of lactation. In this study, peak production is generally achieved between the first and second months and continues to decline until the end of the lactation.

Atabany [6] results, the maximum milk production occurs at the beginning of the lactation and minimum milk production occurs at the end of the lactation. The maximum and minimum milk production of this research at BBPTU Baturraden, showed that the peak of Friesian Holstein (FH) cow's milk production occurs at week 2nd to 5th. This opinion is supported by research conducted at PT. Taurus Dairy Farm, Cicurug-Sukabumi that the peak of milk production occurs in the range of week 2nd to 5th [7].

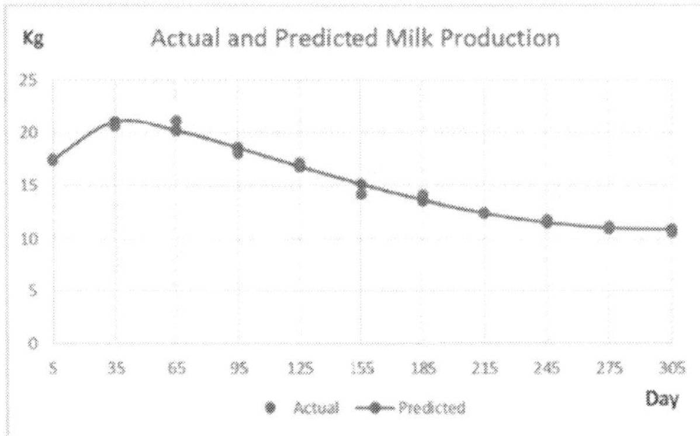

Figure 1. Milk Production Test Day (TD) on First Lactation at BBPTU-HPT Baturraden

The results show that the shape of the Ali-Schaeffer equation curve almost same with the plotting of actual TD milk production data, only a few deviations on the day 65th of milking (TD 3), 95th (TD 4) and 155th (TD 6). This showed that visually based on plotting data, Ali-Schaefer's equation curve is a good estimator curve. Furthermore, this alleged milk production will be used to test the correlation between several different TD recording patterns which results can be seen in Table 3.

Table 3. Correlation of Test Day (TD) Milk Production

Recording Pattern	Test Day Record (TD pattern)	Correlation
Pattern 1	Complete	0.991
Pattern 2	Odd	0.997
Pattern 3	Even	0.995
Pattern 4	1, 2, 3, 5, 7, 9, 11	0.995
Pattern 5	1, 2, 5, 8, 11	0.998
Pattern 6	1, 2, 3, 7, 11	0.994
Pattern 7	1, 2, 6, 10	0.993

Simultaneous or periodic TD recordings show a high correlation, it means all of these recordings can be applied. Complete TD recording requires more costs because milk production should be recorded every month, as well as the physiological trait of livestock are not the same. This will cause unequal production fluctuations and variations, especially during peak production in the second month (TD 3), and when production begins to decline in the third to fifth months (TD 4, 5 and 6). Milk production in the following month is relatively stable decrease form linear function.

The selection of recording patterns is based on the magnitude of the correlation between actual production on a certain time and its prediction. The results showed the recording of milk production Pattern 5, namely the recording of milk production in TD 1st, 2nd, 5th, 8th, and 11th showed the highest correlation value. Pattern 5 can be recommended to be done in smallholder farms that have difficulty recording milk production every month, or livestock companies that have large livestock for the purpose of efficiency.

Other research results to estimate milk production by using Gamma function can be done on a few months only. The recommended combination is the recording of milk production at month 4th and 5th or 4th and 6th, as well as 1st and 8th lactation months combined with record of month 2nd, 3rd, or 4th. The smallest estimate is combinations 3rd and 8th monthly recording or 1st, 3rd and 8th [8].

Test Day Model (TDM) has become the world standard and is widely adopted by various countries with various modifications in the genetic quality evaluation program on a regular basis [9, 10, 11, 12, 13, 14], especially for countries with diverse recordings of milk production, such as in Kenya [15]. In Indonesia, the problem is almost the same, that is the variety of recording system and the absence of database, so that the use of TDM not only become a thing to be considered but has become a requirement in evaluating milk production in dairy cow, so that recording of milk production Pattern 5 can be solved that problem.

4. Conclusion

The Ali-Schaeffer equation can be used to estimate milk production by correlation between the estimated value with the actual value (r) 0.991, the standard error (se) 0.64, and standard deviation (stdv) 0.01. Recording of milk production Pattern 5, namely record milk production on Test Day (TD) 1, 2, 5, 8, and 11, showed the highest correlation value (r=0.998).

References

[1] Attabany A, Purwanto B P, Toharmat T, Anggraeni A 2013 Effisiensi Produksi Sapi Friesian Holstein Pada Generasi Induk dan Generasi Keturunannya. *JIPT.* **1(2)**: 115-121.

[2] Jamrozik J, Schaeffer L R 1997 Estimates Of Genetic Parameters For A Test Day Model With Random Regression For Yield Traits Of First Lactation Holsteins. *J. Dairy Sci.* **80**:762-772.

[3] Dimauro C, Vicario D, Canavesi F, Borlino A C, Macciotta N P P 2006 Analysis of Individual Variability of the Shape of Lactation Curve for Milk Fat and Protein Contents in Italian Simmental Cows. *8th WCGALP*, CD-ROM Communication no. **01**-54.

[4] Indrijani H 2009 Perkembangan Evaluasi Genetik Sapi Perah Berdasarkan Produksi Susu. *Wartazoa* **19(1)**

[5] Pratiwi N, Sudewo A T, Setya A S 2013 Penggunaan Taksiran Produksi Susu dengan Test Interval Method (Tim) pada Evaluasi Mutu Genetik Sapi Perah di BBPTU Sapi Perah Baturraden *Jurnal Ilmiah Peternakan* **1(1)**:267-275.

[6] Attabany A, Purwanto B P, Toharmat T, Anggraeni A 2013 Effisiensi Produksi Sapi Friesian Holstein Pada Generasi Induk dan Generasi Keturunannya *JIPT.* **1(2)**: 115-121.

[7] Indrijani H, Anang A, Noor R R, Talib C 2003 Efektivitas Catatan Test Day untuk Evaluasi Genetik Produksi Susu pada Sapi Perah *Jurnal Pemuliaan Indonesia Zuriat* **14**.

[8] Pallawarukka 1989 Metode Estimasi Produksi Susu 305 Hari dari Pencatatan Beberapa Hari. Pusat Penelitian dan Pengembangan Peternakan *Badan Penelitian dan Pengembangan Pertanian* Departemen Pertanian Bogor 120-131.

[9] Swalve H H 2000 Theoritical Basis And Computational Methods For Different Test-Day Genetic Evaluation Methods *J. Dairy Sc.* **83**:1115-1124

[10] Indrijani H, Anang A, Noor R R, Talib C 2001 Penggunaan Catatan Test Day untuk Mengevaluasi Mutu Genetik Sapi Perah *Jurnal Ilmu Peternakan dan Veteriner* **(4)**

[11] Kaya I, Akbas Y, Uzmay, C 2003 Estimation of Breeding Values for Dairy Cattle Using Test-Day Milk Yiels *Turk J. Vet. Anim Sci.* **27**: 459-464

[12] Cho K H, Na S H, Seo K S, Park B H, Choi J G, Lee Y C, Park J D, Son S K, Kim S, Choi T J, Salces A 2006 Estimation of Genetic Parameters for Change of Test Day Records on the Milk Production and SCS Using Random Regression Model of the Holstein Cattle in Korea. *8th WCGALP*, CD-ROM Communication no. : **01**-51.

[13] Costa C N, Vasconcelos J, Cobuci J A, Thompson G, Carvalheira J 2006 Genetic Parameter of Test Day Milk Yield for Brazilian Holstein Cattle Using an Autoregressive Multiple Lactation Animal Model *8th WCGALP*, CD-ROM Communication no. : **01**-53.

[14] Khani J, Edriss M A, Mehrgardi A A 2006 Estimation of Genetic Parameters for Milk Production Traits Using Test Day Records of Holstein Dairy Cattle of Iran *8th WCGALP*, CD-ROM Communication no. : **01**-63

[15] Ilatsia E D, Muasya T K, Muhuyi W B, Kahi AK 2006 Use of Test Day Milk Yield Records for Genetic Evaluation in Sahiwal Cattle *8th WCGALP*, CD-ROM Communication no. : **01**-61

The Development of Rural Poultry Farming Program to Kampong Chicken Farmers of Treman Village North Sulawesi

S Surtijono[1], R Tinangon[1], J R Leke[1], F N Sompie[1], Z Poli[1] and R Siahaan[2]

[1] Faculty of Animal Husbandry, University Sam Ratulangi, Manado, Indonesia
[2] Faculty of Mathematics and Natural Sciences, University Sam Ratulangi, Manado, Indonesia

E-mail: rinileke@yahoo.com

Abstract. The purpose of research was to empower the members of rural poultry farming to increase productivity of kampong chicken. The development of Science, Technology and Arts (IPTEKS) for society in Treman Village aimed to increase nutrient for society, to help them to build rural poultry industry belongs to community, and to supply kampong chicken to "Damar Merah and El'Shadai" farmer's groups of Treman Village. The materials were kampong chicken supplied by kampong chicken breeder in Treman Village. The research method used is survey method, with the respondents amounted to 20 farmes determined by purposive sampling ie farmers belonging to the group, the development of science and technology for the region. Data analysis used is descriptive analysis. The result show that increasing of skill and competence of "Damar Merah and El'Shadai" farmer's groups after granting new hatching machine and *Day Old Chick*. Both of groups can accept new application in modern poultry farming management through modern breeding system. The development of rural poultry farming program to kampong chicken farmers of Treman Village, North Sulawesi can increase knowledge, skill, and income of the kampong chicken farmers.

1. Introduction

Treman Village is, one of 12 villages in Kauditan Regency, located at the west end of North Minahasa District, North Sulawesi Province. This village is about 3 km north of regency, about 5 km from Bitung City, 20 km from the district capital, and 35 km from the province capital. The total area is 1447 ha. The number of population are 2381 people i.e. 1187 men and 1194 women in 684 families. Education levels of people i.e. elementary school 1231 people, junior high school 665 people, senior high school 522 people, Diploma (D) 27 people, D3 29 people and S1-S3 higher educations (bachelor-master-doctoral degrees) 191 people. The livelihood of the people i.e. 356 farmers, 129 government employees (ASN), 168 "tukang"/skilled labors, and 6 army/police officers. The Kauditan Regency is closed to Bitung City which known as a fishery producer area. Unavailability of employment for productive ages leads to high unemployment rate. In facing economic crisis, economic sectors have resilience such as micro, small and medium sectors and informal businesses. Therefore, the most realistic economic recovery to undertake should be started from those sectors. This strategy can be applied as a tool to strengthen community participation in national and local economic development. The agricultural sectors in North Minahasa include food crops, plantation, animal husbandry, and fisheries. Agriculture crops are dominated by rice, corn, peanuts, soybeans, cassava and sweet potatoes. Coconut plantations are also dominant in this area. Land are under coconut trees are widely planted by farmers with corn, paddy, and bananas. The fishery sector is dominated by fishing areas for supplying North Sulawesi Province. The livestock sector is an important for North Minahasa in addition to agriculture and fisheries. The development of agriculture, livestock and fisheries are mutually supportive and beneficial, so the integrated system provides great benefits to both. On one side, agricultural products such as maize, cassava, tomatoes, agricultural waste such as coconut cake can be used as animal feed to have added value. The fishery sector provides the economic values. The fish waste products can be managed to be fish flour and liquid waste fish oil.

In 2017, Damar Merah and El'shadai farmer's groups were formed as a follow up of RPJMD program of North Minahasa District. Treman Village was a main producer/income source of copra harvested 3 times a year and cloves harvested once a year, made family have a very low income level. Poultry farm traditionally kept under coconut trees, by searching food for their selves, the average ownership of 10 - 20 roosters and hens. The traditional marriage system takes time from laying eggs to raising children 115 days. It has not been able to become a business venture to be income source for people.

Treman Village is closed to Bitung City (Harbour City) and Manado, the Capital of North Sulawesi Province, is very potential for business development of domestic chicken. Agricultural, livestock and fishery development are mutually supportive and beneficial, providing integrated benefits. In one side, agricultural products such as maize, cassava and even rice additional harvests such as bran can be used as animal feed that added value. Fishery waste could also be used as a source of animal protein for livestock. Another impact directly felt by farmers was the increasing of income and welfare which if a week of DOC production can reach 40,000 and eggs reach 10,000 while chicken production can reach 1,500 per day. Selling price for a DOC was Rp 4,000 – Rp 5,000; an egg was Rp 1,200 – Rp 1,500, and chicken meat was Rp 25,000/Rp 35,000 / kg, and life chicken was Rp 28,000/kg.

Treman Village formed farmers/fisheries groups "Damar Merah" and "El'shadai" based on the Decree of North Minahasa Regency No 19a dated January 20, 2005 to develop farming/ livestock business oriented agribusiness and environmentally friendly. The structure of organization "Damar Merah" and "El'shadai" consisted of chairman, secretary, treasurer and 17 members. The farmers' group was originally formed from the "mapalus" group which is "gotong royong' (work together) to prepare the labour to cultivate and to clear the land, to cultivate, to harvest and to raise livestock. This program was determined based on the habits of farmers groups in this area in managing the farming business. The habit is the mutual assistance activities between farmers to manage their farm business where farmers in one village help others in the form of manpower and done in rotation. Some farmers who do the activities of "Mapalus" were recruited in a group of livestock farmers "Damar Merah" and "El'shadai". This activity was supported by Agricultural Extension Workers Penyuluh Pertanian Lapangan (PPL). The groups did short-term group activities including weekly meeting and established social gathering/arisan and mapalus. In long-term activities, The Damar Merah and El'shadai groups are not business-oriented yet.

The low quality of feed given in the form of rice, bran, forage, rice grains, sand, and insects. The nutrient content of the ration was 8.52 - 14.10% protein, 5.70 - 11.63% fat, crude fibre 6.88 - 14.07%, 0.45 - 0.91% phosphorus and 0.02 - 1.04% calcium[1]. These diet are sufficient for basic life and little production only. Introduction of technology through agricultural and fishery product innovations can be utilized as a source of potential livestock rations. Processing can be done through physical, biological and chemical processes and local raw materials techniques. Utilizing the waste of fishery companies of Bitung City both solid and liquid waste. However, the knowledge of waste management needs technology touch to optimize utilization of fishery byproducts to be animal feed in the forms of fish meal and fish oil [2]. Analysis content of fish skipjack flour from head fish waste (HSW) was 45.32% protein, filleting waste (FW) was 49.19% protein, and Arachon waste (SW) was 48%. The processing of liquid boiling waste of fish oil processing contained 10% protein and linoleic fatty acid needed by domestic poultry[3].

The low quality of egg-laying, due to the very strong nature of the incubator caused low production with average time of 115 days. The effective way to eliminate the nature of brooding in domestic poultry with the artificial insemination to kampong chicken. Artificial insemination in poultry is one of the chicken technologies and is a valuable technique in the livestock industry of birds [4].

Conservation of kampong chicken cement using diluents to fertility and fertile spermatozoa period of post artificial insemination. Based on pre-field survey and discussions to local group members and associations, priority issues need to be addressed by members of farmer groups of "Damar Merah" and "El'shadai" with companion from universities are:

1. Livestock business conducted by "Damar Merah" and "El'sahadai" groups is still traditional so that the result is not optimal yet;
2. Lack of knowledge of group members about raising domestic poultry as a source of income that can improve the economy;
3. Lack of knowledge of livestock groups to produce poultry chicken with high productivity by using artificial insemination techniques;
4. Lack of knowledge of group members about the utilization of feed ingredients of the existing rations in the area and how to formulate rations for livestock according to the standard to increase productivity;
5. Lack of ability of group members to adopt appropriate technology to increase livestock population by using simple sedan engines;
6. Lack of team members' ability to adopt technology to reduce mortality rates by using herbal ingredients derived from existing plants in the area as a substitute for synthetic antibiotics;
7. Group members perform their production processes without recording so that activities are performed without accurate data because it is only based on the memories of group members, and
8. Lack of knowledge of farmers on post-harvest technology.

Poultry commodity especially kampong chicken is one of the commodities source of animal protein that fulfil the taste of society, affordable by people's purchasing power widely, high nutritious, fast growing, relatively short productive age, easily developed biotechnology and easy application[5]. Development of kampong chicken North Sulawesi can be achieved by applying product processing technology. Lecturers and students of Sam Ratulangi University can develop community creativity development as a direct target of Science, Technology and Arts (IPTEKS). This effort is a part of university duties to increase the income of the community so that Student Creativity Research/PKM aimed to apply IPTEKS to broaden the society's insight and to increase the income of farmer's groups.

2. Materials and Methods

To achieve the objectives, the consultative method was used in breeding chicken, hatching machine, training the selection of hatching eggs, practicing preparation of rations, and producing of nugget, kampong chicken meatball and ice cream. Preparation of diet/rations use local raw materials. The process of producing skipjack tuna oil as a source of energy, omega-3 and omega-6 used local raw material as a substitute for maize. This Community Partnership Program (CPP/PKM) was held at Village Hall of Treman Village, Kauditan District, Province North Sulawesi. The target audience of this PKM activity was "Damar Merah" and "El'shadai" Farmer's groups of Treman Village (80% of these members were women group/PKK)

2.1. Procedure

Based on the priority problems of livestock farmers group "Damar Merah" and "El'shadai", the group empowering was a necessary. The empowerment was done to handle some priority issues that can be done with several methods including extension materials (institutional strengthening; technology products: producing kampong chicken nugget, kampong chicken meatball, and ice cream; product display, and cash flow analysis) and production preparation (the consulting; training: preparation and designing; purchasing of supported raw materials; finance analysis) (Figure 1).

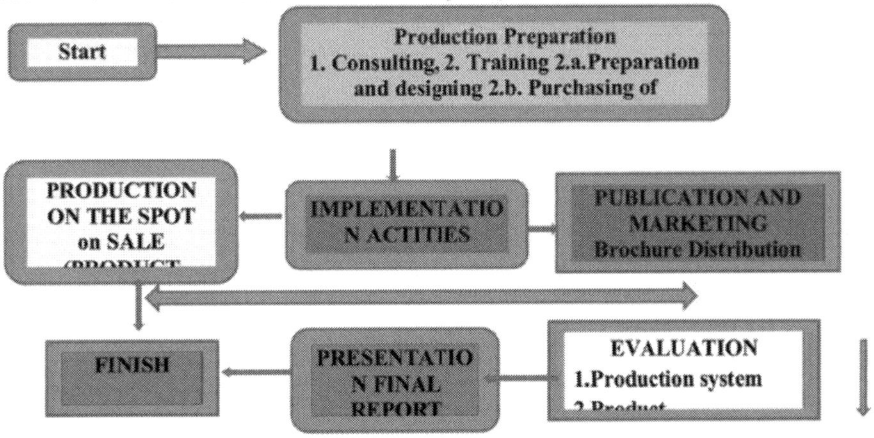

Figure 1. Flow diagram method of rural poultry farming program of Treman Village.

3. Results and Discussion

Consultative and practicing methods of PKM to Damar Merah" and "El'shadai groups were done through IPTEKS implementation. Activities had been implemented include how to breed chickens, making hatching machine (Fig.2.a), preparation of rations by using local raw materials, food processing technology products such as kampong chicken nuggets and meatball and ice cream (Fig.2.b). This activity was attended by 40 - 50 people, 10 lecturers and 10 students and the Head of women group/PKK of Treman Village.

Rural poultry farming using native breeds is being practised in many developing and underdeveloped countries throughout the world[5]. According [6], local or indigenous chicken breeds of Africa and other part of the world include the Bisaya in the Philippines, the indigenous fowl of the Polvaba provinceof Ukraine, the golden –speckled native fowl of Czechoslovakia.

Research in village poultry in different countries has revealed that the genetic potensial of village chicken is generally not the major countries has revealed that the genetic potential of village chicken is generally not the major constraints to their production[7]. The increasing of annual egg production was from 116 eggs to about 140 eggs per hen. The average egg weight to the flock also increased from 43 to 49 g through six generations of selection in a nondescript flock of Indian Desi fowl [8]. The activity was carried out in direct application form. Practicing activities were started from chicken breeding technique, hatching machine operation, feed mixing, nugget and meatball and ice cream producing can be seen in the picture of farming below. The activity of the community activity program was expected to increase the knowledge of chicken breeding ranging from chicken breeding techniques, utilization of local feed, hatching, breeding, feeding, and management of livestock products such as nuggets, chicken meatballs and ice cream.

The evaluation showed that the farmer group's responded to the technology provided through the community partnership program/PKM. This can see through high society enthusiasm and strongly support of government officials of Treman Village to the undertaken activities. According to[9] that food security is an issue that has always been the attention of Indonesian government. The government proclaimed main program of agricultural development such as increasing food security. Increased food security has a very basic development perspective. This is because access to food and balanced nutrition as fulfilment of basic food needs,is the most basic rights for humans. According [10], the consumption of animal protein from livestock is still low so it needs to be strived to provide food products, source of animal protein from livestock in from fresh and processed products[10].

Figure 2. Hatching machine

Figure 3. Ice cream production

Kampong chickens are raised using traditional production techniques by almost every village household. They are a side –line activity and are not considered the main source of family earnings. The members of a family generally work in crop cultivation, as labourers, or as traders. Although some families keep more than 1000 birds, they still work in other activities for their main livelihood. In same cases, farmers have integrated their native chicken operations with freshwater fish farming by contracting the cages above fish-pond.

4. Conclusion

The Community Partnership Program/PKM of Damar Merah" and "El'shadai farmer's groups can be adopted by farmers of Treman Village, This PKM can increase the knowledge, skills and income of kampong chicken Damar Merah dan Elsahadai farmer's groups of Treman Village. The application of breeding system strategy in this program was to improve genetic characters i.e. good genotype and productive phenotype characters. The feeding program was ad libitum feeding that availability of food in food container which composition based on ISO calories and energy and local food resource. This activity (PKM) applied simple hatching machine to increase *Day Old Chick*. The end products were nugget chicken and ice cream production for housewives to provide healthy, safe and halal food for families.

Acknowledgements

Financial support was provided by Research and Community Service Institution/ LPPM of University Sam Ratulangi by IPTEKS Grant from The Ministries of Research, Technology, and Higher Education 2018 and to the farmers of Damar Merah and El'Shadai" of Treman Village for their willingness to become our partner.

References

[1] Sukardi, Mufti 1988 Penampilan prestasi ayam buras di kabupaten Banyumas dan pengembangannya *Procedings Seminar Nasional tentang Unggas Lokal* Semarang **52:** 543 – 551

[2] Leke J R, Achmanu, Sjofjan O, Najoan M, Mandey J S 2015 Organoleptic characteristic of eggs laid by local hens fed Skipjack Fish waste as a source of Omega-3 fatty acids in the diets *Livestock Research for Rural Development* **27(11)** http://www.lrrd.org/lrrd27/11/leke27219.html

[3] Leke J R 2013 Egg Internal Quality n-3 Fatty Acids of Native Chicken Fed on Skipjack Fish (Katsuwonus pelamis l) Industrial Waste Containing Feed *Int. Jour. of Poult.Sci.* **12:** 484 - 488

[4] Ridwan, Rusdin 2008 Konservasi Semen Ayam Buras Menggunakan Berbagai Pengencer Menggunakan Berbagai Pengencer Terhadap Fertilitas dan Periode Fertil Spermatozoa Pasca IB *J.Agroland* **15:** 63 – 67.

[5] Bett H K Peters K J Bokelmann W 2011 Hedonic price analysis to guide in breeding and production of Indigenous chicken in Kenya *Livestock Research for Rural Development* **23(6)** http://www.lrrd.org/lrrd23/6/bett23142htm

[6] Oluyeni Roberts 2003 Poultry production in warm wet climates 2nd ed *Spectrum Books Limited.*

[7] Dolberg F 2003 Review of household poultry production as a tool in poverty reduction with focus on Bangladesh India http:www.fao.org/ag/againfo/programmes/en/pplpi/docare/wp6.pdf.

[8] Khan A G 2008 Indigenous breeds, crossbreds and synthetic hybrids with modifield genetic and economic profiles for rural family and small scale poultry farming in India *World's Poultry Science Journal* **64:** 405 – 415

[9] Nur Y H, Nuryanti Y, Resnia R, Santoso A S 2012 Analysis of factors and projection for national food consumption cases in commodities : rice, soybeans and beef *Scientific Bulletin of R & D Trade* **6:** 38 – 52

[10] Sonbait L Y 2011 Consumer's favorite of processed products for beef in Manokwari City *Agrinimal* **1:** 71-75

IOP Conf. Series: Earth and Environmental Science **334** (2019) 012007 doi:10.1088/1755-1315/334/1/012007

Utilization of Fermented Complete Feed Based on Mixed *Pennisetum Purpureum* and *Indigofera Sp* on Performance of Garut Male Lambs

S Nurachma[1], A A Nurmeidiansyah[1], T Dhalika[2] and D Ramdani[1]

[1]Department of Animal Production, The Faculty of Animal Husbandry, Universitas Padjadjaran, Jatinangor Campus, Sumedang 45363, Indonesia
[2]Department of Nutrition and Feed Technology, The Faculty of Animal Husbandry, Universitas Padjadjaran, Jatinangor Campus, Sumedang 45363, Indonesia

E-mail: sitinurachma206@gmail.com

Abstract. This research was aimed to find out the use of fermented complete feed based on mixed *Pennisetum purpureum* and *Indigofera* sp on performance of Garut male lambs during fattening. A completely randomized design was used to compare 5 different complete feeds containing different *Pennisetum purpureum* (60, 70, 80, 90 and 100%) and *Indigofera* sp (40, 30, 20, 10, 0%) mixtures using 5 replicates. The measurable parameters were dry matter consumption, body weight gain, as well as feed conversion efficiency. The results showed that different mixtures of *Pennisetum purpureum* and *Indigofera sp* had a significant effect ($P<0.05$) on dry matter consumption with the highest dry matter consumption was obtained by mixture of 80% *Pennisetum purpureum* and 20% *Indigofera sp* (1,313.4 gram /head /day). However, the mixtures had no significant effect ($P>0.05$) on body weight gain and feed conversion efficiency of the lambs.

1. Introduction

The ration of ruminant livestock such as sheep is the kind of forage which consists of grass and legumes. Most sheep farmers in Indonesia, especially traditional farmers, in West Java very rely on field grass for the sheep that maintained . The use of field grass to support the development of sheep farming has several constraints, such as the guarantee of the availability of relatively low, can only meet the needs of sheep farming on the scale of traditional livestock farming with the aim as a side business. Field grass is not specifically cultivated to support the development of sheep farms and only obtained on certain land such as rice field, riverbanks, and empty land that has not been utilized or among cultivated plants.

The continuity of sheep farming with a commercial-oriented business objective requires a guaranteed supply of forage feed continuously throughout the year according to the population of the sheep that are kept at a certain time and the projection of later sheep populations when the livestock business is further developed, the existence of forage cultivated into a necessary condition get attention. Various types of grass and cultivation legumes have been widely developed for the purpose of meeting the needs of forage for ruminant sheep such as *Pennisetum purpureum* grass or better known as elephant grass and legume *Indigofera sp.*

Elephant grass or *Pennisetum purpureum* is one type of grass that has good quality and is easy to grow and has a high production [1], it can be relied upon as a substitute forage grass field. But to meet the needs of sheep's nutrients required additional feed that is able to supply deficiencies in the elephant grass. *Indigofera sp* is one alternative that can be utilized as an additional feed, based on its high nutritional content [2]. Indigofera sp protein content based on some research results about 26%, and expected can replace the concentrates that are relatively expensive. Use of *Indigofera sp* as animal feed has been done in some areas and produce good production.

Indigofera sp is a type of legume plants of potential highly developed as forage because it has good nutritional value . According to Hassen [3], *Indigofera sp* legume has a crude protein content of 27% - 31%, crude fiber between 22% - 46% with a high dry matter digestibility value of 72% - 81%. However, the utilization of this plant is constrained by the tannins contained , that can bind to a protein forms a complex "tannin - protein" which resulted in the value of the benefits of protein disrupted due to the strength of the bond certain complex of a protein cannot be digested by the digestive system of ruminants . According to Abdullah and Suharlina [4], the tannin content on *Indigofera sp* legume is 0.027%. Utilization of *Indigofera sp* should be limited to a safe , so the protein in rations can be used efficient. According to Simanihuruk and Sirait [5], the use of Indigofera sp as goat feed ranges from 25% - 75% of total dry matter rations, then Tarigan and Ginting [2], states that *Indigofera sp* as much as 30% - 40% in goat rations low-quality grass yields a consumption response, optimal digestibility and optimal weight gain

The quality of *Indigofera sp* as forage and as additional feed to complete the deficiency on the grass, is expected to support to achieve high productivity. However, in the fattening phase need the addition of concentrate that carcass quality is achieved as expected. As well-known meat quality of sheep is affected by breed, vegetation and feeding practice [6]. Average daily gains in forage finished lambs is usually lower than in lambs finished on high grain diets [7, 8]. The increased of slaughter weight will provide a good profit for farmers, and is expected on an ongoing basis, so it can be a reliable livestock business. The study discover the optimal effect of ration that can be beneficial for development of sheep farms, especially traditional farmers. This study will be basic information for researchers to uncover on the utilization of fermented complete feed based on *Indigofera sp.* and *Pennisetum purpureum* at different balances. Thus, a new theory may be arrived at.

2. Material and methods

2.1. *Experimental Animals*

About 25 Garut male lambs were used in this study and each of them was placed in an individual cage with free access to drinking water. The average age of the lamb are ten months and the average of weight are 23.76 ± 2.08 kg. A completely randomized design was used to compare five (5) different diet treatments based on *Pennisetum purpureum* and *Indigofera sp.* inclusion at various balances: T0 (*Pennisetum purpureum* 100%), T1 (*Pennisetum purpureum* 90% + *Indigofera sp* 10%), T2 (*Pennisetum purpureum* 80% + Indigofera sp 20%) , T3 (*Pennisetum purpureum* 70% + Indigofera sp 30%), and T4 (*Pennisetum purpureum* 60% + Indigofera sp 40%).

Table 1. Composition of feedstuffs and their nutrient contents in each treatment based on dry mater.

Feedstuff	Treatment				
	T0	T1	T2	T3	T4
	%				
Indigofera sp.	0	10	20	30	40
Pennisetum purpureum	100	90	80	70	60
Nutritional content					
Crude Protein	11.72	12.91	14.41	15,29	16.48
Crude Fiber	26.48	26.37	26.26	26,15	26.04
Nitrogen Free Extract	44.47	43.89	43.31	42,73	42.15
Fat	1.19	1.46	1.72	1,99	2.25
Ash	16.14	15.37	14.61	13,84	13.08
Total Digestible Nutrient	59.12	59.30	59.48	59,66	59.84
Gross Energy (Kkal/kg)	3.098	3.089	3.079	3,070	3.060

Each treatment was replicated five time. Concentrates were added every day as much as 500 g/head. Measurable parameters included dry matter intake, body weight gain, feed conversion rate and efficiency of feed utilization. The resulted data were analyzed by Duncan test. In addition, the composition of feedstuffs and their nutrient contents can be seen in Table 1.

2.2. Feeding Trial

Complete feed were prepare containing silage mixed *Indigofera sp.* and *Pennisetum purpureum* were composition according treatment, and concentrate were added. Concentrate is a commercial product with nutrients contents are Dry Matter 84.50%, Crude Fiber 16.50, Crude Fat 7% and Nitrogen Free Extract 34%. The animals were place in individual pens during treatment. The ration were prepared according to the dry matter requirement of 4%. Silage and concentrate were prepare to the experimental lambs individually in plastic tray twice a day. Quantity of ration to lambs given daily was weighed and refusal in the next morning was weighed too. Feed consumption are amount of ration provided is reduced by remaining in dry matter and expressed in gr/head/da., while Body weight gain are final weight minus the first weight . The data were analyzed by Minitab 16 Statistic software.

3. Results and discussion

3.1. Consumption of dry matter

The results of statistical analysis showed that the complete ration based on mixture of *Pennisetum purpureum* and *Indigofera sp.* fermented gave a significant different effect ($P<0.05$) on dry matter consumption of male's Garut lamb fattening phase. Measured nutritional value to determine the efficiency of the use of Garut males age d 8 - 10 months are arranged on various kinds of protein and energy balance, listed in Table 2.

The consumption of rations is physiologically controlled to meet maintenance and production needs that are restricted by the capacity of the gastrointestinal tract, the calculation based on the amount of consumption of dry matter rations [9]. Maintenance need is nutrition to meets the needs of life processes without any activity and production, while production need is the nutrients for growth, milk production and other activities.

Table 2. Efficiency of Ration on Garut Male Lamb of Fattening Phase

Measured Variables	Treatments				
	T0	T1	T2	T3	T4
Consumption of Dry Mater g/head/day	987.06[a]	1,210.90[b]	1,313.38[c]	1,248.80[b]	1,261.07[b]
Daily gain g/head /day	73.20	108.80	124.60	86.80	121.20
Feed conversion	16.15	15.21	11.03	18.21	10.65
Feed efficiency %	7.28	8.76	9.48	6.94	9.66

Information ;
* T0 = Complete Feed Fermentation-Based of Pennisetum and Indigofera sp (100%: 0%), T1 = Complete Feed Fermentation-Based of Pennisetum and Indigofera sp (90%: 10%), T2 = Completely Feed Fermentation Based of *Pennisetum* and *Indigofera* sp (80%: 20%), T3 = Completely Feed Fermentation Based of Pennisetum and Indigofera sp (70%: 30%), and T4 = Completed Feed Fermentation Based of Pennisetum and Indigofera sp (60%: 40%).
** the same letter to the direction lines shows non-significant

Parakkasi [10] states that the feed intake will go well if the animals are in suitable environmental conditions. Environmental factors that have direct impact on feed consumption, temperature, humidity, and sunlight, while not directly influence, such as the weather. At high environmental temperature, feed consumption will be less, as well as in humid conditions, consumption would decreased. According Kearl [11], the dry matter requirements for sheep weighing approximately 30 kg is 1.150 grams / head / day. The results of this study showed that the consumption of dry matter varies between 836.53 to 1.404.00 grams/ head / day. Dry matter consumption in this study is higher than Rianto [12] 730 - 895 grams/head/day on local thin-tailed sheep that given polar, while Sayekti [13] 1,107.9 grams / head / day, on local sheep.The condition indicates that the amount of consumption of dry matter ration in Garut male lamb fattening phase is still in the range of dry matter requirement of sheep.

3.2. *The effect of treatment on growth of body weight*
The result of statistical test showed that complete ration based on mixture of *Pennisetum purpureum* and *Indigofera sp* gave no significant effect on the increase of body weight of Garut male lambs fattening phase. Weight gain is an increase in body size over a period of time as a result of growth, or it is a change in size that includes changes in life weights, shapes, linear dimensions, and body composition [14]. These changes include changes in body components such as muscles, fats, bones, and organs and chemical components especially water, fat, protein and ash in carcasses. Organ changes take place gradually so as to achieve the size and shape characteristics of each tissue.

According to Parakkasi [10], nutrients in the diet affects the growth, especially the energy content, that energy will accelerate weight gain in a phase of fattening, this condition occurs due to increased fat reserves inside the body. In addition to the influence of the energy content, weight gain in fattening also influenced by the protein, especially amino acids in the diet that will affect the formation of tissue proteins. the highest protein content in the ration, the development and increase of protein tissue in the body the faster, so the weight gain increases.

The results of this study indicated that the increase in body weight of male Garut lamb fattening phase ranged between 26 to 169 grams/head /day. Variations in body weight gain relative has a range of values that are very different, this condition occurs due to the influence of individual factors. Acoording Petrović [16], a great number of different factors influence the growth of lambs while nutrition, health condition and genotype belong to the most important ones. Daily weight gain in this study were close to Purbowati [17] , 115.33-128.90 grams/ head/day, and Ekawati [18] , 94 - 106.3grams/head/day.

3.3. *The effect of treatment of feed conversion and efficiency*
The result of statistical test showed that the complete ration based on mixture of *Pennisetum purpureum and Indigofera sp.* gave no significant effect on the conversion value of Garut male lamb on fattening. Feed conversion is closely related to body weight gain. The feed conversion value is the number of rations required to produce 1 kilogram of livestock production, as can be reflected by weight gain . In this study body weight gain non-significant although dry matter consumption was significant different. This may be due to individual livestock factors that are less able to digest , so feed conversion in this study not significant too. The results of this study indicate that the feed conversion value of Garut male lamb on fattening obtained ranged from 9.27 to 35.67 whilst the efficiency of the use of ration on Garut male lamb on fattening phase ranged from 3.11% to 12.70%, while Mathius [19], feed efisiensi on sheep range on 6.78 – 13.72.

4. Conclusion
The results showed that complete ration based on mixture of fermented *Pennisetum purpureum* and legume *Indigofera sp* on Garut male lamb at fattening phase gave effect to the consumption of dry matter with the highest dry ingredient consumption obtained in the provision of complete ration based on mixture of *Pennisetum purpureum* and *Indigofera sp* on ratio 80% vs. 20%, that was equal to 1313.38 gram/head /day, whereas in body weight increase, ration conversion rate and efficiency of ration use did not give real effects.

References
[1] Rahman D, Lasamadi S S, Malalantang, Rustandi S D A 2013 Pertumbuhan dan Perkembangan Rumput Gajah Dwarf (*Pennisetum Purpureum* cv. Mott) Yang Diberi Pupuk Organik Hasil

Fermentasi Em4 *Jurnal Zootek ("Zootek"Journal)* **32(5)**: 158–171, 158-165

[2] Tarigan A, Sirait J, Sheila P G 2013 Produksi Dan Komposisi Nutrisi Indigofera Sp. Pada Intensitas Pemotongan Dan Jarak Tanam Yang Berbeda Di Dataran Tinggi Dengan Curah Hujan Sedang *Seminar Nasional Teknologi Peternakan dan Veteriner* 441-447

[3] Hassen A, North F G R, Van N, Tikh J T 2007 Influence of season/year and spesies on chemical composition and in vitro digestibility of five *Indogofera accessions Anim. Feed Sci. and Technol.* **136**: 312 – 322.

[4] Abdullah L, Suharlina 2010 Herbage yield and quality of two vegetative parts of Indigoferaat differents times of first regrowth defoliation *Media Peternakan* **33(1)**: 44 – 49.

[5] Simanihuruk K, Sirait J 2009 Pemanfaatan leguminosa pohon *Indigofera sp* sebagai pakan basal kambing Boerka fase pertumbuhan *Seminar Nasional Teknologi Peternakan dan Veteriner* 449 – 454

[6] Sibel S Ö, Mürsel Ö, Abdurrahman A, Turgay T 2015 Meat Quality And Fatty Acid Composition Of Chios Male Lambs Fed Under Traditional And Intensive Conditions *Emirates Journal of Food Agriculture* **27(8)**: 636-642

[7] Borton R J, Loerch S C, Clure K E, Wulf D M 2005 Comparison of characteristics of lambs fed concentrate or rygerass to traditional or heavy slaughter weights. I. Production, carcass, and organoleptic characteristics *J Anim Sci* **83**: 679-685.

[8] Aurousseau B, Bauchart D, Faure X, Galot A L, Prache S, Micol D, Priolo A 2007 Indoor fattening of lambs raised on pasture: (1) Influence of stall finishing duration on lipid classes and fatty acids in the longissimus thoracis muscle *Meat Sci* **76**: 241-252.

[9] National Research Council 1985 Nutrient requirements of sheep 6th revised ed. National *Academic Press* Washington DC pp. 45-50.

[10] Parakkasi 1999 *Ilmu nutrisi dan makanan ternak ruminansia.*Universitas Indonesia Press, Jakarta

[11] Kearl L C 1982 Nutrition requirement of ruminant in developing countries *Utah State University* Logan USA.

[12] Rianto E, Haryono E, Lestari C M S 2006 Produktivitas domba ekor tipis jantan yang diberi pollard dengan aras berbeda *Seminar Nasional Teknologi Peternakan dan Veteriner* 431-439.

[13] Sayekti I, Purbowati E, Endi R 2015 Pemanfaatan Protein Pakan pada Domba Lokal Jantan yang Mendapat Pakan pada Siang dan Malam Hari *Animal Agriculture Journal* **4(1)**: 22-27.

[14] Soeparno 2005 Ilmu Dan Teknologi Daging Cetakan 4 *Gadjah Mada University Press.* Yogyakarta. **1**, 35- 45, 51 - 58, 62 - 69, 131- 132, 264, 324 - 327.

[15] Petrović V C, Petrovicl M P, Ružić-Muslić D, Maksimović N, Selionova M I, Aybazov M M, Malyukova M A 2015 Genotype, Sex And Interaction Effect On Lamb Growth Traits *Biotechnology in Animal Husbandry* **31 (1)**: 37-44

[16] Purbowati E, Sutrisno C I, Baliarti E, Budhi S P S 2009 Penampilan Domba Lokal Jantan Dengan Pakan Komplit Dari Berbagai Limbah Pertanian dan Agroindustri *Seminar Nasional Kebangkitan Peternakan* 130-138.

[17] Eni E, Anis M, Sunarso 2014 Efisiensi dan Kecernaan Ransum Domba yang Diberi Silase Ransum Komplit Eceng Gondok Ditambahkan Starter Lactobacillus plantarum *Agripet* **14(2)** : 107-114

[18] Mathius I W, Dwi Y, Eliani W, Budi H, Anderson W, Abu T 2001 Pemanfaatan Energi Terlindungi Untuk Meningkatkan Efisiensi Penggunaan Pakan Pada Domba Induk *Jurnal Ilmu Ternak dan Veteriner* **6(1)**: 7-13

Reproductive Rate Performance of Boer Goat and Its F1 Cross in Indonesia

T Nugroho[1], C Kustiyani[1], A Ratriyanto[1], N Widyas[1], S Prastowo[1]

Department of Animal Science, Universitas Sebelas Maret, Surakarta Indonesia

E-mail: prastowo@staff.uns.ac.id

Abstract. Aim of crossbreeding practice is to increase local or indigenous breeds productivity by introducing exotic breeds. It expected that crossbreed result better performance. For that, this study aims to evaluate the reproductive rate (RR) performance of Boer goat compared to its F1 cross. Kidding record were collected in CV. Kambing Burja, East Java, Indonesia from January 2012 to December 2015. In this study, Boer bucks (purebred) were mated with two doe breeds that are female Boer (100% Boer) and female F1 Cross (50% Boer). For evaluation purposes, litter size, kidding interval and annual pre-weaning mortality were used as components for reproductive rate calculation in both mating system. Result shows Boer does significantly (P<0.05) perform better in litter size at year 2012, 2014 and 2015 when compared to F1 cross. In term of kidding interval, Boer and F1 cross shows different value in each observation year, moreover Boer goat show lower preweaning mortality percentage than F1 cross. According to RR value, we can see that Boer has better performance compared to F1 cross, however its value was varied in every year observation. Based on this result we can concluded that Boer reproductive performance is relatively better than its F1 cross.

1. Introduction

Crossbreeding is a way to increase local or indigenous breeds productivity by introducing exotic breeds that have superior quality [1]. The effect of crossbreeding will bring up the hybrid vigor (HV) or heterosis and it is expected that crossbreed resulted better performance compared to the both parents such as production or reproductive traits [2]. However, it cannot be denied that crossbreeding also have negative effects, such as mating with an unmeasured breed composition which can cause a decreasing of genetic diversity [3].

Indonesia has various local or indigenous goat breeds such as Gembrong, Kacang, Jawarandu, Kosta, Samosir, Etawa Grade and many others [4]. Indonesian local goats are well-known for their robustness towards environment and feed sources, however it has small body size and low production performance [5]. To solve this problem, crossbreeding commonly practiced by introducing new genetic sources with high productivity performance. For example, to improve local goat meat production Boer goat was introduced. Boer goat is a meat type goat originated from South Africa equipped with excellent body conformations, good carcass quality, fast growing and high adaptability to any environmental challenges [6]. Introducing Boer to local goat is able to increase carcass percentage of crossbred goat [7]. Based on that characteristic, Boer become a potential genetic resources in crossbreeding programs for improving Indonesian local goat productivity.

Crossbreeding between Boer and Kacang which aims to produce Boerka goat has been found in Indonesia. Boerka has better capacity to grow, larger body weights than Kacang and relatively adaptive to tropical condition. Compared to other breed, in average, Boerka yielded extra birth and weaning weight of 32% and 35% respectively [8]. Another report shows that crossbred between Boer buck and Etawah does could increase birth weight by 15.6% and weaning weight by 10.07% [9]. The success of a crossbreeding program, could not only measured by the production traits of the crossbreds, however, their reproduction also matters as it relates to the sustainability.

Both fertility and reproductive rate (RR) are the key factors to increase population as well as rearing profit. Reproduction rate is defined as the number of offspring weaned from every doe per year [10,11]. Reproduction rate is estimated by incorporating the elements of litter size, pre-weaning mortality and kidding interval [10]. This parameter measures the productivity of a given population in annual base. The term productivity in this context referred to the ability of a population to survive and produce offspring based on the number of breeding females.

Studies on Boer crosses have been conducted, however none of them concerned in discussing reproduction traits as a unity. Therefore, this study aims to evaluate the RR performance within Boer goat compared to its F1 cross. The term of F1 cross goat is the result of crossbreeding between Boer buck with Jawarandu doe which commonly called Boerja. For the additional information, Jawarandu is an Indonesian local goat which is also a product of crossbreeding between Etawah buck and Kacang does that have been establish many years ago and spread along the archipelago.

2. Materials and Methods

Pedigree and birth records at CV. Kambing Burja, Malang, East Java, Indonesia were used in this study. Data were obtained since the first kidding occur started from January 2012 until March 2016, followed with data classification which based on doe breed and years of kidding. In total, data were retrieved from 18 Boer bucks which mated to 174 Boer does (100% Boer) and 642 F1 cross (50% Boer).

Boer goats were obtained from Australia or which born in Indonesia. Meanwhile, the F1 cross is the result from crossbred between Boer buck and Jawarandu doe. Jawarandu goats were obtained from local market near the farm without parity and age information. Female goats were reared in colony, 20-25 head per flock and didn't separated between breed composition, therefore no different in farm management. Mating process was performed by joining Boer buck in the flock for 45 days.

To evaluate reproductive performance, RR value which consist of litter size (L), pre-weaned mortality percentage (M) and kidding interval in year (I) were calculated using the following formula [10] for every year of observation:

$$RR = \frac{L(1-M)}{I} \qquad (1)$$

Litter size was calculated based on the number of kid from each doe that giving birth. Kidding interval is the distance between two kidding time, calculated in term of year. Moreover, pre-weaned mortality was calculated in % from the number of weaned kid at 77 days divided with number of kids which are born. Data of litter size and kidding interval between breed were compared using *t-test* (α=0.05), while mortality percentage and RR value were analyzed by descriptive.

3. Results and Discussion

3.1. Reproductive Performance of Boer goat and F1 Cross

The yearly reproductive performance of does were shown in Table 1. Boer does significantly ($P<0.05$) shows better performance in litter size at year 2012, 2014 and 2015 when compared to F1 cross. In term of kidding interval, there were no significance ($P>0.05$) differences between breeds in each observation year, moreover Boer goats also shows lower preweaning mortality compared to F1 cross.

Litter size is generally influenced by doe than buck [12]. Boer goat known as prolific bread and has fecundity rates until 210% or litter size for about 2.1 kids born/doe kidded [7]. Boer goat is assumed to be able to ovulate more fertile ova than F1 cross. Moreover, kidding interval in this study shows better compared to the previous study at 0.82 year [8]. In this regards, kidding interval is highly influenced by farm management, where good management practice would decrease kidding interval significantly

IOP Conf. Series: Earth and Environmental Science **334** (2019) 012008 doi:10.1088/1755-1315/334/1/012008

[13] because related to reproductive efficiency in mating management. A strict mating schedule was applied in CV. Kambing Burja, therefore all the breeds are reared under the same condition resulted in relatively uniform of kidding interval. Pre weaning mortality was influenced by management, litter size and mothering ability [8]. It is reported that Boer has better mothering ability than other crossbred [14], Boers are able to produce sufficient milk to raise the kids [7,15] resulted less number of preweaning mortality.

Table 1. Reproductive traits of Boer goat and F1 cross does

Trait	Dam	2012	2013	2014	2015
Litter Size	Boer	1.765	1.578	1.557	1.535
	F1	1.275	1.461	1.419	1.511
	P Value	<0.001	0.046	<0.003	0.631
Kidding Interval (year)	Boer	0.806	0.743	0.745	0.768
	F1	0.736	0.766	0.753	0.776
	P Value	0.124	0.209	0.471	0.590
Preweaning Mortality (%)	Boer	3.333	28.324	14.286	32.653
	F1	8.163	25.959	18.530	43.004
Reproductive Rate	Boer	2.117	1.522	1.791	1.345
	F1	1.591	1.412	1.536	1.109

3.2. Reproductive rate of Boer goat and F1 Cross

According to RR calculation, we can see that Boer has better performance compared to F1 cross in every year of observation (Table 1 and Figure 1). It can be said that Boer has higher number of offspring weaned from every doe per year, thus more efficient in reproductive performance. Moreover, we also see the variation of RR value in all the year observation. Despite breed genetic differences, rearing management or environmental factor seems to be the factor which vary year by year that may influencing the RR variation [12]. In addition to that, the level of reproductive efficiency is also affected by litter size, weaned kid, kidding interval and the length of the reproductive cycle [15]. At this point, we can assume that reproductive efficiency differences between Boer and its F1 cross is related to the genetic potential. This assumption is in line with the finding in the number of litter size as well as preweaning mortality value which represent mothering ability (Table 1). A better genetic value due to long term selection in Boer could be one of plausible explanation. The more number of survived kids per doe per year would be a sign of good farm management practice that leads to higher efficiency. The last but not the least, a good plan along with continues evaluation is a must to achieve better genetic gain in crossbreeding program as well as to maintain the local breed genetic diversity.

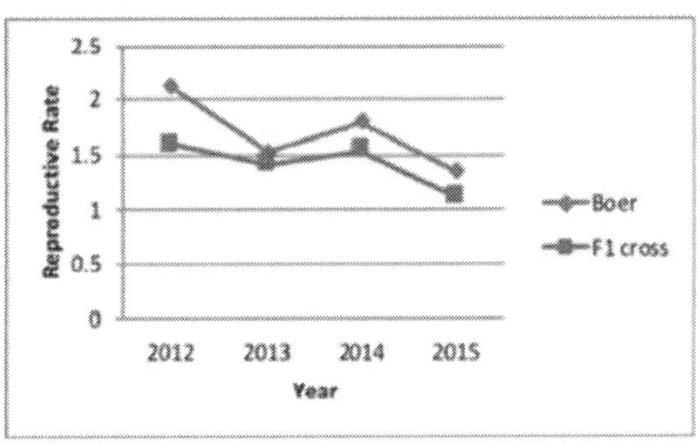

Figure 1. Reproductive rate variation of Boer and F1 cross during observation

4. Conclusion
It is concluded that reproductive performance of Boer goat is better than its F1 cross. A better breeding strategy and the improvement of farm management should be done in parallel to gain success of breeding goal.

References
[1] Widyas N, Nugroho T, Prastowo S 2017 IOP Conference Series: Materials Science and Engineering **193**: 12-37
[2] Prastowo S, Widi T, Widyas N 2017 IOP Conference Series: Materials Science and Engineering **193** 12028
[3] Ribeiro M N, da Rocha L L, de Carvalho F F, Costa R G, Gomes-Arandas J K, Ribeiro N L 2017 Sustainable Goat Production in Adverse Environment: **1** 101–10
[4] Periasamy K, Vahidi S M, Silva P, Faruque M, Naqvi A, Basar M, Cao J, Zhao S, Thi L, Pichler R, Garcia M, Shamsuddin M, Boettcher P, Fernando J, Han J, Ajmone P, Diallo A, Viljoen G J 2017 Small Ruminant Research **148:** 2–10
[5] Budisatria I G S, Udo H M J, Eilers C H A M, Van Der Zijpp A J 2007 Outlook on Agriculture **36** 145–52
[6] Casey N H, van Niekerk W A 1988 Small Ruminant Research **1:** 291–302
[7] Malan S W 2000 Small Ruminant Research **36** 165–70
[8] Elieser S, Sumadi, Budisatria G S, Subandriyo 2012 Journal Indonesian Tropical Animal Agriculture **37:** 15–21
[9] Kostaman T, Sutama I-K 2006 Seminar Nasional Teknologi Peternakan dan Veteriner 2006 pp 522–7
[10] Gatenby R M 1986 Singapore: *Logman Singapore Publishers Ltd.*
[11] Pellerin A N, Browning R 2012 *BMC Veterinary Research* **8(1)**.
[12] Boujenane I, Kansari J 2002 L *Small Ruminant Research* **43** 115–20
[13] Haldar A, Pal P, Datta M, Paul R, Pal S K, Majumdar D, Biswas C K, Pan S 2014 *Asian-Australasian Journal of Animal Sciences* **27** 628–34
[14] Browning Jr R, Leite-Browning M L 2011 *Journal Animal Science* **89** 2696–707
[15] Greyling J P C 2000 *Small Ruminant Research* **36** 171–7

Performance and Income Over Feed Cost of Rambon Duck that Given the Ration Containing *Gold Snail (Pomaceae canaliculata)* and *Noni Fruit (Morinda citrifolia L)* Flour

W Tanwiriah[1], L Nurlina[1], D Garnida[1] and E Sujana[1]

[1]Animal Husbandry Faculty, Universitas Padjadjaran, West Java, Indonesia

E-mail : *wi2ntanwiriah@yahoo.co.id*

Abstract. The golden snail can breed very quickly and could become a pest for the_plants especially rice plants. The Gold snails contains to use as duck feed. One of ducks that developed in West Java, is Rambon duck. In the maintenance of ducks, to improve productivity, usually required additional substances in ration. Noni plant which fruit is very nutritious, and contains anti-microbial that increase the productivity of livestock. This study aims to determine the level of gold snails and the Noni Fruits flour in a ration that gave the best Rambon duck performance and IOFC in the Jatigede Reservoir area. The research was done experimental and all data were analyzed statistically with Anava and further test with Duncan Test. The research used 200 DOD of Rambon duck that given 4 ration treatment containing gold snail flour (K) and noni fruit flour (M). The treatments were P1 (R1+0.45% M), P2 (R1+0.60% M), P3 (R2+0.45% M), dan P4 (R2+0.60% M). Parameters measured were feed consumption, body weight gain, feed conversion, carcass weight and IOFC. The results showed that P2 treatment (gold Snail 5%+0,60% noni fruit) gave the best Rambon duck performance with the value of IOFC Rp. 11,675.74 each bird

1. Introduction

In areas with a lot of water such as rice fields or the area around the reservoir usually live many types of snails. The gold snail (*Pomaceae canaliculata*) is one of many species of snails found in various places and breed very quickly , so in long time that will become pests of plants, especially can damage the rice plants. According to the research gold snails contains high enough protein, so it can be used as a source of protein for duck feed. The content of protein gold snail flour ranged from 38.9% to 51.44% and Metabolizable Energy is 2.394 kcal/kg up to 3.189 kcal/kg [1]. In around the reservoir is also suitable to be maintained waterfowl is like ducks.. One of the ducks that developed in West Java is the Rambon ducks from Cirebon. Rambon duck has a slim body shape, 45 - 50 cm high and upright steps. Male duck Rambon weights average 1.4 kg, while the female 1.2 kg. Rambon ducks have shiny brown male feather in the neck, black head, black beak, and 70% have brown fur and brown thigh [2]. The use of gold snail for duck feed is good because its reduce pests and also reduce the cost of duck feed that very high about 70% of the cost production. According to the research, the use of gold snail flour up to 6% in the ration did not decrease the performance (weight gain, ration conversion, weight and percentage of carcass, abdominal fat) on MA male ducks (Mojosari-Alabio) [3]. The addition of the golden snail flour in the ration to the 9% level also did not affect the duck eggs quality including egg weight, egg white index, egg yolk index, egg yolk, HU value, eggshell weight and eggshell thickness [4]. However, fresh gold snails contain anti-nutritional that can adversely affect livestock growth. These anti-nutritional can be removed by heating or boiling. The gold snail meat flour (previously treated with boiled for 15 minutes) can be used up to 11.25% in the chicken ration without decreasing performance [5]. The snail meat flour (previously treated with soaked lime (Calcium Oxide) for 15 minutes) can be used up to 6% in broiler chicken [6].

Maintenance of ducks in the villages traditionally, rarely use chemical drugs, but replaced with herbal plants that serve as a drug and growth promote. Types of medicinal plants commonly used is noni fruits (*Morinda citrifolia* L). One of the active substances in the noni fruit that c an improve the performance of poultry is Saponin. The use of plants containing saponins in monogastric animals can increase feed production, growth and efficiency, and improve the quality of meat [7,8]. Saponins can increase nutrient absorption in the gut [9, 10]. Noni fruit contains proxeronine alkaloids and proxeronase enzymes that will form xeronine active substances in the intestines. Xeronine's active substances can help proteins perform their functions and work well. These substances are brought to the body's cells through the blood so that the cells work more actively [11]. Xeronine will be absorbed by the body's cells to activate and maximize the function of inactive proteins, regulate the structure and form of active cells. The specific function of this xeronine substance is to protect the cell membrane, so that the cell becomes more active, healthy and there are structural and functional improvements. Including improvements in liver cells [12]. Xeronine is capable of stimulating the immune system, regulating cell function and cellular regeneration of damaged cells [13]. Xeronine can also speed up the absorption of nutrients into the digestive system and harmonize the workings of cells in the body [14]. According to the research, the addition of 3 ml/ kg ration (0.3%) Noni fruit flour in broiler feed can give a real impact in the addition of broiler chicken body weight [15]. In another research, the addition of 4.8 g / kg (0.48%) of non-fruit juice in the feed had no significant effect on final weight and percentage of carcasses, but resulted more efficient rations and lower feed conversion [16]. The aim of the research were to determine the level of gold snail (*Pomaceae canaliculata*) and *Morinda citrifolia* L in the rations that produced the best Rambon duck performance at Jati Gede reservoir area in Sumedang.

2. Experimental Details

This Research used 200 DOD of Rambon duck. Each duck is given a wingtag and then put randomly into the cage and maintained for 8 weeks. Duck is placed on a bamboo cage that a length x width x height is 2 m x 1.5 m x 0.5 m. Each cage is occupied by 10 ducks. Each cage has one round feeder and waterer. For the first 2 weeks of the study each cage was given one 60 watt lamp as a brooder. Feed ingredients are corn, rice bran, soybean meal, fish meal, coconut oil and gold snail powder. The content of nutrients and metabolizable energy of the rations based on the need of pekin duck [17] are ME 2900 kcal / kg and protein 17%. Before used gold snails was boiled for 15 minutes, then dried and ground. Noni fruits was taken in a fairly mature so the fruit is still a bit hard with the color is slightly yellow then dried and ground. The ration of the research can be seen in Table 1.

Research using Completely Randomized Design. There were 4 treatments ration containing gold snails (K) and Noni fruits (M) which is different level. There are 4 treatmens that were P1 (R1 + 0.45% M), P2 (R1 + 0.60% M), P3 (R2+ 0.45% M) and P4 (R2 + 0.60% M), each treatment was repeated 5 times and the number of ducks 10 per cage . The parameters observed were feed consumption, weight gain, feed conversion, carcass weight and Income Over Feed Cost (IOFC). To know the effect of treatment then the data were analyzed statistically with Anava and to know the difference between treatment analyzed by Duncan Test

3. Results And Discussion

The results of research can be seen in Table 2. The feed consumption in all treatment not too differences and variant analyzed indicate that the treatment has no significant effect on the feed consumption. The ration on this research made iso protein and iso energy, so if there is an increase or decrease of consumption caused by treatment of gold snail flour and noni fruits. The result of the research was found that the use of gold snail up to 10% and 0.060% noni flour in the ration give not effect to feed consumption, so it can be said that the ration is still palatable. The palatability of the ration is one of the important factors determining the level of feed consumption that is influenced by smell, taste, texture and color of ration [18]. The use noni fruits too high is not recommended because the noni fruit contains kaporat acid and kapirik acid that gives the typical smell [19]. Indeed, many factors that influence the feed consumption are environmental temperature, genetic, health, ration form, balance of nutritional rations, stress, body size, growth rate and egg production [20].

Table 1. Ration of the Research

Feeds	R1	R2
	..............(%)............	
Corn	51.00	51.00
Soybean Meal	10.00	10.00
Fish Meal	6.00	3.00
Rice Bran	24.00	22.00
$CaCO_3$	1.50	1.50
Cornel oil	2.00	2.00
Premix	0.50	0.50
Gold Snails Flour	5.00	10.00

The contains of the Rations	R1	R2
ME (Kkal/kg)	2835	2839
Crude Protein (%)	17.11	17.31
Crude Fat(%)	8.57	8.56
Crude Fiber (%)	8.35	8.24
Calcium (%)	0.89	0.83
Phosphor (%)	0.72	0.67
Lysin (%)	0.85	0.70
Methionin (%)	0.33	0.27
Met + Systin (%)	0.54	0.47

Table 2. The Effect of Treatment on Rambon Duck Performance

No.	Parameter	P1	P2	P3	P4
1.	Feed Consumption (g)	4,414.81	4,478.44	4,676.84	4,500.68
2.	Body Weight gain(g)	1,169.03 ^{b}	1,236.45 ^{c}	1,132.77 ^{ab}	1,122.91 ^{a}
3.	Feed Conversion	3.78 ^{ab}	3.63 ^{a}	4.14 ^{c}	4.02 ^{bc}
4.	Carcass Weight (g)	665.50 ^{a}	751.30 ^{b}	682.50 ^{ab}	655.80 ^{a}
5.	Percentage of Carcass (%)	54.57	60.97	57.00	55.61
6.	Income Over Feed Cost (Rp)	10,218.80	11,675.74	7,955.15	9,292.75

Note : The different letter at the same line show the treatment significant different

The increase of Rambon duck body weight gain range from 1,122.9 – 1,236.45 g at 8 weeks of age. This result is lower than the results of Ahmad Subhan et al.[7] research in ducks that were given 6% of snail flour that is 1462.63 g. This is because of the different types of ducks namely Mojosari ducks. The highest Rambon duck is achieved by P2 with ration treatment containing 5% gold snail flour and 0.06% noni fruits addition. The treatments of the research gave significant effect ($P< 0.05$) to body weight gain and Duncan test showed that the treatment P2 (5% K + 0.60% M) was significantly higher than P1, P3 and P4. Body weight gain. P1 treatment was significantly higher than P4, but did not differ from P3 and P3 was not different from P4. The best treatment was P2 that was combination of 5% gold snails and 0.06% noni fruits. The use of 0.60% noni fruits has been enough to give all the benefits, while the use of 0.45% not effective to promote growth. This is in line with previous research that the giving of noni fruit of 0.48% in ration did not give a real effect on body weight gain due to lack of dose [16]. Giving noni fruits increase body weight, this is because of the compound of noni fruits. According to [21] active substances compounds in noni fruits including xeronine, proxeronin, proxeronase, serotonin, vitamin C, anti-oxidants, minerals, proteins, enzymes, alkaloids, and fitronutrien others, are very active in strengthening the immune system, improve cell function and accelerate the regeneration of damaged cells. In addition, the use of plants containing saponins in monogastric animals can increase the production, growth and feed efficiency, and improve the quality of meat [7,8]. Saponins can increase nutrient absorption in the gut [9,10]. In this study the best use of gold snails is 5%, because the use of 10% to tend the decrease of body weight. This is because the gold snail contains anti-nutritional substances, so it should be used less than 10%. The use of more gold snails than 6%, feed consumption and body weibht gain decreased and feed conversion increased [5]. Similarly, the results of research in [22] that the replacement of fish meal (10% in ration) with gold snails only 25-75%, that is 2.5% to 7.5%, gives a good influence on individual daily growth rate, feed efficiency, protein retention and fat retention in fish. The results of [3] research argues that the use of 6% golden snail flour in the ration produce a performance that is proportional to feed consumption.

The lowest of Rambon duck feed conversion was achieved by the P2 (5% K+0,60% M) treatment and the highest of feed conversion by P3 treatment. In general, the conversion of rations in ducks is still very large, this happens because of the way of eating of ducks that is after taking the food then immediately drink, so many rations are wasted to the water. As is the case in other studies, the result that the feed conversion of duck is 4.49 – 5.31 [23]. In this study the feed consumption is not different so that the difference is the result of its body weight gain. The highest body weight gain is P2 treatment, so resulting the lowest feed conversion. Feed conversion is the ratio of the amount of feed consumption to body weight gain, when the smaller is more efficient. Feed conversion is influenced by large body and nation, production level, energy ration and environmental temperature [24]. In animal husbandry the value of feed conversion is minimally influenced by three factors of ration quality, feeding management, mortality rate. The key to successful business in broiler farming (livestock) is the feed conversion [25]. According to [26, 27], the size of feed conversion is influenced by feed quality and livestock's ability to convert food into meat, feed balance, body size, environmental temperature, live weight, physical form of feed, strain and sex.

The carcass weight of the research ranged from 665.50 to 751.30 grams. This result is lower than the research by [3] at the 8 weeks old in male ducks that ranged from 854.76 to 954.90 gram. The highest carcass weight in this research is achieved by P2 treatment. This is in line to the final body weight because carcass weight is related to the final body weight. Carcass weight is influenced by age, species of livestock, ration and body weight or body size [28]. The result of ANAVA indicated that the treatment had significant effect ($P< 0.05$) on carcass weight. Duncan test showed that the carcass weight of Rambon duck in P2 treatment was significantly higher than P1, P3 and P4 treatments. This is in line with body weight P2 which also higher, because carcass weight is the largest part of the body of the ducks.

Carcass Percentage of the research ranged from 54.57% to 60.97%. This result is higher than the results of [19] research that ranged from 52.0 to 55.7%, but lower than the results of [3] in male duck that is range from 59.70% to 61.10% from life weight. The results of ANAVA showed that the treatment had no effect on the percentage of carcass. This is because at the age of 8 weeks the ducks grow has not maximal yet, so at this age the grown only just the body frame. Ducks already look big but the muscle in the chest is still thin. So on all ducks non-carcass parts such as head, neck, legs and inside part have developed first. The percentage of carcasses shows the proportion of the body part. Differences in the proportion of carcass occurs when there is a different grow of muscle, especially in the chest area at the same age. According [29] percentage of carcass is influenced by growth rate and quality of feed. Growth rates are indicated by the increase of live weight and the increase percentage of carcass.

The results of the highest Income Over Feed Cost (IOFC) was obtained at the P2 treatment that is Rp. 11,675.74 per head This occurs because of the P2 has high body weight gain, while the amount of feed consumption is relatively the same. IOFC is one of the way to know the economic value in livestock business that is the income of duck sale, reduction with feed cost in one period of duck raising. The more efficient in feed conversion into life weight (its mean the feed conversion is low), the better of IOFC value [24]. IOFC is a barometer to see that the feed cost is the biggest cost in the livestock business [30]. Feed is the primary cost of the livestock business in intensive system where feed costs reaches about 70% of the total production cost, therefore the feed ingredient price greatly affects the production cost [31]. Factors influencing the value of IOFC are feed consumption, final body weight, price per kg of ration and price per kg of live weight [24].

4. Conclusions

The level of gold snails (*Pomaceae canaliculata*) and Noni Fruits (*Morinda citrifolia L*) in the ration had an effect on body weight gain, feed conversion and carcass weight, but had no effect on feed consumption and carcass percentage. The level of gold snails 5% and noni fruit 0.06% in the ration produced the best Rambon duck performance with the value of Income Over Feed Cost Rp. 11,675.74 per head.

Acknowledgements

The authors would like to thank to the Rector of Universitas Padjadjaran who has funded this research through the Internal Grants of Unpad and to Nanda, Rizki, Yusita, Fakhri, Gumilar, Dimas and Fauzan who have assisted the research.

References

[1] Laboratorium Nutrisi dan Makanan Ternak 1995 Fakultas Peternakan Universitas Padjadjaran
[2] Setioko A R, Sopa S, Tine S 2005 Identifikasi Sifat-Sifat Kwalitatif dan Ukuran Tubuh Pada Itik Tegal, Itik Cirebon dan Itik Turi *Seminar Nasional Teknologi Peternakan dan Veteriner* BPT. Bogor.
[3] Ahmad S, Rohaeni E S, Qomariah R 2009 Pengaruh Penggunaan Kombinasi Sagu Kukus dan Tepung Keong Mas Dalam Formulasi Pakan Terhadap Performans Itik Jantan MA Umur 1 – 8 Minggu *Seminar Nasional Teknologi Peternakan dan Veteriner* Balai Pengkajian Teknologi Pertanian Kalimantan Selatan.
[4] Purnamaningsih A 2010 Pengaruh Penambahan Tepung Keong Mas (*Pomaceae canaliculata* Lamarck) Dalam Ransum Terhadap Kualitas Telur Itik *Skripsi Fakultas Pertanian Universitas Sebelas Maret* Surakarta.
[5] Darmawan F 1996 Pengaruh Pemberian Berbagai Tingkat Tepung Daging Keong Murbei (*Pomaceae canaliculata*) sebagai Pengganti Tepung Ikan Dalam Ransum Terhadap Performans Ayam Jantan Tipe Medium *Skripsi* Fakultas Peternakan Unpad Sumedang.
[6] Wirontono A 2001 Pengaruh Pemberian Ransum Yang Mengandung Berbagai Tingkat Tepung Daging Keong Mas (*Pomaceae canaliculata*) Yang Diolah Dengan Kapur Sirih, Terhadap Performan Ayam Broiler *Skripsi* Fakultas Peternakan Unpad Sumedang.
[7] Cheeke P R 2001 Actual and Potential of Yucca Schidigera and Quillaja Saponaria in Human and Animal Nutrition *Recent Anv Animal Nutr*. Aust. **13**: 115-126.
[8] Miah M Y, Rahman M S, Islam M K, Monir M M 2004 Effects of Saponin and L-carnitine on The Performance and Reproductive Fitness of Male Broiler *Int.J. Poultry Sci* **3**:530- 533.
[9] Jhonson I T, Gee J M, Price K, Curl C, Fenwick G R 1986 Influence of Saponin on Gut Permeability and Active Nutrien Transport in Vitro *J Nurt*,**115**: 2270-2277
[10] Onning G, Q Wang, B R Westbrom, N Asp, B W Karlson 1996 Influence of Oat Saponins on Intestinal Permeability in vitro and in vivo in The Rat Br *J. Nurt*. **76**: 141-145
[11] Heinicke 1994 Xeronin and Cell Regenerration in scientific Research on Noni Fruit *Alexandra Dittimor* Frangfrut.

[12] Wijayakusuma H M 1992 Tanaman Berkhasiat Obat di Indonesia. Jilid 1 *Pustaka Kartini* Jakarta pp. 9
[13] Winarno R G 2003 Rahasia Morinda Citrifolia atau Noni. www. kompas.com/harian/0110/rgmI.html diakses tanggal 20 Maret 2016.
[14] Djauhariya E, Rahardjo M, Ma'mun 2006 Karakterisasi Morfologi dan Mutu Buah Mengkudu.*Bul. Plasma Nutf.* **12(1).**
[15] Sujana E, Syafril D, Garnida D 2007 Efek Pemberian Ransum Mengandung Tepung Buah Mengkudu (*Morinda citrifolia* Linn) Terhadap Performa Ayam Broiler *Seminar Nasional "Pengembangan Sistem Produksi dan Pemanfaatan Sumberdaya Lokal untuk Kemandirian Pangan Asal Hewan"* Fakultas Peternakan Unpad Sumedang.
[16] Bintang L A K, Sinurat A P, Purwadaria T 2007 Supplementation of Morinda citrifolia waste as bioactive compound on the Performances of Broiler *JITV* **12(1)**: 1-5
[17] Sinurat A P 2000 Penyusunan Ransum Ayam Burat dan Itik. *Pelatihan Proyek Pengembangan Agribisnis Peternakan* Dinas Peternakan DKI Jakarta
[18] Church D C 1979 Livestock Feed and Feeding *Durhan and Cowney, Inc.* Portland Oregon
[19] Winarti C 2005 Peluang Pengembangan Minuman Fungsional Dari Buah Mengkudu (*Morinda citrifolia* L.) *Jurnal Litbang Pertanian* **24(4)**: 149-155
[20] Ensminger 1980 Poultry Sience Second Edition *The Interstate Printers & Publisher Inc.* Danville Illinois.
[21] Djauhariya E, Tirtoboma 2002 Mengkudu (Morinda citrifolia L.) Tanaman Obat Tradisional Multi Khasiat *Warta Penelitian dan Pengembangan Pertanian* **7**: 1-7
[22] Sulistiono 2007 Keong Mas Sumber Pakan dan Obat-obatan. *Departemen Manajemen Sumberdaya Perairan* Fakultas Perikanan dan Ilmu Kelautan Institut Pertanian Bogor
[23] Bayu W, Heny Setyo P, Nuryadi 2015 Pengaruh Penambahan tepung Buah Mengkudu (Morinda citrifolia L.) Dalam Pakan Terhadap Penampilan Produksi Itik Hibrida *Jurnal Ilmu-Ilmu Peternakan* Fapet UB. **25(2)**: 28-35. http://jiip.ub.ac.id/
[24] Rasyaf M 2004 Beternak Ayam Pedaging *PT Penebar Swadaya*, Jakarta
[25] Abidin Z 2002 Meningkatkan Produktifitas Ayam Kampung Pedaging *Agromedia Pustaka*
[26] Kamal M 1997 Pengaruh penambahan DL metionin sintetis ke dalam ransum fase akhir terhadap perlemakan tubuh ayam broiler *Bull Peternakan* **18**: 40 – 46.
[27] Zuprizal 1993 Pengaruh penggunaan pakan tinggi protein terhadap penampilan, karkas dan perlemakan ayam pedaging fase akhir *Bull. Peternakan* **17**: 110 – 118.
[28] Iskandar S 2005 Pertumbuhan dan Perkembangan Karkas Ayam Silangan Kedu x Arab Pada Dua Sistem Pemberian Ransum *JITV* **10(4)**: 253-259
[29] Soeparno 2005 Ilmu dan Teknologi Daging Cetakan Keempat *Gadjah Mada University Press* Yokyakarta.
[30] Prawirokusumo S 1990 Ilmu Gizi Komparatif *BPFE* Yogyakarta.
[31] Supriyati, Z D, Kompiang I P, Soekanto P, Abdurachman 2003 Peningkatan Mutu Onggok Melalui Fermentasi dan Pemanfaatannya Sebagai Bahan Pakan Ayam Kampung *Pros. Seminar Nasional Teknologi Peternakan dan Veteriner* Bogor.

Evaluation of Microorganism from Integrated Processing Mixed Sheep Feces and Rice Straw as Starter Culture

Y A Hidayati[1], **S Nurrachma**[1] **and W Juanda**[1]

[1]Faculty of Animal Husbandry, Universitas Padjadjaran, 45363 Sumedang, Indonesia

E-mail : yuli.astuti@unpad.ac.id

Abstract. This study aims to determine the potential of microorganism in processing of mixed sheep feces and rice straw, used as a starter in the preservation of forage feed. The method is, in the 1st phase with 3 treatments of carbon and nitrogen ratio with 6 repetitions and the 2nd phase with 3 treatments of molasses in the base medium, which are grouped in dry matter content with 6 replications. The results of the 1st phase showed the preferred treatment of T2 (C/N ratio 30) yielded an initial number of bacteria of 2.87 x 10^{14} cfu /g and fungi 9.6 x 10^{13} cfu/g, and decreased into 4.2x 10^{10} cfu/g and 3.7 x 10^{11} cfu/g, respectively. At the end of initial degradation, bacteria identified were *Bacillus* sp and fungi identified were *Aspergillus niger, Aspergillus glaucus, Aspergillus terreus, Rizophus* sp. The results of the 2nd phase showed that the highest bacterial colonies were achieved in T2 with DM3 of 38 x 10^6 cfu / ml and the bacterial consortium identified were *Bacillus* sp and *Lactobacillus* sp. The number and types of bacterial consortium isolated and identified in the liquid medium can be used as a starter in the preservation of forage feed.

1. Introduction

In the sheep farming, waste is produced in the form of feces, urine and the remaining feed in the cage. The amount of feces produced 5 - 10% per head per day, a considerable amount if concentrated in one location. Feces contain a number of microorganisms that can be utilized and pathogenic microorganisms, in addition feces contain organic materials that can be exploited by microorganisms to grow. Based on this, feces have the potential as a source of nutrients and sources of microorganisms that can be utilized further through integrated processing.

Integrated processing is a system of processing waste gradually and produces a wide range of products, one of which inoculum a bacterial consortium that can be used as a starter in the preservation of forage feed. This bacterial consortium can be obtained by isolating and identifying bacteria and fungi from the substrate (sheep's feces and rice straw) through an integrated treatment that begins with the initial degradation stage of the substrate. The number of bacteria and fungi growing in the substrate, in the later stages is extracted to obtain the source of inoculum and nutrients in the form of single cell proteins, and the next step adds carbon source as a source of energy for the growth and activity of bacterial consortium to be used as starter for pickling of forage feed.

Forage feed is needed in its availability, through the processing of forage feed preservation is guaranteed its availability. In the processing of preservation of forage feed is needed starter that has been obtained by buying a starter with a price that is quite expensive, so it is necessary to make starter by utilizing available resources that utilize sheep feces with integrated processing. To ensure the quality of starter produced, it is necessary to evaluate the content of bacterial consortium to be used as starter.

This research is needed to evaluate the potential of bacteria in starter, obtained from sheep feces through integrated processing. The results of this study, in addition to producing inoculants of a bacterial consortium that can be used as a starter in preserving forage feed, is also friendly to the environment, because in the process of processing can kill pathogenic microorganisms in the feces.

2. Material and Methods

2.1. Materials

The material used in this research is sheep feces and rice straw, and materials for bacterial and fungal isolation (500 g Nutrient Agar medium, 500 g Potato Dextrose Agar, 500 ml NaCl, distilled water, materials for identification of bacteria and fungi, carbol gentiana violet, lugol, 250 ml alcohol 95%, 100 ml air fuchsin, immersion oil, alcoholic acid, H_2SO_4, methylene blue, carbol fuchsin

The research method used in this research, the 1st phase is experimental method in laboratory using complete randomized design with 3 treatments (T1 = C / N ratio 25, T2 = C / N ratio 30, T3 = C / N ratio 35 with 6 repetitions and the 2nd phase, using a randomized block design with 3 treatments (T1 = addition of 2.5% molasses in the base medium T2 = addition of 5% molasses in the base medium T3 = 7.5%), which are grouped in basic media with dry matter content of each DM1 = 3.08%, DM2 = 2.49%, DM3 = 2.26%, DM4 = 1.08% with 6 replications. The variables observed in the Phase 1 were the number and types of indigenous bacteria and fungi that grown at the beginning and end of the initial degradation process of integrated treatment of sheep feces and rice straw mixture. The variables observed in the 2nd phase were the number of colonies and types of bacteria grown on liquid media resulting from the integrated processing extraction of mixture of sheep feces and rice straw.

2.2. Research procedure

This research consisted of two phases. First phase was isolation and identification bacteria and fungi on initial degradation process. Sheep feces with 60% moisture content were used as a sample. Sample of sheep feces and rice straw were analyzed C and N content through Walkley Black and Kjedahl method respectively. In the first phase, treatments were based on C/N content of sheep feces and rice straw mixture: 25, 30, and 35. After substrate for each treatment obtained, substrate incubated for 1 week (initial degradation). Substrate aerated until the moisture content reaches 16%. Fungi and bacteria grew on before and after initial degradation were isolated and identified. Second phase was addition of molasses in the base medium (T1 = 2.5%, T2 = 5%, T3 = 7.5%) into mixture of sheep feces and rice straw to produce starter for forage preservation. Substrate from initial degradation was extracted using hot water solvent with a ratio of 1: 4 (1 kg of substrate: 4 kg hot water) to DM1 (DM = 1: 8 for DM2, 1:12 for DM3 and 1: 16 for DM4; (DM1 = 3.08%, DM2 = 2.49%, DM3 = 2.26%, DM4 = 1.08%). Growing indigenous bacteria from initial research into basic media (DM1, DM2, DM3, DM4) in liquid form which added molasses according to treatment, then incubated anaerobically for 2 weeks. Then isolated calculated the amount of bacterial consortium and identified bacteria growing at each treatment. Treatments that produce the most bacterial colonies that will be used as starter in the preservation of forage feed.

3. Result and Discussion

3.1. Temperature and pH on the initial degradation process of the sheep feces and rice straw substrate with various C/N ratio

Based on observations, the temperature and pH achieved in the initial degradation process can be seen in Figure 1 and Figure 2.

Observations on the initial degradation process of the substrate of sheep and rice straw, at all treatments, on the initial substrate, the temperature reached 65-62 ° C on days 1 and 2, then tended to decrease until day 7 to a temperature of 32 ° C, all treatments yielded a pattern decrease the same temperature, this describes the decomposition of organic matter on the substrate. This is in line with the opinion [1, 2], which states on decomposition of aerobic organic material in composting will be followed by high temperature.

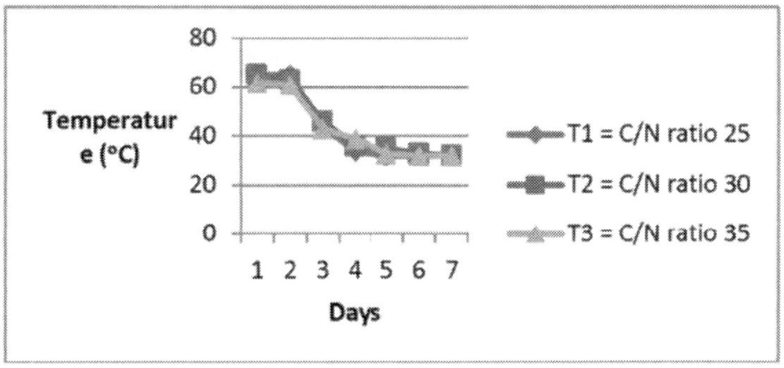

Figure 1. Temperature of the initial degradation process of the sheep's stool substrate and rice straw with various C / N ratios

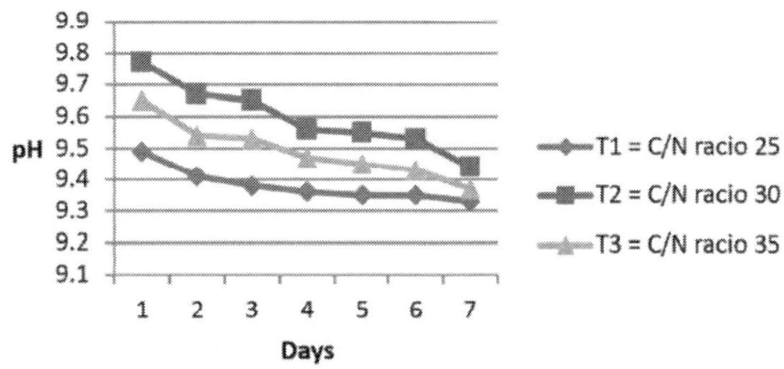

Figure 2. pH in the initial degradation process of the sheep's stool substrate and rice straw with various C / N ratios

Likewise at the beginning of the process, the pH values reached 9.77 - 9.49, then tend to decrease until the 7th day pH reaches 9.33-9.44. All treatments produce the same pattern of pH reduction, this illustrates that the decomposition of organic ingredients will reduce the pH value. This is in line with the opinion [2], which states the pH value of the substrate will change according to the content of organic material present in the substrate.

3.2. Number of bacteria and fungi in the early degradation process on the substrate of sheep feces and rice straw

Based on observations, the number of bacteria and fungi that were successfully isolated can be seen in Figure 3 and Figure 4.

The results of observations on various treatments, at the beginning of the degradation process, the number of bacteria isolated 241 x 10^{10} cfu / g to 287 x 10^{10} cfu / g, then at the end of the initial degradation process the number of isolated bacteria 42 x 10^{10} cfu / g to 49 x 10^{10} cfu / g, in all treatments showed a decrease in the number of bacteria, however the highest decrease was achieved in the treatment of T2 (C / N ratio 30) which reached 85.37%, this illustrates that in the treatment of T2 degradation process of organic substrates is good, this produces high temperatures and such conditions can kill pathogenic bacteria. Presumably in the treatment of T2, the nutrient balance contained in the substrate corresponds to the growing need for indigenous bacteria and activity on the substrate, this is in line with the opinion [3, 4], which states that bacterial growth is influenced by nutrients and the amount of bacterial inoculum. However, the number of isolated bacteria at the end of the degradation process is still possible to be used as a starter in the subsequent process in the integrated processing of sheep and rice straw. Because in the waste treatment process, there is decomposition of organic material that causes heat and causes the temperature at the beginning of the waste treatment to be high (50 °C) and results in the death of pathogenic bacteria contained in the waste, the pathogenic bacteria live at 37 °C. This is supported by the opinion [5, 6] which states the number of bacteria that can be used as a starter from 10^6 to 10^8 cfu / g.

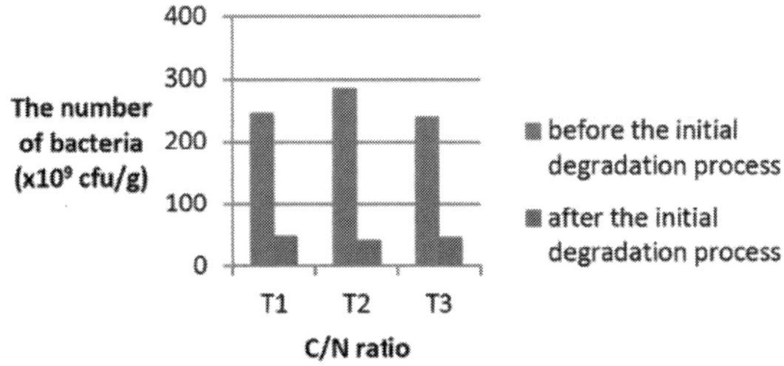

Figure 3. The bacteria isolated before and after the initial degradation process on various C / N ratio.

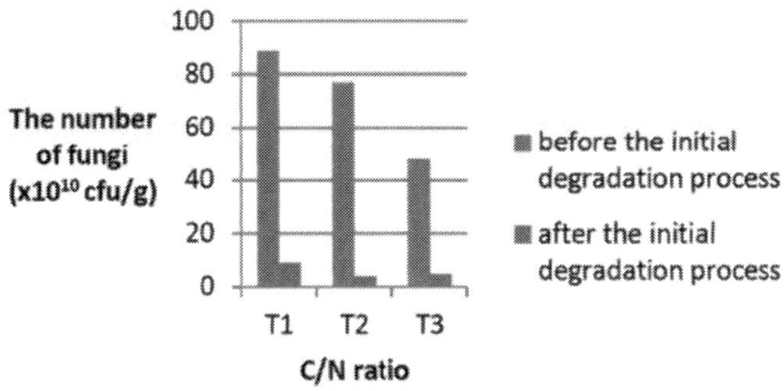

Figure 4. The fungi isolated before and after the initial degradation process on various C / N ratio.

Similarly, the number of fungi at the beginning of the degradation process was 48 x 10^{10} cfu/g to 89 x 10^{10} cfu/g, then at the end of the initial degradation process the number of bacteria isolated was 4 x 10^{10} cfu/g to 9 x 10^{10} cfu/g. the highest decrease was achieved in T2 (C / N ratio 30) treatment which reached 94.8%. The amount of bacteria and fungi at the end of the degradation process allows to be used as a starter in the subsequent process in the integrated processing of sheep feces and rice straw.

3.3. Identification of bacteria and fungi in the initial degradation process on the substrate of sheep and rice straw

Based on observation result, bacteria and fungi identification result on early degradation process on sheep feces substrate and rice straw can be seen in Table 1.

The observations showed that the bacteria identified before the initial degradation process were *Bacillus* sp, *Escherichia coli*, *Enterobacter* sp, *Pseudomonas* and after the initial process of bacterial degradation identified only *Bacillus* sp. This shows that bacterial pathogens die during the initial degradation process, because in the process of degradation a high temperature (65 °C) is formed which can kill pathogenic bacteria. This is in line with the opinion [7], which states that *Escherichia coli, Enterobacter sp, Pseudomonas* will live at temperature 5 – 30 °C.

Fungi that were identified before the initial degradation process were *Aspergillus* sp, *Neoprenpora sitopila, Rhizophus* sp, after the initial degradation process of the identified fungi were *Aspergillus niger*, *Aspergillus glaucus*, *Aspergillus terreus*, *Rizophus* sp. During the initial degradation process the fungus *Neurospora sitopila* die, because the fungus only live at 20 – 30 °C environmental temperature [8], While the fungi that remain alive until the initial degradation process ends, will serve as a single cell protein source in the next process.

Table 1. Identification of bacteria and fungi in the initial degradation process on the substrate (sheep feces and rice straw).

Bacteria Identified		Fungi Identified	
Day 1	Day 7	Day 1	Day 7
Bacillus sp	*Bacillus* sp	*Aspergillus terreus*, *Aspergillus* sp	*Aspergillus niger*
Escherichia coli			*Aspergillus glaucus*
Enterobacter sp			*Aspergillus terreus*
Pseudomonas sp.		*Neurospora sitopila*, *Rhizophus* sp	*Rizophus* sp

3.4. Number of Bacteria Consortium on liquid media with various treatments.

Based on the observation result of the isolation of the number of bacterial consortium growing on each treatment can be seen in Figure 5.

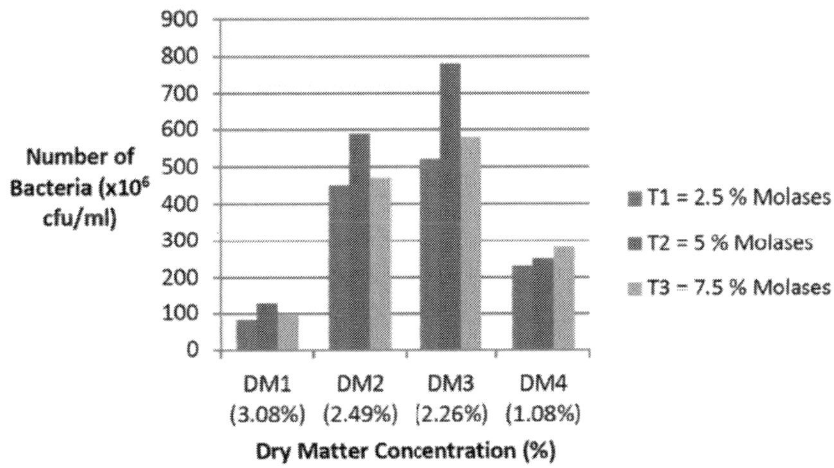

Figure 5. The number of bacterial consortiums grown in each treatment

The results showed that the highest growth of bacterial consortium was obtained at treatment of T2 (T2 = addition of 5% molasses) and at DM3 = 2.26%, i.e. 780 x 107 cfu/ml. This is predicted because the amount of dry matter 2.26% and 5% molases in T2 DM3 support the growth of indigenous bacterial consortium on liquid media resulting from integrated processing of sheep feces and rice straw. This is in line with the opinion [3, 4],which states that the number of growing bacteria is affected by the balance of inoculum and the availability of nutrients and environmental conditions.

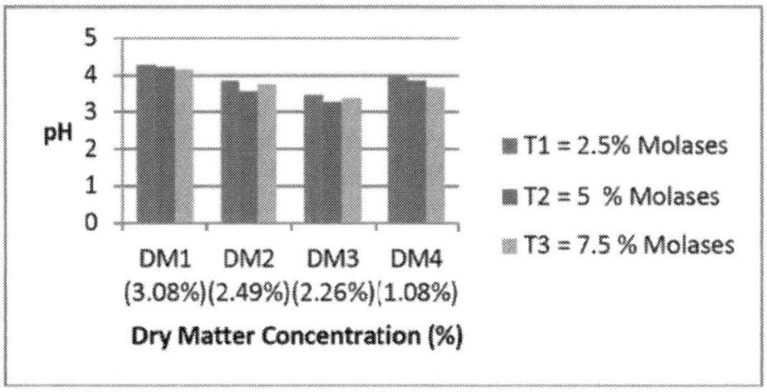

Figure 6. Average pH at various treatment

Average pH achieved in T2DM (addition of 5% molasses in all percentages of dry matter) treatment is 3.26 describes the acidic media conditions and suspected bacteria that grow is a group of lactic acid bacteria. This is in line with the opinion [9] which states that lactic acid bacteria grow at acid pH conditions.

3.5. Identification of Bacterial Consortium on liquid media with various treatments

Based on observations, found some bacteria with rod-shaped, gram positive, facultative anaerobic, nonspore-forming, can ferment carbohydrate by forming lactic acid and identified are *Bacillus sp, Lactobacillus sp.* Almost all Lactic Acid Bacteria (LAB) only get energy from sugar metabolism, so their growth habitats are limited to environments that provide enough sugar. Nutritional requirements of LAB complex include amino acids, vitamins, purines and pyrimidines. Lactic acid produced from LAB can provide bacteriocidal effect for other bacteria because the pH of the environment can decrease to 3-4.5. At that pH LAB remains alive, while other bacteria, including bacteria, decompose food will die [9].

This study will help researchers to uncover the number and type of bacteria in the starter for the preservation of forage feed that many researchers do not know. Thus a new theory of bacterial consortium as starter for forage preservation can be obtained.

4. Conclusion

Based on the result of the initial degradation, the highest number of bacteria and fungi was obtained in T2 treatment (C / N ratio 30) which was 42 x 10^9 cfu / g and 4 x 10^{10} cfu / g and identified *Bacillus* sp and *Aspergillus niger*, *Aspergillus glaucus*, *Aspergillus terreus*, *Rizophus* sp. While, the highest number of bacteria in the substrate from second phase was obtained at T2 treatment (DM3), which was 38 x 10^6 cfu / ml and identified *Bacillus* sp, *Lactobacillus* sp. *Bacillus sp* and *Lactobacillus sp* are lactic acid bacteria that can function as forage preservatives (silage).

Acknowledgments

Thanks to Rector Universitas Padjadjaran who has provided financial assistance and facilities to the author for the sake of this research, in the University Grant Program through the University Lecturer Competence Research 827/UN6.3.1/LT/2017.

References

[1] Neugebauera M, Sołowieja P, Piechockia J, Czekałab W, Janczakb D 2017 *International Journal of Smart Grid and Clean Energy* **6(1).**

[2] Işıl A T E, Ayhan Ü, Murat T 2015 *International Journal of Environment and Waste Management* **18(2)**

[3] Kathryn E R, Davis, Shayne J J, Peter H J 2005 *Appl. Environ Microbiol* Feb **71(2)**: 826–834

[4] Thomas E 2015 *Front Microbiol.* **6** 287
[5] Wakil S M, Ayenuro O T, Oyinlol K A 2014 *Food and Nutrition Sciences* **5:** 495-506
[6] Ranadheera C S, Vidanarachchi J K, Rpcha R S, Cruz A G, Ajlouni S 2017 Fermentation **3(67)** doi: 10.3390/fermentation3040067
[7] Manyi-Loh C E, Mamphweli S N, Meyer E L, Makaka G, Simon M, Okoh A I 2016 *Int. J. Environ. Res. Public Health* **13(9)**: 843
Kanti A, Sudiana I M 2018 Production of Phytase, Amylase and Cellulase by Aspergillus, Rhizophus and Neurospora on Mixed Rice Straw Powder and Soybean Curd Residue *Earth and Environmental Science* **166** 012010 doi :10.1088/1755-1315/166/1/012010
[8] Ni K, Wang Y, Li D, Cai Y, Pang H 2015 https://doi.org/10.1371/journal.pone.0121967.

Increased Amylopectin Content Potential in Corn Grains of Quality Protein Maize (QPM)

Edy[1], S Numba[1] and B Ibrahim[1]

[1]Faculty of Agriculture University of Muslim Indonesia Makassar, Makassar, Indonesia

Email: nuhungedy63 @ yahoo.com

Abstract. Maize can be developed as an alternative staple food in anticipation of dependence on rice consumption which continues to increase along with the increase of population. In order for increase corn consumption it needs to be assembled to improve corn nutrition and taste at an affordable price. The objective of this research was to increase amylopectin content in QPM corn grains without reducing the potential of protein quality and production potential.The method used in this research was Back Cross Breeding. This research was done until the formation of Genotype F2. Stages of the method that has been done were started by hybridization between corn Variety of Srikandi Putih (Recipient's parents) and corn Local Waxy Corn (Donor Parents). Recipientparents were planted in 3 rows and donor parents in 1 row. Each row consisted of 40 plants. The crosses resulted in the genotype F1. Genotype F1 planted in 2 rows and then selfing 21 plants to yield Genotype F2. The research results showed that amylopectin content of corn grains in F1, F2, Parent of Srikandi Putih Variety and Local Waxy Corn were 92,57%, 91,31%, 81,92% and 97,80%, respectively. At this stage of the study, it is concluded that there was a change of amylopectin content of grains in Genotype F1 and Genotype F2 due to crosses compared to their parents.

1. Introduction

Maize can be developed as one of the basic staple foods in anticipation of the dependence on rice consumption which is increasing until now. The productivity of corn produced by breeders is already high, including Srikandi Putih Variety. Generally corn crops are used for animal feed, only a small portion is used for food and food industries. Along with the increasing number of population then the staple food needs must also be met. Therefore it is necessary to diversify staple food, one of them by utilizing corn as an alternative staple food. To make corn as an alternative staple food, it is necessary to improve its quality and taste to be tasty and nutritious. The merging of two good properties of two different corn varieties can be done if their genetic diversity is available.

High quality protein maize (Quality Protein Maize, QPM) is available and available on the market, but its flavor is not pulverized (so rough) to human consumption, as well as Local Waxy Corn is also available and available in the South Sulawesi Region. This Local Waxy Corn is one of South Sulawesi's typical corn with a taste that is very tasty and popular people, both young corn and processed old seeds because it contains high amylopectin, but the lack of low production and protein content is less. Disadvantages of both types of corn need to be removed with certain plant breeding techniques to obtain high protein corn and feels pulverized. Corn with special properties can be established through repetitive and programmed plant breeding programs [1]. In waxy corn there is a wx recessive gene in homosigot (wxwx) that affects the chemical composition of starch, causing tasty and savory taste. Backross breeding methods can be applied to integrate the donor genes from high amylopectin-specific corn to high quality protein corn and high productivity. Thus, corn will be obtained which has the desired special properties.

Amylopectin content in Waxy Corn nearly 100%. Ordinary corn endosperms consist of a mixture of 72% amylopectin and 28% amylose [2]. According to [3], the endosperm content of pulut corn is almost entirely amylopectin. In Waxy Corn there is a wx recessive gene in homosigot (wxwx) that affects the chemical composition of starch, causing tasty and tasty flavors. The yield of waxy corn is generally low, only 2-2.5 t/ha and can not stand the disease of downy mildew. Waxy corn is highly related to the high content of amylopectin in starch. Starch consists of two glucose polymer compounds, namely amylose and amylopectin. The weight of amylose and amylopectin molecules depends on the amylose botanical source which is a component with a straight chain, while amylopectin with branched chain. Amylose is a straight-line helical-shaped polysaccharide with α-1.4 glycosidic bond. The branching point of amylopectin is an α-1.6 bond. The number of glucose molecules in the amylose chain ranges from 250 to 350 units [4]. Starch is composed by at least three main components, namely amylose, amylopectin, and intermediate materials such as lipids and proteins. These components affect the functional properties and amylographic of corn flour [5]. The results showed that the protein content of Waxy corn Takalar and Waxy corn Gorontalo was relatively the same, 0.78% and 0.79% respectively. Comparison of amylopectin and protein content between maize Variety of Srikandi Putih and Local Waxy corn showed amylose content in corn of Srikandi Putih 30.60% (amylopectin 69.40%), protein 1.22%. Waxy corn Takalar contains 5.79% amylose (94.21% amylopectin) and its protein content is only 0.7%% [6]. These two properties need to be combined to obtain high amylopectin and high protein corn as well. The objective of this research is to increase amylopectin levels in QPM corn kernels without reducing the potential of protein quality and production potential.

2. Materials and methods

The study was conducted in Bajeng Gowa Regency, South Sulawesi Province. This study is a follow-up of the first study. The materials used in this research are: corn seed of Varieties Srikandi Putih, corn seed of Local Variety Waxy corn, F1 seed, and F2 seed. This research only reached the formation stage of Genotype F2. Stages of the method that has been done is started by hybridization between corn Variety Srikandi Putih (parent Receiver) and corn Local Pulut (parent Donor). The resulting crosses result in the genotype F1. Genotype F1 planted 2 lines then selfing 21 plants and yielded Genotype F2. Design research in the field as follows: This research consists of 2 processes. Process 1 using cross breeding cross-pollination method, which is Hybridization Method between Varieties of Srikandi Putih and Local Waxy corn.Srikandi Putih Variety as female parent (P1) planted 3 rows while Local Waxy corn as male parent (P2) planted 1 row. Each line consists of 40 plants and repeated 3 times. The crosses result will produce F1 generation. Then proceed to Process 2 designed with Block Random Design. Treatment is Plant Genotype, ie Srikandi Putih Variety, Local Waxy corn Variety, F1 and F2 Generation. Repeated 3 times.

3. Result and discussion

3.1. The length and diameter of Cobs (cm)

Table 1. The length and diameter of the cobs in the various genotypes tested.

Treatment	Length of the cobs	Diameter of the cobs
Srikandi Putih Variety (parent ♀)	17.8a	4.6a
Local Waxy Corn (parent ♂)	16.1ab	3.5b
F1	14.7b	3.6b
F2	16.8ab	4.7a

Note: the numbers folloed by the same letter are not different at the LSD 0.05 test level

Table 1 shows the longest length of cobs in Srikandi Putih Variety and is significantly different from Local Waxy Corn Variety while the genotype F1 and F2 are not significantly different. Furthermore, the cob's diameter on Srikandi Putih Variety and F2 genotype were not significantly different but both were significantly different from Local Waxy Corn and F1 genotype. This shows that the character of length and diameter of cob has started to change the nature of the positive direction. It is known that the length and diameter of the ear is influenced by many genes.

3.2. Weight of 100 seeds and weight of seeds per plant (g)
Table 2 shows the weight of 100 seeds in Srikandi Putih Variety highest and significantly different with all genotypes tested while Local Waxy Corn Variety with genotype F1 were not significantly different but both were significantly different with genotype F2. The highest seed weight per plant in the Srikandi Putih Variety was significantly different from all genotypes tested. Furthermore Genotype F1 and F2 were not significantly different but both were significantly different from Local Waxy Corn. This shows that in general there has been a change in the nature of the descendants of F1 and F2 but it seems still inconsistent between the weight of 100 seeds with the weight of seeds per plant. This is due to segregation in the crossing process so that production in each generation varies. Differences in genetic makeup is one of the factors causing plant appearance diversity [7]. [8] describes that genetic variability is the diversity of individual traits in a population arising from genetic factors. Genetic variation can occur due to mixing of breeding materials, genetic recombination as a result of crossing and mutation or polyploidization.

Table 2. The weight of 100 seeds and weight of seeds per plant in the various genotypes.

Treatment	Weight of 100 seeds	Weight of seeds per plant
Srikandi Putih Variety (parent ♀)	36.0a	79.0a
Local Waxy Corn (parent ♂)	31.8b	19.0c
F1	32.7b	26.8b
F2	23.4c	24.3b

Note: the numbers folloed by the same letter are not different at the LSD 0.05 test level

3.3. Amylopectin and Seed Protein levels (%)
Table 3 shows the highest levels of amylopectin in Local Variety of Waxy Corn and significantly different from all genotypes tested, but the amylopectin levels between the genotype F1 and F2 were not significantly different but both were higher and significantly different from the Srikandi Putih Variety. Furthermore, the highest protein content in the Srikandi Putih Variety and significantly different with all genotypes tested. The protein content in Genotype F1 did not differ significantly with the Genotype F2 but both were significantly higher than the Local Waxy Corn Variety.

Table 3. Levels of Amylopectin and Seed Protein of various genotypes tested.

Treatment	Amylopectin	Protein
Srikandi Putih Variety (parent ♀)	81.92c	10.4.a
Local Waxy Corn (parent ♂)	97.80a	8.60c
F1	92.57b	9.3b
F2	91.31b	9.4b

Note: the numbers folloed by the same letter are not different at the LSD 0.05 test level

This indicates that there has been a positive nature change in F1 and F2 but not yet stable. The amylose and amylopectin composition in the genetically controlled seed kernels. In general, Dent corn and pearl corn of endospermic types contain amylose 25-30% and amylopectin 70-75% of total starch. Waxy Corn has a starch content of nearly 100% amylopectin. The presence of a single waxy (wx) gene that is recessive to epistasis located on chromosome nine affects the chemical composition of starch, resulting in very little amylose accumulation [9]. Another study showed that QPM varieties released for the first time in Indonesia were composite maize of Srikandi Kuning-1 and Srikandi Putih-1 with productivity of 7.0 t/ha [10]. Srikandi Kuning Variety contain 10.38% protein, lysine 0.477% and tryptophan 0.093%, while Srikandi Putih Variety contains protein 10.44%, lysine 0.410% and tryptophan 0.087% [11]. The results showed that the protein content of Waxy Corn Takalar and Waxy Corn Gorontalo was relatively the same, 0.78% and 0.79% respectively. Comparison of amylopectin and protein content between maize variety of Srikandi Putih and Waxy Corn showed amylose content in corn of Srikandi Putih 30.60% (amylopectin 69.40%), protein 1.22%. Waxy Corn Takalar Corn contains 5.79% amylose (94.21% amylopectin) and its protein content is only 0.7%% [6]. The results of research [12] that the F1 protein (genotype crossed by Srikandi Putih Variety x Local Waxy Corn Variety) increased 8.1% compared to the parent of Local Waxy Corn Variety.

4. Conclusion

- Character of production of Srikandi Putih Variety has started to appear on Genotype F1 and F2 but not yet stable.
- Character of amylopectin content Local Waxy Corn Variety has been seen in Genotype F1 and F2 but not yet stable.
- Character of protein content of Srikandi Putih Variety has been seen in Genotype F1 and F2 but not yet stable.

Advanced research is needed to obtain new genotypes with high production characters with high levels of amylopectin and stable proteins to be prepared into new variety.

Acknowledgements
This research is supported financially by Directorate of Research and Community Service (DRCS) Research Department, Technology, and Higher Education (Ristek-Dikti) Republic of Indonesia in the form of Research Scheme of Superior University. Therefore, we would like to thank the director and staff of DRPM Ristek-Dikti, Head of Research and Development Institute of LP2S and Dean of the Faculty of Agriculture of Muslim University of Indonesia (UMI) and the students for their assistance so that this Phase II research could Finished successfully

References
[1] Azrai M., Mejaya M J, Yasin M H G 2008 Pemuliaan Jagung Khusus (Special Maize Breeding) *Balai Penelitian Tanaman Serealia (Cereal Crops Research Institute)* Maros.
[2] Jugenheimer R W 1985 Corn Improvement, Seed Production, and Uses *Robert E Krieger Publishing Company* Malabar Florida
[3] Alexander D E, Creech C 1977 Breeding special nutritional and industrial types In: Corn and Corn Improvement *The American Society of Agronomy Inc.*
[4] Dziedzic S Z, Kearsley M W 1995 The technology of starch production In Dziedzic S Z, Kearsley M W (Eds) *Handbook of Starch Hydrolysis Products and Their Derivatives Blackie Academic and Professional* London
[5] Suarni, Aqil M, Firmansyah I U 2008 Starch characterization of several maize varieties for industrial use in Indonesia *Proceeding of The 10th Asian. Regional Maize Workshop* pp. 74-78.
[6] Suarni, Firmansyah I U, Aqil M 2013 Keragaman Mutu Pati Beberapa Varietas Jagung (Variety of Starch Quality Some Varieties of Maize) *Penelitian Pertanian Tanaman Pangan (Agricultural Research of Food Crops)* **32(1)**.
[7] Sitompul S M, Guritno B 1995 Analisis Pertumbuhan Tanaman (Plant Growth Analysis) *Gadjah Mada University Press* Yogyakarta.
[8] Mangoendidjojo W 2003 Dasar-Dasar Pemuliaan Tanaman (The Basics of Plant Breeding) *Penerbit Kanisius (Kanisius Publisher)* Yogyakarta
[9] Fergason V 1994 High amylase and waxycorn In: Halleur A R (Ed) Specialty Corns *CRC Press Inc.* USA.

[10] Yasin H G, Syuryawati M, Kasim F 2010 Varietas Unggul Jagung Bermutu Protein Tinggi (Superior Varieties of Maize High Quality Protein) *Iptek Tanaman Pangan Crops science and technology* **5(2)**: 146-158.

[11] Azrai M 2004 Penampilan Varietas Jagung Unggul Baru Bermutu Protein Tinggi di Jawa dan Bali (Appearance of New Superior Varieties of Maize of High Quality Protein in Java and Bali *Buletin Plasma Nutfah (Bulletin of Germplasm)* **10(2)**: 49-55.

[12] Edy, Sudirman N, Baktiar I 2017 Increased Potential of Protein Content of Waxy Corn *International Journal of Environment, Agriculture and Biotechnology (IJEAB)* **2(4)**: 1990 - 1993.

Luteinizing Hormone Receptor Gene Polymorphism of Pasundan Cattle in Ciamis

D N Arifin[1], S Said[2], C Sumantri[3], W P B Putra[2] and D N Hadi[4]

[1]Post Graduate School Bogor Agricultural University, Bogor, Indonesia
[2]Research Centre for Biotechnology-Indonesian Institute of Sciences, Cibinong, Indonesia
[3]Faculty of Animal Science Bogor Agricultural University, Bogor, Indonesia
[4]Cattle Breeding and Artificial Insemination Development Centre, Ciamis, Indonesia

E-mail: dnarifin9@gmail.com

Abstract. Pasundan cattle is a local livestock that has lived and well adapted in West Java. In 2015, the population declined by 20.96% caused by the high selling prices and changes in land functions. The impact of declining of the population can causes genetic degradation indirectly. One of the efforts to increase the population is selection of cattle based on luteinizing hormone receptor gene polymorphism that can used as molecular selection for reproductive traits. Thirty seven heads of Pasundan cattle with artificial insemination calving records was used in this study. The exon 11 of LHR gene was amplified and subsequently analysed by sequencing method. The sequencing result were found three genotypes of CC, CT and TT in SNP g.1337C>T. The frequency of T allele (0.527) shows more dominant over C allele (0.473). The chi square test showed that Pasundan cattle were in Hardy-Weinberg equilibrium. The heterozygosity value of Ho (0.513) indicating that the heterozygosity value of the LHR gene belongs to high category (>50%). The high genetic diversity of the Pasundan cattle population indicates that there has not been selection program yet. The results of this study can be used as a reference to continue the selection program improving genetic quality in the reproductive traits of Pasundan cattle.

1. Introduction

Pasundan cattle live spreads throughout the South Coast of West Java and buffer zone area of protected forests along the North Priangan region[1]. In 2013, the population of Pasundan cattle reaches about 50,000 heads [6]. In 2015, the population decreased by 20.96%, the most common in the buffer zone area. It causes by highly selling price and changing of land function. The impact of declining of the population can causes genetic degradation indirectly, such as decrease of genetic quality, effective population, inbreeding increment and purity degradation [4]. Conservation of cattle pasundan important to be implemented, to overcome the threat of the population base area experiencing some limitations that can lead to extinction [1].

Reproduction traits plays an important role in increasing livestock populations. Reproduction of Pasundan cattle potency needs to be studied more deeply through genotypic studies which is a modern method of selecting superior offspring. According to [9] one approach to increase the informativeness of genetic information is to identify the markers for genes that have direct effects on specific aspects of fertility such as the luteinizing hormone/choriogonadotropin receptor (LH/CGR). LHR is one of reproductive genes related to reproduction. These gene affect the endocrine system that plays a major role in controlling reproductive function regulated through the pituitary-gonad hypothalamus. Mutations in the LHR genes, affect the regulation of gonadotropin hormones that are potentially linked to reproductive phenotypes such as early puberty and calving interval [12].

2. Materials and Methods

2.1. Procedure

Thirty seven heads of Pasundan cattle with artificial insemination calving records was used in this study. The sample is a productive female cattle in Cattle Breeding and Artificial Insemination Development Center (BPPIBTSP) Ciamis, West Java.

2.2. DNA extraction

DNA extraction refers to the reference of [14]. Blood samples were collected from a coccygeal vein into a vacutainer tube containing K3EDTA. Blood samples were stored under refrigerator temperature (4 – 5 °C) until DNA isolation is performed. DNA isolation was performed using Genomic DNA mini kits (Geneaid Biotech Ltd., Taiwan) according to the procedures available from the manufacturer. The obtained DNA product was stored at -20 °C until the polymerase chain reaction (PCR) process is performed.

2.3. DNA amplification

The PCR assay for LHR gene was using with a primer set (forward 5'_CAA ACT GAC AGT CCC CCG CTT T_3' and reverse 5'_CCT CCG AGC ATG ACT GGA ATG GC_3') proposed by [10] and [3]. The extracted DNA samples were taken as 2µl, then added 0.5µl primer F, 0.5µl primer R, 5 µl PCR mix and 2 µl destilation water. Mixtures of these were incubated using a thermocycler PCR machine. A mixture of PCR LHR gene was performed, following the protocol: initial denaturation at 94 °C for 5 min. Continued with 37 cycles starting at 94 °, 60 ° and 72 °C (45 seconds respectively) followed by the final extension at 72 °C for 10 minutes. Digested PCR products were electrophoresed on 1% agarose gels and then visualized under a UV transilluminator. After PCR products were visualized, volume of mixtures converted into 30 µl for sequencing analysis. The data of sequences obtained were processed and analyzed by using FinchTV and MEGA7.

3. Results and Discussion

3.1. DNA sequence

The sequencing of LHR gene in Pasundan cattle yields 303 bp sequence and only one SNP is found. The single nucleotide polymorphism found lies in the position of SNP g.1337C> T. This position coincides at the point of restriction by *HhaI* enzyme in RFLP method [3, 10]. Based on GenBank (NC_037338.1), Pasundan cattle undergo a mutation at that base position. There are three genotypes found, labelled CC, CT and TT by identified its chromatogram using FinchTV. The type of mutation occurring in the LHR gene of Pasundan cattle is a transition mutation that changes between the pyrimidine base (C) and pyrimidine (T). This SNP was analysed as non-synonymous mutation that altered amino acid Arginine to Tryptophan.

Sequencing analysis is used initially to be expected to find many SNPs. In a previous study, using two different primer pairs,[8] found other SNPs from LHR gene analysis results in exon 11. There were found three SNPs at position g.1401G> T, g.1470C> T and g.1581G> T in *Bos taurus*.

3.2. Genotypic and allelic frequencies

The LHR gene in Pasundan cattle is polymorphic. [7] states an allele is polymorphic if one of its allele frequencies is less than 0.99. The frequency of LHR gene genotype in Pasundan cattle shows that the frequency of CT genotype (0.513) is generally the highest in the observed population or the dominant allele. The genotype frequency of CC (0.216) is the lowest and the genotype TT (0.270) which is also not much different. The frequency of T allele shows higher than C allele (Table 1).

Table 1 . Frequency genotype and allel of LHR gene in pasundan cattle

Frequency	Genotype Freq. (n)			Allele Freq.		N
	CC	CT	TT	C	T	
Observed	0.216 (8)	0.513 (19)	0.270 (10)	0.473	0.527	37
Expected	0.224	0.498	0.278			

At the mixing of *Bos taurus* × *Bos indicus* cattle in Europe, identification of the LHR gene in the PCR-RFLP method with the *HhaI* enzyme also produced three genotypes, CC, CT and TT. The frequency of the TT genotype originates from 0 – 0.091, while the frequency of CT genotypes comes between 0.366 – 0.888 and CC belongs to 0.151 – 0.574. The frequency of C alleles is higher than the T allele [10]. In Turkish dairy cows (*Bos taurus*), the frequencies of CC, CT and TT genotypes are 0.928, 0.047 and 0.023, respectively. The frequency of C allele (95.20) is higher than the T allele (4.80) [2]. In addition, feed on Nellore (*Bos indicus*) cattle in Brazil, high frequency of TT genotype (0.540), low frequency of CC genotype (0.030) and T allele frequency (0.755) higher than C allele (0.245) [11].

Based on these findings, each type of cow has a different genetic diversity. The genetic diversity of Pasundan cattle (*Bos sondaicus*) is close to the diversity possessed by Nellore cattle as *Bos indicus* which has a higher T allele frequency than C alleles. This may be due to morphological and physiological similarities, and Pasundan cattle also have *Bos indicus* blood which is a living and adapts well in the tropical climate.

3.3. Genetic parameters

Genetic diversity can be determined by the value of heterozygosity, Hardy-Weinberg Equilibrium and Polymorphism Information Content(PIC) as genetic parameters shown in Table 2.

Table 2 . Genetic parameters in LHR gene of Pasundan cattle

Heterozygosity		Chi-square (χ^2)	PIC
Ho	He		
0.513	0.498	0.0027	0.374

Heterozygosity observation (Ho) of LHR gene in Pasundan cattle have value that is 0.513, while heterozygosity expectation (He) is 0.498. The value of Ho is greater than He. The heterozygosity value above 0.5 (> 50%) indicates the variation of a gene in a population is high [7]. This is because at the beginning of the procurement of Pasundan cattle, BPPIBSPT Ciamis bring cattle pasundan from various breeders so that the pattern of random mating and not the occurrence of inbreeding.

The value of chi-square Hardy-Weinberg Equilibrium shows there's no significant, thus the diversity of cattle population pasundan still in a balanced state. This condition indicates that the selection program has not been implemented in Pasundan cattle. In this case the application refers to the diversity of LHR genes.

The PIC value is one of the parameters that indicates the level of information of a genetic marker in livestock [5]. The value of PIC Pasundan cattle obtained is 0.374. According to [15], the PIC value is high if the value of PIC> 0.50, if PIC 0.25 – 0.50 is categorized as moderate and when PIC <0.25 is categorized as low. Based on these parameters, the PIC value of LHR gene in Pasundan cattle is included in the moderate category. It is also found in Sudanese cattle that have relatively similar PIC values [13]. It is assumed that the LHR gene in *Bos indicus* is still diverse. Although not in a high category, this gene is still potential to serve as assist in selection (marker assisted selection).

Other studies conducted by [8] indicating that the combinations of the three SNPs g.1401G> T, g.1470C> T and g.1581G> T in *Bos taurus* were significantly associated with variations in the fertility characteristics of cattle, particularly calving intervals and days to first service. In this case the diversity of LHR genes as genetic markers in Pasundan cattle has potency and applicable as marker assist selection (MAS).

Luteinizing hormone is a hormone that plays an important role in ovulation. In ovulation, increasing of LH will cause the ovum cells out of the ovary to be fertilized by sperm. Ovulatory conditions that are not optimal can cause no ovulation that affects pregnancy failure in cattle. This has an impact on the decrease of reproduction performance, which is increasing the amount of service per conception, prolonged the length of days open and calving interval. Economically it will increase its cost in the business of beef cattle, both longer maintenance for the expenditure of feed and maintenance costs to improve the condition of livestock under normal circumstances.

4. Conclusion

LHR gene in the Pasundan cattle shows diversity with the discovered of three types of genotypes. Based on the genetic parameters obtained, the LHR gene has high heterozygosity, the population belongs to Hardy-Weinberg equilibrium, and the moderate PIC value that still have potency as marker assist selection. So the results of this study can be used as a reference to continue the selection program improving genetic quality in the reproductive traits of Pasundan cattle.

Acknowledgements

We would like thanks to Research Centre for Biotechnology, Indonesian Institute of Sciences for funding the research. And thanks also to Cattle Breeding and Artificial Insemination Development Center (BPPIBTSP) Ciamis for the research permit.

References

[1] Arifin J, Komar S B, Setyowati E Y, Yunasaf U, Anang A, Indrijani H, Sulasmi 2015 Sebaran gen, keseimbangan populasi dan ukuran populasi efektif sapi pasundan pasca migrasi di Majalengka *Jurnal Ilmu Ternak* **15**

[2] Arslan K, Akyüz B, Akçay A, Ilgar E G, Macun H C, Çınar M U 2017 Association of number of artificial inseminations per pregnancy in holstein dairy cows with polymorphism in luteinizing hormone receptor and follicle stimulating hormone receptor genes *Slov Vet Res* **54** p 91

[3] Arslan K, Akyüz B, Agaoglu O K 2015 Investigation of STAT5A , FSHR , and LHR gene polymorphisms in Turkish indigenous cattle breeds (East anatolian red , South anatolian red, Turkish grey , anatolian black , and zavot) *Russ. J. Genet* **51** 1088–1095

[4] [BP3IPTEK] Badan Penelitian Pengembangan dan Penerapan IPTEK Provinsi Jawa Barat 2016 *Panduan Praktis Pembibitan Sapi Pasundan* (Bandung: BP3IPTEK)

[5] Bostein D, White R L, Skolnick M, Davis R W 1980 Construction of a genetic linkage map in man using restriction fragment length polymorphisms *Am. J. Hum. Genet***32** p 314

[6] Dwitresnadi R, Sulaeman M, Arifin J 2015 Kinerja usaha pembibitan sapi potong pasundan pada pemeliharaan semi ekstensif *Jurnal Ilmu Ternak* **31** p 1

[7] Hartl D L, Clark A G 1997 *Principles of Population Genetics* (Sunderland: Sinauer associates)

[8] Hastings N, Donn S, Derecka K, Flint A P, Woolliams J A 2006 Polymorphisms within the coding region of the bovine luteinizing hormone receptor gene and their association with fertility traits *Animal Genetic* **37** p 583

[9] Huhtaniemi I P 2002 The role of mutations affecting gonadotropin secretion and action in disorders of pubertal development *Best Practice and Research Clinical Endocrinology and Metabolism* **16** p 123–38

[10] Marson E P, Ferraz J B S, Meirelles F V, Balieiro J C C, Eler J P 2008 Effects of polymorphisms of LHR and FSHR genes on sexual precocity in a Bos taurus × Bos indicus beef composite population *Genetic Molecular Research* **7** p 243

[11] Milazzotto M P 2001 Mutações no gene do receptor do hormônio luteinizante (LHR) bovino e associação com precocidade sexual em fêmeas Bos primigenius indicus (Nellore) [thesis]

Botucatu (BRA): Campus de Botucatu, Instituto de Biociências, Universidade Estadual Paulista

[12] Milazzotto M P, Rahal P, Nichi M, Miranda N T, Teixeira L A, Ferraz J B S, Eler J P, Campagnari F, Garcia J F 2008 New molecular variants of hypothalamus – pituitary – gonad axis genes and their association with early puberty phenotype in Bos taurus indicus (Nellore) *Livestock Science* **114** p 274

[13] Omer N N, Gornasb N, Rahmatallac S A, Ahmed M K A 2016 Genetic characterization of indigenous Sudanese cattle using FSHR and LHR genes *Am. sci. res. j. eng., technol., sci.* **24** p 1

[14] Sambrook J, Fritsch E F, Maniatis T 1989 *Molukler Cloning: A Laboratory Manual* (USA: CSH Laboratory Press)

[15] Xin-Sheng W, Tian-Wen W, Hui-Ling Z, Guang-Long C X Q, Jin-Hua C, Xiu-Bai Z, Guo-Hong C 2008 Correlation analysis of wool yield in wan line angora rabbits using microsatellite DNA markers *J. Biol. Sci* **8** p 679

IOP Conf. Series: Earth and Environmental Science **334** (2019) 012013 doi:10.1088/1755-1315/334/1/012013

Single Nucleotide Polymorphism on Exon 2 Leptin Gene of Pasundan Cattle

N Hilmia[1], D Rahmat[1], Dudi[1] and D N Hadi[2]

[1]Animal Production, Faculty of Animal Husbandry Universitas Padjadjaran
[2] Balai Pengembangan dan PerbibitanTernak Sapi Potong Cijeungjing Ciamis

E mail : nena.hilmia@unpad.ac.id

Abstract. Single Nucleotide Polymorphism on exon 2 of leptin gene which changes encoding from Arginine to Cysteine could change the function of Leptin as an obese gene which regulated feed intake and energy expenditure. The aims to analyze point mutation based on SNP exon 2 of Leptin gene of Pasundan cattle. This research used 49 DNA of Pasundan cattle from BPPT SP Cijeungjing. Multiplication of DNA was done by Polymerase Chain Reaction (PCR). Analyze of SNP on exon 2 Leptin gene based on nucleotide sequence from direct sequencing of PCR product which was alligned with leptin gene sequence from gene bank using Bioedit and MEGA 5.2 program. The results showed nucleotide sequence of exon 2 Leptin gene in Pasundan cattle is different from Leptin gene sequence of *Bos indicus* and *Bos taurus*. It was found one synonymous SNP that did not change amino acids Serine encoding on S17S, and two non-synonymous SNP which altered amino acids encoding, i.e R25C and R25H. In Pasundan cattle, the frequency of C allele (42.86%) was higher than A allele (31.64%) and T allele (25.50%). Six genotypes were identified i.e. CC (24.49%), CT (32.65%), CA (4.08%), TT (8.16%) and TA (2.05 %), AA (28.57%).

1. Introduction

In order to overcome increasing of food demand especially in meat, the empowerment of genetic resources of local cattle as the main producer of red meat, becomes an appropriate choice, since local cattle has been integrated and highly adaptive to the environment and farmer culture life. Although their productivity is still lower than imported cattle and their crossing, local cattle could survive and be productive with minimum inputs, more resistant to disease and more resistant to climate change. One of the local cattle in West Java, that has been established as a new breed with the Minister of Agriculture Decree no. 1051/kpts/SR.120/10/2014, namely is Pasundan cattle.

The molecular genetic technology was evolved rapidly, it is, a new breakthrough to be able to explore the genetic potential of beef cattle, quickly and precisely as a marked assisted selection (MAS). The Single Nucleotide Polymorphisms (SNP) analysis is one of technique to evaluate point mutation at nucleotide sequence level. SNP can be used discover and gene mapping, candidate gene analysis, and genetic disorder diagnosis, estimates of environmental stimulation responses, xenobiotic and diet, physiological diversity and genomic testing.

Productivity as a quantitative trait is influenced by many gene, one of the potential genes that influence this trait is the leptin gene. The leptin gene is called the obess gene that acts as an anti-obesity gene. Leptin is a hormone, the product of obese gene, that was involved in physiological processes such as food intake, energy metabolism [1], regulation of body weight, reproduction, and immune system function [2]. It was assumed that point mutations at nucleotide sequence in this gene will affect the action of the leptin hormone as a leptin gene product in regulating of energy metabolism, Disorder on energy metabolism may influence livestock productivity, especially on feed intake, body weight, and fat deposition.

One of point mutation on exon 2 leptin gene that have been reported associated with productivity in beef cattle is SNP R25C. Several studies showed that SNP in exon 2 of Leptin gene contributed to fat accretion which was responsible for carcass fat quality, fat deposition, backfat thickness and butter fat [3-9]. Three of SNP on Leptin gene (promotor and exon 2) in Nellore cattle reported that there were differences in longissimus dorsi area among different genotypes [10]. Point mutation on exon 2 leptin gene is an important SNP that may influence productivity of beef cattle. The objective of this research was to identify Single Nucleotide Polymorphism on Leptin gene of Pasundan Cattle.

2. Materials and Methods

2.1. Material and DNA Amplification

The identification of SNPs on Pasundan cattle Leptin gene was evaluated using 49 DNA samples which was taken from *Balai Pengembangan Perbibitan Ternak Sapi Potong* (BPPTSP) Cijeungjing, Ciamis district West Java. The sequence target of Leptin gene was amplified through Polymerase Chain Reaction (PCR). Each PCR reaction contained buffer 10x, dNTP, primer, taq polymerase and dH20. Leptin gene sequence with 620 bp of length, was amplified using forward primer 5'CTCACTGCTGCGTGGTCTAC3'; revers 3' GCACTAGGATTCCGGTCTGG 5'cover a part of intron 1, exon 2, and a part of intron 2. The initial denaturation at 95°C for 5 minutes, followed by 33 cycles of denaturized at 95 °C for 45 seconds, 1 minute annealing at 58 °C, extension at 72 °C for 1 minute followed by polymerase at 72 °C for 5 minutes.

2.2. Data Analysis

Genotyping of Leptin gene was analyzed by direct sequencing and was alignment using Bioedit and MEGA 4 program. The heterozygosity, alleles, and genotypes frequency of leptin gene based on SNPs R25C and R25H were calculated with Nei formula [11]. This analyzes was operated using POPGENE.32 program.

3. Results and Discussion

Amplification of the 620 bp leptin gene includes part of intron 1, exon 2 and intron 2 partially. Multiplication of DNA target on exon 2 Leptin using PCR machine. PCR product was evaluated by agarose electrophoresis, as shown in Figure 1 below

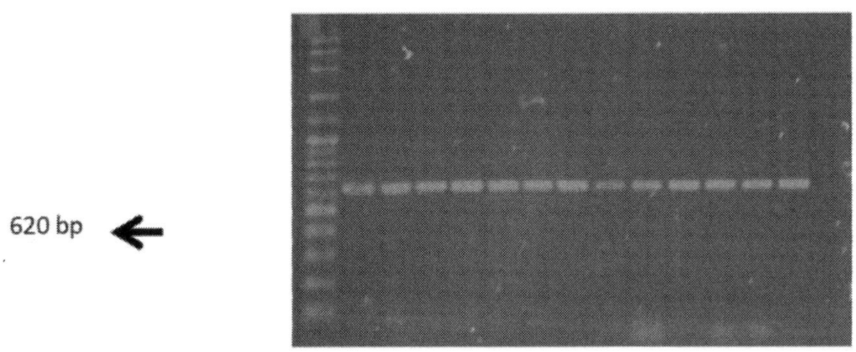

Figure 1. PCR product of Leptin gene along 620 bp

The Single Nucleotide polymorphism (SNP) consists of two, i.e. synonymous and non synonimous SNP. Synonymous mutation is a change of one nitrogen base at codon that was not alter the coding of amino acids, thus have not an affect protein formed, also called silent mutation. Non synonimous mutation is point mutation on nucleotide sequence that alters the coding of amino acids. Non synonymous mutation is divided into missense mutation, a mutation in the codon that affects previous amino acid coding and nonsense mutation changes in the codon into termination codon thus cutting off protein formation [12].

In these research were found three SNP on exon 2 leptin gene Pasundan cattle, one synonymous SNP and two non synonimous SNPs. Synonymous SNP was detected at g.1025T>C/g1158T>C (access no NCBI EU313203.1 and U50365.1). There was a transition from tymin to cytosin, that was not altered encoded 17th amino acid (Serine). There were found two non synonimous mutation on R25C/g.1047C>T/g1180C>T and R25H/g.1048G>A/g.1181G>A (NCBI acces no. EU313203.1 and U50365.1). This SNP altered amino acid 25th Arginine to Cysteine and Arginine to Histidin. There were base nitrogen transition from cytosine to thymine and guanin to adenine. There two kind of substitution mutation, transitions that changed base nitrogen A (adenine) to G (guanin) and C (cytosine) to T (thymine), or vice versa, (pyrimidine) and transversions are substitution mutation between purine and pyrimidine [12]. The aligned result of exon 2 leptin gene and allele determination using bioedit and MEGA 4, were showed at Figure 2. The SNP S17S was not included on allele determination, because, it was a silent mutation.

```
        10        20        30        40
....|....|....|....|....|....|....|....|....
Bos_indicus_leptin
TATCTGTCTTACGTGGAGGCTGTGCCCATCCGCAAGGTCCAGGATGACA
Bos_taurus _leptin
TATCTGTCTTACGTGGAGGCTGTGCCCATCCGCAAGGTCCAGGATGACA

Allele A      ...........................A...............
Allele T      .......C.................T...............
Allele C      .......C.................C...............
```

Figure 2. Alignment result of nucleotide sequence exon 2 leptin gene : SNP S17S/g.1025T>C/g1158T>C, R25C/g.1047C>T/g1180C>T and R25H/g.1048G>A/g.1181G>A(acces no. NCBI EU313203.1 and U50365.1)

Those non synonymous mutation in line with [13] who revealed that in 78 head of the local cattle Ciamis that were taken from Tambaksari and Cijulang, Ciamis District, leptin gene was polymorphic. There were found non synonymous SNP R25C and R25H. The non-synonymous SNP R25H was specific mutation on Ciamis local cattle, based on there was not found in *Bos indicus* as well as in *Bos taurus*. Those point mutation were assumed had an effect to productivity, because alteration of amino acids on gene may change their protein function. The SNP R25C in leptin gene is causative mutation which could alter leptin function as an anti-obese hormone in the physiological processes [14].

Table 1. Heterozygosity, Allele and Genotype Frequency based on SNP R25C (Arg25Cys/g.1047C>T) and R25H (Arg25His/g.1048G>A) on Pasundan Cattle

Breed	n	Allel Frequency			Genotype Frequency					
		C	T	A	CC	CT	CA	TT	TA	AA
		42	25	31	12	16	2	4	1	14
Pasundan	49	0,429	0,255	0,316	0,245	0,327	0,041	0,082	0,020	0,286

He = Expected Heterozygosity

Previous study on leptin gene showed that SNP R25C is a non-synonymous mutation that can change the biological function of leptin gene, its demonstrated by highly level of leptin mRNA in homozygous cattle (TT), that has an point mutation on R25C [4]. Based on those SNPs, on Pasundan cattle, leptin gene was polymorphic, as presented on Table 1 bellow.

The frequency of C allele//no mutation (42.9%) was higher than T allele (25.5%) and A allele (31.6%). Hereafter, CT genotypes (32.7%) were higher than CC genotype (24.5%), CA (4.1%), TT (8.2%), AA (28.6%), and TA (2.0%). The alleles frequency on this research in line with the previous on 78 DNA sample of Ciamis local cattle, indicated there were three alleles, C allele as the highest frequency (55.5%), than T (30%) and H/A allele (15%) [13]. The result of this study were different from leptin gene study on *Bos taurus* as well as on *Bos indicus* as is a new mutation SNP R25H. However the allele no mutation on this research was higher than allele mutation. These results were in accordance with other study which analyzed SNP in Leptin gene exon 2 (Arg25Cys) showing the frequency of C allele was higher than T allele. The study on 323 Fries Holland cattle, showed the frequency of C and T alleles were 67% and T 33% respectively [4]. The research on 169 individuals of Charolais X Fries Holland bull, indicated that C allele frequency was 65%, while T allele was 35% [15]. The SNP analysis on 55 Charolais cattle and 17 Simmental cattle found 34% and 32% for T allele respectively whereas C allele were 66% and 68% respectively[3]. In contrast, they reported different results on British cattle (Angus and Hereford) that the frequency of T allele (0.58% and 55 %), was higher than C allele (0.42% and 0.45%). Others study reported that there was low frequency of T allele and there was not found TT genotype in Nelore (Bos indicus) cattle, whereas it was found TT genotype with low frequency (19%) in Nelore X Bos taurus cattle [14]. There were mutation on leptin gen in Turkey cattle. The frequencies of T and C allele were 0.54±0.06 and 0.46±0.06 for East Anatolian Red cattle, and 0.48±0.05 and 0.52±0.05 for Anatolian Black [16].

Several studies revealed there were significant association between Leptin genotype based on SNP (Arg25Cys) with carcass grade fat, carcass yield grade, and lean meat yield. Thymine homozygotes (TT) had more carcass grade fat compared to heterozygotes and to animals homozygous for the cytosine allele, heterozygotes in turn had more carcass grade fat compared to cytosine homozygotes. Additionally, thymine homozygotes had lower lean meat yield and yield grades compared to heterozygotes and cytosine homozygotes [6]. Cattle with TT genotype had higher carcass grade compared to those with CT and CC genotypes [8]. This SNP on 154 bull is reported there were association between T allele with increasing fat deposition and higher levels of Leptin mRNA, the T allele was associated with fatter carcasses and conversely the C allele was associated with leaner carcasses[3]. The longissimus dorsi muscle tenderness from TT genotype was more tender than that from CC and CT genotypes [7]. There were significant association between Leptin genotype based on SNP Arg25Cys with carcass grade fat, carcass yield grade, and lean meat yield [3,6,8].

4. Conclusion

Leptin gene in Pasundan cattle was polymorphic. There were found one synonymous SNP that did not change amino acids Serine encoding on S17S, and two non-synonymous SNP which altered amino acids encoding, i.e. R25C and R25H. These non-synonymous mutations changed amino acids encoding from Arginine to Cysteine and Arginine to Histidine respectively. In Pasundan cattle, the frequency of C allele (42.86%) was higher than A allele (31.64%) and T allele (25.50%). Six genotypes were identified i.e. CC (24.49%), CT (32.65%), CA (4.08%), TT (8.16%) and TA (2.05 %), AA (28.57%).

References

[1] Hossner K L 1998 Cellular molecular and physiological aspect of leptin: Ptential application in animal production *Can. J. Anim. Sci.* pp. 463-472.

[2] Van der L T, te Pas M F W, Veerkamp R F, Liefers S C 2005 Leptin gene polymorphisms and their phenotypic associations. *Vitamins and Hormones* **(71)**: 373 – 404 DOI: 10.1016/S0083-6729(05)71013-X

[3] Buchanan F C, Fitzsimmons C J, Van Kessel A G, Thue T D.Winkelman-Sim D C, Schmutz S M 2002 Association of a missense mutation in the bovine leptin gene with carcass fat content and leptin mRNA levels *Genet Sel Evol.* **34**: 105-116 doi: 10.1051/gse:2001006.

[4] Liefers S C, Veerkamp R F, Te Pas M F W, Delavaud C, Chilliard Y, Platje M, Van der Lende T 2002 Associations between leptin gene polymorphisms and production, live weight, energy balance, feed intake, and fertility in holstein heifers *J. Dairy Sci.* **85**:1633–1638.

[5] Buchanan F C, Van Kessel A G, Boisclair Y R, Block H C. McKinnon J J 2007 The leptin arg25cys affects performance, carcass traits and serum leptin concentrations in beef cattle *Can. J. Anim. Sci.* **87**:153–156.

[6] Nkrumah J D, Li C, Basarab J B, Guercio S, Meng Y, Murdoch B, Hansen C, Moore S S 2004 Association of a single nucleotide polymorphism in the bovine leptin gene with feed intake, feed efficiency, growth, feeding behaviour, carcass quality and body composition *Can. J. Anim. Sci.* **84**: 211–219.

[7] Schenkel F S, Miller S P, Ye X, Moore S S, Nkrumah J D, Li C, Yu J, Mandell I B, Wilton J W, Williams J L 2005 Association of single nucleotide polymorphisms in the leptin gene with carcass and meat quality traits of beef cattle *J. Anim. Sci.* **83**: 2009-2020.

[8] Kononoff P J, Deobald H M, Stewart E L, Laycock A D, Marquess F L S 2005 The effect of a leptin single nucleotide polymorphism on quality grade, yield grade, and carcass weight of beef cattle *J. Anim. Sci.* **83**:927-932.

[9] DeVuyst E A 2010 The Economics of Gene Testing Cattle *Oklahoma Cooperative Extension Service* http://osufacts.okstate.edu.

[10] Da Silva R C, Ferraz J B, Meirelles F V, Eler J P, Balieiro J C, Cucco D C, Mattos E C, Rezende F M, Silva S L 2012 Association of single nucleotide polymorphisms in the bovine leptin and leptin receptor genes with growth and ultrasound carcass traits in Nellore cattle.

[11] Nei M 1987 Molecular Evolutionary Genetics New York (US) *Columbia University Press.*

[12] Graur Dan 2003 Single base mutation *Nature Ensiclopedia of The Human Genomes Mcmillan Publisher Ltd. Nature Publishing Group* p. 287-290.

[13] Hilmia N 2013 Karakterisasi Fenotipe Dan Potensi Genetik Serta Hubungannya Dengan Produktivitas Dan Kualitas Daging Sapi Lokal Di Ciamis Jawa Barat *Disertation* Institut Pertanian Bogor Bogor.

[14] Fortes M R S, Curi R A, Chardulo L A L, Silveira A C, Assumpção M E O D, Visintin J A, de Oliveira H N 2009 Bovine gene polymorphisms related to fat deposition and meat tenderness *Genet. and Mol. Biol.* **32(1)**:75-82

[15] Lagonigro R, Wiener P, Pilla F, Woolliams J A, Williams J L 2003 A new mutation in the coding region of the bovine leptin gene associated with feed intake [Short Communication] *Anim Genet.* **34**: 371–374

[16] Kaygisiz A, Bengi C, Cilek S 2011 Investigation of leptin gene polymorphisms in East Anatolian Red Anatolian and Black cattle and determination of genetic distance from Brown Swiss Cattle *J. Anim. Plant. Sci.* **21**: 121-125.

Localization of Plasma Membrane Calcium ATPase 2 and 4 (PMCA2 and PMCA4) in Chicken Sperm

R Setiawan[1,2], C Priyadarshana[1], A Tajima[3], N Ishikawa[3] and A Asano[3]

[1]Graduate School of Life and Environmental Sciences, University of Tsukuba, Ibaraki, 305-8577, Japan
[2]Department of Animal Production, Faculty of Animal Husbandry, Padjadjaran University, Bandung, 45363, Indonesia
[3]Faculty of Life and Environmental Sciences, University of Tsukuba, Ibaraki, 305-8577, Japan

E-mail: asano.atsushi.ft@u.tsukuba.ac.jp

Abstract. Calcium ions play important roles in several cellular processes including signalling pathway, membrane fusion and cell adhesion. In sperm, it is known that intracellular calcium concentration ($[Ca^{2+}]_i$) regulates acrosome exocytosis and flagellar motility. $[Ca^{2+}]_i$ is generally regulated by influx and efflux of calcium ion. Plasma membrane calcium ATPase (PMCA) pumps the $[Ca^{2+}]_i$ excess, providing a fine-tuning of $[Ca^{2+}]_i$ concentration for sperm functions. To date, information related PMCA family in chicken sperm is still not documented yet. Therefore, this study was subjected to characterize expression and localization of PMCA family in chicken sperm. In this study, we selected PMCA2 and PMCA4 as representative to a specific tissue and a housekeeping isoform, respectively. Immunohistochemistry of PMCA2 and PMCA4 in chicken testis showed that both transmembrane proteins are expressed at the edge adluminal of the seminiferous tubule, indicating that these proteins are present in morphologically matured sperm that completed spermatogenesis. This was confirmed by immunocytochemistry, showing that PMCA2 and PMCA4 are highly enriched in the midpiece of chicken sperm. Furthermore, sub-cellular fractionation, combined with western blotting, showed that both isoforms are exclusively found in the membrane fraction of the sperm, together suggesting both PMCA are localized to the plasma membrane overlaying midpiece.

1. Introduction

According to the Food and Agriculture Organisation's (FAO) statistics, the global demand for poultry meat and eggs tend to increase in the next decades exponentially as a result of population growth and changes in dietary habits. Total poultry meat demand in developing countries is projected to double (or approx. 93.5 million tonnes) by 2030. This is a challenge for our ability to develop the livestock production efficiently.

To date, many efforts have been made to develop chicken production systems to maximise production ability. Recent developments in the field of poultry reproductive technology have led to the importance of fertility as a main chord in sustainable chicken production. For example, sperm cryopreservation and artificial insemination have been widely used for genetic make-up and efficiency use of the male genetic resource. However, a significant problem with this kind of applications is the fertility rates remain unreliable for use in commercial production. This is mostly because it is still unclear the natural state of mechanisms by which chicken sperm undergo deterioration of fertilizing ability underlying sperm preservation.

For the successful fertilisation, sperm require to travel in the female reproductive tract and undergo physiological changes before reaching the female egg. Moreover, after reaching the target, the sperm must interact with the egg's extracellular matrix and perform acrosomal exocytosis. Calcium is considered playing roles in most of these processes. It is well known that the elevation of intracellular calcium ($[Ca^{2+}]_i$) plays a vital role as a second messenger in the physiology of cells. In sperm, $[Ca^{2+}]_i$

responsible for flagellar movement as well as activation of signalling pathways leading to acrosome exocytosis that pre-requisite to fertilise the oocyte [1]. Previous studies documented that the increase of basal $[Ca^{2+}]_i$ level ranged from 25-75 nM to 500 nM leads to membrane fusion and acrosome reaction [2,3]. The functional importance of $[Ca^{2+}]_i$ is also supported by the evidence that the fertilising ability of cryopreserved mammalian sperm is impaired by extreme elevation of $[Ca^{2+}]_i$ due to the compositional changes in sperm membrane during cryopreservation [4,5]. Even though the elevation of $[Ca^{2+}]_i$ is essential for sperm functions, but the sustained elevation of $[Ca^{2+}]_i$ is toxic for the cell [6].

Plasma membrane calcium ATPase (PMCAs), transmembrane proteins acts as a primary channel involved in ATP-dependent expulsion $[Ca^{2+}]_i$, providing a fine-tuning $[Ca^{2+}]_i$ for sperm functions. PMCAs consist of four different isoforms, where PMCA1 and PMCA4 are expressed in several tissues including testis, while PMCA2 and PMCA3 are expressed in very limited tissues [7], suggesting the functional distinction between isoforms. In fact, PMCA1 and PMCA4 are highly conserved in sperm from different species [7,8]. In murine sperm, PMCA4 has been found in sperm tail [9], while in bull sperm it is enriched in the midpiece [10]. Although some researches have been carried on the functional importance of PMCA in sperm from different animals, no single study exists which focus on avian sperm. Therefore, this study investigates the subcellular location of PMCA2 and PMCA4 as a foundation to dissect regulating mechanisms of $[Ca^{2+}]_i$ in chicken sperm.

2. Materials and Methods

2.1. Localisation of PMCA2 and PMCA4 in chicken testis

We firstly deparaffinized and rehydrated paraffin embedded testis in 100% xylene, 100% ethanol, and 70% ethanol for 15, two, and one minute, respectively. Heat-induced epitope retrieval for 10 minutes in 0.01 M citrate buffer, pH 6.0 was performed and followed by cooling down at room temperature (RT). Blocking solution (10% of goat serum in PBS-) incubated the slides for one hour at RT. The slides were then incubated in PMCA2 or PMCA4 polyclonal antibody [1:50 rabbit serum immunoglobulin (IgG) antibody in PBS-)] at 4 ^{0}C for overnight. After three times washing in PBS-, anti-rabbit IgG Alexa Fluor 488 antibody (1:200) was applied to incubate the slides for one and half hour at RT. Coverslips were mounted on to the glass slides using VECTASHIELD mounting medium with 4',6-diamidino-2-phenylindole (DAPI) (Vector Laboratories, Burlingame, CA, USA). Histological sections of each sample were used as negative controls, substituting primary antibody with PBS-. All procedures after antigen retrieval were performed in a humid chamber to avoid the slides to dry out.

2.2. Localization of PMCA2 and PMCA4 in chicken sperm

Sperm were allowed to settle on anti-liquid bordered coverslips at 39 ^{0}C for 30 minutes, followed by fixation in 4% paraformaldehyde for 15 minutes at RT. The sperm samples were then washed, and permeabilized in 0.5% Triton X-100 for 1 min. Blocking solution (10% goat serum in PBS-) was applied on to the samples for 1 hour at RT. The samples were then incubated in a primary antibody (chicken PMCA2/PMCA4 polyclonal antibody 1:150 PBS-) at 4 ^{0}C for overnight. After three times washing in PBS-, the samples were then incubated in goat anti-rabbit IgG Fluor 488 antibody (1:200). VECTASHIELD mounting medium with 4'6-diamidino-2-phenylindole (DAPI) (Vector Laboratories, Burlingame, CA, USA) mounted the coverslips on to glass slides. Sperm samples without a primary antibody were used as negative controls. All the procedures were performed in a humid chamber.

2.3. Sub-cellular fractionation of PMCA2 and PMCA4 in chicken sperm

Sub-cellular fractions of sperm were prepared from chicken sperm according to [11], with slight modifications. In brief, sperm samples were washed by PBS- for 5 minutes at 1000 x g. The resulting pellet was then resuspended in PBS- containing protease inhibitors (Free EDTA protease inhibitors cocktail, Roche, Basel, Swiss) and 0.5% Triton X-100. The supernatant was then homogenised, five times burst sonicated, and centrifuged at 4 °C, 10000 x g, for 10 minutes. The resulting pellet was designed as a "cytoskeleton" fraction; the supernatant was then centrifuged at 4 °C, 20000 x g, for two hours. The pellet and supernatant were subjected as "membrane (Me)" and "cytosol (Cs)" fractions, respectively. All the sperm fractions were then quantified using Bradford Assay Kit (Bio Rad, US). 70 µg of protein from each fraction was subjected to SDS-polyacrylamide gel electrophoresis (SDS-PAGE).

3. Results and Discussion

3.1. Localisation of PMCA2 and PMCA4 in chicken testis

PMCA is known as a transmembrane protein which has the main task in Ca^{2+} clearance [12]. The protein consists of four isoforms, PMCA1-4, encoded by four different genes and have tissue-specific expression. PMCA1 and 4 are expressed in most tissues and act as housekeeping isoform, whereas PMCA2 and 3 are more restricted, mostly in brain and striated muscle [13]. Okunade et al. (2004) reported that PMCA4 is more abundant isoform in mouse testes compared to PMCA1, making PMCA4 is the major isoform in the tissue, even though PMCA1 can compensate its absence. In contrast to housekeeping isoforms, a previous study observed in mice that PMCA2 acts as the primary controller of milk calcium concentration as 60% of milk calcium decline is due to the absence of PMCA2 [14]. Moreover, the study also revealed that the null PMCA2 mutation of mice cannot be fully compensated by other calcium-transporting channels, secretory pathway Ca^{2+}-ATPase (SPCA) and sarco/endoplasmic reticulum Ca^{2+}-ATPase (SERCA). Considering the functional importance of PMCA2 and PMCA4 in regulating $[Ca^{2+}]_i$, these isoforms might play roles in chicken sperm functions, since the sperm require calcium-dependent events to acquire their fertilizing abilities. However, no previous study has investigated PMCA2 and PMCA4 in chicken sperm, making calcium mechanisms underlying sperm functions are still unclear. Therefore, this study is firstly subjected to identify the localisation of PMCA2 and PMCA4 in chicken sperm to gain further insights regarding the specific biological functions of calcium-dependent sperm functions.

In this study, two different polyclonal primary antibodies-against chicken PMCA2 and PMCA4 were used to determine the origin of the proteins in chicken testes. Both proteins were detected in chicken seminiferous tubules from the basal to the edge adluminal compartment of seminiferous tubules (Fig.1). This localisation indicates that both PMCA2 and PMCA4 might be synthesised and embedded into sperm membrane during spermatogenesis. The similarities between PMCA2 and PMCA4 protein localization indicate the requirement of Ca^{2+} clearence at late spermatogenic cells. The fluorescent intensity emitted from the edge of adluminal compartment is stronger than that from the basal compartment. A possible explanation for the stronger intensity at the edge adluminal seminiferous tubules was that the cytosolic content in mature sperm is less than non-mature sperm. At this compartment, the spermatids undergo extensive remodeling, whereby their cytoplasm content is removed by approximately 70% during spermiation [15,16], resulting in the cell more compact and dense. Moreover, numerous proteins have also been localized to specific sites during spermiation [17].

3.2. Localisation of PMCA2 and PMCA4 in chicken sperm

In order to investigate the distribution of PMCA2 and PMCA4 in particular in subcellular compartment of chicken sperm, immunocytochemistry on chicken sperm was performed. Our results identify that chicken sperm express the family members of plasma membrane calcium ATPase, PMCA2 and PMCA4 (Fig. 2). These both proteins are localized on the specific subcellular compartment at the level of the midpiece. This localization is the same with that in bull sperm [10], and is direct vicinity to the high concentration Ca^{2+} stores in mammalian sperm [18]. In the midpiece of sperm where mitochondria are abundant, the production of ATP is generated by an increase in $[Ca^{2+}]_i$ via Ca^{2+} sensitive dehydrogenases and NADH production [19].

IOP Conf. Series: Earth and Environmental Science **334** (2019) 012014 doi:10.1088/1755-1315/334/1/012014

It has been known that the increase of $[Ca^{2+}]_i$ is essential for sperm to acquire the fertilizing abilities. For example, it plays roles in sperm capacitation in bull [20], acrosome reaction in bull and ram [21], and sperm-oocyte binding [22]. However, the sustained high level of $[Ca^{2+}]_i$ leads to coagulative necrosis due to the loss of plasma membrane's ability to maintain a large gradient of calcium ions [6]. PMCA family acts as $[Ca^{2+}]_i$ pumps to expel excess $[Ca^{2+}]_i$ which makes sperm to excite fertilization ability collectly. This was proved by a previous study using mutant mice showing that deletion of PMCA4 results in double $[Ca^{2+}]_i$ concentration [9] and failure of sperm capacitation and hyperactivated motility [8]. Although the functional relevance of PMCA2 and PMCA4 in chicken sperm have not been established, it is attempting to predict that the PMCA family members located in the midpiece are responsible for $[Ca^{2+}]_i$ maintenance underlying sperm functions, in particular in the flagellar movement.

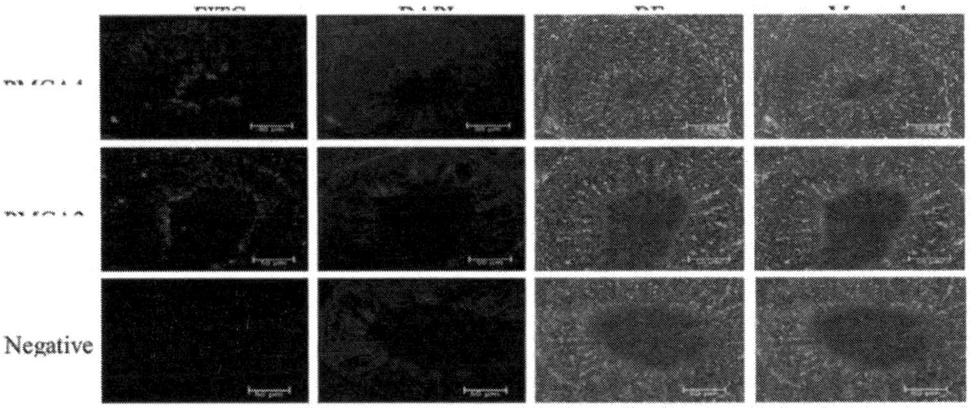

Figure 1. Immunohistochemistry in paraffin-embedded chicken testes using polyclonal antibody against PMCA2 and PMCA4. Immuno-positive signals of both PMCA2 and PMCA4 were detected at the basal to the edge of adluminal

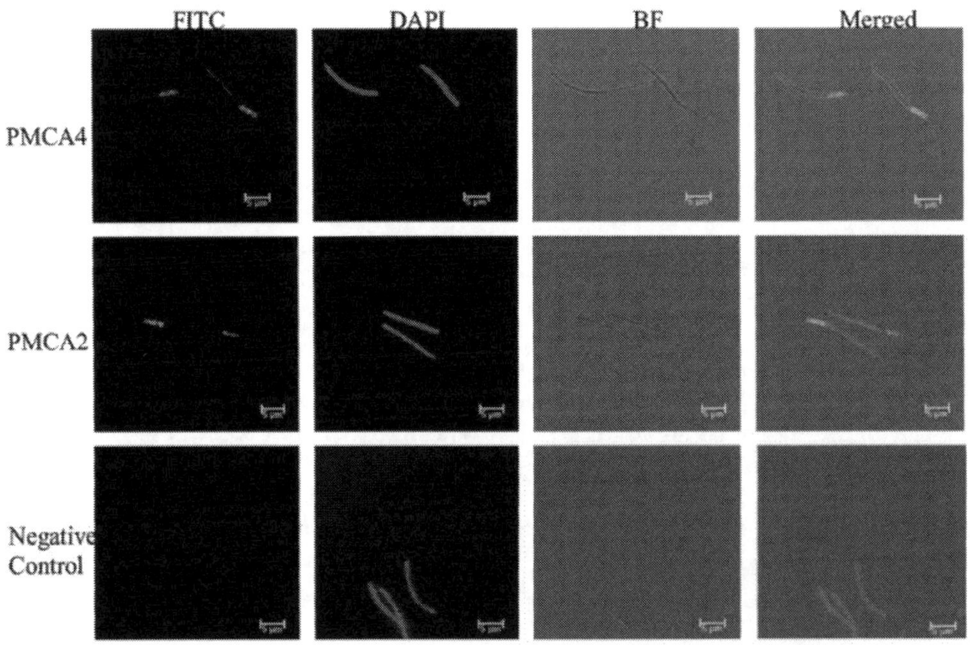

Figure 2. Immunocytochemistry of PMCA2 and PMCA4 in chicken sperm. PMCA2 and PMCA4 were identified in the midpiece of the sperm (*green*). Nuclei were fluorescence-labelled using DAPI (*blue*). Bars: 5 μm.

3.3. Sub-cellular fractionation of PMCA2 and PMCA4 in chicken sperm

Membrane fraction of chicken sperm was prepared using Triton X-100, a non-ionic detergent. Western blotting analysis was performed using a polyclonal antibody against each for PMCA2 and PMCA4. PMCA2 and 4 showed their distribution into both cytosol and membrane fractions. This is consistent with a previous study showing the abundance of PMCA4 in membrane fraction isolated from bull sperm [10].

Lipid rafts are specific membrane domains that play a important roles in diverse cellular processes. In general, PMCA tends to be concentrated into caveolae, a particular type of lipid rafts compartment to govern various signal transduction pathways [23]. This type of lipid rafts is surrounded and enriched by acidic phospholipids such as cholesterol (CHOL), sphingomyelin (SPM), phosphatidic acid (PA), phosphatidylserine (PS), phosphatidylinositol(PI), and phosphatidylinositol 4,5-bisphosphate (PIP_2) which stimulate the velocity maximum (V_{max}) and decrease enzyme's affinity (K_m) of PMCA [24,25,26]. The stimulated-route can also be noticed from calmodulin-stimulated enzymes. The binding event of acidic phospholipid and polyunsaturated fatty acids to calmodulin and the first cytosolic loop of PMCA activate the pump to decrease $[Ca^{2+}]_i$ by 100 nM [27,28]. Intrestingly, recent our study demonstrated that chicken sperm possess lipid rafts in sperm head region same as where locallization of PMCA2 and 4 are found in chicken sperm, suggeting future investigation for a potential mechanism for regulation of PMCA function in sperm.

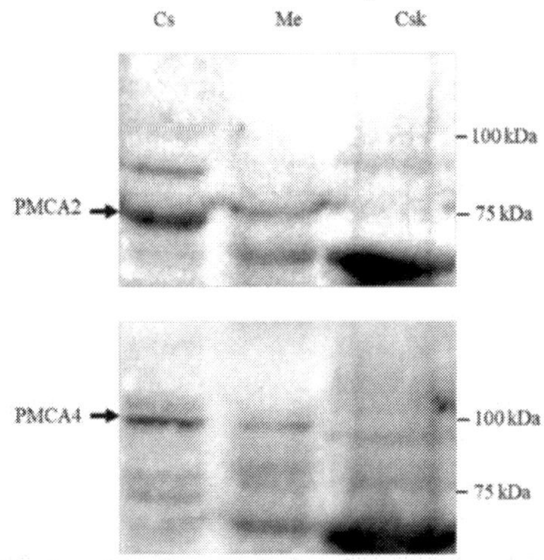

Figure 3. SDS-PAGE and Western blotting analysis of PMCA2 and 4 in chicken sperm membrane fractions. Protein (70 µg per lane) were probed with polyclonal antibody against chicken PMCA2 and 4. PMCA2 and 4 were localized in cytosol (Cs) and membrane (Me) fractions. Cytoskeleton fraction is abbreviated as CSK.

Evidence presented in this paper indicates potential important of PMCA2 and 4 maintaining the basal $[Ca^{2+}]_i$ and provides a foundation to dissect a molecular mechanism involved in regulation of flagellar motility. In addition, it has been shown that lipid rafts have direct interaction with a number of Ca^{2+}-activated signalling molecules in caveolae, such as protein kinase C (PKC) and phosphatidylinositol 4,5-bisphosphate (PIP_2) [29,30], which suggest potential roles of PMCA in sperm functions. Therefore, this study needs to address identification of the signaling molecules regulated by lipid-protein interaction to unravel the downstream events of PMCA regulating $[Ca^{2+}]_i$ for chicken sperm functions.

4. Conclusion

The localization of PMCA2 and PMCA4 in chicken testes and spermatozoa have similar localization which were strongly expressed at the edge of adluminal seminiferous tubules and the midpiece, respectively. Also, our findings identified the subcellular distribution of these $[Ca^{2+}]_i$ pumps were found in both the cytosol and membrane.

Acknkowledgements

The first author would like to acknowledge the Indonesian Endowment Fund for Education (LPDP), Ministry of Finance, and the Directorate Generale of Higher Education (DIKTI), Ministry of Research, Technology and Higher Education, Republic of Indonesia for financial support.

References

[1] Zhang J, Xiao P, Zhang X 2009 Phosphatidylserine externalization in caveolae inhibits Ca^{2+} efflux through plasma membrane Ca^{2+}-ATPase in ECV304 *Cell Calcium* **45**: 177–184

[2] Breitbart H 2002 Role and regulation of intracellular calcium in acrosomal exocytosis *Journal Reprod Immunol* **53**: 151–159

[3] Brewis I A, Morton I E, Mohammad S N, Browes C E, Moore H D 2000 Measurement of intracellular calcium concentration and plasma membrane potential in human spermatozoa using flow cytometry *Journal of Andrology* **21**: 238–249

[4] Robertson L, Bailey J L, Buhr M M 1990 Effects of cold shock and phospholipase A2 on intact boar spermatozoa and sperm head plasma membranes *Molecular Reproduction and Development* **26**: 143–149

[5] Zhao Y, Buhr M M 1995 Cryopreservation extenders affect calcium flux in bovine spermatozoa during a temperature challenge *Journal of Andrology* **16**: 278–285

[6] Farber J L 1990 The role of calcium ions in toxic cell injury *Environmental Health Perspectives* **84**: 107–111

[7] Strehler E E, Zacharias D A 2001 Role of alternative splicing in generating isoform diversity among plasma membrane calcium pumps *Physiological Reviews*: **81** 21–50

[8] Okunade G W, Miller M L, Pyne G J, Sutliff R L, O'Connor K T, Neumann J C, Andringa A, Miller D A, Prasad V, Doetscman T, Paul R J, Shull G E 2004 Targeted ablation of plasma membrane Ca^{2+}-ATPase (PMCA) 1 and 4 indicates a major housekeeping function for PMCA1 and a critical role in hyperactivated sperm motility and male fertility for PMCA4 *Journal of Biological Chemistry* **279**: 33742–33750

[9] Schuh K, Cartwright E J, Jankevics E, Bundschu K, Liebermann J, Williams J C, Armesilla A S, Emerson M, Oceandy D, Knobeloch K P, Neyses L 2004 Plasma membrane Ca^{2+} ATPase 4 is required for sperm motility and male fertility *Journal Biol Chem* **279**: 28220–28226

[10] Post H, Schwarz A, Brandenburger T, Aumüller G, Wilhelm B 2010 Arrangement of PMCA4 in bovine sperm membrane fractions *International Journal of Andrology* **33**: 775–783

[11] Travis A J, Foster J A, Rosenbaum N A, Visconti P E, Gerton G L, Kopf G S, Moss S B 1998 Targeting of a germ cell-specific type 1 hexokinase lacking a porin-binding domain to the mitochondria as well as to the head and fibrous sheath of murine spermatozoa Molecular *Biology of the Cell* **9**: 263–276

[12] Wennemuth G, Babcock D F, Hille B 2003 Calcium Clearance Mechanisms of Mouse Sperm *The Journal of General Physiology* **122**: 115–128

[13] Brini M 2009 Plasma membrane Ca^{2+}-ATPase: From a housekeeping function to a versatile signaling role *Pflugers Archiv European Journal of Physiology* **457**: 657-664

[14] Reinhardt T A, Lippolis J D, Shull G E, Horst R L 2004 Null mutation in the gene encoding plasma membrane Ca^{2+}-ATPase isoform 2 impairs calcium transport into milk *Journal of Biological Chemistry* **279**: 42369–42373

[15] O'Donnell L 2014 Mechanisms of spermiogenesis and spermiation and how they are disturbed *Spermatogenesis* **4** e979623

[16] Sprando R L, Russell L D 1987 Comparative study of cytoplasmic elimination in spermatids of selected mammalian species *American Journal of Anatomy* **178**: 72–80

[17] Vogl A W, Young J S, Du M 2013 New Insights into Roles of Tubulobulbar Complexes in Sperm Release and Turnover of Blood-Testis Barrier *International Review of Cell and Molecular Biology* **303:** 319–355

[18] Costello S, Michelangeli F, Nash K, Lefievre L, Morris J, Machado-Oliveira G, Barrat C, Kirkman-Brown J, Publicover S 2009 Ca^{2+}-stores in sperm: Their identities and functions *Reproduction* **138:** 425–437

[19] Ren D, Xia J 2010 Calcium Signaling Through CatSper Channels in Mammalian Fertilization *Physiology* **25:** 165–175

[20] Parrish J J, Susko-Parrish J L, Graham J K 1999 In vitro capacitation of bovine spermatozoa: Role of intracellular calcium *Theriogenology* **51**: 461–472

[21] Dragileva E, Rubinstein S, Breitbart H 1999 Intracellular Ca(2+)-Mg(2+)-ATPase regulates calcium influx and acrosomal exocytosis in bull and ram spermatozoa *Biol Reprod* **61:** 1226–1234

[22] Xu Y, Yang W 2017 Calcium influx and sperm-evoked calcium responses during oocyte maturation and egg activation *Oncotarget* **8:** 89375–89390

[23] Pang Y, Zhu H, Wu P, Chen J 2005 The characterization of plasma membrane Ca^{2+}-ATPase in rich sphingomyelin-cholesterol domains *FEBS Letters* **579:** 2397–2403

[24] Hope H R, Pike L J 1996 Phosphoinositides and phosphoinositide-utilizing enzymes in detergent-insoluble lipid domains *Molecular Biology of the Cell* **7**: 843–851

[25] Pike L J, Casey L 1996 Localization and turnover of phosphatidylinositol 4,5-bisphosphate in caveolin-enriched membrane domains *Journal of Biological Chemistry* **271:** 26453–26456

[26] Pike L J, Han X, Chung K N, Gross R W 2002 Lipid rafts are enriched in arachidonic acid and plasmenylethanolamine and their composition is independent of caveolin-1 expression: A quantitative electrospray ionization/mass spectrometric analysis *Biochemistry* **41:** 2075–2088

[27] Brodin P, Falchetto R, Vorherr T, Carafoli E 1992 Identification of two domains which mediate the binding of activating phospholipids to the plasma-membrane Ca^{2+} pump European *Journal of Biochemistry* **204:** 939–946

[28] Enyedi A, Flura M, Sarkadi B, Gardos G, Carafoli E 1987 The maximal velocity and the calcium affinity of the red cell calcium pump may be regulated independently Journal of *Biological Chemistry* **262:** 6425-6430

[29] Czeredys M, Samluk Ł, Michalec K, Tułodziecka K, Skowronek K, Nałęcz K A 2013 Caveolin-1 - A novel interacting partner of Organic Cation/Carnitine Transporter (Octn2): Effect of protein kinase C on this interaction in rat astrocytes *PLoS ONE* **8** e82105

[30] Morén B 2014 Caveolae associated proteins and how they effect caveolae dynamics *Umeå University*

Effect of Sucrose on in vitro Bud Multiplication of Torch Ginger (*Etlingera elatior*)

L Chaidir[1], S Hasani[1], A Diana[1], M Subandi[1] and N Wicaksana[2]

[1]Department of Agrotechnology, Faculty of Science and Technology, UIN Sunan Gunung Djati Bandung, Indonesia
[2]Department of Agrotechnology, Faculty of Agricultura, Padjadjaran University, Indonesia

Email : libertychaidir@uinsgd.ac.id

Abstract. Torch Ginger (*Etlingera elatior* (Jack) R. M. Smith) is plant material that can be used as a medicine, food, cosmetics, and bio-pesticides. The existence of torch ginger in nature is becoming more limited due to the utilization of health care services mainly as a medicine continues to increase. In conventionally, plant propagation of torch ginger is derived from a rhizome, but seedlings produced tend to have the level of low proliferation and susceptible to infections of soil pathogens. This research aimed to know the influence of sucrose and BAP concentration on bud growth of torch ginger in in vitro. The research was conducted in the Tissue Culture Laboratory Universitas Islam Negeri Sunan Gunung Djati Bandung. Research method used was a descriptive design with experiment treatments i.e. sucrose (20 g/L, 40 g/L, and 60 g/L) and BAP (0 mg/L and 1 mg/L). The result showed that the use of sucrose 40 g/L + BAP 1 mg/L gave the best response to the shoot growth on 6 WAI and capable to produce green shoots on 8 WAI.

1. Introduction

Torch ginger (*Etlingera elatior* (Jack) R.M. Smith) is a native Indonesian herbal plant from the family Zingiberaceae which is widely used by people, as medicinal material, food, ornamental plants, cosmetic material, also as bio-pesticides. At this time, the use of torch ginger especially as medicinal material continues to increase. Like most other medicinal plants, torch ginger is also harvested directly from nature, so its existence is getting limited. This condition encourages the need of cultivation activity as an effort to preserve torch ginger so that its existence is maintained.

Conventionally, torch ginger propagation material comes from the rhizome, which is the part of the plant that is in the soil. Seedlings that are propagated through this method tend to have low proliferation rates and are very susceptible to pathogenic infections such as rhizome rots caused by Phytium species and leaf spot due to Coletotrichum species [1]. Such propagation techniques can lead to accumulation of pathogens in the seedlings, especially viruses that will be inherited through generations. This causes the supply of healthy torch ginger seedlings to be limited.

Effort that can be done to obtain disease-free torch ginger in large number and uniform plants with a relatively shorter time is by conducting plant propagation in in vitro [2]. The appropriate use of culture media and growth regulating substances become important factors as an effort to obtain optimum result in in vitro propagation.

In the use of medium, carbon sources is a very essential component because carbon functions as a source of energy needed by cells for growth [3]. Sucrose as an organic component consisting of glucose and fructose must be present in the culture media as an energy source for non-photosynthetic explants, so that they can perform growth [4]. In the culture medium, the addition of carbon sources becomes very important, because plants that are cultured in a bottle cannot obtain energy from the photosynthesis process as do plants that are conventionally cultivated. In order to meet the nutritional and energy needs, it is necessary to provide carbon sources that can be directly used by plants in their growth. Sucrose is an important and abundant compound in plants. Sucrose as a carbohydrate, serves as a source of energy for explant growth.

Besides carbon sources, torch ginger propagation also requires plant growth regulators (PGR). According to PGR which plays a role in shoot growth, especially is cytokinin. The purpose of adding cytokinin is to stimulate shoot growth, encourage metabolism and cell division, also reduce apical dominance to encourage lateral shoot initiation. BAP (Benzyl Amino Purine) is one of the groups of synthetic cytokinin which is most often used in plant propagation in in vitro culture, because it has a high enough effectiveness for shoot propagation, easily obtained and relatively cheaper.

2. Materials and Methods

A descriptive method was used in this research, which is the description of the growth response of torch ginger explants to the concentrations of sucrose and BAP given. Treatment design consisted of two factors, the first factor was the concentration of sucrose (s1 = 20 g/L, s2 = 40 g/L, s3 = 60 g/L), the second factor was the concentration of BAP (b0 = without BAP and b1 = 1 mg/L).

3. Results and Discussion

The results showed that explants began to show a growth response at 2 week after initiation (WAI). The response did not show the formation of shoot directly, but preceded by tissue swelling. The addition of sucrose and BAP on the culture medium greatly affect the growth pattern experienced by torch ginger explants. Of the total grow explants, the time required to demonstrate the first growth response was 2 weeks. Shoot formation is the main objective of this study. Shoots are a prospective new plants that grows from plant parts. Growth of shoots is very important in tissue culture, because shoots act as materials in the next propagation.

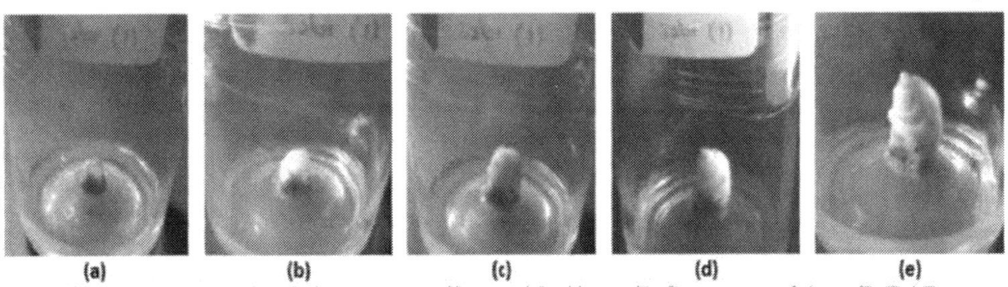

Figure 1. Growth of shoot on medium with 40 mg/L Sucrose and 1mg/L BAP

a. Condition of explant at 1 WAI
b. Condition of explant at 2 WAI
c. Condition of explant at 4 WAI
d. Condition of explant at 6 WAI
e. Condition of explant at 8 WAI

Based on the observation results, culture medium which was capable on growing the best shoot was s2b1 (40 mg/L Sucrose and 1 mg/L BAP). In this medium, explant was able to grow a bud at 6 WAI. The formation of buds began with the occurrence of tissue swelling in explants. Swelling was seen starting at 2 MSI, where the explants experienced an increase in size. This swelling pattern mainly occurred vertically and continued to occur up to 5 WAI. Entering 6 WAI, explants showed different conditions, the exterior was yellowish white with a stronger structure (Figure 1d). This was thought to be a previously swollen tissue had developed into a bud. The same condition was still seen in 7 WAI observations, a change occurred only in the slightly bud size increase. At 8 WAI, a significant change had occurred, where the previously formed buds changed color to green became a bigger shoot (Figure 1e).

Green color change that occurred in these shoot indicated the presence of chlorophyll in the tissue. As explained that the more green color produced, the higher the chlorophyll content. Chlorophyll is a green pigment located in the chloroplasts, which is the place where photosynthesis occurs in plants. Characteristics of green shoot indicated that shoot growth occurs well. Sucrose and BAP concentrations used had shown optimal shoot growth.

Treatment of S2B1, namely 40 g/L sucrose and 1 mg/L BAP was the best combination treatment that could produce optimal growth of torch ginger shoots. The use of 40 g/L sucrose had been able to meet the needs of carbon as an energy source for explants in its growth. BAP of 1 mg/L was a suitable concentration for regeneration of torch ginger explant tissue in shoot formation in in vitro culture. With a proper combination of sucrose and BAP concentrations, so the best shoot growth of torch ginger explant can be generated.

In tissue culture, plants could not conduct photosynthesis processes as plants that are cultivated on open field, so they need additional carbohydrates as the energy source. [5] explained that the differences in growth regulators concentration in tissue culture cannot control the development of plants without the addition of carbohydrates. These carbohydrates have a fundamental role in the process of shoot proliferation and affect shoot growth and survival. The presence of sucrose in culture media in this study greatly influenced shoot growth. The use of sucrose with optimal concentration will ensure the availability of energy sources for the cell to grow [3]. If the carbon source is sufficient, the cells will form quickly, so that the explant is able to grow shoots.

Addition of cytokinin is also very important in the regeneration process of torch ginger shoots. Cytokinin is used to stimulate cell division, cell enlargement and shoot formation. [6] determined that the addition of cytokinin into culture medium is expected to overcome the problem of the low rate of cell division in plant shoots. In this study, the use of BAP was combined with sucrose.

4. Conclusion

The best concentration of sucrose and BAP for the growth of torch ginger shoot in in vitro was in the combination of 40 mg/L sucrose and 1mg/L BAP which had been able to produce healthy and green shoot.

Acknowledgments
We would like to thank the Chairman of Research Center UIN Sunan Gunung Djati Bandung, Dr. Yudi Darmalaksana who had provided research grant in the BOPTAN UIN research grant in 2018.

References

[1] Abdelmageed A H A, Faridah Q Z, Norhana F M A, Julia A A, Kadir M A 2011 Micropropagation of Etlingera elatior (*Zingiberaceae*) by using axilarry bud explants *Journal of Medicinal Plants Research* **5(18)**: 4465-4469.

[2] Ibrahim H, Setyowati F M 1999 Alphabatical Treatment of Species In de Guzmna C, Siemonsma J S (Eds) : Plant Resources of South-East Asia Spices *Backnuys Publisher* Leiden the Netherlands **(13)**: 123-126.

[3] Sitorus E N, Hastuti E D, Setiari N 2011 Induksi Kalus Binahong (*Basella rubra* L.) secara in vitro Pada Media Murashige & Skoog dengan Konsentrasi Sukrosa yang Berbeda *BIOMA* **13(1)**: 32 - 37

[4] Kafindra, Khumaida N, Ardie, S W 2015 Induksi Rimpang Mikro kaemferia parviflora secara in vitro dengan Penambahan BAP dan Sukrosa *J. Hort Indonesia* **6(1)**: 54-63.

[5] Ramage C M, Williams R R 2001 Mineral nutrition and plant morphologis. *In vitro cell Biol. plant* **38**: 116-124.

[6] Rajthar J, Viehmannova I, Cepkova P H, Fernandez E, Milella L 2014 In vitro propagation of Drosera intermedia as influenced by cytokinins, pH, sucrose, and nutrient concentration *J. Food Agric.* **26(6)**: 558-564.

Diversity of morphological characters of 30 local torch ginger accessions from Pangandaran of West Java of Indonesia

C Hidayat[1], L Chaidir[1], A Supriadin[1], O T Kurahman[1], and A M Mulya[1]

[1]Department of Agrotechnology, Faculty of Science and Technology, UIN Sunan Gunung Djati Bandung, Indonesia

E-mail: cephidayat62@uinsgd.ac.id

Abstract. Torch ginger (*Etlingera* spp.) is one of the herbal plants native to Indonesia that has long been known and used as medicines. Exploration on torch ginger plant needs to be done to conservate and determine genetic diversity. Study of genetic diversity torch ginger in the district of Pangandaran is important because as long as there has been no study in depth the potential of genetic diversity and phylogenetic relationship germplasm of torch ginger Genetic diversity based on this research had a proportion of 47% of the total of 100% in the PC1-PC3 was based on the analysis Principal Component Analysis (PCA) with the character influencing that form clumps, line leaf midrib, plant height, stem diameter, leaf length, color scales rhizome, stem color, rhizome and the leaf base forms. Based on the results of the genetic diversity cluster analysis found that narrow torch ginger euclidean indicated by coefficient of 0.79 and was divided into six clusters. Accession which had the closest kinship i.e. accession N1 from Cikadu subdistrict of Cintakarya and accession N2 from Cikoang subdistrict of Cintakarya, while accession with the farthest kinship namely accession N29 from Jangraga subdistrict of Mangunjaya with range euclidean of 0.79.

1. Introduction

Torch ginger plants consist of green torch ginger (*Etlingera elatior*) and red torch ginger (*Etlingera hemisphaerica*), which have the potential as functional food material that have various functions, namely antibacterial, natural medicines, cosmetics, beverages and food flavoring [1] so its existence is quite searchable to be utilized by the community. Pangandaran Regency has the potential for torch ginger exploration because due to extensive land and supporting environmental factors. Breeding on torch ginger plants in Pangandaran Regency needs to be done because so far there has been no study of the potential for genetic diversity and kinship relationships of torch ginger germplasm. The existence of germplasm is increasingly narrowed due to genetic erosion caused by human and natural factors. By narrowing the germplasm, the genetic diversity of plants will be smaller and can cause extinction.

The breeding process carried out on torch ginger plants can be done by exploration, identification and characterization through morphological and agronomic characters. Pangandaran Regency is a location that can be used to explore torch ginger plants because it is included to the production center, moreover the community uses the torch ginger commodity itself. Germplasm collections can be used better if the characteristics of the plant are known [2]. The germplasm collection aims to study the level of existing diversity to maintain genetic diversity. Factors that influence the differences in plants are not only because of differences from internal factors, but also external factors. Differences in the appearance of plants can be caused by differences in the nature of plants (genetic) or environmental differences, or because they influence one another [7].

Efforts that need to be done to maintain genetic diversity such as primitive species, wild species, also specific and superior species, namely the preservation of germplasm by means of in situ and ex situ. The existence of high genetic diversity is one of the guidelines that must be noticed to obtain superior cultivars. If there is a high genetic diversity there is a greater chance of the best character selection when compared to characters which have a low genetic diversity.

The kinship relationship between two individuals or populations can be known by using the similarity of some characters, with the assumption that different characters are caused by differences in genetic array. Kinship is important to know because it has benefit of facilitating the search for substitute varieties if a variety has problems in its cultivation process.

2. Material and Methods

The research was carried out by exploration in the field by collecting torch ginger accessions in Pangandaran District. Exploration activity in Pangandaran District was carried out in several villages which was expected to obtain more collection of torch ginger plants. The method used was a roaming method with purposive sampling method. In this study, objects that are in accordance with the objectives of the study were taken, where the study was carried out in the area in Pangandaran District which found torch ginger plants. Research conducted on individual plants using direct observation for the measurement of morphological characters and using descriptive, namely in examining an object or a class of events in the present with the aim to be able to make characters systematically, factually and accurately about the natures or characteristics and relationships of phenomena observed.

3. Result and Discussion

Principal Component Analysis (PCA) is an analytical method that is often used to determine the diversity in plant collections. PCA is a technique used in order to find out how much a character contributes to diversity so that the results are commonly used to identify traits that become a characteristic of a variety [3]. In this study, the PCA aimed to determine the characters that can affect differences or variations in 30 accessions of torch ginger plants (*Etlingera* spp.) observed in Pangandaran District.

Table 1 Eigenvalue and Variability Values of 20 Characters in 30 *Etlingera* spp accessions in Pangandaran Regency

Component	Eigenvalue	Percent *(%)*	Cumulative *(%)*
1.	3.90	19.52	19.52
2.	3.49	17.44	36.96
3.	2.04	10.18	47.14

The calculation of the PCA was analysed based on 20 characters of torch ginger plant (*Etlingera* spp.), namely: clump shape, rhizome shape, color of rhizome scales, stem color, upper leaf color, lower leaf color, bud color, wax layer, leaf midrib line , leaf texture, leaf tip shape, leaf texture, leaf tip shape, leaf edge shape, leaf edge color, leaf base shape, leaf shape, plant height, stem diameter, leaf length, leaf width and leaf node distance.

The results of the principal component analysis (PCA) showed the presence of eigenvalue, percent and cumulative values. The principal component used was a component that had eigenvalue more than two so that it produced three principal components of eigenvalue and variability values.

Table 1 shows the Eigenvalue of each variable with the percentage in ordered values. These three components were components that had more than two eigenvalue. In the Table 1 it can be seen that the principal component (PC1) had an eigen value of 3.90 with a percentage of 19.51% of the variation of 30 accessions of torch ginger plants (*Etlingera* spp.) from all locations with red and green types. The percentage of PC1 was obtained from the clump shape, leaf midrib, plant height, stem diameter and leaf length. The second component (PC2), the eigenvalue owned was 3.49 with a percentage of 17.44% obtained from the color characteristics of the rhizome scales and stem color. The third component (PC3), the eigen value owned was 2.04 with a percentage of 10.18% obtained from the character of the rhizome form and the base leaf shape.

Table 1 describes the components of PC 1 - PC 3 having a total of 47% of the total 100% proportion of the torch ginger plants variation character that appear and. PC 1 - PC 3 had eigenvalue 2.04-3.90 where the eigenvalue was above two so that PC 1 - PC 3 could be called as the characters that affected variations occurred in 30 accessions of torch ginger plants (*Etlingera* spp .) which consisted of clumps, leaf midrib lines, plant height, stem diameter and leaf length, color of rhizome scales and stem color, rhizome shape and base leaf shape.

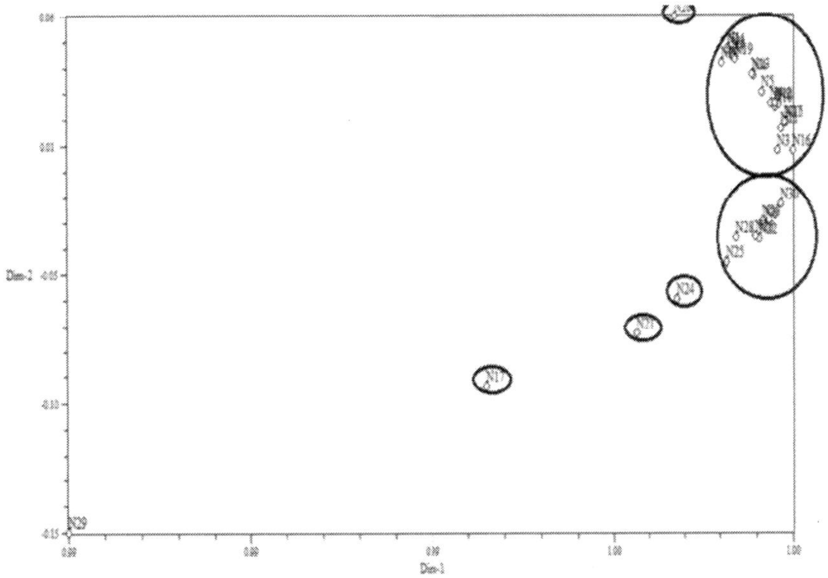

Figure 1 Biplot Graph of Spreading Pattern among 30 Accessions of Torch ginger Plants (*Etlingera* spp.) in Pangandaran District

Explanation:N1= Cikadu Cintakarya; N2= Cikoang Cintakarya; N3= Ciawi Cintakarya; N4= Panglanjan Cintaratu; N5= Bontos Cintaratu; N6= Burujul; N7= Sukarenah; N8= Patinggen; N9= Harjaresik; N10= Kedungwuluh; N11= Sindangwangi; N12= Ciganjeng; N13= Sirungwatang; N14= Cintamaju; N15= Karangsari; N16= Ciawi Cintakarya; N17= Maruyung Sari; N18= Paledah; N19= Sukanagara; N20= Emplak; N21= Pejaten; N22= Karang Torch ginger; N23= Sidamulih; N24= Karang Benda; N25= Karang Jaladri; N26= Sinargalih; N27= Sukamaju; N28= Mangunjaya; N29= Jangraga; N30= Sindangjaya.

In Figure 1, the biplot graph of the torch ginger crop distribution pattern based on 20 characters observed there were six groups. Accession of torch ginger plants (*Etlingera* spp.) which were in the same circle were accessions that had plants with similar character. The first group consisted of only 1 accession, the N20. The second group consisted of 19 accessions namely N14, N13, N19, N1, N2, N3, N4, N5, N7, N8, N9, N18, N10, N11, N16, N12, N15, N27, N29 which clustered close together and were in the same circle and were accessions with similar morphology of clump shape, color of rhizome scales, upper leaf color, wax layer, leaf midrib lines, leaf texture, leaf tip shape and edge color, as for agronomic similarities seen from plant height, stem diameter, leaf length and leaf width.

The third group consisted of 7 accessions namely N30, N23, N6, N26, N22, N28, N25 with similarities in clump morphology, stem color, lower leaf color, wax layer, leaf midrib lines, leaf texture and edge shape while agronomic similarities were shown from leaf length and leaf width. The fourth group consisted only of accession N24, the fifth group consisted only of accession N21 and the sixth group only consisted of accessions N17, each of which had the character of plant height and stem diameter that was different (smaller) than other accessions.

Cluster analysis is an analysis that is used to classify elements that are similar as research objects to be used as clustering that is different and mutually exclusive or special. Cluster analysis is an analysis included in the multivariant statistical analysis of interdependent methods. The purpose of cluster analysis was to explore the torch ginger plant, which was to classify objects based on similar characteristics among the objects. The objects are classified in one or more clusters or groups so that the objects in one cluster will have similarities with one another.

In the cluster analysis, which can state the distance of non-conformity, is Euclidean Coefficient. The euclidean distance which has a range of more than one indicates a large non conformity coefficient. The small non conformity coefficient states that each genotype with each other has a narrow variation and vice versa if the larger non conformity coefficient, the wider variation in each genotype. The euclidean distance of torch ginger plants in Pangandaran Regency was in the range of 0.05 to 0.79 (Figure 2) which states that the non-conformity coefficients in torch ginger plant populations in all locations had a narrow non conformity stating that the variation contained in the population was narrow..

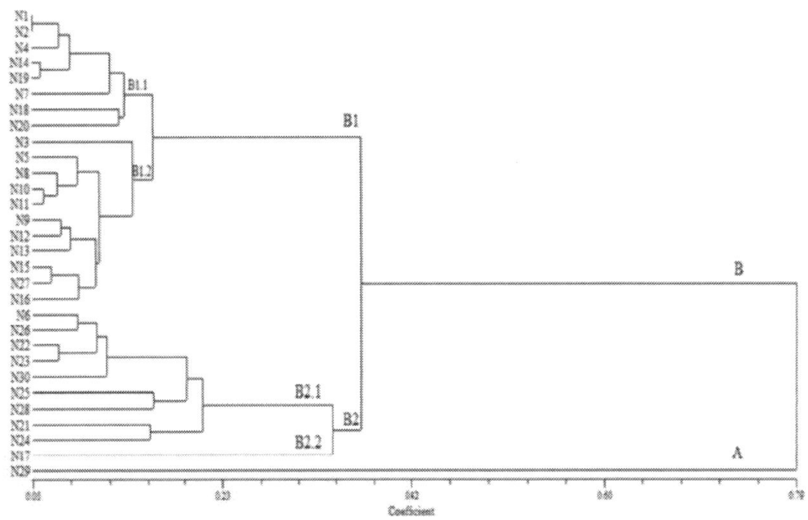

Figure 2. Dendrogram of 30 Accessions Torch ginger

In Figure 2, a dendrogram of a total of 30 accessions torch ginger plants in Pangandaran District had formed two clusters, cluster A and cluster B. Cluster A consisted of only one accession, namely accession N29, namely accession from Jangraga Village, Mangunjaya Subdistrict with an euclidean range of 0.79 which had a phenotypic appearance with quite striking differences, namely from brownish green of rhizome scales and only had 119.1 cm plant height with a stem diameter of 1.36 and leaf length of 3.87, but that did not mean that other characters were very different from other accessions .

Cluster B was a cluster consisting of 2 clusters, namely clusters B1 and B2, each of which also formed sub-clusters. The total accessions contained in cluster B1 were 19 accessions. The B1 cluster consisted of sub-cluster B1.1 and sub cluster B1.2. Sub cluster B1.1 consisted of 8 accessions, namely N1, N2, N4, N14, N19, N7, N18, N20 while the sub cluster B1.2 consisted of 11 accessions namely N3, N5, N8, N10, N11, N9, N12 , N13, N15, N27, N16. Cluster B2 consisted of sub cluster B2.1 and sub cluster B2.2. Sub-cluster B2.1 consisted of 9 accessions namely accession N6, N26, N22, N23, N30, N25, N28, N21, N24 while sub cluster B2.2 consisted of only 1 accession namely N17.

Cluster B had many accessions that had a parallel non conformity distance that means these accessions tended to have morphological appearances that were closely related. Accessions that had the closest kinship were the plants that exist in the accession N1 originating from Cikadu Village, Cintakarya District and N2 accession, originating from Cikoang Village, Cintakarya District with the closest coefficient distance, which was 0.05. This could be due to the location of the accession which is quite close because it is in the same sub-district with relatively similar environmental factors.

If more and more characters are observed, the magnitude of the differences or even similarities between accessions will be increasingly visible [14]. This torch ginger plant in Pangandaran Regency had a kinship pattern of relationships between accessions and diversity that had been shown based on the results of exploration that can be used for breeding purposes, assembling of new varieties and subsequent superiority.

4. Conclusion

There was a close relationship between torch ginger plants (*Etlingera* spp) with euclidean distance ranges from 0.05 to 0.79 which resulted in two clusters namely cluster A and cluster B which showed variation in torch ginger plants. The farthest kinship relationship was in accession 29, namely accession from Jangraga Village, Mangunjaya Subdistrict with euclidean range of 0.79, while the closest kinship to N1 accession was from Cikadu Village, Cintakarya District and N2 accession, originating from Cikoang Village, Cintakarya District with the closest distance coefficient, which was 0.05 because the locations of the accessions are in the same sub-district with relatively similar environmental factors.

Acknowledgements
We would like to thank the Chairman of Research Center UIN Sunan Gunung Djati Bandung, Dr. Yudi Darmalaksana who had provided research grant in the BOPTAN UIN research grant in 2017.

References

[1] Jackie T, Haleagrahara N, Chakravarthi S 2011 Antioxidant effects of Etlingera elatior flower extract against lead acetate-induced perturbations in free radical scavenging enzymes and lipid peroxidation in rats *BMC Res. Notes* **67:** 1–8
[2] Setyowati M, Hanarida I, Sutoro 2007 Karakteristik Umbi Plasma Nutfah Tanaman Talas (*Colocasia esculenta*) *Bul. plasma nutfah* **13:** 49–55
[3] Afuape S O, Okocha P I, Njoku D 2011 Multivariate Assessment of The Agromorphological Variability and Yield Components among Sweet Potato (*Ipomoea Batatas* (L.) Lam) Landraces *African J. Plant Sci.* **5:** 123-132.
[4] Baihaki A 2000 Teknik Rancangan dan Analisis Penelitian Pemuliaan Fakultas Pertanian Bandung: *Universitas Padjadjaran.*
[5] Barros L M, Paiva J R, Crisostómo J R, Cavalcanti J J V 2007 Cajueiro In : Bruckner C.H. (ed.) : Melhoramento de Fructeiras Tropicais Editoria UFV, *Vicosa* pp. 159-176.
[6] Chaidir L, Hidayat C, Supriadin A, Aina A, Nuryan, Ramdhani M A 2018 Morphological Characterization of Etlingera Elatior (Jack) Explorated in Sukabumi, West Java, Indonesia *IOP Conf. Series: Materials Science and Engineering* **288(2017)** 012091.
[7] Chaidir L, Yuliani K, Qurrohman B F T 2016 Eksplorasi dan Karakterisasi Tanaman Genjer (*Limnocharis flava* (L.) Buch) di Kabupaten Pangandaran berdasarkan Karakter Morfologi dan Agronomi *Jurnal Agro* **3(2)**.
[8] Gardner F P, Pearce R B, Mitchell R L 1991 Physiology of Crop Plants Diterjemahkan oleh Susilo H *Universitas Indonesia Press*: Jakarta.
[9] Habsah M, Lajis N H, Sukari M A, Yap Y H, Kikuzaki H, Nakatani N, Ali A M 2005 Antitumour-Promoting and Cytotoxic Constituentss of Etlingera Elatior *Malaysian Journal of Medical Sciences* **12:** 6-12.
[10] Ibrahim H, Mood J 2001 The Torch Gingers: Sentinels of The Forest *Garden Asia* pp. 12-19.
[11] International Union for the Protection of New Varieties of Plant 1996 Guidelines for the Conduct of Tests for Distinctness, Univormity and Stability: Ginger *(Zingiber officinale* Rosc.)
[12] Lim C K 2001 Taxonomic notes on Etlingera Giseke (*Zingiberaceae*) in Peninsular Malaysia: the "Achasma" taxa and supplementary notes on the "Nicolaia" taxa *Folia Malaysiana* **2:** 141–178
[13] Luc-Cayol F, Fereol L 1997 *Alpinia martinica (Zingiberaceae):* An Intergeneric Hybrid Between *Alpinia Purpurata* and *Etlingera elatior HortSci* **32:** 914- 915
[14] Rohlf F J 2001 NTSYS-pc: Numerical Taxonomy and Multivariate Analysis Sy.

Correlation Between Fruit Characters with Capsaicin Content in F_2 Population of Chili

R F Yenny[1], N Rostini[2], Hersanti[2], Anas [2]

[1]Postgraduate Program of Plant Science, Faculty of Agriculture, Universitas Padjadjaran Jl. Raya Bandung Sumedang km 21, Jatinangor, 45363, West Java, Indonesia

E-mail: neni.rostini@unpad.ac.id

Abstract. Chili is one of the most important vegetable spice crop valued for its aroma, flavor, and pungency for Indonesian food. An experiment on chili (*Capsicum annuum* L) was undertaken to study the genetic correlation and relationship of morphological and fruit characters on capsaicin content using path analysis. The F_2 population was developed from the cross between the high capsaicin content of cayenne pepper CF02 inbred line and curly chili K15 inbred line. Capsaicin content was negatively associated with fruit surface, fruit length, fruit diameter, fruit weight, and fruit weight per plant. Based on path analysis, these characters had direct effect and indirect effect on capsaicin content. Direct selection may be executed considering these traits as the primary selection criteria to reduce the indirect effect of other characters during the development of high yielding chili variety.

1. Introduction

Chili or pepper (*Capsicum annuum* L.) belongs to family Solanaceae. The center origin of chili is Mexico, with secondary centers in Guatemala and Bulgaria [1]. Chili is one of the most important vegetable spice crop valued for its aroma, flavor, and pungency [2]. The pungency of chili is due to a crystalline acrid volatile alkaloid capsaicin ($C_{18}H_{27}NO_3$) present in the placenta of fruit.

The plant breeder tries to partition the correlation coefficients into components of direct and indirect effects by employing the path coefficient analysis, which involves measurement of the influence of one trait upon the set of other traits through standardized partial regression coefficient to increase the efficiency of selection. Correlation and path analysis was proposed by Wright [3] to organize the intercourse between predictor variables and the response variable. Correlation measures association between capsaicin and fruit characters. This method is very useful in plant breeding programs to determine whether there is a relationship between fruit characters and the character of capsaicin content so that it is helpful to improve correlated responses [4]. Therefore the present investigation was carried out to determine the nature and degree of association among the fruit characters with their capsaicin content and their direct and indirect effects to chili.

2. Methods And Materials

The experiment was carried out in 2016-2017 at the farmer field Ciparanje, Jatinangor district. Totally 300 F_2 chili plants obtained from the cross CF02 X K15 type and it was evaluated for high yield and quality characters during the year 2016. Observations were recorded on single plant wise for each of F_2 populations. Cultivation techniques used in the research location are standard cultivation techniques in chili. The seeds are sown first in a sterile seedling tray until the age of three weeks after planting (WAP). The data was observed from nine quantitative characters, which includes fruit surface, fruity brightness, resilience, fruit length, fruit diameter, number of fruits, fruit weight per plant, weight per fruit, and capsaicin content. The data were statistically analyzed for estimation of the correlation coefficient among different characters. The path coefficient of various fruits characters towards capsaicin content was calculated according to [5].

3. Results and Discussion

High correlation only shows the closeness of the relationship between traits but cannot prove a causal relationship. Path analysis can be used to determine the causal association and can sort it into direct and indirect influences [6]. Fruit characteristics included in the path analysis were fruit surface, fruit length, fruit diameter, fruit weight per plant, and weight per fruit, because these characters correlate with capsaicin levels (P> 0.05) based on the *Pearson correlation test*. The results of this study indicate that the type of cayenne pepper has the spiciest taste which is characterized by high capsaicin contents. The level of capsaicin strains tested ranged from 0.09 - 1.06. The highest yield was achieved by lines 212 and 249 respectively 1.05 and 1.06. The results of the correlation coefficients was presented in Table 1.

Table 1. The result of Correlation Coefficients between Fruit Characters in F_2 chili Plants

Traits	Z1	Z2	Z3	Z4	Z5	Z6	Z7	Z8	Y
Fruit surface	1	.355**	.070	.677**	.055	-.057	.397**	.647**	**-.238****
Fruit Bright	.355**	1	-.050	.312**	.051	-.028	.208**	.350**	-.094
Resillience	.070	-.050	1	.046	.044	.016	.111	.040	.006
Fruit Lenght	.677**	.312**	.046	1	.044	-.085	.548**	.871**	**-.290****
Fruit Diameter	.055	.051	.044	.044	1	.237**	.227**	.096	**-.176***
Number of fruit	-.057	-.028	.016	-.085	.237**	1	.678**	-.151*	-.074
Fruit weight per plant	.397**	.208**	.111	.548**	.227**	.678**	1	.525**	**-.215****
Weight per fruit	.647**	.350**	.040	.871**	.096	-.151*	.525**	1	**-.299****
Capsaicin content	-.238**	-.094	.006	-.290**	-.176*	-.074	-.215**	-.299**	1

Information: Z1: Fruit surface; Z2: Fruit Brightness; Z3: Resilience; Z4: Fruit Length; Z5: Fruit diameter; Z6: Number of fruit; Z7: Fruit weight per plant; Z8: Weight per fruit; Y: Capsaicin content.

The results of correlation analysis on the F_2 generation of the cross between the high capsaicin content of cayenne pepper CF02 inbred line and curly chili K15 inbred line showed that the significant highly correlation with capsaicin content. This indicates that chili genotypes that have high capsaicin levels tend to have a character. The results of correlation analysis showed that the characters of fruit surface, fruit length, fruit diameter, fruit weight per plant, weight per fruit were highly significantly correlated with capsaicin levels, with a correlation coefficient (r) of -0.238, -0.290, - 0.176, -0.215, and -0.299. This result indicated that chili genotypes which have low capsaicin levels tend to have small fruit length, small fruit diameter and fruit weight per plant and low weight per fruit. The correlation between capsaicin levels and other characters is not significant or negative.

This shows that the character most influence the capsaicin content levels. According to [7], the independent variable that has the most significant impact on the dependent variable is determined as a direct influence. Direct influence can be interpreted if each increase in one standard deviation in an independent value will increase the standard deviation C of a dependent variable. There is a correlation between morphological characters with the character of capsaicin contents, giving opportunities are used as morphological characters selection criteria to get characters fruit components required.

The path-coefficient analysis can explain the relative contribution of fruit characters to capsaicin levels, both directly and indirectly (Table 2). Five morphological characters contributed 34.4 % to the variation of capsaicin content. However, there characters significantly contributed to the capsaicin content of chilli (F <= 0.05).

Table 2. Direct and indirect effects for five quantitative traits in chilli (*Capsicum annuum* L.).

Variable	Direct influence	Indirect influence					Total
		Z1	Z4	Z5	Z7	Z8	
Fruit surface	-.053	0	-0.07582	-0.00825	-0.01032	-0.09058	-.238
Fruit length	-.112	- 0.03588	0	-0.00656	-0.01425	-0.12194	-.290
Fruit diameter	-.149	- 0.00293	-0.00493	0	-0.0059	-0.01344	-.176
Fruit weight per plat	-.026	- 0.02104	-0.06138	-0.03382	0	-0.0735	-.216
Weight per fruit	-.140	- 0.03429	-0.09755	-0.0143	-0.01365	0	-.299

Path diagrams are used to clarify further the description of path analysis put forward (Figure 1). By combining diagrams geometric and algebraic equations, then in-depth statistical analysis study the causal-effect correlation between variables to be more weighted in the sense that the results are easier to understand. Path diagrams are arranged based on the underlying knowledge about causal relationships based on the hypotheses made [8].

Efforts to determine the characters that can be used as criteria of effective selection can be seen from the magnitude of direct influence, the correlation between characters and the difference between associations between independent variables and influences directly the variable [9]. Thus the character which matches the selection criteria is characters had influence values directly positive marked equal to the correlation value and had a difference small. The fruit diameter has a value of -0.176 is the character that gives the most direct influence on the capsaicin content.

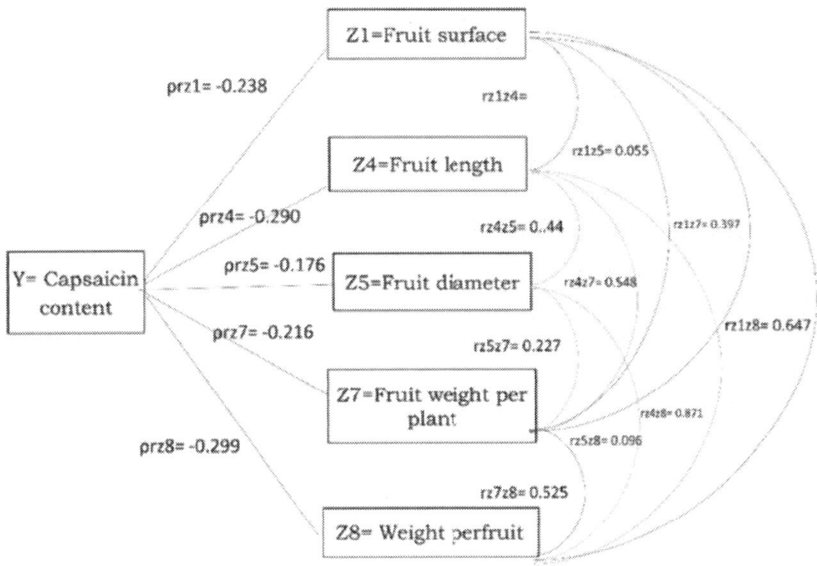

Figure 1. Path diagram quantitative character on the F_2 generation chili.

4. Conclusion

Based on the results of the experiment can be concluded that the characters that have a direct influence on capsaicin content is fruit surface, fruit length, fruit diameter, fruit weight per plant, and weight per fruit. Morphological characters that have direct influence can be used as selection criteria for improving spiciness character.

Acknowledgments

The authors appreciated to Universitas Padjadjaran for funding the research through Academic Leadership Grant (ALG) 2017-2018.

References

[1] Salvador M H 2002 Genetic Resources of Chili (*Capsicum anuum* L.) in Mexico *Proceedings of the 16th Int. Pepper Conf.* Tampico Tamaulipas Mexico November pp. 10-12

[2] Hazra P, Chattopadhyay A, Karmakar K, Dutta S 2011 Moddern Technology in Vegetable Production *New Indian Publishing Agency* New Delhi India pp. 478.

[3] Wright S 1921 Correlation and causation *Journal of Agricultural Research* **20**: 557-587.

[4] Falconer D S, Mackay T F C 1996 Introduction to Quantitative Genetics England (DE): *Longman.*

[5] Dewey O R, Lu K H 1959 A correlation and path coefficient analysis of components of crested wheat grass seed production *Agron. J* **51**:515-518.

[6] Li C C 1956 The concept of path coefficient and its impact on population genetics *Biometrics* **12(2)**: 190-210.

[7] Munawar M, Shahbaz M, Hammad G, Yasir M 2013 Correlation and Path Analysis of Grain Yield Components in Exotic Maize (*Zea mays*. L) Hybrids *International Journal of Sciences Basic and Applied Research (IJSBAR)* **20(1)**: 22-27.

[8] Singh R K, Chaudhary B D 1985 Biometrical methods in quantitative genetics analysis New Delhi (IN): *Kalyani Publisher.*

[9] Yunianti R, Sastrosumarjo S, Sujiprihati S, Surahman M, Hidayat S H 2010 Kriteria seleksi untuk perakitan varietas cabai tahan *Phytophthora capsici* L *Jurnal Agronomi Indonesia* **38(2)**: 14-21.

Inheritance Pattern of Capsaicin Content of Indonesian Chili Landraces (*Capsicum annum* L.)

N Rostini[1], R Fitry Yenny[2], Hersanti[1], Anas[1] and S Amien[1]

[1]Agriculture Faculty, Universitas Padjadjaran Jl. Raya Jatinangor Km 21 Bandung, Indonesia
[2] Postgraduate Program of Plant Science, Agriculture Faculty, Universitas Padjadjaran

E-mail: neni.rostini@unpad.ac.id.

Abstract. Capsaicin is a compound that produces a spicy taste in chili. Inheritance of capsaicin is valuable information for breeders to determine the selection methods and when the selection will be conducted. The objective of this experiment is to observe the inheritance pattern of the capsaicin content of Indonesian chili landraces. The experiment was conducted at Ciparanje Experimental Station of Universitas Padjadjaran. One hundred sixty-seven F_2 plants were analyzed for capsaicin content in the laboratory. The results showed the distribution of capsaicin content was continuous using the Kolmogorov Smirnov test. Many genes might control capsaicin content in Indonesia chili landrace. The capsaicin content of the male parent was 4,565 ppm, and the female parent was 8,540 ppm. The capsaicin content among F_2 plants ranged from 621 ppm to 14,348 ppm. The average of capsaicin content in the F_2 population was 4,814 ppm. The capsaicin content of some F_2 progenies showed higher than their male parents. However, some F_2 progenies showed also low capsaicin content compared with their female parent. Segregation transgressive of capsaicin content occurred in F_2 generation.

1. Introduction

The spiciness taste in chili due to capsaicinoid content, including dihydro-capsaicin and capsaicin. Inheritance of capsaicin content is essential for researchers because it determines the effectiveness of the breeding process in chili. Heterosis obtained on F_1 chili hybrid from the crossing between spicy chili pepper and non-spicy chili pepper, and also observe the transgressive variation in F_2 [1]. The spiciness in the genus Capsicum is the qualitative character that is affected by the function of the *pun1* allele. On the non-spicy chili 'Nara Murasaki,' function of *pun1* alleles was lost due to a mutation of a single nucleotide adenine insertion with on allele *pun1* and turned into a *pun14* allele [2]. Capsicum chinense hybrid of 'No. 3441' and 'Habanero' shows the dominant gene controlled of the spiciness and non-spicy flavor is recessive genes controlled [3]. The pungency of *C. chinense* peppers was controlled by *Pun1* locus. On the other hand, non-pungency was controlled by a recessive allele, named $pun1^2$[4]. The level of spiciness in spicy chili was determined by capsaicin content. In this experiment, we would examine the inheritance pattern of the level of spiciness Indonesian chili landrace.

2. Materials and Methods

Segregation population of the progenies crossing among Indonesian chili landrace was used in this experiments. The experiment was conducted in the field at Ciparanje experimental station in Universitas Padjadjaran (UNPAD). Capsaicin content of 167 of F_2 plants was analyzed in the Pharmacy Laboratory.

Capsaicin content of 167 progenies was analyzed the normality distribution by Kolmogorov Smirnov test. The normal distribution of data shows the polygenic trait and have a quantitative variation. The number of F_2 plants has similar capsaicin content with a parent could be counted using equation according to [5]: $(1/4)^n$, where n is the number of gene pairs controlling capsaicin. The phenotypic class in F_2 was estimated by equation (2n+1). The phenotypic ratio in F_2 population was calculated using

$(a+b)^{2n}$ [5]. If the discreet distribution data was observed in the F_2 population, the data then was computed using the Chi-square test. The result was then compared with the Mendelian ratio [6].

3. Result and Discussion

Performance of parents and the F_2 plants could be seen in Figure 1 and Figure 2. The F_2 population from crossing two landrace Indonesian cultivars called curly chili (male) and cayenne pepper (female). This chili was grouped in to *Capsicum annum* (L.). Curly chili is a chili with a dropping fruit position, fruit more than 10 cm long with a diameter of less than 1 cm. While cayenne pepper usually has an upright fruit position, the length of the fruit is about 2-4 cm with a diameter of less than 1 cm.

In the F_2 population, there was an apparent variation in fruit length, and fruit position. In Figure 2, the most appearance of fruits looked like cayenne pepper (Figures 2a and 2b), but there were also variations such as curly chili with an upright fruit position (Figure 2c) or cayenne pepper fruit with a dropping fruit and about 6 cm long, longer than usual (Figure 2d). Variations occur because of the gene segregation in the second generation of these.

The result of the Kolmogorov-Smirnov Test in Table 1. showed that the data was normal distribution or the capsaicin content is a quantitative character, controlled by many genes. The result is similar to the research of an interspecific hybridization of *Capsicum annum/ C. chinense*. This crossing showed significant additive gene effect for the capsaicinoid [7]. The existence of transgressive variation showed that the Indonesian chili landrace parent did not have all active genes controlling the spicy flavour. From the data, it was found that there were three plants of 167 F2 plants showed the lowest performance for capsaicin content, and three plants showed the highest capsaicin. The estimated number of genes controlling the capsaicin content was three pairs of genes at least (n=3) and phenotypic classes 2n+1=7 with distribution as shown in Table 2 and Figure 3. The lowest capsaicin content was 621 ppm, and this plant was considered without effective allele (0 genes effective), and the highest capsaicin content of plant was 14348 ppm and was considered have six effective alleles (Table 2).

A unit used to express pungency called the Scoville heat unit. Symbol, SHU [8] in Table 2, capsaicin content was converted to SHU by multiplying the ppm value by 16 because the pure capsaicin is 16,000,000 SHU. Habanero was a hot chili cultivar with 150,000 – 350,000 SHU. In this population, the capsaicin content of transgressive variation more than 150,000 SHU and could be classified as hot pepper. The study of capsaicin content and pungency level of different fresh and dried chilli peppers showed six Habanero varieties (Habanero Red Savina (HRS), Habanero Yellow (HY), Habanero Maya Red (HMR), Habanero Tasmania (HT), Habanero Paper Latern (HPL), and Habanero Red (HR), HRS has the highest capsaicin content of 147,870 SHU among other Habanero [9].

The hot chili progeny higher than the parents in the F_2 population, enriching the Indonesian chili collection and it is also possible to obtain a higher spiciness because the genes controlling spiciness could be more than three pairs. The breeders could hybrid another Indonesian landrace, to get spiciness variations with a combination of another spicy gene in another Indonesian chili landrace.

(a) (b)

Figure 1. Performance of Parent : (a) female, (b) male

Figure 2. Performance of F_2 Progeny Crossing of Indonesian Landrace Chili Cultivars, (a) male type, (b) female type, (c) and (d) variation

Table 1. Result of Kolmogorov-Smirnov Test

		Data
N		167
Normal Parameters[a,b]	Mean	0.4695
	Std. Deviation	0.17487
Most Extreme Differences	Absolute	0.061
	Positive	0.061
	Negative	-0.039
Kolmogorov-Smirnov Z		0.726
Asymp. Sig. (2-tailed)		0.668

Test distribution is **Normal**.

Table 2. Segregation of Capsaicin Content

The number of genes effective	The Number of Plant	Capsaicin content (ppm)	Scoville Heat Units (SHU) (ppm x 16)	Parent
0	2	621 - 2,582	9,936 - 41,312	
1	17	2,583 - 4,543	41,328 - 72,688	
2	40	4,544 - 6,504	72,704 – 104,064	Male
3	54	6,501 - 8,464	104,016 – 135,424	
4	39	8,465 – 10,425	135,440 – 166,800	female
5	13	10,426 – 12,386	166,816 – 198,176	
6	2	12,387 – 14,348	198,192 – 229,568	

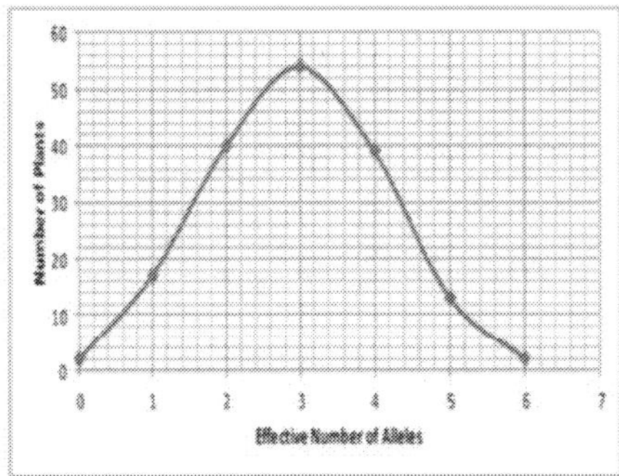

Figure 3. Segregation curve of capsaicin content in F_2 Population

Transgressive variation of capsaicin content were the homozygote recessive and the plants having one effective gene for negative transgressive, while homozygote dominant and the plants having five effective genes for positive transgressive (Figure 4).

P Male AAbbcc X Female aaBBCC1

F_1 AaBbCc

F_2

The effective gen	0	1	2	3	4	5	6
Ratio of number of plants (expected)	1	6	15	20	15	6	1
Number of plants (observed)	2	17	40	54	39	13	2
Scoville Heat Units (SHU)	9936 - 41312	41328 - 72688	72704 - 104064	104016 - 135424	135440 – 166800	166816 - 198176	198192 - 229568
Genotypes	1 aabbcc	2 Aabbcc 2 aaBbcc 2 aabbCc	1 AAbbcc 4 AaBbcc 4 AabbCc 1 aaBBcc 4 aaBbCc 1 aabbCC	2 AABbcc 2 AAbbCc 2 AaBBcc 8 AaBbCc 2 AabbCC 2 aaBBCc 2 aaBbCC	1 AABBcc 4 AABbCc 1 AAbbCC 4 AaBBCc 4 AaBbCC 1 aaBBCC	2 AABBCc 2 AABbCC 2 AaBBCC	1 AABBCC

Figure 4. The illustration of hybridization diagram from parent to F_2 generation.

4. Conclusion

The results of the distribution analysis of capsaicin content showed normal data distribution using the Kolmogorov Smirnov test. Capsaicin segregation in the F_2 populations was polygenic inheritance controlled by three gene pairs. The capsaicin content of the female parent was 1535 ppm, and the male parent was 7340 ppm. Capsaicin content on the F_2 population ranges between 622 ppm – 11,348 ppm. The middle value of F_2 capsaicin was 4,814 ppm. There was transgressive variation shown by higher capsaicin content in F_2 progeny compared to the male parent and lower than female.

Acknowledgement

The authors appreciated to Universitas Padjadjaran for funding the research through HIU-RHPU 2017-2018.

References

[1] Garcés-Claver A, Gil-Ortega R, Álvarez-Fernández A, Arnedo-Andrés M S 2007 Inheritance of capsaicin and dihydrocapsaicin, determined by HPLC-ESI/MS, in an intraspecific cross of Capsicum annuum L *Journal of Agricultural and Food Chemistry* **55(17):** 6951–6957.

[2] Kirii E, Goto T, Yoshida Y, Yasuba K, Tanaka Y 2017 Non-pungency in a Japanese Chili Pepper Landrace (*Capsicum annuum*) is Caused by a Novel Loss-of-function Pun Allele *The Horticulture Journal* **86(1):** 61–69.

[3] Koeda S, Sato K, Takisawa R, Kitajima A 2015 Inheritance of Non-pungency in "No.3341" (*Capsicum chinense*) *The Horticulture Journal* **84(4):** 323–326.

[4] Stewart C Jr, Mazourek M, Stellari G M, O'Connell M, Jahn M 2007 Genetic control of pungency in C. chinense via the Pun1 locus *J. of Experimental Botany* **58(5):** 979–991.

[5] Rothwell N V 1979 Understanding Genetics *Oxford Univ. Press* New York in Burns G W 1980 The Science of Genetics An Introduction of Genetics *Macmillan Pub Co., Inc.* New York.

[6] Burns G W 1980 The Science of Genetics An Introduction of Genetics *Macmillan Pub Co., Inc.* New York.

[7] Zewdie Y, Bosland P W 2000 Capsaicinoid Inheritance in an Interspecific Hybridization of *Capsicum annuum* x *C. chinense J. Amer.Soc.Hort.Sci.* **125(4):** 448–453.

[8] Scoville W L 1912 Note on Capsicums *Journal of the American Pharmaceutical Association* **1:** 453.

[9] Popelka P L, Jevinová P, Šmejkal K, Roba P 2017 Determination of capsaicin content and pungency level of different fresh and dried chilli peppers *Folia Veterinaria* **61(2):** 11—16.

IOP Conf. Series: Earth and Environmental Science **334** (2019) 012019 doi:10.1088/1755-1315/334/1/012019

The Role of Rumen Fluid as A Source of Methanogenic Bacteria on Coalbed Methane Activation

E T Marlina[1], Y Ardiansyah[1], T B A Kurnani[1], Y A Hidayati[1], K N Rahmah[1], I M Joni[2] and E Harlia[1]

[1]Faculty of Animal Husbandry, Universitas Padjadjaran, Indonesia
[2]Department of Physics, Faculty of Mathematics and Sciences, Universitas Padjadjaran, Indonesia

E-mail: eulis.tanti@unpad.ac.id

Abstract. CBM is highly potential as a source of renewable energy that can be activated with methanogenic bacteria. The rumen content is a waste of slaughterhouse can be utilized as a source of methanogenic bacteria. The aims to study the role of rumen fluid as a source of methanogenic bacteria in activating coal bed methane. Descriptive research method using 2 types of media consist of standard media (A) and enriched media (B). The application of rumen fluids for CBM activators are lignite, sub-bituminous, and bituminous coal in biogenic fermentation. The inoculum dose was observed at levels 6% of rumen media. The technique for counting microbes is through Total Plate Count in anaerobic Hungate tube. VFAs production was measured by titration method and methane production was analysed by gas chromatography (GC). All of parameter observed was measured on days 0, 2, 10, and 15. The highest number of anaerobic bacteria was found on day 10 and methane production was found on day 15. The highest production of VFA and methane occurred in biogenic fermentation of lignite. The results showed that the contents of rumen have the potential for the activation of coal bed methane in biogenic fermentation.

1. Introduction
Indonesia is one of the countries with the largest coal reservoir in the world. Coal mining is scattered in Sumatera, Kalimantan, Sulawesi, and Papua islands. However, 60 percent of total coal reservoirs in Indonesia consist of low quality coal because they have low calorific value (sub-bituminous) [1,2]. Clean coal technology in coal mining will be indispensable in the future and Indonesia is expected to be actively involved in the process as one of the main players in the coal mining sector. One of the activities in this clean technology is the development of coalbed methane (CBM), which has a lot of potential in Indonesia, has begun to get attention lately. CBM is one of the alternative energy sources that can replace fossil fuels that generate during coal formation process [3, 4, 5]. CBM can be formed biogenic through microbial activity such as methanogenic bacteria [6, 7, 8].

Rumen content has potentially used as methanogenic bacteria source [9, 10]. It can be obtain from slaughterhouse waste. The number of bacteria in rumen reach 10^9 CFU/ml while 2-4% of total bacteria is methanogens [11, 12]. Methanogens is methane generated bacteria in its metanbolism. It live in strictly anaerobic condition such as rumen.

2. Material and methods

2.1. *Sample and media preparation*
Rumen fluid was taken from the content of rumen from slaughterhouse. Cultured media for anaerobic bacteria was consisted of 2 types of dilution: standard media (A) and enriched media (B). Diluent solvent no. 14 [13]. was used to dilute samples. Standard media (A) consisted of nutrient agar, lactose broth,

resazurin, distilled water, and rumen fluid, while enriched media consisted of K_2HPO_4, NaCl, $(NH_4)_2SO_4$, KH_2PO_4, $CaCl_2$, $MgSO_4.7H_2O$, glucose, cellobiose, resazurin, bacto agar, rumen fluid, and distilled water.

2.2. Evaluation on total plate count of anaerobic bacteria and methane production

Each sample was cultivated in Hungate tube by using roll tube technique so that the melted media spread evenly in tube wall. After media solidified, all samples were incubated at 37 °C for 15 days. The growth of bacteria was observed on day 2, 5, 10, and 15. Coal used was lignite, subbituminous, and bituminous. Production of methane was measured using gas chromatography (GC). Concentration of VFAs was measured using titration method.

2.3. Statistic Analysis

The research analaysis was used descriptive methods. Data was calculated by the average of samples obtained at various sample with triplicates. Data obtained was calculated by using SPSS 17.

3. Result and discussion

Cultured media is a nutrients source that support optimal growth. The quality and quantity of nutrients can determine the growing rate of bacteria. In this study, anaerobic bacteria from rumen content were grown on standard media (A) and enriched media (B). Measurement of the number of anaerobic bacteria was performed on days 2, 5, 10, and 15 (Table 1). The number of anaerobic bacteria on media A and media B increases with time of incubation. However, on day 15 the number of anaerobic bacteria decreases. Based on the bacterial growth curve, it can be explained that the growth of anaerobic bacteria on day 2 is still experiencing an adaptation period in terms of nutrition, pH and temperature of the environment. Furthermore, bacteria experienced a rapid growth process on the 5th day until day 10 so that will produce the largest number of bacteria and reached peak on the day 10 (Fig. 1).

Table 1. Growth of Anaerobic Bacteria and Methane Production from Rumen Fluid on Media (A) and Media (B)

Days	Media A		Media B	
	Anaerobic Baceria (10^6 cfu/ml)	Methane (ppm)	Anaerobic Baceria (10^6 cfu/ml)	Methane (ppm)
2	36.67±14.33	361.26±101.52	33.33±7.18	676.55±142.72
5	96.67±18.98	1443.42±122.14	65.23±22.87	1720.94±155.12
10	133.33±50.21	11196.67±605.12	188.33±39.92	42171.69±5431.21
15	98.33±7.18	14730.85±930.55	136.67±32.53	75358.37±4630.99

Media A= Standard Media; Media B= Enriched Media

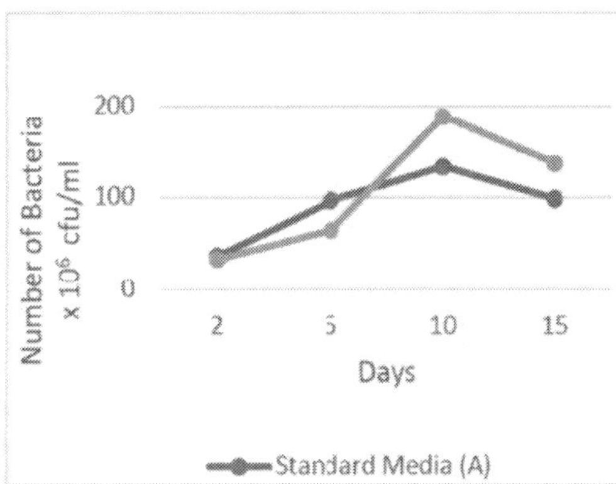

Figure 1. Growth of anaerobic bacteria from rumen fluid on different media

Table 2 . Concentration of VFAs and methane production of different coal type

Type of Coal	Variable	Days			
		2	5	10	15
Lignite	Methane (ppm)	617.54±181.39	1514.51±139.11	53820.24±6275.52	90307.66±7895.55
	VFA (mM)				
	Acetate	11.73	14.67	17.67	30.62
	Propionate	2.59	3.56	3.32	6.29
	Iso Butirate	0.42	0.56	0.68	1.25
	N butirate	0.54	1.15	1.15	1.77
	Iso valerat	0.22	0.35	0.37	0.62
	N valerat	-	-	-	0.03
Sub-bituminus	Methane (ppm)	644.25±158.72	1211.35±115.22	13610.46±805.44	54831.92±6271.71
	VFA (mM)				
	Acetate	12.89	14.68	17.69	23.08
	Propionate	1.87	2.58	4.20	5.43
	Iso butirate	0.52	0.55	0.71	1.13
	N butirate	0.50	1.35	1.26	1.75
	Iso valerate	0.19	0.26	0.34	0.74
	N valerate	-	-	-	-
Bituminus	Methane (ppm)	505.90±125.11	1240.77±120.12	17861.22±828.25	53182.92±6157.22
	VFA (mM)				
	Acetate	11.70	13.72	15.45	19.85
	Propionate	1.21	1.89	2.74	4.18
	Iso butirate	0.40	0.48	0.72	1.81
	N butirate	0.48	1.27	1.31	1.68
	Iso valerate	0.21	0.28	0.33	0.81
	N valerate	-	-	-	-

On the day 15, nutrient available decrease because of the increasing population of bacteria. This condition is similar to bacteria growth curve [14, 15]. Bacteria needs nutrient to maintain its cell. In this study, the utilization of two different media (standard and enriched media) is to determine the optimal nutrient requirement for anaerobic bacteria.

Enriched media contains more complete nutrint than standard media [13, 16]. The main composition of enriched media is bacto agar and rumen fluid. Bacto agar is purified agar by reducing the content of impurity pigments, salt content (NaCl), and the foreign content (organic and inorganic) as low as possible. Various minerals are added to support the growth of anaerobic bacteria on enriched media [17, 18]. As a result, growth of anaerobic bacteria on enriched media that has more complete nutrients is higher than standard media (Fig. 1).

Measurements of volatile fatty acid (VFA) concentration and methane production in various coal types such as lignite, sub-bituminous, and bituminous were performed on enriched media (Table 2). Methane production was directly proportional to the increase in the number of anaerobic bacteria (Table 2). This illustrates that anaerobic bacteria from rumen fluid grown on enriched media were capable of producing methane while the highest VFAs production after a 15-day incubation period was acetate. Acetate is the main precursor for methane production during anaerobic digestion of organic matter [19]. Volatile fatty acids (VFAs) are important mid-products in the production methane. Their concentrations affect the efficiency of fermentation [20]. High VFAs concentration increases the methane yield [21]. The highest acetate concentration was achieved on 15th day lignite coals (Table 2). Lignite, sometimes called "brown coal", is a low-calorie coal type, with a carbon content of 25-35% and 30-60% water content [22]. The lignite coal response in producing VFAs and methane better than other coal types is unexplainable. However, anaerobic bacteria from rumen fluid can adapt and grow better in lignite than other type.

4. Conclusion

Rumen fluid has a high potential in providing anaerobic bacteria to produce coalbed methane in the biogenic fermentation process. Lignite coal produces the most favorable response in the production of VFAs and methane compared to sub-bituminous and bituminous coal. The 15-day incubation time produces the highest amount of VFAs and methane production, both in lignite, sub-bituminous, and bituminous.

Acknowledgments
The authors thank the Rector of Universitas Padjadjaran for the financial support under the Academic Leadership Grant (ALG).

References

[1] Amrullah D, Rahmat S B, Simatupang D P 2018 The Center of Coal and Geothermal Mineral Resources Geological Agency *Indonesian Ministry of Energy and Mineral Resources* http://psdg.bgl.esdm.go.id
[2] Agency for the Assessment and Application of Technology 2015 Outlook Energy of Indonesia 2015 *BPPT* Jakarta.
[3] Moore T A 2012 Coalbed methane: A review *International Journal of Coal Geology* **101:** 36-81.
[4] Pan Z, Wood D A 2015 Coalbed methane (CBM) exploration, reservoir characterization, production, and modeling: A collection of published research (2009-2015) *Journal of Natural Gas Science and Engineering* **26**: 1472-1484.
[5] Li S, Tang D, Pan Z, Xu H, Tao S, Liu Y, Ren P 2018 Geological conditions of deep coalbed methane in the eastern margin of the Ordos Basin, China: Implications for coalbed methane development *Journal of Natural Gas Science and Engineering* **53:** 394-402.
[6] Park S Y, Liang Y 2016 Biogenic methane production from coal: A review on recent research and development on microbially enhanced coalbed methane (MECBM) *Fuel* **166**: 258-267.
[7] Fuertez, Cordoba J G, McLennan J D, Adams D J, Sparks T D 2018 Potential application of developed methanogenic microbial consortia for coal biogasification *Internationla Journal of Coal Geology* **188**: 165-180.
[8] Bucha M, Jedrysek M O, Kufka D, Plesniak L, Marynowski L, Kubiak K, Blaszczyk M 2018 Methanogenic fermentation of lignite with carbon-bearing additives inferred from stable carbon and hydrogen isotopes **186:** 65-79.
[9] Zhang H, Zhang P, Ye J, Wu Y, Fang W, Gou X, Zeng G 2016 Improvement of methane production from rice straw with rumen pretreatment: A feasibility study *International Biodeterioration & Biodegradation* **113**: 9-16.

[10] Zou Y, Xu X, Li L, Yang F, Zhang S 2018 Enhancing methane production from *U. lactuca* using combined anaerobically digested sludge (ADS) and rumen fluid pre-treatment and the effect on the solubilization of microbial community structures *Bioresource Technology* **254:** 83-90.

[11] Yokoyama M T, Johnson K A 1988 Microbology of the rumen and intestine *The Ruminant Animal* D.C. Church ed. Simon and Shuster New York p. 125-144

[12] Millen D, De Beni A M, Lauritano P R D 2016 Rumenology XVI **314** p.67 ISBN: 978-3-319-30531-8.

[13] Ogimoto K, Imai S 1981 Atlas of Rumen Microbiology *Japan Scientise Press Tokyo* Japan.

[14] Paulton R J L 1991 The bacterial growth curve *Journal of Biological Education* **25(1991)**: 92-94

[15] Al-Qadiri H M, Al-Alami N I, Lin M, Al-Holy M, Cavinato A G, Rasco B A 2008 Studying of the bacterial growth phases using fourier transform infraredspectroscopy and multivariate analyses *Journal of Rapid Methods and Automation in Microbiology* **16:** 73-89

[16] Ozbayram E G, Kleinsteuber S, Nikolausz M, Ince B, Ince O 2017 Effect of bioaugmentation by cellulolytic bacteriaenriched fromsheep rumen on methane productionfrom wheat straw *Anaerobe.* **46:** 122-130.

[17] Mueller B 2015 Experimental Interactions between clay minerals and bacteria: A review *Pedosphere* **25(6)**: 799-810.

[18] Larsen J, Sanchez H P, Arellano I D, Morales V C, Montoya L C, Moreno J V 2017 Interactions between microbial plant growthpromotors and their effects on maize growth performance in different mineral and organic fertilization scenarios *Rhizoxphere* **3(1)**: 75-81.

[19] Karakashev D, Batstone D J, Trably E, Angelidaki I 2006 Acetate oxidation is the dominant methanogenic pathwayfrom acetate in the absence of Methanosaetaceae *Applied and Environmental Microbiology* **72(7)**: 5138-5141 DOI: 10.1128/AEM.00489-06.

[20] Wang Y, Zhang Y, Wang J, Meng L 2009 Effect of volatile fatty acid concentration on methane yield and methanogenic bacteria *Biomass and Bioenergy* **33(5)**: 848-853 DOI: 10.1016/j.biombioe.2009.01.007.

[21] Li H, Cao F, Wang Y 2014 Effect of volatile fatty acids on methane production of dry anaerobic digestion *Chinese Journal of Environmental Engineering* **8(6)**: 2572-2578.

[22] Bowen B H, Irwin M W 2008 Coal Characteristics *The Energy Center* at Discovery Park Purdue University Indiana Center for Coal Technology Research *CCTR* Potter Center 500 Central Drive West Lafayette.

IOP Conf. Series: Earth and Environmental Science **334** (2019) 012020 doi:10.1088/1755-1315/334/1/012020

Coastal Conservation for Ecotourism Activity in Kiluan Island Coast Lampung Province

A Akbar [1], D J Prihadi [2], Y Dhahiyat [3] and W Pamungkas[2]

[1] Marine Science Study Program, Universitas Padjadjaran
[2] Department of Marine Science, Universitas Padjadjaran
[3] Department of Water Resources, Universitas Padjadjaran

E-mail : ali.tensai @ gmail.com

Abstract. Kiluan island located in the waters area of the Kiluan bay, Kiluan Negeri Village, Tanggamus District, Lampung Province. Ecotourism is a development of tourism concept which based on nature conservation, and because of its magnificent natural potential and its beautiful scene, Kiluan Island coast has become a popular ecotourism spot. However, the conservation management never been better. The purposes of this research is to identify any Potentials, and every important aspects for Kiluan island coast's conservation and ecotourism spot development. This research used survey methods, which is water quality measurements, and questionnaire. The condition of Kiluan island's water qualities are still accordant with the standard of Indonesian constitutions specified by the Ministry of Environment Number 51 of 2004, section 1 about Sea Water Quality Standard for Marine Tourism, with the result that the water is very suitable to be used as marine ecotourism spot. Tourist perception about the condition of the Kiluan island coast are very good, that 67% scores good for the flora and fauna ecosystem, 73% scores very good for the potentials of natural tourism, 73% scores very good for the Natural Scenery, 63% scores good for the water activities enjoyment, and 53% scores very good for beach recreation tours.

1. Introduction

Indonesia is the biggest Archipelago country in the world, in an area 7.7 million km^2 wide, and consist of 17,504 islands with 95,181 km^2 coastal line, from Sabang to Marauke, [1]. Ecotourism concept is an integrated tourism concept which combine conservation, and empowerment of local communities, with the result that local communities can also acquire the benefits directly from the tourism activities, because they manage it by them self, and they develop and use all the potentials by themselves [2]. Kiluan Bay which located in Lampung Province, is one of the famous bay in Lampung and has been utilized as tourism spot with ecotourism concept. This area have many natural resources, and unique marine biota, which are Bottle nose dolphins (*Turciops truncates*), long nose dolphins (*Stenella longirostris*), Hawksbill turtle (*Eretmochelys imbricata*), also coral reefs in the Kiluan Island [3].

Inside the Kiluan Bay area, there is a small but famous island, within 2 hectares, named Kiluan Island. This Kiluan Island coast has been utilized as an ecotourism area because of its fabulously beautiful Scenery, and it has coral reefs area that can be used as snorkeling spot. Until this moment, there are so many visitors that came to Kiluan Island for sightseeing for the Scenery, and doing recreation activities, such as beach recreation, swimming and snorkeling. Because of the ecotourism concept, is a tourism concept that empower the local communities, so the society in Kiluan Bay has established an organization called Tourism Awareness Society Group (Pokdarwis). This local communities organization has a role to manage all of the tourism activities and management in Kiluan Bay, Including Kiluan Island.

However, the ecotourism management in Kiluan Bay area which managed by Pokdarwis Kiluan still never been better, and has to be improved, especially in the Kiluan Island because its already popular. So,the purposes of this research is to identify any Potentials, and every important aspects for Kiluan island coast's conservation and ecotourism spot development.

2. Materials and Methods

2.1. Formulation of the problem
Kiluan Island has been utilized as an ecotourism spot for a long time ago. However, the Pokdarwis's management was still lack of attention in managing conservation in the Kiluan Island Coast. So, it needs to do a research to identify the potentials, and every important aspect that Kiluan Island have. Hopefully, this research can give an imaginary output for Pokdarwis to improve their ecotourism management and also have an attention in the conservation management.

2.2. Design of Research
this research were carried out by identifying the potentials by measuring the oceanographic conditions of Kiluan Island Coast, and assess the Visitor Perceptions using survey methods in descriptive and quantitative analysis.

2.3. Sample Selection and Data Retrieval
This research using primary data, or the data that taken directly from the research site, which is Kiluan Island Coast. The retrieval data is divided into 3 station that represent three section of Kiluan Island Coast, there are West section as 1st station, North section as 2nd station, and East section as 3rd station. The measurement of Oceanographic conditions was conducted by water quality data retrieval from 5 parameters, in accordance with sea water quality standard index specified by the Ministry of Environment, Indonesian constitutions number 51 of 2004, section 1 about Sea Water Quality Standard for Marine Tourism as in Table 1.

Table 1. Indonesian constitutions specified by the Ministry of Environment Number 51 of 2004, section 1 about Sea Water Quality Standard for Marine Tourism

No.	Parameters (Physics and Chemical)	Unit	Quality Standard
1.	Clarity	Meters	>6
2.	Temperature	^{o}C	Natural
3.	pH	-	7 – 8.5
4.	Salinity	$^{o}/_{oo}$	Natural
5.	Dissolve Oxygen	mg/l	>5

Source: Attachment to Decree No. 51 / MENLH / 2004 concerning Sea Water Quality Standards for Marine Tourism

Besides that, also done a research about visitor perceptions about the conditions of Flora and Fauna Ecosystem, Natural Potentials, the Scenery, Water Activities enjoyment, and Beach Recreation tours. The visitor perceptions research were done using random sampling questionnaire methods, towards 30 visitors as respondent randomly in 12 may 2018. The scoring scores were divided into 5 scales, Very good, Good, Not good, Very not good, and do not know. Then, the scores will be presented in graphics.

3. Result

3.1. General Condition of Research Location

Kiluan Island is located in the middle of Kiluan Bay's waters area, which belong to Pekon Village Kiluan Negeri, Tanggamus Distric, Lampung Province. Since a long time ago, this island is actually has been famous as coastal ecotourism spot among Lampung Province people. Thee visitor needs to rent a small boat (max. capacity 8-10 persons) to cross the Kiluan Bay through the Island, and the boat rentals were managed by Pokdarwis. There is only one family that live inside this small island as his livelihood, it's a family of Mr. Dirham, also as a member of Pokdarwis

The sector of tourism management in Kiluan Bay, were organized by local Communities organization, there are Pokdarwis and a NGOs called *Yayasan Ekowisata Cikal* that have a role to accompanying Pokdarwis Kiluan Bay, so they can stick to the Ecotourism Concept in every program. It has been said that empowering local communities, is the most fundamental principle to implement ecotourism concept [4]. This Pokdarwis establishment is a form of government effort in directing local communities to be independent in organize their own areas, by utilizing the area as an ecotourism spot. Its also been said that preservation of natural resources and local culture will ensure the sustainability of development [4]. Hopefully, with this independent Pokdarwis functionate can organized every utilizations, constructions, and protection of tourism in Kiluan Bay, including Kiluan Island.

One of the main ecotourism management in Kiluan Bay is Dolhins watching tour, because as the result of interview with the chair of Yayasan Ekowisata CIKAL NGOs, that the Kiluan Bay area is the only ocean area in West Indonesia that became dolphin habitat as dolphins seeking for food in this area. So there are plenty much of dolphins appeared on the sea. Besides that, another active activities that managed by Pokdarwis is an ecotourism activity in Kiluan Island, there are beach recreations, swimming, fishing, and snorkeling, with Pokdarwis managing the boat rentals to cross through the Island from the Bay. This island has a coral reef area within 250 m^2, that used to be snorkeling spot so the tourist can see corals and coral fishes.

Figure 1. (A) South Part of the Island (B) North part of the Island
Source : Personal Documentation.

Besides Pokdarwis, there were also formed a local communities Organization, that have a purpose to keep watch on, and controling every activities that used and utilized natural resources in Kiluan's area including the ecotourism activities, it called Supervisory Community Group (Pokmaswas). The implementation of pokmaswas's control that makes them to take actions are handling of stranded mamals that often founds in Kiluan Island areas, such as Whales, Dolphins. If this happens, mr. Dirham as the residents and the member of Pokdarwis will report it to Pokmaswas so they will take actions to solve the problems.

The East part of the island is directly adjacent with Hindian Ocean, so the island positions blocking the ocean current from Hindian Ocean to the bay [5] said that. A coast that facing directly to the ocean will be marked with steep cliff, and the beach materials is rough because it received massive wave from the ocean huge current level. This geophysical illustrates the East part of Kiluan island that directly faced the Hindian Ocean. It has a wide range of Coral Platform along the coastal line in the east, with the big waves delivers coral shards, so the beach materials were rough sand. Otherwise, the west part of the island has a very low current level, with beach materials were fine white sand, without any coral shards. The north part of the island is very unique because it is the transitions area between east and west, so this area is the one that has beautiful coral reefs, with many kind of reef fishes. The south wart of the island has no beach, It is only a steep rocky terrain, with massive waves. The conditions of Kiluan Island can be seen in Figure 1.

Kiluan Bay is only accessible by 1 road from Bandar Lampung city within 2-3 hours, by car or motorcycle. Then, it needs 5-10 minutes to cross the bay through the island by rented boat. From the interview results with Mr Dirham as member of Pokdarwis, there are 90% of the visitors come to Kiluan Island, only to enjoy the beauty of Kiluan's Scenery from the island and doing some beach recreations, swimming or snorkeling.

4. Discussion

There are 5 parameters according to the sea water quality standard index by Indonesian constitutions number 51 of 2004, section 1 about Sea Water Quality Standard for Marine Tourism, which is, temperature, salinity, pH, Dissolve Oxygen, and Water Clarity. The measurement result shows that the Oceanographic conditions in Kiluan Island coast has convenient score to be used as a marine tourism activity spot, as shown in table 2 below.

Table 2. Oceanographic Measurement Result of Kiluan Island Coast

Station	Observed Parameters				
	Temperature (^{0}C)	Salinity (ppt)	pH	DO (mg/l)	Clarity (Meter)
1	28	28	7.57	7.63	100% (7m)
2	28	29	7.33	7.67	100% (7m)
3	28	28	7.4	7.60	100% (7m)

The sea surface water temperature at all 3 data retrieval sections, using a thermometer range about 27 °C-29 °C. this range of sea surface water temperature s in accordance with the natural value of seawater quality standard specified by the Ministry of Environment Decree No. 51 / MENLH / 2004 on the Quality Standard of Sea Water is about 26 °C – 30 °C. Salinity of seawater at all 3 data retrieval sections taken using a refractometer ranged from 28 $^0/_{00}$-29 $^0/_{00}$. The range of salinity values is also in accordance with the standard natural value of seawater quality standards by the Ministry of Environment Decree No. 51 / MENLH / 2004, with permissible changes up to <5% of average seasonal salinity.

The waters clarity that measured using a secchi disk can be seen through the bottom of the waters, indicating that the water's brightness is 100%. This waters clarity is a very important aspect in coastal tourism activities, so that visitors may feel satisfied doing the water activities, such as swimming, and snorkeling. Besides, water clarity also has a high aesthetic value as ecotourism spot. The degree of acidity (pH) obtained using pH meter at all 3 sections of data retrieval ranged from 7.4-7.57. The value of the degree of acidity (pH) is included in the standard value of seawater quality standard by the Ministry of Environment Decree No. 51 / MENLH / 2004 regarding the Sea Water Quality Standard is about 7-8.5. This condition allows the visitors swimming at the beach without getting hurt and irritation.

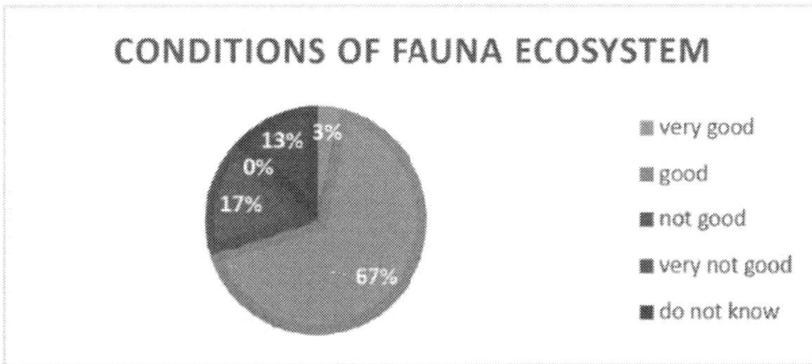

Figure 2. Visitor Perseptions of Conditions of Fauna Ecosystem

The Dissolve oxygen (DO) measured using DO meters ranged from 7.60-7.67 mg/l, has exceeded the standard value of seawater quality standard by Ministry of Environment Decree No. 51 / MENLH / 2004 concerning the Sea Water Quality Standard is >5 mg/l. This condition evidenced by the large number of coral fishes can be seen in the waters, because the Dissolve Oxygen is an important oceanographic factor for the survival of aquatic biota.

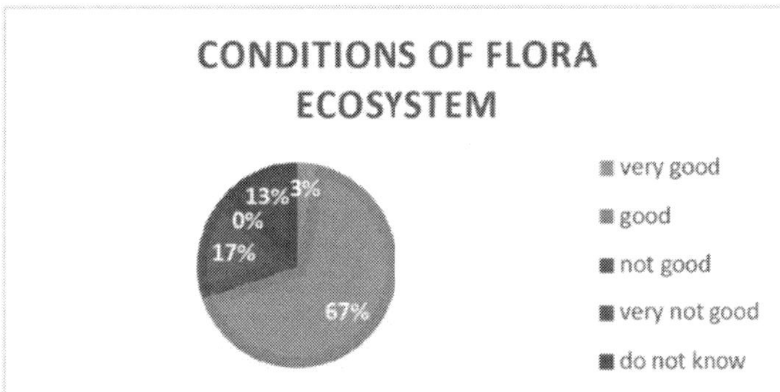

Figure 3. Visitor Perception Conditions of Flora Ecosystem.

The percentage of visitor interviews about Conditions of Fauna Ecosystem shows 3% rate Very Good, 67% rate Good, 17% rate Not good, and 13% rate Do not Know as can be seen in Figure 2 below. The ecosystems in this area of Kiluan Island are the reef fish ecosystem, and the common pelagic fish are the target the visitors to do fishing. Visitors can see that there are still many fish and other animals such as Starfish in the coral reef ecosystem contained in Station II. Visitors can also see reef fish while swimming due to the clarity of the water that reaches 100%, so this condition became a special attraction for the visitors. The concept of Ecotourism ran by Pokdarwis are by selling the view of fauna that remain in its natural integrity, one of them to preserve the ecosystem of coral reefs with reef fishes.

In addition to Fauna, Flora also became one aspect that must be maintained and managed for its sustainability, while the management done by Pokdarwis, must be in accordance with the concept of Ecotourism. According to the percentage of visitor perceptions ratings, that 3% rate Very Good, 67% rate Good, 17% rate not good, and 13% rate Do not know, as can be seen in Figure 3. There is no mangrove trees in the area around Kiluan Island, only coastal vegetation, and other trees.

Tourism potential of Kiluan Island Beach to be an ecotourism destination is very magnificent, and of course, attract all of the visitors. 73% of visitors rated Very Good, and 27% of visitors rated Good.

In accordance with the statement [6], that coastal sand is one of the potential categories and the main selling power for a beach to become a tourism area. Coral reef at station 2 is also still occupied by various reef fish so it still has the potential to do snorkeling activity over there [7], that coral reef ecosystem is one of the main tourism potentials in coastal area. Fishery resources can be a potential factor for coastal tourism, and many visitors are fishing and directly roast the fish they catch and eat it in the Island [8]. The Scenery is one of the visitor attractions which is also one of the extraordinary potentials of the island of kiluan [9]. The beauty and aesthetics of the landscape, the clarity of the water, the mountain scenery, and the sunset [8], that all of those factors will be the memories for visitors who come to this Kiluan Island. The diagram of the result of visitor's assessment on tourism potential can be seen in Figure 4.

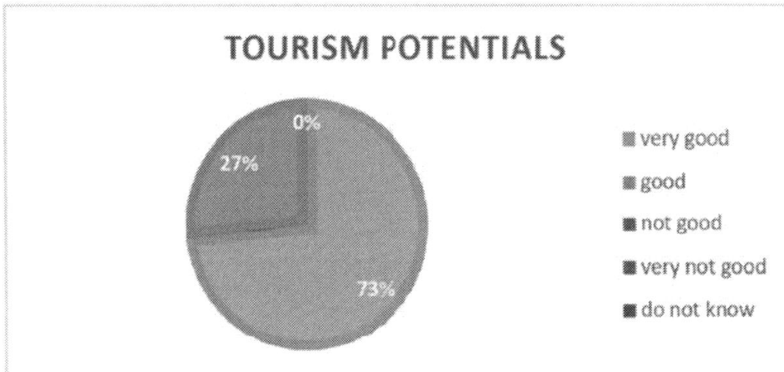

Figure 4. Visitor Perceptions of Tourism Potentials

The aesthetics of a landscape is determined by the beauty of nature with its different forms of earth face, and there are 6 factors that affect it [10]. Among them are, Regularity, Texture, Familiarity with the environment, Widespread viewing space, excitatory plurality, and Value of confidentiality. The natural Scenery on Kiluan Island beach is very beautiful, because the natural beauty of Kiluan Island Beach contains 6 factors [10], so 73% of respondents give very good, and 27% of respondents give good. The Scenery on the beach of Kiluan Island is rated higher [11], which gets 42% Good, and 58% is very good at the beauty of Manalusu beach, Pangandaran. Diagram of the assessment results The coastal landscape of Kiluan Island can be seen in Figure 5 below.

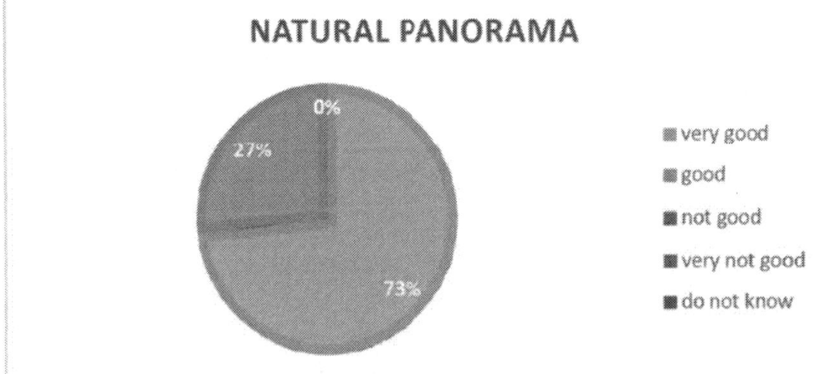

Figure 5. Visitor Perceptions of Natural Scenery

Figure 6. (A) View in Station 1 (B) View in Station 2 (C) View in Station 3(D) View in Station 4

Swimming and any water activities included in the activities of marine tourism Contact Activities, because in direct contact with sea water [8]. Referring to the Ministry of Environment Decree No. 51 / MENLH / 2004 Concerning the Sea Quality Standards, Subsection 1, about Marine Water Quality Standards for Marine Tourism, tells that there are physical and chemical indicators of the sea water for visitors to do water activities and swimming. It has been discussed Previously, that the Kiluan Island's sea waters conditions of has been in accordance with all the minimum parameters in the quality standard, so it can be concluded that water Contact Activities such as swimming, or snorkeling, are suitable and safe to do in sea waters in Kiluan Island. The results of the assessment show that 34% rate Very Good, 63% rate Good and 3% rate not good due to the geophysics condition in station 3 is not suitable for swimming activities, as the diagram shown in Figure 6.

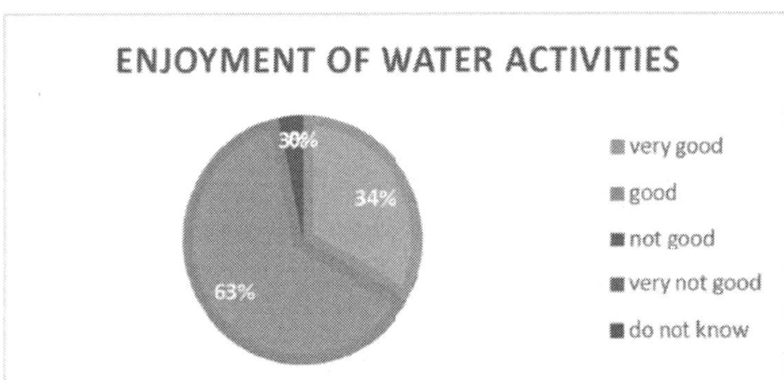

Figure 7. Visitor Perception of the Enjoyment of Water Activities.

Beach sand picnic activity is categorized as Litoral Activities, or activity on sand at low tides [8]. The basic material of sand beach beaches has a great affect to the amenities of beach sand picnic activities, such as sitting on the sand, playing sand, and sunbathing [11]. 36% of visitors rate Very Good, 57% rate Good and 7% rate not good as can be seen in Figure 7. 7% of visitors rate noy good because at station 3 there is a lot of coral shard that blend with the sands, so that visitors have less interest to walk in the station 3. unlike the station 1 as the most crowded spot, because of the beauty of white fine sand.

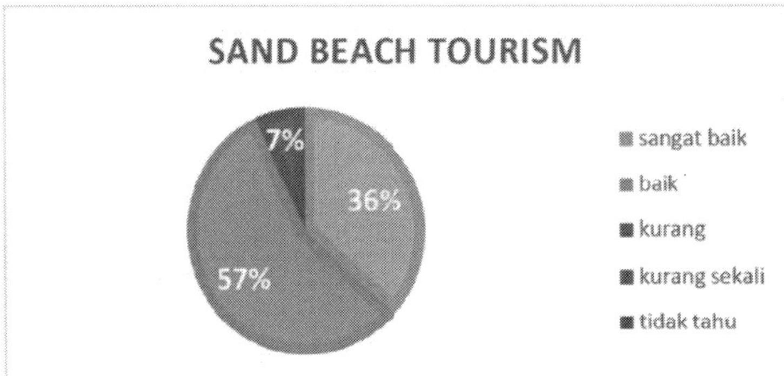

Figure 8. Visitor Perceptions of the Enjoyment of Sand Beach Tourism

5. Conclusion

The oceanography condition of Kiluan Island is still in accordance with the quality standard according to the Ministry of Environment Decree No. 51 / MENLH / 2004 Concerning the Quality of Sea Water, Subsection 1 about Marine Water Quality Standards for Marine Tourism, so it is very safe for visitors to do water tourism activities, and this aspect becomes one of nature's potential that Kiluan Island coast have. Besides, other natural potentials owned by Kiluan Island Coast are the beaty white fine sands and the natural Scenery, because both of those elements is most precious selling power for a beach ecotourism destination, that 67% scores good for the flora and fauna ecosystem, 73% scores very good for the potentials of natural tourism, 73% scores very good for the Scenery, 63% scores good for the water activities enjoyment, and 53% scores very good for beach recreation tours.

References

[1] Prihadi D J 2015 Keberadaan Ikan Kodok di Pulau Nusa Penida Provinsi Bali. *Jurnal Akuatika* **VI (2)**: 187-197
[2] Murkayanti, Saraswati A 2005 Pengembangan ekowisata sebagai pendekatan pengelolaan sumberdaya pesisir berkelanjutan. Kasus desa Blendung – Kabupaten Pemalang. *Jurnal Teknik lingkungan P3TL-BPPT* **6(2)**: 391-396.
[3] Marpaung L A, Ardiyansyah, Ritonga, R 2010 Analisis yuridis sumberdaya alam dalam rangka pembangunan kawasan wisata teluk kiluan berbasis kearifan lokal. *Keadilan Progresif* **7(2)**: 106
[4] Tuwo A 2011 Pengelolaan Ekowisata Pesisir dan Laut : Pendekatan Eklogi, Sosisal-Ekonomi, Kelembagaan, dan Sarana Wilayah. Perpustakaan Nasional RI Katalog dalam Terbitan (KDT). Brilian Internasional, Surabaya.
[5] Hantoro W S 2002 Pengaruh Karakteristik Laut dan Pantai Terhadap Perkembangan Kawasan Kota Pantai. (Prosiding). Pusat Penelitian Geoteknologi LIPI, Bandung.
[6] Fandeli C M 2002 Perencanaan Kepariwisataan Alam, Fakultas Kehutanan Universitas
[7] Supriharyono 2000 Pengelolaan Ekosistem Terumbu Karang. Jakarta: Djambatan.
[8] Hudisaputra A K 2011 Pengelolaan wilayah pesisir teluk kiluan kabupaten tanggamus melalui pengembangan ekowisata. Artikel. Program Studi Magister Ilmu Lingkungan. Program Pascasarjana Universitas Padjadjaran. Bandung.

[9] Armos N H 2013 Studi kesesuaian Wisata Boe Desa Mappakalompo Kecamatan Galesong Ditinjau Berdasarkan Biogeofisik.. Skripsi. Fakultas Ilmu Kelautan dan Perikanan. Jurusan Ilmu Kelautan. Universitas Hassanudin, Makassar.

[10] Deliyanto B 2000 Lahan Dalam Perspektif Lingkungan. *Jurnal Pengembangan Wilayah dan Kota* **2(2)**: 45-49

[11] Christina E 2013 Kajian Kesesuaian Lokasi Wisata Pantai Dalam Pengembangan Pantai Pulau Kiluan Sebagai Kawasan Wisata Pantai yang Berkelanjutan di Kabupaten Garut Provinsi Jawabarat. Skripsi. Fakultas Perikanan dan Ilmu Kelautan. Prodi Ilmu Kelautan. Universitas Padjadjaran, Jatinangor.

Study of Water Balance in Arjasari Agricultural Land (A case study of Intercropping System of Corn and Chilies)

G F Furqan[1], E Suryadi[1] and S Dwiratna NP[1]

[1]Alumnus of Agricultural Engineering and Biosystem Department of Universitas Padjadjaran, Cibaduyut Raya No.35/200A, Bandung 40235, Indonesia
[2]Lecturer of Agricultural Engineering and Biosystem Department of Universitas Padjadjaran, Jl. Raya Bandung-Sumedang Km 21, Jatinangor 40600, Indonesia

E-mail : ghanifaliq @ gmail.com

Abstract. Development of food crops is not only addressed to a momentary increase in production but also attention to the aspect of land conservation in order for the system of production can be sustainable. Intercropping system is one way of managing agricultural land which can minimize the risk in the utilization of dry land for the development of food crops. The availability of water on dry land are generally affected by rainfall and the ability of the soil to retain water. The potential water resources available in the ground is indispensable in agricultural hydrology and water management in the framework of developing food crops. This study aims to determine the moisture content of the soil and water balance of the land in the region of Arjasari, Bandung Regency. The methods used in this research is a descriptive analysis. The research was carried out from March until November 2015. The results of this study showed that the moisture content of the soil in the study ranged from 314.2 mm/m up to 472 mm/m with the moisture content of the soil which can be utilized by the plant a maximum of 185 mm/m.

1. Introduction

Seeing the extent of dry land in Indonesia, dry land has the potential to be developed into productive agricultural land. Data of 2013 states that Indonesia has dry land used as a garden or moor of 11,876,881 ha of about 148 million ha [1]. Data show that dry land utilized is still less than 10% of available land. Potential water resources in the form of water savings available in the soil are indispensable in agricultural hydrology and water management in the framework of developing food crops, especially since the water demand for most agricultural cultivation in Indonesia is still dependent on rainwater. Given this, then the water management must be done optimally.

Multiple cropping system is one way of managing agricultural land that can minimize the risks in dry land use for food crop development as food crop development is not only aimed at instantaneous production improvement but also attention to land conservation aspect so that production system can be sustainable. In addition, according to [2], the combination of legume and non-legume crops in the intercropping system can generally increase the productivity of agricultural land.

One of the plants that has considerable national demand is chili. According to [3], the average consumption of chili per capita of Indonesia population reached 14,235 kg. Maize is one of the strategic commodities for Indonesia. Corn is a staple food in some areas such as Madura and Nusa Tenggara. The national corn demand for food, feed and industry continues to increase every year. The current use of corn commodities is dominated by animal feed industry by 51% [4]. With a large national demand and tends to increase every year, its fulfilment must be sought from domestic production.

Increased productivity can be pursued by maximizing the growth and development of plants, so the plant is expected to give maximum results. Crop water requirements are one of the most important aspects for plant growth and development. Crop water requirement is the amount of water needed by plants to grow optimally which can also be interpreted as the amount of water used to meet the evapotranspiration process of plants. With the fulfilment of water needs of plants, plants are expected to grow and develop to the maximum so as to increase productivity.

By analysing the groundwater content, water use can be optimized so that irrigation water supply for plants can be optimized. Water balance (agriculture), ie analysis with respect to the nature and behaviour of the soil to the atmosphere, and as a support required physical data soil, especially water content at the level of field capacity and permanent wilt point [5]. Water balance is closely related to groundwater availability. This becomes important for agricultural activities because groundwater can be utilized to meet the needs of plant water.

Arjasari Sub-district is geographically located at 107o 30 'to 107o 40' east longitude and 7o 30 'to 7o 50' south latitude. The area of Arjasari sub-district is 49,35 km2 or 4935,30 ha. The area is divided into several categories such as the area of wetland farmland, the area of non-rice farming land and the area of non-agricultural land. Based on the topography of most areas in the district of Arjasari is an area of plains and slopes or ridges with varying heights from 700 m to 1000 m above sea level. Arjasari Subdistrict is flooded by river Cipalasari and Citalugtug river. According to BPS (2014) data, Arjasari rainfall in 2013 is 1885 mm / year with an average of 5.08 mm / per day, the highest rainfall recorded in April, November and December. The climate is very cool temperatures ranging between 18o -26o Celsius [6].

2. Methods

This research is done by descriptive analysis method. By analysing and describing the state of the object being studied. This research was conducted from March until November 2015 at SPLPP field of Universitas Padjadjaran located in Arjasari Village, Arjasar Subdistrict, Bandung Regency. The tools used are, Ring Sample, Palu, Hoe, and Notebook. The materials used are rainfall data from 3 weather stations that is Cileunca Station, Cisondari and Cipaku. In addition to rainfall data, also used climatological data obtained from BMKG Station Bandung for the last 10 years as well as maps of Citarum watershed area and plant data referring to FAO Irrigation and Drainage Paper no. 56 of 1989.

2.1. Soil Sampling and Testing

In relation to water balance calculations, an analysis of soil samples was performed. The method used in sampling is the Undisturbed Samples method. After soil sampling is done, soil physical properties analysis to find the value of field capacity, the value of permanent wilting point, infiltration rate and the level of soil dryness at the beginning of the planting period. Sample of soil taken as many as 6 pieces around the land used for cultivation activities. The stages of soil sampling are as follows;

- First ground surface is cleaned first from grass and garbage. Dig the soil up to 30-40 cm deep or until the soil is free from the root zone of the plant.
- Place the first sample ring on the ground with the pointed portion below. Immerse the ring to approximately ½ tall rings using hammers and wooden planks.
- Place the second sample ring at the top. Then immerse it until the second ring goes down to ¼.
- Take both sample rings from the ground and split the part between the two samples carefully. This is done to avoid any disturbance that can damage the soil sample.
- Close the sample ring and label the soil sample. Then the sample was tested in the laboratory.
- Sample of soil that has been taken from the field of research was further tested in the laboratory of Indonesian Ministry of Agriculture Land Research Centre in Bogor, West Java.

2.2. *Data collection*
The required data is climatological data obtained from the Water Resources Management Agency, West Java Province, the data used is the rainfall data station closest to the research location, such as Cipaku station, Cisondari station and Cileunca Station for the last 10 years. In addition, temperature data, humidity, radiation time, evaporation of each of the last 10 years and value of plant evapotranspiration coefficient are also needed. And also required data in the form of maps to measure the extent of polygon area for the calculation of data Rainfall region.

2.3. *Data processing*
The collected data is then processed to obtain potential evapotranspiration value. The climatology data used is the regional rainfall data. To calculate the area rainfall data used method polygon Thiessen. This method requires at least 3 stations to work on.
After the rainfall data obtained region, then the data is processed in conjunction with the climatological data of Bandung obtained from the Meteorology, Climatology and Geophysics. The several stages of data processing are as follows;

1. Calculate the average rainfall value of the region each month to obtain 80% chance rainfall using the following equation;
$$R80= n/5+ 1 \quad (1)$$
Where: R80 = Rain Opportunity 80% n = amount of data
2. Calculate potential evapotranspiration value by using Cropwat 8.0 for Windows software by entering climatological data in accordance with the available column of software, until finally the potential evapotranspiration value is automatically obtained.

2.4. *Determination of Water Holding Capacity Value*
Determination of WHC value is based on groundwater content that has been tested previously in the laboratory of Soil Physics, Land Research Center, Ministry of Agriculture, Republic of Indonesia. The WHC value of soil is the average value of the difference between soil moisture test results at pF 2.54 and pF 4.2

2.5. *Needs of crop water*
Crop water requirements are defined as the amount of water provided to offset the water lost by evaporation and transpiration. Evapotranspiration is a combination of evaporation from the soil surface or evaporation and evaporation from plant leaves or transpiration. The magnitude of the evaporation value is influenced by climate, variety, type and age of the plant. By incorporating plant efficiency (kc), consumptive use of plants is a function of potential evapotranspiration of plants [7]. Consumptive usage can be calculated by the following equation:
$$Etc = Eto \ x \ Kc$$
Where: Etc = Water Requirement Plants (mm / day),
Eto = potential evapotranspiration (mm / day),
kc = coefficient of plant.

2.6. *Water Balance Sheet Analysis*
Water balance calculation procedures according to [8] use the bookkeeping system with the following steps;

2.6.1. Rainfall column. Filling with the monthly rainfall data or rainfall with a certain chance that can represent the entire land. In this study used rainfall with 80% chance.

2.6.2. Potential Evapotranspiration Column (ETP). Populate with ETP value from local station in ETP priority sequence with empirical formula (Penman-Monteith) in Software Cropwat 8.0.

2.6.3. CH-ETP column. Filled with CH value difference with ETP.

2.6.4. Potential Accumulated Water Loss Columns for Evaporation (APWL). Fill with the sum of the negative CH-ETPs sequentially month after month.

2.6.5. Groundwater content column (KAT). Filling the kAT value where APWL occurs with the formula;

$$KAT = TLP + [[1,00041 - (1,07381 / AT)] \,|APWL|\, xAT]$$

Information ;
TLP : Permanent Late Point
KL : Available Water Field Capacity
|APWL| : Absolute Value of APWL

2.6.6. Column Change of Ground Water Level (dKAT). The DKAT value of the month is the KAT of that month minus the previous month's KAT. A positive value denotes a change in groundwater content that takes place at CH> ETP (rainy season), the addition stops when dKAT = 0 after the KL is reached. Conversely, if CH <ETP or dKAT is negative, then all CH and some KAT will be evapotranspirated.

2.6.7. The actual Evapotranspiration column (ETA). When CH> ETP then ETA = ETP because ETA reaches maximum; If CH <ETP then ETA = CH + |dKAT| because all CH and dKAT will all be evapotranspirated.

2.6.8. Deficit Column (D). Deficit means reduced water for evapotranspiration so that, D = ETP-ETA

2.6.9. Surplus Column (S). Surplus means excess water when CH <ETP so, S = CH-ETP-dKAT

3. Result and study

The research was conducted at *Sanggar Penelitian, Latihan, dan Pengembangan Pertanian* (SPLPP) Univeraitas Padjadjaran, located in Arjasari Village, Arjasari Subdistrict, Bandung Regency. SPLPP Arjasari stands on an area of 213.4 ha, with a type of latosol soil and has an elevation of about 950 masl. Arjasari village is located about 25 km from Bandung city. This research was conducted on an area of 226.8 m^2 with the length of research land used 75.6 m and width 3 m. The beginning of planting was done on the second ten day in April, and harvested on the second ten day in July 2015. The research was conducted by planting corn and chili crops intercropping with the distance between corn and pepper plant 70 cm. The distance between corn plants is 25 cm, while for the distance of pepper plant is 50 cm.

Hydrological data which is the daily rainfall data from 3 weather stations that is Cisondari Station, Cipaku Station and Cileunca Station. Daily rainfall data used is the data of the last 10 years, the period 2005-2014. Rainfall data obtained from the Department of Water Resources Development West Java Province. While the data used climatology taken from the Meteorology, Climatology, and Geophysics Bandung. The data used include the maximum temperature data, minimum temperature, average temperature, wind speed, air humidity and the duration of solar radiation for the last 10 years. Selection of climatology station Bandung is because it is the nearest station in the location of research. Some parameters are calibrated with certain methods so that it is expected to represent the situation in the research location.

The result of rain data measurement is a rainfall data of a point, whereas for the purpose of analysis is needed rain data of a region so that the data must be converted into the rainfall area first. According to [9], one of the methods used to calculate the rainfall of this region is the Thiessen method. Thiessen method is determined by making polygons between the rain heading in a watershed area, then the average rainfall flow rate is calculated from the number of multiplications between each polygon area and the rainfall height divided by the entire area of the watershed.. The following is the rainfall data opportunity 80% of the research location.

IOP Conf. Series: Earth and Environmental Science **334** (2019) 012021 doi:10.1088/1755-1315/334/1/012021

Table 1. Rainfall Opportunity 80% chance at the study site. The data obtained from the processing of rainfall data for 10 years. The above data is calculated monthly using the equation;R80 = n / 5 + 1

Month	jan	Feb	mar	apr	may	jun	jul	aug	sep	oct	nov	dec	TOTAL
Rainfall (mm)	145.5	200.3	239.4	185.6	154.4	79.2	7.4	20.2	21.9	170.3	247.2	164.1	1635.4

The determination of climatic classification is based on the Schimdt-Ferguson and Oldeman Classifications. The use of two methods in this analysis is because the Oldeman climate classification is directed to food crops such as rice and secondary crop, while the use of the Schmidt-Ferguson climatic classification is used because other climatic classification methods are less suitable for the situation in Indonesia, especially regarding the techniques used to assess rainfall. Rainfall at the study sites based on Schmidt-Ferguson climatic type classification is included in Climate Type C, since it has a Q value of 37.5%. Meanwhile, according to the Oldeman climate classification, the rainfall at the study site falls into the E4 climate type. according to Oldeman this area is generally too dry, may only be once planted secondary crop, and even then depending on the rain.

The existence of limited climatological data in the District Arjasari require the use of climatological data from BMKG station Bandung City. Climatology data from BMKG Station Bandung City needs to be modified to be more in accordance with the conditions in the field research, modification of climatological data is done by using Braak method. Braak method in its use requires data in the form of reference temperature data, as well as elevation (elevation) station reference temperature and elevation of where the temperature value is sought. As for Braak formula approach ie;

$$T = X - 0.0061\ h \tag{2}$$

Where T is temperature (°C), h is the difference between the reference station elevation and the searched station (masl) and X is the reference temperature value (°C). High reference station BMKG Bandung is 791 masl, while the elevation of SPLPP Unpad in the village of Arjasari is 950 masl. The following is the modified temperature data using the Braak method:

Table 2. Modified Temperature Data

Month	Minimum Temp (°C)	Maximum Temp (°C)	Average Temp (°C)
Jan	19.41	27.78	23.6
Feb	19.18	27.79	23.49
Mar	19.15	27.75	23.45
Apr	19.15	28.25	23.7
may	19.01	28.29	23.65
jun	18.65	28.19	23.42
jul	17.63	28.4	23.02
aug	17.64	28.43	23.04
sep	17.81	29.08	23.45
oct	18.57	27.5	23.04
nov	19.02	28.07	23.55
dec	19.27	28.09	23.68

As shown in table 2, the data at the study sites ranged from 17.63°C to 29.08°C with an average temperature of about 23°C. The lowest temperature was recorded in July at 17.63°C while the highest temperature was recorded in September at 29.08°C. The data is the average temperature for the period of 2005-2014.

Data of air humidity, duration of solar irradiation, and also used is air humidity data from BMKG station of Bandung City. This is done because of limited data at research sites where there is no climatology station available. All climatology data used is the average data during the period 2005-2014. Air humidity ranges from 73% to 84%. The highest air humidity, at 84%, occurred during the peak of the rainy season between November and February. The average wind velocity value of 1.68 m /s with the lowest wind speed of 1.5 m/s occurred in May and the greatest wind velocity occurred in January at 1.78 m/s. In general, the duration of solar irradiance ranges from 4.68 hours to 6.81 hours that occurred in August, while for the shortest irradiation occurred in January.

This water balance research required potential evapotranspiration (ETP) data, plant evapotranspiration (ETc), as well as actual evapotranspiration (ETA) values. The potential evapotranspiration value is calculated by Penman-Monteit method using Cropwat 8.0 software. The calculation of Penman-monteith evapotranspiration value requires data of maximum temperature value, minimum temperature, wind speed, solar irradiance time, and humidity value of air and coordinate research field. Here are the results of potential evapotranspiration calculations on Cropwat 8.0 software;

Table 3. Potential Evaporation Value

Month	Jan	Feb	mar	Apr	mei	Jun	Jul	Agt	Sep	Okt	Nov	Des
1st ten day	38.5	37.8	34.3	34.9	33.1	33.6	36.9	40.7	42.8	39.3	36.1	37.1
2nd ten day	38.5	37.8	34.3	34.9	33.1	33.6	36.9	40.7	42.8	39.3	36.1	37.1
3rd ten day	42.35	30.24	37.73	34.9	36.41	33.6	40.59	44.77	42.8	43.23	36.1	40.81

The value of plant evapotranspiration (ETc) is the water required by plants. ETc value is the result of multiplication of evapotranspiration value with coefficient value of plant. Each plant has a different Kc value. The value of KC in plants is also different at each growth phase. The value of Kc in the intercropping pattern, in the development phase of Kc value is calculated using the equation 104 in FAO irrigation and Drainage Paper No. 56:

$$KC\ Mid = \frac{(f1\ x\ h1\ x\ K1) + ((f2\ x\ h2\ x\ K2)}{(fi\ x\ h1) + (f2\ x\ h2)}$$

Where ; F1 = area of land cover by plant 1
F2 = area of land cover by plant 2
H1 = plant height 1 K1 = Kc value of plant 1
H2 = plant height 2 K2 = value of plant Kc 2

The result of calculation using equation 1 for com and chili plants obtained value of Kc equal to 1.13. This value is obtained by calculating the maximum height of corn crop as high as 1.5 m, while 0.7 m of chili pepper. The maximum planting height refers to FAO Drainage and Irrigation Paper No.56. For land cover by maize by 67% and pepper plant by 33%. Data of water requirement of plants can be seen in figure 1 below;

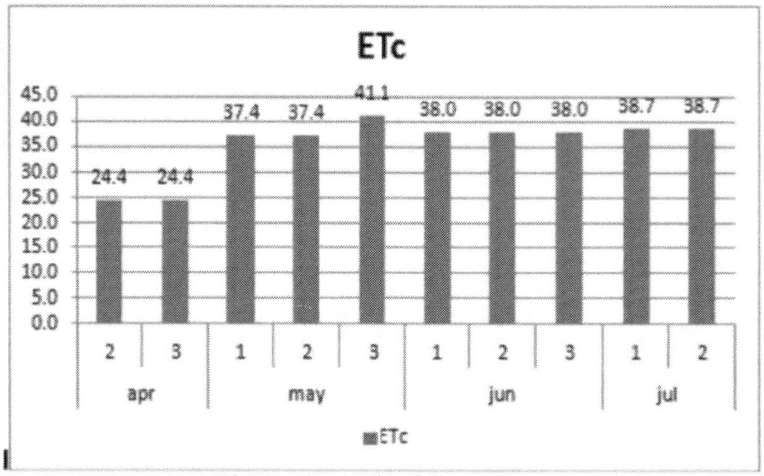

Figure 1. Needs of Crop Water

Crop Water Requirements closely related to potential evapotranspiration values and plant coefficients (Kc). Corn crops intercropped with pepper plants have a water requirement of 353.52 mm. This value is obtained by summing the crop water requirements from the beginning of planting, ie on the second ten day of April to the time of harvesting of plants in the second ten day in July. The need for water of this plant is the requirement of water of corn and chilli plants in the intercropping system conducted in Arjasari, Bandung regency.

Classification of soil types in research sites based on triangle of soil texture issued by USDA belong to clay type with composition 73% clay, 4% dust fraction and 23% sand fraction. This data is obtained from the results of examination of soil physical properties of land research conducted in Soil Physics laboratory, Soil Research Institute, Ministry of Agriculture, Republic of Indonesia in the city of Bogor.

Data of soil physical characteristic which is ground water content at each different phase is the result of sample test taken from the research field. The samples taken are as many as 3 pieces of samples to be taken average value on each phase. The average field capacity value in the research field was 47.2% or 472 mm / m. Field capacity value was obtained from soil moisture test at pF 2.54, while test of permanent wilting point value was done at pF 4.2 with average result 28.7% or 287.3 mm / m. Test result data can be seen in the following table;

Table 4. Summary of Laboratory Test Results of Soil Sample

Sample	KL		TLP		AT	
	%	mm/m	%	mm/m	%	mm/m
1	48.5	485	30.9	309	17.6	176
2	44.9	449	26.2	262	18.7	187
3	48.3	483	29.1	291	19.2	192
Average	47.2	472.3	28.7	287.3	18.5	185.0

Water balance calculation is a combination of evapotranspiration value data with rainfall data, plant data, and soil physical properties data. Water balance analysis is done by using Thornthwaite-Mather method. The water balance analysis is conducted to observe the water conditions in the land on each ten day. This becomes important because information is needed whenever there is a water deficit, and whenever there is a surplus of water on the land. This can be useful as a handle to determine the beginning of the growing season so that the water deficit is not too large. One of the most influential factors to the surplus and the negative water in the research field is the high rainfall. As shown in Figure

3, the water deficit in the field of research occurs in the third ten day period from July to the third ten day of September. The water deficit reached 65.3 mm in the third ten day of September. The largest water surplus in the research field occurred on the second ten day in November at 126.9 mm. The following general description of the water field research can be seen in Figure 2 below;

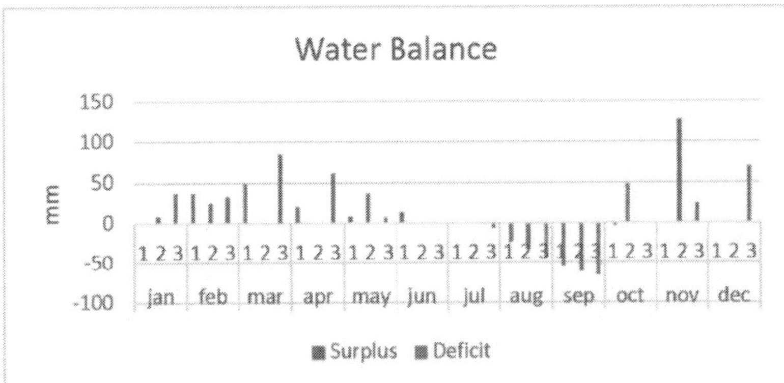

Figure 2. Water Balance of Land Research

Result of soil moisture test on pF 2.54 and pF 4.2, got average value equal to 47.23% and 28.73%. So that the value of ground water content in the field capacity condition of 472 mm/m, while at the condition of permanent wilting point, ground water content of 287 mm/m. The difference between the groundwater value in the field capacity condition and the permanent wilting condition is the value of the groundwater content, meaning that the water that can be utilized by the plant is 185 mm/m. During the study period the water readily available in the soil is maintained at a value of 50%, from the available groundwater value of 185 mm, so the land considered to be water shortage if water is readily available in soil less than 92.5 mm.

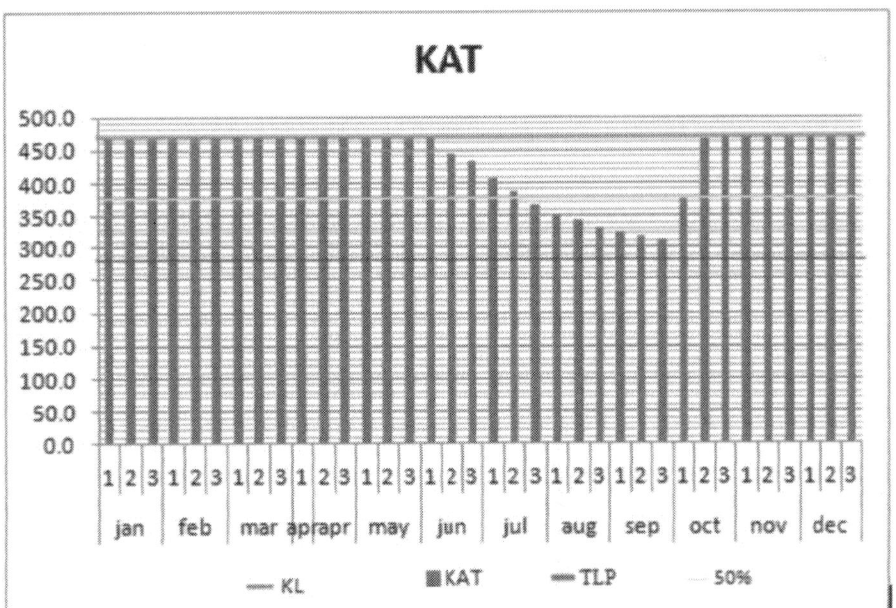

Figure 3. Graph of Earth Water Content

Readily available water content is maintained at 50%, this means groundwater content in new research areas is considered to reach a deficit value when readily available water content has fallen by

50% from its original value of 185 mm or as much as 92.5 mm value of field capacity. The value of the groundwater content in the research field is experiencing a deficit when groundwater content has been less than 379.5 mm. The result of water balance analysis by using Thornthwaite-Mather method shows that water deficit in research field occurred in third period from July to third ten day in September.

Measurements on corn plants were conducted to collect some data such as plant height, stem diameter, length of segment, and number of leaves. Sample used for each parameter used 5 plants which will then be measured also the result of its production. In general, the uniformity of plant growth, in the second plot is better than the first plot. This can be seen from the growth of plant height and the number of leaves that plants have on the second replication is more uniform, but the development of corn crop is better in the first plot. This is seen from the height of the plant, the diameter of the stem, and the number of leaves owned by the plant on the first plot is large when compared with the second plot plant.

The productivity of the plants produced in each replication was measured by several parameters, namely the number of cobs, the weight of the ear, and the sugar content in the corn produced. The yield of corn plant has not reached its maximum potential. According to the description of the plant from the Seed Technology Laboratory Faculty of Agriculture, Padjadjaran University, this cultivated corn cultivar has a potential yield of 137.5 gr per cob because most have only weights of about 80-120 grams only, although there are 2 plants that produce corn that exceeds its potential. The use of sweet corn as the research object make sugar in corn as one of the parameters measured to determine the harvest both in quality and quantity. The corn produced has a sugar content ranging from 9.1 to 13.4 brix.

Table 5. Results of Measurement & Production of Maize Plant

Parameter	**R 1**	**R 2**
Crop Height (cm)	179	159.8
Crop Diameter (cm)	16.6	14
Number of Leaves	8.6	8
Number of Cobs	1.2	1.2
Cobs Weight (gr)	1040	900
Cobs Weight w/o Skin (gr)	560	600
Sugar level (Brix)	11.56	10.78

In addition to corn crops, the yields produced by pepper plants are also measured. The harvest time on both plots is done simultaneously. In some harvest time, there are plants that still have not produced chili fruit at harvest time. Chillies grown are Chilli varieties of Shakira obtained from Seed Technology Laboratory, Faculty of Agriculture, compared with chili varieties Rimbun 3 production of chili pepper varieties of this Shakira can compete with varieties Rimbun 3. Results from pepper plant of 4 times the harvest produces average chili weighing 1062 grams or about 265 grams each time the harvest.

References

[1] Departemen Pertanian 2014 Statistik Lahan Pertanian Tahun 2009-2013 *Pusat Data dan Sistem Informasi Pertanian* Kementrian Pertanian.

[2] Vandermeer J 1989 The Ecology on Intercropping *Cambridge University Press* New York.

[3] Departemen Pertanian 2014 Survey Sosial Ekonomi Nasional *Departemen Pertanian Nasional* Jakarta

[4] Ditjentan Pangan 2010 Mempertahankan Swasembada Jagung Menuju Kemandirian Pangan *Sinar Tani* Edisi 20-18 Oktober 2010 **3376** Tahun XLI.

[5] Zulkarnaini 1995 Penentuan Pola Tanam Berdasarkan Neraca Air Pada Beberapa Daerah Kering di D.I. Aceh *Tesis Pascasarjana (S2) Fakultas Pascasarjana* Institut Pertanian Bogor Bogor.

[6] Badan Pusat Statistik 2014 Statistik Kecamatan Arjasari http://bandungkab. bps.go.id/publikasi_ bps/2014/statda_kec/150Arjasari/files/search/searchtext.xml 28 Juli 2015.
[7] Doorenboos, Pruitt W O 1997 Guidelines for Predicting Crop Water Requirement *FAO Irrigation and Drainage Paper* **24** Rome.
[8] Thornthwaite C W and Mather J R 1957 Instruction and Tabels for Computing Potensial Evaporatranspiration and The Water Balance *Drexel Institute of Technology Laboratory of Climatology* **10(3)** Centerton. New Jersey.
[9] Dwiratna S 2012 Penuntun Praktikum Hidrologi *FTIP Universitas Padjadjaran*.

Remote Sensing Data, Landsat 8 Oli / Tirs For Identification of Evapotranspiration Spatial Distribution in Sub DAS Cisangkuy West Java Indonesia

D R Kendarto[1], C Asdak[1] and A Hapsari[2]

[1] Soil and water Engineering Laboratory, Agricultural Engineering, Universitas Padjadjaran, Sumedang, Indonesia
[2] Graduate Scholl, Agricultural Engineering, Universitas Padjadajaran, Sumedang, Indonesia

E-mail: dwirustamkendarto@gmail.com

Abstract. The Cisangkuy sub-watershed area, West Java is mostly up land farming. In the dry season the region often experiences drought. Efforts to anticipate regional drought can be done by calculating the balance of water availability of the region. One of the most important water balance calculation parameters is evapotranspiration. Direct evapotranspiration measurements in the field represent actual circumstances, but for large areas require large and long-term costs. One way to get regional evapotranspiration data is by reducing evapotranspiration information from remote sensing data. LANDSAT 8 OLI / TIRS satellite imagery is one of the resource satellites commonly used in agriculture. This study aims to determine the estimated value of evapotranspiration in the sub area of Cisangkuy watershed using LANDSAT 8 OLI / TIRS satellite imagery. The method used in this research is the energy balance model with descriptive analysis. The results showed that the highest value of evapotranspiration on August 17 in Cisangkuy Sub-Basin area in mixed forest was 4,190 mm / day. The results of data reliability analysis using root mean square (RMS) error indicate that satellite image can be used to identify evapotranspiration in Cisangkuy sub watershed with RMS error value of 1.124. The RMS value is still smaller than the average evapotranspiration field measurement difference, so it can be said that the evapotranspiration estimation is accurate.

1. Introduction

Evapotranspiration is one of the important information in agricultural planning, especially with regard to establishing cropping patterns and irrigation planning [1]. Evapotranspiration is the amount of water lost from the land due to evaporation of the land surface and transpiration by plants that take place simultaneously [2,3], so that it can be known the amount of water lost and must be replaced through the purchase of water to the land or known as irrigation water. Evapotranspiration data are also widely used in water balance analysis, groundwater availability, runaway water analysis and flood analysis plans, especially with regard to watershed management [4,5].

The actual measurement of evapotranspiration is usually carried out through field measurements or calculations based on environmental parameters usually using the Penman-Montied method as a reference [6], but the weakness of this conventional method is that evapotranspiration information is only for homogeneous regional characteristics [6]. Information on actual evapotranspiration for large areas with diverse variations in land use requires enormous time and costs. Therefore technology is needed that is able to provide information on evapotranspiration as a whole in a wide range of spatial diversity.

Remote sensing data is one of the recording data by a remote sensing vehicle that provides an integral view of the earth's surface, so as to provide a comprehensive picture of the earth's surface information. Remote sensing data is potentially used as a spatial data source because it has advantages in temporal, spectral and spatial resolution referring to data requirements.

The use of remote sensing data is expected to provide evapotranspiration information accurately for large areas. Mass balance approach to reduce evapotranspiration information through image analysis using the SEBAL method [2,3].

Cisangkuy Sub-watershed with an area of 34,024 Ha is part of the upper Citarum watershed in Bandung Regency [5]. Most of the Cisangkuy sub watershed area is rainfed agricultural land, so that when entering the dry season, agricultural areas that mostly rely on rainwater will experience drought. Cropping and irrigation pattern planning in the Cisangkuy Das area is needed so that irrigation water needs and water availability can be calculated, so that cropping planning provides accurate results. One of the information needed in analysing water availability in cropping planning is land evapotranspiration.

The spatial information of evapotranspiration provided in the study area is still lacking because in the area around the study only represented by one weather station. Research on the reduction of spatial information on evapotranspiration using remote sensing data is necessary because of the superiority of remote sensing images in spatial and temporal resolution. This study aims to determine the spatial distribution of evapotranspiration in the Cisangkuy sub-watershed area using LANDSAT 8 OLI / TIRS satellite imagery data and determine the ability of satellite image analysis in estimating the spatial distribution of evapotranspiration values in the Cisangkuy watershed.

2. Methods

This research was conducted in the Cisangkuy watershed, Citarum Hulu, West Java. Data processing was carried out at the Land and Water Engineering Laboratory, Faculty of Agricultural Industrial Technology, Padjajaran University. Analytical descriptive method is used to reduce spatial information on evapotranspiration from remote sensing data using an energy balance model. The remote sensing data used is the Landsat 8 OLI / TIRS image obtained from the USGS page. The root mean square error is used to analyse errors that occur between predictive data and actual data as a measure of research reliability. Image processing module is used as a series of remote sensing digital data processing to obtain spatial distribution of region evapotranspiration.

2.1. Initial image processing of Landsat 8 OLI/TIRS

The remote sensing data used is the LANDSAT 8 OLI / TIRS recording image on August 17, 2016, which was downloaded from the USGS page. Cutting the radiometric correction image and geometric correction is done before the process of identification of soil moisture. Image cutting is done to determine the boundary of the study, adjusting the area in the image with the location of the study and for reducing the size of the data. Radiometric correction is useful for reducing errors due to interference with reflection values due to interference and sharpening the image. The technique used in radiometric correction of this image is linear contrast stretching. Geometric correction is used to correct projection errors and add image coordinate accuracy. Geometric correction is done by referring several points using the coordinates of the earth map as a reference for Ground Control Point with the maximum root mean square error value is 1.

2.2. Image processing to identify evapotranspiration spatial distribution

Calculation of evapotranspiration using satellite imagery in this study uses an energy balance model by passing several calculation stages, namely brightness temperature (Tb) calculation, land surface temperatures (Ts), net radiation (Rn), G soil heat fluxes, air heating flux (H), latent heat flux (LE) and evapotranspiration (ET) using a digital image processing module. Calculation of surface temperature estimates using remote sensing data uses the equation:

$$\text{Tb (brightness temperatures)} : K2\ln(k1l1+1) \quad (1)$$

Where: K1= 774,8853 wm-2sr-1μm-1 dan K2= 1321.08 K, so that the surface temperature (Ts) is determined by the equation:

$$\text{Ts (land surface temperatures)}: Tb1+(\lambda Tb\eth)Ln\varepsilon$$

Where: Tb= brightness temperatures, λ= average wavelength of emission radiation 11,5μm, ε= object emissivity, δ= 1,438 10-2 Mk *Formatting author affiliations*

2.3. Energy balance methods to extract evapotranspiration spatial distribution

Determination of evapotranspiration values using the formulation of the method of the energy balance concept The main equation model is: Rn = H + G + λE with the derived equation is as follows:

$$Rn = Rs\ in - Rs\ out - RL\ out$$
$$Rs\ in = Rs\ Out/\alpha$$
$$\alpha = \pi L\lambda d2/ESUN\lambda\cos\theta$$
$$Rs\ out = \pi \times L\lambda \times d2 \times 1/band$$
$$RL\ out = \varepsilon\ \sigma\ Ts4$$
$$G/Rn = Ts\ (0.0038\alpha+0.0074\alpha2)(1\text{-}0.98NDVI4)$$
$$H = \beta(Rn-G)/1+\beta$$
$$LE = Rn - G - H$$
$$ET = LE \times 0.0864MJ/m2/hari * 0.408\ mm/hari$$

where: Rn= net radiation, G= soil heat fluxs (Wm-2), H= sensible heat fluxs (Wm-2), Rsin= short radiation in, Rsout= short wave radiation out, RL out= long radiation out, Lλ= bands spektral radiance, d2= astronomy distance sun to earth, ESUN= averages solar spectral irradiance, Θs= sun zenith angle, 1/band= median wave length of band used, ε= emisivitas, σ= Stefan-Bolztman constant (5.67 x 10- 8 Wm-2K-4), , β= water body, α= surface albedo n
NDVI= Normalized Difference Vegetation Index, LE = energy latent

Data reliability test is done by comparing the actual soil moisture data and soil moisture data from the LANDSAT image digital analysis process. Reliability test uses the Root Mean Square equation.

3. Result and discussion

3.1. Surfaces temperatures, net radiation, and albedo in Cisangkuy sub watershed

Surface temperature (Ts) is the temperature that comes from the outermost part of an object that gets energy from sunlight. Surface temperature is affected by emissivity, object heat and thermal conductivity. The highest surface temperature in the study area occurred in the corn field (Banjaran) 25.25° C, the lowest temperature occurred in mixed forest land 21.38° C. Increased emissivity of corn plantations (Banjaran) is higher than other research locations with low conductivity values, resulting in an increase in higher temperatures. Mixed forest (Pangalengan) has an increase in emissivity lower than other sample points, but the conductivity value is higher, so the increase in temperature is the lowest.
Net radiation is the difference between short and long waves that come to the surface of the earth with short waves and long waves coming out of the earth's surface. The Cisangkuy sub-watershed area is dominated by high net radiation, marked in red in most areas of each sub-district in Figure 1. Areas with low net radiation values are found in several locations scattered in several districts, such as Pangalengan, Banjaran and Cimaung subdistricts. in blue. The highest net radiation value on satellite imagery in the study area is mixed forest (Pangalengan) with a value of 388,798 W/m2 and the lowest net radiation value from grassland (Banjaran) with a value of 346,527 W/m2.

The smaller the value of reflected or albedo radiation in a field, the greater the net radiation value of the land, such as in mixed forests (Pangalengan) with the highest Rn value among other sample points. In the grasslands in Banjaran, the albedo value is greater than the value of radiation absorbed compared to other sample points, so the Rn value is the smallest. In addition, mixed forests with higher plant heights compared to other regions make the net radiation value produced also greater. Small albedo values on vegetated land cover are also caused by multilevel canopies that can absorb solar radiation at several levels of canopy [7]. Net radiation distribution, soil heat fluxes in the area of the Cisangkuy watershed is presented in Figure 1

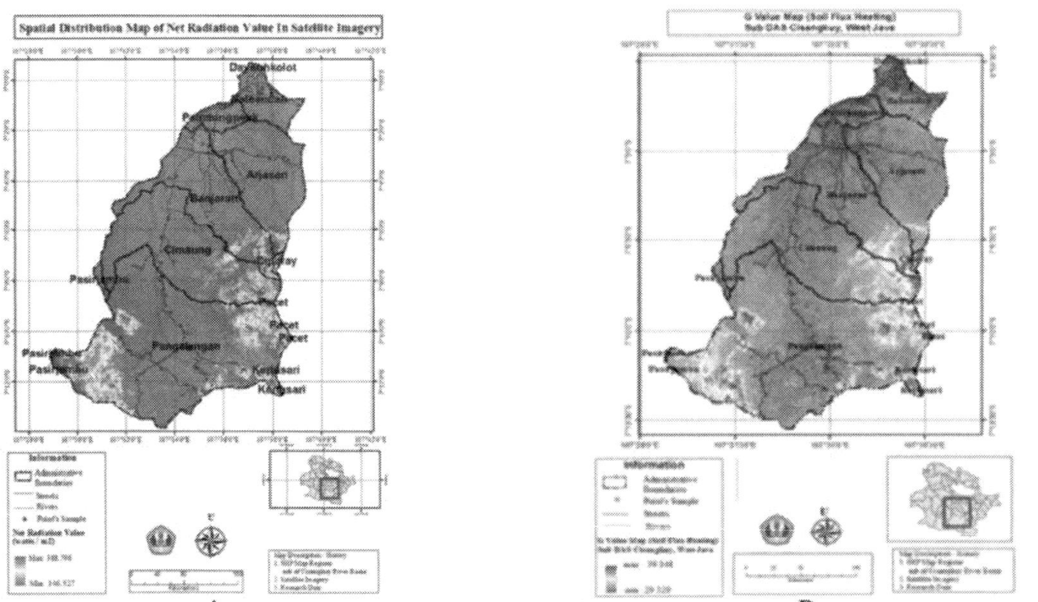

A B

Figure 1. Nett Radiation Spatial Distribution (A) and Soil Heat Fluxes Spatial Distributions (B).

3.2. Distribution soil heat fluxes in Cisangkuy sub watershed

Soil Heat Flux (G) is heat that reaches the surface of the soil and is used for physical and biological processes of the soil. The sub-watershed area of Cisangkuy is dominated by high soil heating flux values, characterized by brown color in almost all sub-districts in Figure 1 and low color markings scattered in several points such as Pangalengan, Cimaung, Banjaran and Arjasari districts. The measurement results show the highest results in the G value among the sample points taken located in the corn field in Banjaran with a value of 39,348 W/m2 and the lowest value is located in mixed forest in Baleendah worth 29,329 W/m2.

Soil heat fluxes (G) are directly proportional to surface temperature (Ts), can be seen at the highest point at G and surface temperature (Ts) which is at the same point, namely the corn field. Vegetation density also affects the G value because land with moderate vegetation cover such as gardens gets a greater percentage of G value, because the amount of net radiation that is passed to the soil and absorbed into the soil is greater than that of tightly closed vegetation.

3.3. Distribution spatial sensible heat fluxes in Casangkuy Watershed

Sensible heat flux (H) is the energy used to move heat from the surface to the air due to the temperature difference between the two, usually it will feel like an increase or decrease in temperature in the atmosphere. The lowest value in H calculation using satellite imagery is found at the river point in Pamengpeuk with a value of 105.045 W/m2 and the highest point is found in the cornfields in Banjaran with a value of 254.777 W/m2. Formula H shows that there are two factors that influence the calculation of H, namely surface temperatures (Ts) and air temperatures (Ta). This is because in this measurement the Ts value is higher than other factors, so it can be estimated that Ts is directly proportional to H, it can be seen at the highest value of Ts and H calculations at the same point.

Figure 2 shows that the sub-watershed area of Cisangkuy is dominated by the value of H with a moderate value, which is marked in yellow. The highest and lowest values for H are only spread in several regions in each district. The widest area with a high distribution of air flux values is found in Baleendah and Pamengpeuk districts, which are marked in red. Clouds found in some sub-districts make lower values due to cloud-covered surfaces, so that the digital value of the image becomes lower or can be a minus, this is marked in blue.

3.4. Distribution evapotranspiration in Cisangkuy sub watershed

The process of evaporation of water in evapotranspiration is carried out using latent heating flux (LE), which is the energy used to evaporate some water from the soil surface to the atmosphere. LE can be converted to evapotranspiration by multiplying by the conversion factor of evapotranspiration value. The calculation results show that the most ET value in satellite imagery is located in mixed forest in Pangalengan with a value of 4.190 mm / day, while the lowest value in this calculation is 1.113 mm/day in paddy field. The highest ET value point with the highest radiation point Rn is located at the same point, i.e. mixed forest, so that it means that net radiation is directly proportional to evapotranspiration.

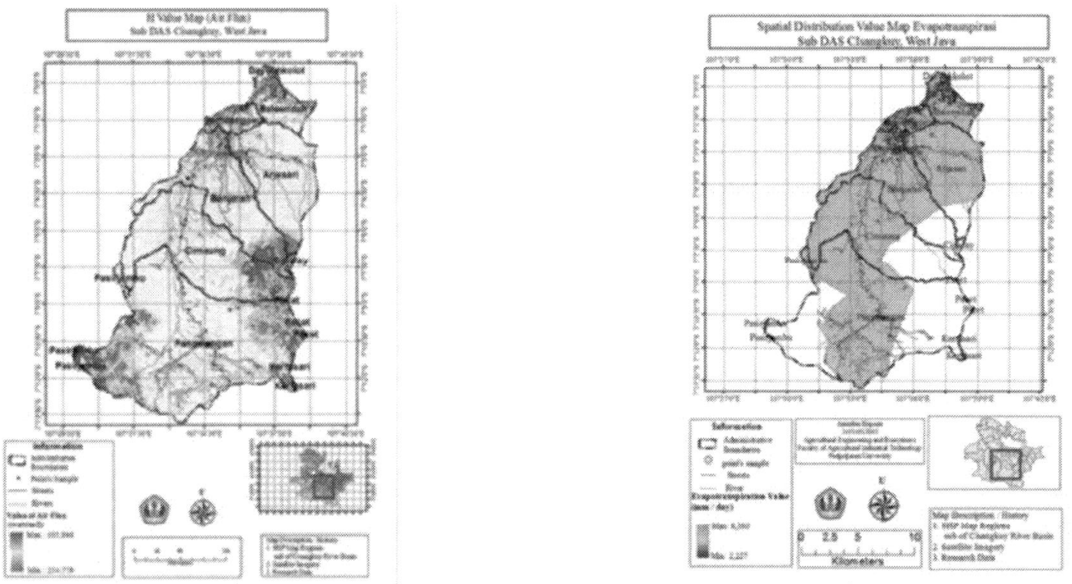

A B

Figure 2. Sensible Heat Fluxes Spatial Distributions (A) and Evapotranspiration Spatial Distributions (B).

The results showed that mixed forests (Pangalengan) had the highest evapotranspiration values, because the Rn value at that point was higher among other sample points, but the values of G and H were low when compared to other sample points. The use of paddy fields in Banjaran has the lowest evapotranspiration value compared to the net radiation values of other samples, but the value of G and air heating flux (H) are high compared to other sample points. Land use with different cover can also be a cause of high or low evapotranspiration values. The value of land use in industry and urban areas has a lower evapotranspiration value than other land uses[8].

Table 1. Evapotranspiration in region area distribution

Number	Region	Evapotranspiration (mm/day)
1	Baleendah	0.17 – 5.5
2	Banjaran	0.55 – 4.88
3	Cimaung	0.89 – 4.94
4	Pamengpeuk	0.47-4.78
5	Arjasari	0.44-4.88
6	Pangalegan	1.29-5.77

The Cisangkuy watershed on average experiences evapotranspiration with high values, indicated by the green color in most of the sub-districts on the map, while the regions that experience lower evapotranspiration are marked in blue in Figure 4. The image shows the blue color usually caused by the existence of clouds covering the land, so that the reading by the image becomes less accurate. Of the 6 sub-districts, Pangalengan experienced the highest evapotranspiration with a range of 1.2680 - 5.7663 mm / day. The spatial distribution of average evapotranspiration value per district can be seen in Table 1.

4. Conclusion

The spatial distribution of Cisangkuy sub DAS evapotranspiration values using satellite images of LANDSAT 8 OLI /TIRS showed that the highest evapotranspiration values occurred in mixed forest land in Pangalengan with a value of 8.380 mm /day and the lowest evapotranspiration values in paddy fields in Banjaran with a value of 2.23 mm /day. High Evapotranspiration values occur because the value of net radiation (Rn) is higher among other sample points, but the value of soil heating flux (G) and air (H) heating flux is lower than other sample points. LANDSAT 8 OLI / TIRS satellite imagery is able to estimate the distribution of evapotranspiration in the Cisangkuy sub DAS, West Java with a test value of RMS Error 1,794. This value is accurate because the value of the RMS Error does not exceed the average value of the squared difference between the calculation of satellite imagery and the field. Based on the results of satellite images, the Pangalengan area has the highest daily evapotranspiration with a value of 1,268 - 5,766 mm / day.

References

[1] Priyonugroho A 2014 Analisis Kebutuhan Air Irigasi (Studi Kasus Pada Daerah Irigasi Sungai Air Keban Daerah Kabupaten Empat Lawang) *J. Tek. Sipil dan Lingkung.* **2(3)**: 457–470.

[2] Irmak A, Ratcliffe I, Ranade P, Hubbard K G, Singh R K, Kamble 2011 Estimation of Land Surface Evapotranspiration With a Satellite Remote Sensing Pro *Gt. Plains Res. Spring* **21(1)**.

[3] Nouri H, Beecham S, Anderson S, Hassanli A M, Kazemi F 2015 Remote sensing techniques for predicting evapotranspiration from mixed vegetated surfaces *Urban Water J.* **12(5)**: 380–393.

[4] Haryanto E T, Herwanto T, Kendarto D R 2007 Perubahan Bentuk Penggunaan Lahan dan Implikasinya terhadap Koefisien Air Larian DAS Citarum, Jawa Barat *Bionatura* **9(1)**.

[5] Kendarto D R 2008 Penggunaan Teknik Penginderaan Jauh Dan Sistem Informasi Geografis Untuk Menentukan Potensi Ketersediaan Air Tanah Kasus di Sub DAS Cisangkuy, DAS Citarum Bandung *Jurnal Teknotan.* **2(3)**.

[6] Tsouni A, Kontoes C, Koutsoyiannis D, Elias P, Mamassis N 2008 Estimation of actual evapotranspiration by remote sensing: Application in Thessaly plain, Greece *Sensors* **8(6)**: 3586–3600.

[7] Setiawan R 2007 Metode Neraca Energi Untuk Perhitungan Indeks Luas Daun Menggunakan Data Citra Satelit Multi Spektral **21(2)**: 27–38.

[8] Khomarudin R 2014 Penentuan Evapotranspirasi Regional Dengan *P. Bidang, P. Penginderaan, D. Geofisika, and M. Ipb.*

The Ability of Phosphorhizobacteria Isolates to Produce Organic Acid and Promote Phosphatase Activity to Increase The Growth of Maize (*Zea mays* L.) in Selected Medium

F H Khumairah[1], A Nurbaity[1], M R Setiawati[1], B N Fitriatin[1] and T Simarmata[1]

[1] Department of Soil Science, Faculty of Agriculture, Universitas Padjadjaran, West Java, Indonesia

Email: fiqriahhanum@gmail.com and tualar.simarmata@unpad.ac.id

Abstract. Utilization of phosphorhizobacteria has the potential to increase the growth of maize by producing organic acid and promoting its phosphatase activity. The research to investigate the capability of phosphorhizobacteria isolates to produce organic acids to convert organic P into inorganic P so that P is available and can increase the nutrient uptake by plants. The experiment was conducted at the Soil Biology Laboratory in Department of Soil Science, Faculty of Agriculture, Universitas Padjadjaran. It was arranged as complete randomized factorial design, consisted of 16 treatments (combination of singular and consortia of phosphorhizobacteria isolates: J1M, J3T and J5H and two concentrations) with 3 replications. The suspension of phosphorhizobacteria isolates were inoculated into media Murphy then planted the maize seedlings into it. The produced organic acid, phosphatase enzyme and seedling growth were measured at 14 days after planting. Results showed that phosphorhizobacteria isolates has the ability to produce oxalic acid, tartaric acid, acetic acid, lactic acid and maleic acid along with maize seedlings. The interaction of consortia phosphorhizobacteria isolates (J1M+J3T) capable to increase maize biomass 21.87% higher than control and P uptake by 3.13 mg/100 g. The J5H isolate can significantly promote the highest phosphate solubilizing index and phosphatase activity

1. Introduction

Maize (*Zea mays* L.) is an important food crop in Indonesia as a source of carbohydrates. From 2011 to 2015, the productivity and production of maize crops have increased. Maize production is still lower than the government's program production target of 20.33 million tons [1]. Production of maize plants that are not optimum yet and one of the reason is due to the decreasing of harvested maize crop area. In 2015, the maize crop harvesting area decreased by 1.29% compared to the previous year, due to the change of maize crops into land for other food commodities and the conversion of agricultural land to non-agricultural [2].

The effectiveness of inorganic P fertilizer in Ultisols is very low i.e. only 10-30% of P fertilizer is given so that 70-90% of P fertilizer remains in the soil and it is difficult to absorb for the plant [3]. Phosphorus is very easy to bind with metals present in Ultisols especially with Al and Fe which form insoluble Al-P and Fe-P bonds. These bonds cause the P element to be unavailable to the crop and fertilizer becomes inefficient [4].

The low availability of P nutrients in Ultisols becomes a barrier to the cultivation of maize in this soil. Utilization of phosphorhizobacteria can be done to dissolve P nutrients becomes available to plants. Phosphorhizobacteria dissolves phosphate through two mechanisms by producing phosphatase enzymes that convert P organic into P inorganic and by removing organic acids that can form stable complexes with binding cations of P in the soil and breaking the Al- P or Fe-P bond [5, 6], so that the P element becomes available and can be absorbed by the plant.

The utilization of phosphorhizobacteria can be done through single inoculation or consortium. Inoculation of the phosphorhizobacteria consortium was able to promote better plant growth and P uptake compared to single inoculation [7]. Therefore, compatibility test of phosphorhizobacteria isolates with plants was required for phosphorhizobacteria inoculum consortium formulation. Some of the previously selected phosphorhizobacteria isolates were obtained from rhizosphere in several districts of West Java. J1M obtained from maize rhizosphere of Majalengka, J3T from maize rhizosphere of Tasikmalaya and J5H from rhizosphere of natural forest of Garut are expected to be compatible and capable of supporting plant growth.

2. Materials and Method

2.1. Design and procedure of experiment
The experiment was conducted from October 2017 until March 2018 at the Soil Biology Laboratory in Department of Soil Science, Faculty of Agriculture, Universitas Padjadjaran, Bandung, Indonesia. The experiment was arranged as completely randomized block design, consisted of 16 treatments (one control and fifteen combination of singular and consortia of PSB isolates with two concentrations). The three selected isolates (J1M, J3T and J5H) were maintained in Pikovskaya media and stored inside the Laboratory of Soil Biology Department of Soil Science and Land Resources, Faculty of Agriculture, Universitas Padjadjaran, Bandung. Phosphaterhizobacteria were inoculated in Pikovskaya medium and incubated in 150 rpm rotary shaker for 72 h at 28 °C to obtain the population density about 10^8 CFUs m/L. The TPC method (Total Plate Count) was used to enumerated the bacteria population. Subsequently, the suspension of phosphorhizobacteria isolates was applied to maize seedlings in liquid planting medium Murphy.

Before treatment, maize seeds (BISI 2) were sterilized in 0.1% mercuric chloride solution for 1 min, then sterilized in 70% ethanol, washed with 4 times with sterilized water and allowed to dried (air-dried). Preparation of maize seedlings is done by planting seeds of BISI 2 variety on paper towel. Two sheets of paper towel were moistened using aquadest and planted maize seeds with a wide spacing and then covered with other paper and then rolled with a plastic then kept in dark conditions for four days to grow roots then used for bio-test (bioassay).

The planting of maize seedlings on Murphy media was carried out on test tubes of 20 mm diameter and 300 mm sterilized sizes. Adding liquid media Murphy according to each treatment. In the control treatment, add the Murphy media as much as 100 ml into the test tube without adding a bacterial suspension. At a 5% phosphorhizobacteria concentration treatment, added a liquid medium of Murphy as much as 95 ml and a bacterial suspension of 5 ml. At the phosphorhizobacteria treatment of 10%, added a media of Murphy as much as 90 ml and a bacterial suspension of 10 ml. Then plant the maize seedlings by soaking the roots of maize plants into the liquid planting media Murphy and the top of the plant is supported by plastic and floxoc pipe. Then arranged and stored in a greenhouse.

2.2. Determination of organic acid and phosphatase activity
The type and quantity of released organic acid into liquid planting media Murphy was analyzed by high performance liquid chromatography (Singapore), in the Laboratory Central, Universitas Padjadjaran. The organic acid standard to be used in this study was analyzed by reversed phase chromatography using Grace smart RP 18.5 μ column, and read at $\lambda = 210$ nm. Before injection into HPLC, organic acid standards and samples were first filtered with 0.18 μm Whatman paper. The analysis was carried out at isocratic conditions at 40 °C using 50 mM phosphate solution as a mobile phase (dissolving 6.8 g of potassium dihydrogen phosphate in 900 ml water, pH value adjusted by adding phosphoric acid to 2.8 using pH meters, then adding water up to 1000 ml) , then filtered with Whatman paper 0.45 μm. The flow rate of the mobile phase is set to 0.7 ml/min. Prior to injection, the device must be stabilized first[8].

Phosphatase activity was determined according to Eivazi and Tabatabai method by giving p-nitrophenyl substrate so that p-nitrophenol compound formed by enzyme activity was then stained by sodium hydroxide solution which could be detected by 400 nm spectrophotometer [9, 10].

2.3. *Statistic data analysis*

Statistical analysis were processed by SPSS software ver. 21 (Information Technology Service, Universitas Padjadjaran). Data were analyzed using one-way analysis of variance (ANOVA) and the difference between treatments was tested by Duncan's multiple range test at $p \leq 0.05$.

3. Results and Discussion

3.1. *Organic acid production*

Phosphorhizobacteria produces organic acids that react with $AlPO_4$, $FePO_4$, and $Ca_3(PO_4)_2$ in releasing a P bound. From these reactions are formed organic chelates of Al, Fe, and Ca so that P is liberated and soluble and available for plants [11]. Organic acids such as acetic acid, formic acid, lactic acid, oxalic acid, malic acid and citric acid are organic acids produced by the microbes. The order of organic acid ability in dissolving phosphate is: citric acid>oxalic acid=tartrate acid=malic acid>lactic acid=formic acid=acetic acid [12]. Organic acids that form a more steady complex with metal cations will be more effective in removing Ca, Al and Fe minerals so the soil will release a larger P. Ease of phosphate released following the order of $Ca_3(PO_4)_2$> $AlPO_4$> $FePO_4$ [13-15].

The results of the analysis were not statistically processed. Based on the result indicated that average of all isolate of phosphorhizobacteria either single or consortium at concentration 5 and 10% able to produce organic acids, which are oxalic acid, maleic acid, tartaric acid, acetic acid and lactic acid with different amount in each treatment (Table 1). Consortium inoculum of the phosphorhizobacteria isolate produces more organic acids that are more effective to dissolve P [16]. The organic acids were produced by phosphorhizobacteria from glucose as the primary metabolite used for cell survival [17]. Production of organic acids by phosphorhizobacteria is also capable of lowering pH, as an anion exchanger with phosphate and as chelating agent P absorbent [18].

Table 1. Production of Organic Acid in Liquid Planting Media Murphy due to Provision of Various Isolates and Concentrations

Treatments	Oxalic acid (ppm)	Maleic acid (ppm)	Lactic acid (ppm)	Tartaric acid (ppm)	Acetic acid (ppm)
b0k1	7.33	0.15	10.86	12.89	n/d
b1k1	9.48	0.16	18.00	n/d	n/d
b2k1	9.20	0.16	116.19	2.77	n/d
b3k1	8.95	0.15	10.20	n/d	1207
b4k1	9.86	0.16	7.61	n/d	0.18
b5k1	7.72	0.16	42.79	n/d	1.40
b6k1	15.21	0.18	42.17	n/d	n/d
b7k1	7.83	0.15	7.30	n/d	1.41
b0k2	24.95	0.16	6.68	n/d	n/d
b1k2	14.76	0.15	21.68	n/d	n/d
b2k2	9.64	0.21	7.52	n/d	n/d
b3k2	15.98	0.16	6.49	0.53	55.91
b4k2	11.15	0.16	7.35	n/d	3.53
b5k2	7.61	0.15	6.56	n/d	62.69
b6k2	7.86	0.16	9.17	n/d	54.97
b7k2	18.27	0.16	6.62	n/d	17.88

n/d = number of organic acids not detected.

3.2. Phosphatase activity

The results of statistical analysis showed that there were no interaction between single or consortium phosphorhizobacteria isolates with phosphorhizobacteria concentrations but showed independent effect on each treatment (Table 2). The use of single isolate and consortium of phosphorhizobacteria was not significantly different from control, but in treatment b3 the use of single isolate J5H was significantly different from the treatments of b2, b5 and b6 which showed higher phosphatase activity value and followed by different notation of Duncan's Multiple Range Test.

Enzyme phosphatase is an enzyme that will be produced when the availability of phosphate is low [19]. Phosphatase enzymes can break phosphate bonds with organic compounds into available forms and can be absorbed by plants. Phosphatase activity will work as the number of P organic, the high value of phosphatase activity is suspected because phosphorhizobacteria works by actively hydrolyzing organic P [3].

Table 2. Phosphatase Activity in Liquid Planting Media Murphy Resulting from Provision of Various Isolates and Concentrations

Treatments	Phosphatase Activity (μg pNP/g/h)
Phosphorhizobacteria	
b_0 = control	1.735 ab
b_1 = J1M	1.814 ab
b_2 = J3T	1.526 a
b_3 = J5H	2.103 b
b_4 = J1M + J3T	1.885 ab
b_5 = J1M + J5H	1.576 a
b_6 = J3T + J5H	1.463 a
b_7 = J1M + J3T + J5H	1.760 ab
Concentration	
k_1 = 5%	1.600 a
k_2 = 10%	1.865 b

Figures followed by the same notation are not significantly different based on Duncan Multiple Range Test at 5% real level.

Table 3. Plant Height and Root Length of Maize Seedlings due to Provision of Various Isolates and Concentrations

Treatments	Plant Height (cm)	Root Length (cm)
Phosphorhizobacteria		
b_0 = control	20.917	10.883
b_1 = J1M	23.783	10.867
b_2 = J3T	23.558	10.567
b_3 = J5H	25.483	11.083
b_4 = J1M + J3T	22.383	11.667
b_5 = J1M + J5H	25.275	10.067
b_6 = J3T + J5H	21.383	9.800
b_7 = J1M + J3T + J5H	20.817	9.567
Concentration		
k_1 = 5%	22.396	10.550
k_2 = 10%	23.504	10.575

Figures followed by the same notation are not significantly different based on Duncan Multiple Range Test at 5% real level.

3.3. Maize seedlings growth promotion

Single or consortium inoculation and phosphorhizobacteria concentrations did not have a significant effect on plant height and root length and no interaction between them on plant height and root length of maize seedlings at 14 days after planting (Table 3). The use of phosphorhizobacteria isolates with different concentrations provides various plant growth responses.

The growth of plant height and root length is influenced by the availability of nutrients in the planting medium. The increase in plant height and root length is the effect of bacteria that can convert unavailable elements into available for plants [20]. The phosphate solubilizing bacteria isolated from the maize plant rhizosphere is capable of producing plant growth promoting substances and is capable of producing phytohormones in assisting plant growth [21, 22]. The average plant height and root length of maize crops with 10% phosphorhizobacteria concentration during the experiment was better than the phosphorhizobacteria concentration of 5%, this is because the phosphorhizobacteria population contained larger.

4. Conclusion

All isolates of phosphorhizobacteria either single or consortium at concentration 5 and 10% able to produce organic acids, which are oxalic acid, maleic acid, tartaric acid, acetic acid and lactic acid with different amount in each treatment. The use of single isolate J5H showed higher phosphatase activity value than other isolates either in single or consortium. The use of phosphorhizobacteria isolates with different concentrations provides various plant growth responses. Therefore, further research are needed to develop and formulate these phosphorhizobacteria of biological agent or biofertilizer to improve the growth and yield of maize.

Acknowledgments

We would like to give our biggest appreciation to Universitas Padjadjaran for funding this research through ALG (Academic Leadership Grant).

References

[1] Indonesian Ministry of Agriculture 2015 Peraturan Menteri Pertanian Republik Indonesia Nomor 14/Permentan/OT.140/3/2015.

[2] Chafid M 2016 Outlook komoditas pertanian tanaman pangan jagung *Pusat Data dan Sistem Informasi Pertanian* Kementrian Pertanian.

[3] Fitriatin B N, Yuniarti A, Mulyani O, Fauziah F S, Tiara M D 2009 Pengaruh mikroorganisme pelarut fosfat dan pupuk P terhadap P tersedia, aktivitas fosfatase, populasi mikroorganisme pelarut fosfat, konsentrasi p tanaman dan hasil padi gogo (*Oryza sativa* L.) pada ultisols *Jurnal Agrikultura* **20.**

[4] Elfiati D, Delvian, Hanum H 2016 Indeks pelarutan fungi pelarut fosfat dengan menggunakan empat sumber fosfat *Prosiding Seminar Nasional Lahan Suboptimal* ISBN: 979-587-659-7.

[5] Saparatka N 2003 Phosphatase activities (ACP, ALP) in agroecosystem soils *Doctoral thesis* Swedish University of Agricultural Sciences Uppsala.

[6] Rao S 1994 Mikroorganisme tanah dan pertumbuhan tanaman *UI-Press* **2**

[7] Minaxi J S, Chandra S, Nain L 2013 Synergistic effect of phosphate solubilizing rhizobacteria and arbuscular mycorrhiza on growth and yield of wheat plants *Journal of Soil Science and Plant Nutrition* **13:** 511-525

[8] Nour V, Trandafir I, Ionica M E 2010 HPLC organic acid analysis in different citrus juice under reversed phase condition *Notulae Botanicae Horti Agrobotanici Cluj-Napoca* **11:** 42-48

[9] Eivazi F, Tabatabai M A 1977 Phosphatase in soils *Soil Biol. Biochem.* **9:** 167-172

[10] Schinner F E, Kandeler E, Ohlinger R, Margesin R 1996 Method in Soil Biology *Springer-Verlag*

[11] Sharma S B, Sayyed R Z, Trivedi M H, Gobi T A 2013 Phosphate solubilizing microbes: sustainable approach for managing phosphorus deficiency in agricultural soils *SpringerPlus* **2** 587

[12] Ryan P R, Delhaize E, Jones D L 2001 Function and mechanisms of organic anion exudation from plant roots Annl. Rev. *Plant Physiol Plant Mol. Biol.* **52:** 527-560

[13] Hue N V, Craddock G R, Adamet F 1986 Effect of organic Acids on aluminium toxicity in subsoils *Soil.Sci.Am.J.* **50:** 28-34.

[14] Gupta N, Sabat J, Parida R, Kerkatta D 2007 Solubilization of tricalcium phosphate and rock phosphate by microbes isolated from chromite, iron, and manganese mines *Acta Bot. Croat* **66:** 197-204

[15] El-Azouni I M 2008 Effect of phosphate solubilizing fungi on growth and nutrient uptake of soyben (*Glycine max* L.) plants *J.Applied Sci.Res* **4:** 592-598

[16] Lacobazzi V, Infantino V, Convertini P, Vozza A, Agrimi G, Palmieri F 2009 Trancription of the mitochondrial citrate carrier gene: identification of a silencer and its binding protein ZNF224 *Biochem. Biophysic Res. Comm.* **386:** 186-191.

[17] Rodríguez H, Rossolini G. M, Gonzalez T, Li J, Glick B R 2000 Isolation of a gene from *Burkholderia cepacia* is-16 encoding a protein that facilitates phosphatase activity *Current microbiology* **40:** 362-366.

[18] Chen Y P, Rekha P D, Arun A B, Shen F T, Lai W A, Young C C 2006 Phosphate solubilizing bacteria from subtropical soil and their tricalcium phosphate solubilizing abilities *Appl. Soil Ecol.* **34:** 33-41.

[19] Joner E J, Aarle I M, Vosatka M 2000 Phosphatase activity of extra-radical arbuscular mycorrhiza hyphae: a review *Plant Soil* **226:** 199-210

[20] Zaidi A, Khan M S 2006 Co-inoculation effects of phosphate solubilizing mico-organism and glomus fasciculatum on green gram - bradyrhizobium symbiosis *Turk J Agric For* **30:** 223-230.

[21] Zao Ke, Pentiigen P, Zhan X, Do X, Lio M, Yu X, Chen Q 2013 Maize rhizospehere in sichuan china, host plant growth promoting burkholderia cepacia with phosphate solubilizing and antifungal abilities *Microbiological Research* **169:** 76 – 82.

[22] Wua B, Caob S C, Lib Z H, Cheunga Z G, Wonga K C 2005 Effects of biofertilizer containing n-fixer, p and k solubilizers and am fungi on maize growth *Ganoderma* **125:** 155-162.

Effect of P-Fertilizer and Organic Matter to Value of pH and pHo, P-avaibility, and Yeild of Mieze in Andisols from Pangalengan West Java

M Arifin[1,2], A Sandrawati[1], A Setiawan[1], R Devnita[1], E Trinurani[1] and A Yuniarti[1]

[1]Lecturer at Department of Soil Science and Land Resources, Faculty of Agriculture, Universitas Padjadjaran

E-mail: mahfud.arifin@unpad.ac.id.

Abstract. One of the problems in the utilization of Andisols for agriculture is their high retention of phosphate by allophane mineral which has variable charge. Such phenomenon can be reduced by giving high phosphate fertilizer or increasing organic matters. This research aimed to understand the effect of dose of P fertilizer and types and dose of organic matter fertilizers to reduce P-Retention and soil pH_0 on Andisols planted with sweet corn, so the efficiency of phosphate fertilizer utilization could be-achieved. This research was conducted in glasshouse experiments using Completely Randomized Design respectively. The result showed there were no interaction effects between phosphate and organic matters on P retention, P total, P availability, P uptake, pH and pH_0. Optimal dose of phosphate fertilizers for sweet corn was 45 kg/ha P_2O_5 for all organic matter fertilizers.

1. Introduction

Andisols or Andosol in Indonesia, according to the National Soils Classification System, has an area of around 5.4 million ha or 2.9% of total mainland in Indonesia [1]. Andosol are soils with ABwC horizons, have a molic or umbric horizon aboved B cambic horizon, and have one or both of requirements in the depth of $\geq$ 35 cm: (a) bulk density < 0.90 g/cm and dominated by amorphous materials, (b) > 60% volcanic ash or pyroclastic materials [2]. Andisols generally located in the agroecosystem zones in which they are used for tea plantation or vegetable farming in highland. The problem that often arises in this soil type is low plant productivity due to its chemical characteristics, such as high P-retention, high leaching of bases, and its physical characteristics, such as susceptible to erosion [3].

Low phosphate content in the Andisols can be caused by strong fixation of P by hydrous oxide or silicate clay and suspended by Fe, Al, and Mn ions [4]. High P fixation is also due to the presence of dominated allophane and imogolite minerals which strongly retain P in their exchange complex. Allophane minerals have P sorption capacity almost hundreds fold than gibbsite [5,6]. This type of mineral is also amphoteric and they are able to retain phosphate in huge amount [7].

The decrease of phosphate retention can be done by lowering zero point of charges (ZPC) or pH_0 value. ZPC emerge due to amphoteric characters in their oxides and or iron/Al oxides and Allophane or other amorphous materials in clay fractions which are the main character of clay with variable charges [8]. The decrease of ZPC can increase the surface of negative charge and increase cation exchange capacity after phosphate application. In addition, an increase of organic matter content is also one of the ways to decrease ZPC [9]. These processes occur because organic anions are absorbed to the surface of particles, then they cover some of the positif charges on the surface, so cations retention increase.

Organic matter affected in plant growth with its effect to chemical, physic, and biological soil properties [10]. A high dose of organic matter application would increase the release of cations to the soil solution, so it could maintain or even increase pH [11]. Humus as a result of decaying organic matter is one of the components with variable charge in which the release of cations and anions from mineralization of organic matter can increase electrolyte concentration [10]. This process has an effect to the decrease of potensial surface or pH_0, yet increase soils' cation exchange capacity.

The crops in allophanic soil need higher of labile P, it is important to manage the P sorption capacity of allophane and other Al compound to increase the supply of labile P [12]. Addition of organic manure along with inorganic P could increase the available of P in soil [13]. Phosphorus (P) availability from manure (all type of animal production sources) is high (> 70%) [14]. Poultry manure more readily supplies P to plants than other organic manure sources [12].

In term to increase crop production, the integration of mineral fertilizers with organic amendments (e.g. animal manure and compost) is recommended [15]. However, this is a challenge for the farmer that fertilizers are expensive is expensive. It will urge the needs of researches to find alternative doses and types of organic matter in order to have an efficiency manure usage. This research aimed to understand the effect of dose of P fertilizer and types and dose of organic matter fertilizers to reduce P-Retention and soil pH_0 on Andisols planted with sweet corn, so the efficiency of phosphate fertilizer utilization could be-achieved.

2. Materials and Methods

The research was greenhouse experiment that conducted in experimental farm of Agriculture Faculty, Jatinangor, Sumedang District, West Java. Sweet corn *(Zea mays saccharata Sturt)* was used as an indicator plant. The greenhouse experiment used Completely Randomized Design with factorial pattern consisting of 2 factors and 3 replicates. Treatments design consist with: P_2O_5 doses as first factor, and type of organic mater as second factor. First Factor : a_0 = 0 kg/ha P_2O_5, a_1 = 45 kg/ha P_2O_5, a_2 = 90 kg/ha P_2O_5, a_3 = 135 kg/ha P_2O_5, a_4 = 180 kg/ha P_2O_5. Second Factor : b_0 = without organic matter; b_1 = 4 t/ha cow manure, b_2 = 4 t/ha chicken manure, b_3 = 4 t/ha sheep manure, b_4 = 4 t/ha rice straw compost.

Data were analysed with analysis of variance. The difference of treatment means was analyzed with F-test at the level of signifcancy of 5%. If the results of F-test showed real difference between treatments, they were further tested by the smallest real difference-test (BNT). Observation in the greenhouse experiment included soil pH, P-retention, P-potential, P-Bray, P-uptake, pHo, and weight of sweet corn.

3. Result and Discussion

3.1. P-retention, P-total, P-availability, and P-adsorption

Table 1 showed that application of P fertilizer up to the dose of 180 kg/ha P_2O_5 did not affect P-retention. However, P-retention tended to increase as an increase of P fertilizer doses. This situation can be related to the balance of P in the soils. Some of the factors that affect P-retention are the amount of allophane and imogolite minerals [12,16,17] and asosiated Al or Fe oxide with humic substance [18].

Organic matter has a high content of humic acid [19], this humic substance resulted from biochemichal evolution during decomposition of organic substrates [10]. Humic substance can increase the soluble and extractable P [20]. The role of organic matter can affect available of through its organic acid content which can increase the number of negative charges in soil colloids [21]. Organic anions are known to compete with P and occupy fixation locations in which can decrease the ability of soils to retain P [22-24]. Based on Table 1, organic matter application affected to P-retention. Application of 4 ton/ha organic matter could decrease the P-retention from 78.9% to 65.9-73.2%, depend on organic matter type. Chicken manure can decrease P-retention up to 16.5% rather than other organic matter types.

Application of cow manure with dose of 4 ton/ha was not significantly different with control (without organic matter) in P-total, but was significantly different with other organic mater types (Table 1). Availability of Phosphate and its quantities are different between animal manures [25, 26]. In acidic soil, cow manure could change soil pH and increase P availability [27]. In addition, P fertilization also did not increase P-availability in Andisols. This is in line with the result of P-total.

Table 1. Independenly Effect of P Fertizer and Organic Matter on P-retention, P-total, and P-availability

Treatments	Mean P-retention	Mean P-total	Mean P-availabilty
	%	m/100gr	ppm
a_0 (0 kg/ha P_2O_5)	70.5 a	168.98 a	4.85 a
a_1 (45 kg/ha P_2O_5)	71.6 a	160.06 a	5.16 a
a_2 (90 kg/ha P_2O_5)	73.9 a	168.37 a	5.68 a
a_3 (135 kg/ha P_2O_5)	69.3 a	178.38 a	4.09 a
a_4 (180 kg/ha P_2O_5)	75.9 a	163.33 a	4.09 a
b_0 (Without organic fertilizer)	78.9 a	173.38 ab	6.11 a
b_1 (4 ton/ha cow manure)	73.2 ab	184.76 a	6.92 a
b_2 (4 ton/ha chicken fertilizer)	65.9 b	160.92 b	4.67 a
b_3 (4 ton/ha sheep fertilizer)	70.4 b	164.18 b	4.15 a
b_4 (4 ton/ha rice straw compost)	68.9 b	155.88 b	4.03 a

The numbers followed by the same letter are not difference according to the Duncan Multiple Range Test at 5%

3.2. *pH dan pH_0*

Application of organic matter did not influence soil pH in sweet corn field environment (Table 2).

Table 2. Independent Effect of P Fertilizer and Organic Matter to pH, pH_0

Treatments	Mean of pH	Mean of pH_0
a_0 (0 kg/ha P_2O_5)	5.13 a	4.51 a
a_1 (45 kg/ha P_2O_5)	5.17 a	4.51 a
a_2 (90 kg/ha P_2O_5)	5.11 a	4.51 a
a_3 (135 kg/ha P_2O_5)	5.13 a	4.48 a
a_4 (180 kg/ha P_2O_5)	5.21 a	4.52 a
b_0 (Without organic fertilizer)	5.15 a	4.51 a
b_1 (4 ton/ha cow manure)	5.15 a	4.49 a
b_2 (4 ton/ha chicken manure)	5.16 a	4.47 a
b_3 (4 ton/ha sheep manure)	5.11 a	4.52 a
b_4 (4 ton/ha rice sraw compost)	5.17 a	4.53 a

The numbers followed by the same letter are not difference according to the Duncan Multiple Range Test at 5%

Andisols have a higher buffering capacity because of dominancy of allophane materials and their colloids, beside that application of organic matter could increase the buffer capacity [28]. Meanwhile, C-organic content naturally had negative correlation with pH in Andisols soils in tea plantation, West Java [6].

3.3. *Plant Yields (Weight of Stem and Stem + Cornhusk)*

Application of P fertilizer and organic matter types did not influence plant yield (Table 3). Nevertheless, application of P fertilizer could decrease the yield while organic matter tended to increase the yield. This can be expected due to low dose materials in the application (P fertilizer and organic matter), so P-uptake did not have a significant effect. However, P fertilizer is needed for the assimilation process and building up energy in the photosintate formation.

Table 3. Independent Effect of P Fertilizer and Organic Matter to Stem Weight and Stem + Dry Cornhusk

Treatments	Mean Stem (g)	Mean Stem +Dry husk (g)
a_0 (0 kg/ha P_2O_5)	164.32 a	203.85 a
a_1 (45 kg/ha P_2O_5)	168.87 a	206.71 a
a_2 (90 kg/ha P_2O_5)	158.76 a	201.16 a
a_3 (135 kg/ha P_2O_5)	162.51 a	202.31 a
a_4 (180 kg/ha P_2O_5)	162.93 a	202.05 a
b_0 (Without organic fertilizer)	160.31 a	202.29 a
b_1 (4 ton/ha cow manure)	166.91 a	202.87 a
b_2 (4 ton/ha chicken manure)	164.77 a	205.82 a
b_3 (4 ton/ha sheep manure)	163.21 a	203.71 a
b_4 (4 ton/ha rice sraw compost)	162.19 a	201.39 a

The numbers followed by the same letter are not difference according to the Duncan Multiple Range Test at 5%

P-uptake tended to increase with increasing level of dose of P fertilizer compared to no P fertilizer (control). This condition shows that sweet corn plants give respond toward P, but it is not sufficient to causea difference in the P-uptake. It relates with low of P-availability (Table 1) in supplying P nutrients to the growth and development of roots.

4. Conclusions

P fertilizer and organic matter have no interaction on the P-retention, P-total, P-availabilty, P-uptake, and corn yield. Application of P fertilizer independently did not affect all the response variables while organic matter only affects on P-retention and soil P-total. There is a need of further research related to the effect of types of organic materials in various series of Andisols

References

[1] Subagyo H, Suharto N, Siswanto A B 2004 Tanah pertanian di Indonesia. dalam pengembangan dan manajemen tanah – tanah di Indonesia Bogor: *Pusat Penelitian Tanah dan Agroklimat* p.30 – 61

[2] Subardja D, Ritung S, Anda M, Sukarman, Suryani E, Subandiono R E 2014 Petunjuk Teknis Klasifikasi Tanah Nasional Bogor: *Balai Besar Penelitian dan Pengembangan Sumberdaya Lahan Pertanian*

[3] Inoue K 1986 Chemical properties In Ando Soil in Japan. ed K Wada Kyushu: *Kyushu University Press.*

[4] Brady N C, Weil R R 2016 The Nature and Properties of Soils 15th edition New York: *Pearson Education.*

[5] McLaughin J R, Ryden J C, Syers J K 1981 Sorption of inorganic phosphate by iron- and aluminium-containing components *Jour Soil Sci* **32:** 365-377

[6] Arifin M 1994 Pedogenesis Andisol Berbahan Induk Abu Volkan Andesit Basal pada Beberapa Zona Agroklimat di Daerah Perkebunan Teh Jawa Barat Institut Pertanian Bogor: *Ph.D Thesis.*

[7] Tan K H 2011 Principles of Soil Chemistry 4th Edition Florida: *CRC Press.*

[8] Tan K H 1984 Andosols New York: *Reinhold Company.*

[9] Uehara G, Gillman G 1981 The Mineralogy, Chemistry, and Physics of Tropical Soils with Variable Charge Clays Colorado: *Westview press/Boulder.*

[10] Stevenson F J 1982 Humus Chemistry; Genesis, Composition, Reaction New York: *A Wiley Interscience Publication.*

[11] Sufiadi E 1999 Variasi Titik Muatan Nol, pH, Retensi Fosfat, dan Kapasitas Tukar Kation Andisols Tanjungsari serta Hasil Kentang Sebagai Efek Takaran Bokashi dan Fosfat Universitas Padjadjaran: *Ph.D Thesis.*

[12] Parfitt R L 2009 Allophane and imogolite: role in soil biogeochemical processes *Clay Mineral* **44:** 135-155

[13] Shipra G, Bahl G S 2008 Phosphorus availability to maize as influenced by organic manures and fertilizer P associated phosphatase activity in soils *Bioresource Technology* **99:** 5773-5777

[14] Eghball B, Wienhold B J, Gilley J E, Eigenberg R A 2002 Mineralization of manure nutrients *Jour. Soil and Water Conserv* **57:** 470–473.

[15] Negassa W, Gebrekidan H, Friesen D K 2005 Integrated use of farmyard manure and NP fertilizers for maize on farmers' fields *J. Agric. Rural Dev. Tropics Subtropics* **106:** 131–141

[16] Devnita R 2010 Kajian Karakteristik Mineralogi Muatan Permukaan dan Ameliorasi Andisol di Daerah Jawa Barat Universitas Padjadjaran: *Ph.D. Thesis.*

[17] Arifin M, Nurlaeny N, Devnita R, Fitriatin B N, Sandrawati A, Supriatna Y 2018 The variable charge of andisols as affected by nanoparticles of rock phosphate and phosphate solubilizing bacteria *AIP Conference Proceedings: The 1st International Conference and Exhibition on Powder Technology Indonesia* (ICePTi 2017) 030035-1 - 030035-7

[18] Vistoso E, Theng B K G, Bolan N S, Parfitt R L, Mora M L 2012 Competitive sorption of molybdate and phosphate in Andisols *J Soil Sci Plant Nutr* **12:** 59–72

[19] Rupiasih N N, Vidyasagar P 2009 Analytical study of humic acid from various sources commonly used as fertilizer: emphasis on heavy metal content *International Journal of Design & Nature and Ecodynamics* **4:** 32-46

[20] Guppy C N, Menzies N W, Moody P W, Blamey F P C 2005 Competitive sorption reactions between phosphorus and organic matter in soil: A review *Aust J. Soil Res.* **43:** 189–202

[21] Huang P M 2004 Soil mineral–organic matter–microorganism interactions: fundamentals and impacts *Adv Agron* **82:** 391–472

[22] Tan K H 1982 Principle of Soil Chemistry New York: *Marcel Dekker, Inc.*

[23] Fiantis D 2000 Colloid-surface characteristics and amelioration problems of some volcanic soils in West Sumatra, Indonesia Universiti Putra Malaysia: *Ph.D. Thesis.*

[24] Devnita R, Joy B, Arifin M, Setiawan A, Rosniawaty S, Meidina F S 2018 The phosphorus status of andisols as influenced by nanoparticles of volcanic ash and rock phosphate. *AIP Conference Proceedings: The 1st International Conference and Exhibition on Powder Technology Indonesia* (ICePTi 2017) 030033-1 - 030033-7

[25] Li G, Li H, Leffelaar P A, Shen J, Zhang F 2014 Characterization of Phosphorus in Animal Manures Collected from Three (Dairy, Swine, and Broiler) Farms in China *PLoS ONE* **9:** 1-8

[26] Turner B L, Leytem A B 2004 Phosphorus compounds in sequential extracts of animal manures: chemical speciation and a novel fractionation procedure *Envi Sci Tech* **38:** 6101–6108

[27] Shen J, Yuan L, Zhang J, Li H, Bai Z, Chen X, Zhang W, Zhang F 2011 Phosphorus dynamics: from soil to plant *Plant Physiology* **156:** 997 – 1005

[28] Naidu R, Morrison R J, Janik L, Asghar M 1997 Clay mineralogy and surface charge characteristics of basaltic soils from Western Samoa Clay Minerals **32:** 545-556

Microaggregate and Macroaggregate of Andisol Affected by Arbuscular Mycorrhizal Fungi and Rhizobacteria

C Hidayat[1] , DH Arief [2], J Sauman[2], Anne Nurbaity [2]

[1]Agrotechnology Dept, UIN Sunan Gunung Djati Bandung, Jl. AH. Nasution No. 105, Bandung 40614,Indonesia
[2]*Faculty of Agriculture, Universitas Padjadjaran,Jl. Bandung-Sumedang Km.21, Sumedang 45363, Indonesia*

E-mail: cephidayat62@uinsgd.ac.id

Abstract. The formation of soil microaggregate and macroaggregate involves different microbes. A Screenhouse experiment to determine the effect of Arbuscular Mycorrhizal Fungi (AMF) and rhizobacteria on soil microaggregate stability and soil macroaggregate stability (soil microaggregate and macroaggregate stability) had been carried out. The treatments included AMF (control and *Glomus sp.)* and rhizobacteria (*Pseudomonas diminuta, P. diminuta + Bacillus alvei, P diminuta + P malei, P malei*). The results showed that inoculation *Glomus* sp. and rhizobacteria didn't increase the stability of microaggregates (0.2 mm – 0.3 mm and < 0.2 mm). Inoculation *Glomus* sp and rhizobacteria increased water stable aggregate (2,83 mm – 4,8 mm, 0,50 mm – 1,00 mm, and 0,30 mm - 0,50 mm) and inoculation *Glomus* sp increased water stable aggregate (2,00 mm – 2,83 mm dan 1,00 mm – 2,00 mm). The experiment indicated that rhizobacteria didn't effect soil microaggregate due to silt fraction domination and *Glomus* sp was involved in the formation of macroaggregate

1. Introduction

Aggregates are the basic unit of soil structure, consisting of primary particles, organic materials, and living organisms all bound together in clusters ranging in size less than 2 μm to greater than 2 mm [1]. Soil aggregates are essential components of soil structure that fundamentally affect soil quality, fertility, and sustainability [2]. For Andisol that has unstable sizes [3] and located in rugged terrain in the mountain, [4] soil aggregates become the determining factor of successful plant cultivation, moreover the aggregates stability processes are soil dependent [5].

Soil aggregation formation follows a hierarchical model where primary particles (<53 μm diameter) form microaggregates (<250 μm diameter) via physicochemical forces and persistent binding, whereas microaggregates are assembled into macroaggregates (>250 μm diameter) together with organic debris [2]. Soil microbiota plays important roles in the formation and the stability of soil aggregates [6]. Bacteria plays role in the formation and stabilization of microaggregate [7] through production of extracellular polymeric substances (EPS) that act as binding agent of soil particles [8]. AMF are involved in the formation of macroaggregates through external hyphae [9]. From the results of previous studies found that the rhizobacteria play a role in the formation of microaggregates and FMA involved in the formation of macro aggregates, in this study attempted to combine the microbes was to improve micro aggregates and macro aggregates simulant. The aim of the research was to find out the effect of arbuscular mycorrhizal fungi and different rhizobacteria on the formation of soil microaggregate and macroaggregate of Andisol.

2. Material and Method

The experiment was conducted at the Screen House of Kayu Ambon Lembang 1000 m above sea level using randomized block design (RBD) in factorial pattern with 2 factors and 3 replications. Factor 1:

AMF inoculation (f_o = without *Glomus* sp and f_1 = *Glomus* sp) and factor 2: rhizobacteria inoculation (r_o = without MHB, r_1 = *P. diminuta*, r_2 = *P. diminuta* + *B.* alvei, r_3 = *P. malei*, r_4 = *P. diminuta* + *P. malei*). The parameters observed in this study were aggregate stability index, waterproof stable aggregate. Those observation parameter were analyzed by analysis of variance 5% using DSAASTAT statistical program.

Andisol from Lembang was mixed with fine goat manure with a dose of 20 ton ha^{-1}, then sterilized at temperature 100°C for 8 hours as much as 2 x with 24 hour interval. Once cooled the soil was put into polybag 30 cm x 20 cm as much as 6 kg. Seed of Granola (G_3) potato plant was planted in the middle. Microbial inoculation was done by placing 50 spores of *Glomus* sp. within 5 cm under potato seed and rhizobacteria inoculation using 80 ml (106 CFU ml^{-1}) dissolved in 800 ml of sterile aquades to spread throughout the media. Dosage of fertilizer used: N 300 kg ha^{-1}, 75 kg P_2O_5 ha^{-1}, 100 kg K_2O ha^{-1}. N fertilizer was used at half dose and the remainder at 30 days after planting. P and K fertilizers were given entirely at planting time. Maintenance consisted of irrigation carried out as needed, weed removal, and control of pests and diseases in the event of an attack using pesticides.

3. Result and Discussion

3.1 *Microaggregate*

Inoculation *Glomus* sp. and rhizobacteria either single or double had no interaction and main significant effect on micro aggregates (0.2 mm - 0.3 mm) (Table 1). Aggregate sizes of 0.2 mm - 0.3 mm and <0.2 mm are categorized as micro aggregates, inoculation *Glomus* sp. and rhizobacteria did not simultaneously increase the stability of micro aggregates. Both single and double rhizobacteria also did not significantly increase the value of micro aggregate stability. Similarly, the inoculation of Glomus sp. did not significantly improve the stability of micro aggregates. Based on the opinion of [8], the formation of micro aggregates is influenced by rhizobacteria. Rhizobacteria releases EPS that will glue the soil particles together [11].

Table 1. The Effect of AMF dan Rhizobacteria on Microaggregate (<0.2 mm and 0.2 mm – 0.3 mm) at last Vegetative Phase

Microbe	< 0.2 mm*	0.2 – 0.3 mm*
Rhizobacteria inoculation :		
No rhizobacteria	8.42 a	20.30 a
P. diminuta	8.88 a	23.22 a
P. diminuta+ B. Alvei	8.07 a	21.73 a
P. malei	7.93 a	24.20 a
P. diminuta+ P. Malei	7.78 a	19.20 a
AMF inoculation :		
No *Glomus* sp.	8.11 a	21.25 a
Glomus sp.	8.32 a	22.21 a

Means in the same column with the same letter are not significantly different (Duncan Multiple Range Test α = 0.05).

From the data in Table 1, it can be seen that the rhizobacteria used in this experiment, *P. diminuta, P. diminuta + B. alvei, P. malei, P. diminuta + P. malei* had no effect on micro aggregate formation. This is related with the dominance of the silt fraction (58.61%) on Andisol used in this experiment resulting low electrical charges (electrostatic/ electrochemical forces), thus less support for aggregate soil formation. EPS of rhizobacteria is an organic material rich with negative charges that will binds chemically with a rich allophane mineral clay with positive charge to form a stable aggregate. Such conditions do not occur on the experimental soil used due to the dominance of the silt fraction or slightly electrically charged.

3.2 Macroaggregates

Water stable aggregates (WSA) as an aggregate measurement which is not destroyed by wetting is an indicator that is used to measure macro aggregate stability. WSA are macro aggregates (> 0.25 mm) [11]. In this study, the WSA consist of aggregate size of 0.3 mm - 0.5 mm, 0.5 mm - 1 mm, 1 mm - 2 mm, 2 mm - 2.83 mm, and 2.83 mm - 4.8 mm.

The influence of inoculation *Glomus* sp. and rhizobacteria on WSA was divided into two groups. First, *Glomus* sp. and rhizobacteria gave interaction effect on water stable aggregate at aggregate sizes of 2.83 mm - 4.8 mm, 0.50 mm - 1.00 mm, and 0.30 mm - 0.50 mm. Second, independent effect of *Glomus* sp. on aggregate size 2.00 mm - 2.83 mm and 1.00 mm - 2.00 mm was occurred. Inoculation *Glomus* sp. and rhizobacteria showed interaction effect on the WSA 2.83 mm - 4.8 mm, 0.50 mm - 1.00 mm, and 0.30 mm - 0.50 mm in the last vegetative phase (Table 2).

Table 2. The Effect of AMF dan Rhizobacteria on WSA 2.83 mm – 4.8 mm, 0.50 mm -1.00 mm, and 0.30 mm – 0.50 mm at Last Vegetative Phase

	No Rhizobacteria*	*P.diminuta**	*P.diminuta + B.alvei**	*P.malei**	*P.diminuta + P.malei**
			aggregate size 2.83 mm – 4.8 mm --- % ---		
No *Glomus* sp.	79.90 a B	64.70 a AB	69.27 a AB	56.73 a A	68.83 a AB
Glomus sp.	69.57 a AB	89.03 b B	78.47 a AB	81.93 b AB	67.60 a A
	aggregate size 0,50 mm - 1,00mm ---%---				
No *Glomus* sp.	8.57 a B	14.60 a C	4.70 a A	14.77 b C	17.47 a C
Glomus sp.	17.93 b B	25.70 b D	17.90 b B	2.97 a A	22.70 b C
	aggregate size 0,30 mm - 0,50 mm ---%---				
No *Glomus* sp.	25.27 a A	27.13 a A	27.80 a A	27.13 a A	24.93 b A
Glomus sp.	26.33 a B	29.50 a B	27.27 a B	25.93 a B	14.90 a A

*Means in the same column with the same letter are not significantly different (Duncan Multiple Range Test α = 0.05).

Data from Table 2 showed that rhizobacteria inoculation either single or double on aggregate size of 0.30 mm - 0.50 mm and 0.50 mm - 1.00 mm generally increased WSA compared with no microbial inoculation. However, at the largest aggregate size (2.83 mm - 4.8 mm) rhizobacterial inoculation actually lowered WSA compared with no microbial inoculation. Inoculation of rhizobacteria together with *Glomus* sp. non significantly increased the WSA value, except inoculation of *Glomus* sp. + *P. diminuta* on aggregate size 0.50 mm - 1.00 mm. These data reflect that *Glomus* sp plays a role in macro aggregate formation, in line with the results of [12] which stated that FMA plays a role in the formation of macro aggregates. The data in Table 2, further reinforces the belief that what increases the macro aggregate is *Glomus* sp, since it can increase WSA 1.00 mm - 2.00 mm and 2.00 mm - 2.83 mm which belongs to the macro aggregate category.

Inoculation *Glomus* sp. and rhizobacteria did not always show interaction effect on all sizes of WSA. Rhizobacterial inoculation had no significant effect on aggregate size 1.00 mm - 2.00 mm and

2.00 mm - 2.83 mm. While inoculation *Glomus* sp. gave significant effect (Table 3). Inoculation *Glomus* sp. increased WSA 1 mm - 2 mm by 15,20% and WSA 2 mm - 2,83 mm by 14,65% than no *Glomus* sp inoculation (Table 3).

Tabel 3. The Effect of AMF dan Rhizobacteria on WSA 1.00 mm – 2.00 mm and 2.00 mm – 2.83 mm at Last Vegetative Phase

Treatments	WSA	
	1.00-2.00mm	2,00-2,83 mm
	--%--	
FMA :		
No *Glomus* sp.	71.23 a	68.56 a
Glomus sp.	86.06 b	78.60 b
Rhizobacteria:		
No rhizobacteria	79.12 a	73.68 a
P. diminuta	77.63 a	77.95 a
P. diminuta + B. Alvei	79.37 a	76.47 a
P. malei	70.53 a	67.93 a

In macro aggregate formation, in this experiment > 0.3 mm, inoculation *Glomus* sp. significantly affected the WSA value. The increased in stability of macro aggregates as a result of *Glomus* sp inoculation due to the external hyphae possessed by the AMF species [13]. In this case *Glomus* sp. provides physically skeletal structures: 1) binds the soil particles together, and 2) the external hyphal acts as organic and inorganic binders in binding micro aggregates to macro aggregates.

From data of macro aggregate stability and stability of micro aggregate it can be concluded that *Glomus* sp. and rhizobacteria together do not play a role in the formation of macro aggregates, only *Glomus* sp. independently plays a role in the formation of macro aggregates. As for the formation of micro aggregates *Glomus* sp. and rhizobacteria either jointly or independently does not give effect. Specifically about rhizobacteria, according to [14], it plays an important role in the formation of micro aggregates, but in this experiment it was not so. The formation of micro aggregates determines the quality of soil aggregate stability because it lasts longer, so that if the micro aggregates are formed steadily, then the macro aggregates formed will also be stable. In this experiment, micro aggregate stability without microbial inoculation and inoculated microbial *Glomus* sp. and rhizobacteria was not significantly different. *Glomus* sp plays role in formation of the macro aggregate, although resulted unstable aggregate stability index. So it can be concluded, in order to produce a solid soil aggregate should start with the formation of a stable micro aggregate. Thus to improve the quality of agricultural soil to support maximum plant growth, formation of stable micro aggregate is the most important.

4. Conclusion

Rhizobacteria and *Glomus* sp jointly or independently had no effect on the formation of micro aggregates, while *Glomus* sp was involved in the formation of macro aggregates even resulting unstable aggregate.

Acknowledgements
This work was supported by a grant research from General Directorate of Higher Education, Ministry of Education and Culture. Indonesia. Authors thank to Yati Setiati for her contribution in conducting the experiment and formatting the manuscript for publication and Sofiya Hasani for English revision.

References

[1] Frey S, 2005 Aggregation. In *Microbial aspects*. Elsevier.
[2] Lehmann A, Leifheit, EF, & Rillig MC 2017 *Mycorrhizas and Soil Aggregation*, 241–262.

[3] Tan KH 2008 *Soil in the humid tropics and monsoon region of Indonesia* Bocca Raton, FL. USA: CRC Press.

[4] Nurbaity A, Hidayat C, Hudaya D, & Sauman J 2013 Mycorrhizal fungi and organic matter affect some physical properties of andisols *Soil Water Journal*, **2(2)** 639–644.

[5] Amézketa E, Aragüés R, Carranza R & Urgel B 2003 Macro- and micro-aggregate stability of soils determined by a combination of wet-sieving and laser-ray diffraction 1 83–94.

[6] Lehmann A, & Rillig MC 2015 Soil Biology & Biochemistry Understanding mechanisms of soil biota involvement in soil aggregation : A way forward with saprobic fungi ? *Soil Biology and Biochemistry* (June), 1–5.

[7] Davinic M, Fultz LM, Acosta-Martinez V, Caldero ´n FJ, Cox SB, Dowd SE, Moore-Kucera J 2012 Pyrosequencing and mid-infrared spectroscopy reveal distinct aggregate stratification of soil bacterial communities and organic matter composition. *Soil Biology and Biochemistry* **46** 63–72.

[8] Chenu C, & Cosentino D 2011 Microbial regulation of soil structural dynamics. In K. Ritz & I. Young (Eds.), *The architecture and biology of soils: life in inner space* (pp. 37–70).

[9] Curaqueo G, Acevedo E, Cornejo P, Seque A, Rubio R, & Borie F 2010 Tillage effect on soil organic matter, mycorrhizal hyphae and aggregates in a Mediterranean agrosystem. *Journal of Science and Plant Nutrition*, **10(1)** 12–21.

[10] Miller R, & Jastrow J 2000 Mycorrhizal Fungi Influence Soil Structure. In Y. Kapulnik & D Douds (Eds.), *Arbuscular Mycorrhizas: Physiology and Function* (pp. 3–18). Springer, Dordrecht.

[11] Nurida N, & Undang K 2009 Perubahan agregat tanah pada ultisols jasinga terdegradasi akibat pengolahan tanah dan pemberian bahan organik. *Jurnal Tanah Dan Iklim* **30** 37–46.

[12] AlamiY, Achouak W, Marol, C., & Heulin, T. (2000). Rhizosphere soil aggregation and plant growth promotion of sunflowers by an exopolysaccharide-producing Rhizobium sp. Strain isolated from sunflower roots. *Applied and Environmental Microbiology*, **66(8)** 3393–3398.

[13] Andrade G, Mihara KL, Linderman RG., Bethlenfalvay, G. J., Londrine, U. E. De, & Microbiologia, D. De. 1998 Soil aggregation status and rhizobacteria in the mycorrhizosphere, 89–96.

[14] Rillig MC, & Mummey DL 2006 Mycorrhizas and soil structure *New Phytologist*, **171** 41–53.

The Effect of Rice Husk Silicate Extract on Plant Height, Electrical Conductivity and pH of Paddy Hydroponics

B Frasetya[1], M Subandi[1], S Ginandjar[1] and Y Setiati[1]

[1]UIN Sunan Gunung Djati Bandung, Jl. A.H. Nasution No. 105 Bandung 40614 Indonesia

E mail: budyftq1682@gmail.com

Abstract. Fluctuations in rice production due to conversion of paddy field to other various land use defected the rice production in Indonesia, and in turn it may endanger the food security. Efforts to utilize hydroponics as a technology of vegetable production in limited land availability has developed. However, the practice of hydroponics in rice plants is still limited due to the availability of special nutrient. This study aims to determine the benefits of silicate rice husk extract on the growth of rice hydroponics. The research design used was complete randomized block design. The observed growth parameters are : plant height, electrical conductivity of nutrients and pH. The observed data were analyzed using variance analysis. If the results of variance analysis are significant, the analysis will be proceed with Duncan multiple range test. The results showed that application of rice husk silica extract of 20 ml/L through leaves gave the same response with sodium silicate (Na_2O_3Si) application through nutrient solution, but had significant effect on the change of electrical conductivity value and pH of nutrient solution. The results showed that silicates derived from organic materials can replace the role of silicates derived from inorganic materials on paddy hydroponic.

1. Introduction

In Indonesia, rice is a staple food source for most of Indonesia daily meal. Based on statistical data released by Center for Agricultural Data and Information Systems [1] that the growth of paddy field in Indonesia achieve saturation point with approximately 0.31% in Java Island. The future of rice production challenges are not only constrained by wetland availability, but climate change increase gap between rice production and demand [2]. Imbalance between rice production and supply require technological innovation as a step to maintain rice supply [3]. The application of hydroponics technology to increase rice production by utilizing unproductive lands is still rare due to economic and technical value considerations.

Hydroponics of rice in terms of economic aspects are still less economical when compared with cheaper and faster rice imports. According to [4] rice import policy is less appropriate if still taken as a long-term policy to overcome fluctuations in rice production. Community competence of rice hydroponic is rare when compared to hydroponic vegetable and fruit. Characteristics of paddy crops as silicate accumulators (Si) should be considered in hydroponic nutrient generation [5]. Application of hydroponic nutrients adjusted to planted varieties may increase yield even though nutrient application is not the only one factor of hydroponic cultivation success [6]. Si content in paddy field is abundant but in hydroponic silicate source must be supplied through nutrition application in the form of AB Mix. Nutrition of hydroponic plants that have been distribution to the farmer generally for vegetables without Si enrichment.

Element of Si in paddy plants as functional micro elements [7]. The growth and yield of paddy crops with added Si fertilizer is higher than that without the addition of Si (control) fertilizer [8]. In conventional rice cultivation the Si requirement is entirely supplied from the soil, but in the hydroponic system Si element have to add to the rice nutrient formulation. Utilization of local wisdom as a source of Si fertilizer is by utilizing rice husk as Si source. Si potency in rice husk according to Chandra *et al.* [9] about 17%, the yield of rice husk as a byproduct of rice mill 20-22% [10]. An abundant Si source from rice husks is an opportunity to replace Si fertilizer derived from synthetic fertilizers.

The utilization of rice husk as a source of Si can be done by various methods such as by composting, fermentation and extraction process. The process of composting and fermentation of rice husks takes longer time due to the high Si content and the large area due to rice husks was voluminous. The method of extraction of Si is chosen with the consideration of time required more quickly, Si extract produced in the form of concentrate to facilitate the distribution and application by farmers.

Si content from rice husk silicate extract according to laboratory analysis of Laboratory of Soil and Plant Nutrition [11] Universitas Padjadjaran is 0.68 %. Utilization of rice husk silicate extract on hydroponics rice potential as an alternative source of Si which is cheap and availability to increase the growth of paddy plants.

2. Materials and Methods

This research has been conducted in April-June 2018 in Cileunyi District, Bandung Regency (± 689 m above sea level). Materials and tools used in research, namely: nutrient container, netpot, Styrofoam, paddy seed (Ciherang varieties), sprayer, EC meter, pH meter, KOH 1.5%, rice husk and rockwool.

The research design used was a complete randomized design. Factor Si source consist of two levels, namely: A = Silica rice husk extract 20 ml/L and B = inorganic silica (Na_2O_3Si) Each treatment was repeated 10 times. Parameters measured, i.e. plant height, electrical conductivity (EC) of nutrients and pH of nutrients. The observed data were then analysis using analysis of variance at 5% level, if the F test result significant, the data then analysis with Duncan multiple range test at 5% level.

The research procedure stage, respectively: making of rice husk silicate extract, making of plant nutrient AB Mix, making hydroponics installation (Fig 2), seeding, transplanting and maintenance. The stage of preparation of silicate extract and hydroponic nutrient preparation is an important stage of all stages of research. The preparation of the silicate extract of paddy (Fig 1) begins by smoothing the rice husk size into flour, then add 50 g of husk into the pot with addition KOH 1.5% as much as 100 ml, add water 500 ml then heated to boiling [12] . Cooling process of rice husk extract then filtered so that macro particles do not destruct the sprayer at the time of silicate application.

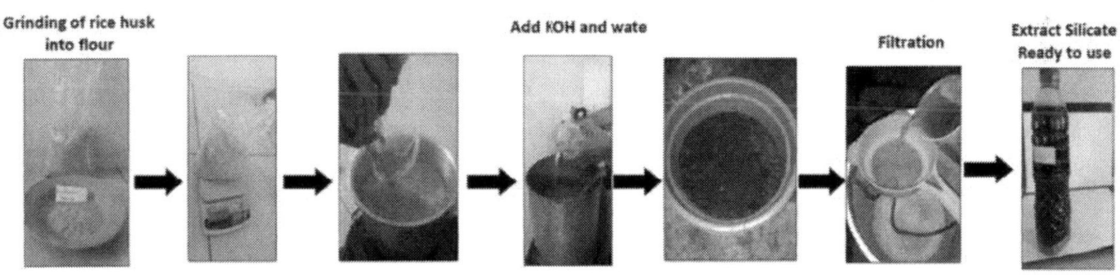

Figure 1. Stages of Silicate Rice Silicate Extraction Process

The AB Mix nutrient preparation carried out in this study consisted of two formulations i.e. formulation without Si (-Si) and formulation with the addition of Si (+Si). Nutrient formulation of Peckenpaugh (2004) for paddy crops was used as the basis for making nutrient AB Mix paddy (in units of ppm), namely: N (249); P (58); K (80); Mg (65); Ca (317); S (87); Si (100); Fe (5); Mn (3); B(0.7); Zn (0.4); Cu (0.07) and Mo (0.05). Based on the calculation of Qurrohman (2017) [6] based on the reference of nutrient formulation obtained calculation of fertilizer requirement (in grams unit), namely: Fertilizers A ($Ca(NO_3)_2$ (1.444 g); KNO_3 (213 g); Fe-EDTA (38 g)), Fertilizer B ((KH_2PO_4 (260 g); $MgSO_4.7H_2O$ (673 g); $MnSO_4.4H_20$ (12 g); $CuSO_4.5H_20$ (0.28 g); $ZnSO_4.7H_2O$ (1.8 g); H_3BO_3 (4 g); $(NH_4)_6Mo_7O_{24}$ (0.1 g); Na_2O_3Si (329 g)).

Figure 2. The Process of Transplanting and Installation of Paddy Hydroponic (Wick System modification with auto pot)

3. Results and Discussion

3.1. Plant Height

Based on the results of the F test, silicate rice husk extract and synthetic silicate (Na_2O_3Si) were not significant effect with the plant height parameters (Table 1). The result of paddy height hydroponic (Ciherang variety) modified wick system at 21 day after planting (DAP) is lower than Amrullah *et al.* [8] on rice (Ciherang variety) aged 12 DAP (48.67 cm) and 28 DAP(73.40 cm) in the control treatment without silicate application. The height of paddy plants in this study is shorter compared to conventional paddy growth.

Application of paddy plant nutrition in accordance with the growth phase in this research have not gain expected growth results. Hydroponic technology according is one technology to increase growth and yield [14, 15]. The cultivation of hydroponic paddy according to Köhl has considerable difficulty compared to vegetable hydroponics [16]. Basically paddy plants have a high adaptability to low oxygen content and high $Fe2^{+}$ content. In this study modification of wick system installation and silicate application has not resulted an optimum growth response.

Table 1. Average Height Parameter at 21 DAP

Treatments	Average (cm)
A	19.96^{ns}
B	21.85^{ns}
Coefficient of Variance (CV)	12.17 %

Remarks: ns = non-significant at 5 % significant level F test

The visual appearance of the plants in this study (Fig.3) the leaves showed symptoms of chlorosis, stunted growth and roots that grew with "browning symptoms" of brownish roots. Paddy plants with healthy growth have a dark green leaves appearance and white roots. Based on diagnostic key for identifying nutrient deficiencies in rice from Dobermann & Fairhurst paddy plants show symptoms of N and Fe deficiency [17]. The symptoms of N deficiency in this study were suspected because N and Fe uptake were inhibited, the nutrient source used in this study used a N-total formulation of 249 ppm lower than that of rice nutrition used by Júnior, *et al.* N-total formulation of 263 ppm [18]. The inhibition of N uptake in this study was influenced by the average pH of the treated nutrient A (6.17) and B (6.71). Recommended pH optimum for hydroponic rice growth according to Ying-Hua *et al.* [19] the level of acidity of nutrient maintained at pH 5.5.

The results of this study confirm the results of previous studies show that plant height is affected by pH and EC value. Hydroponic technology has many advantages over conventional cultivation but not automatically produce quality plant [20].

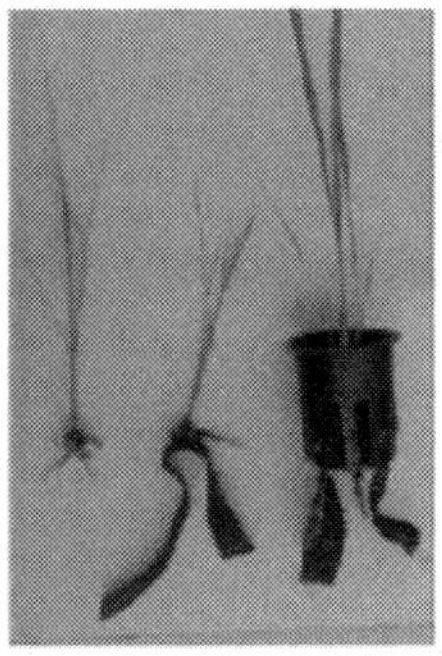

Figure 3. The condition of the hydroponic rice plant

The growth of paddy crops in this study has not gain expected results. The use of rockwool as a planting medium in paddy plants has not been optimal in supporting plant growth. Utilization of organic planting medium is expected to help plant growth. The research results show that composting and rice husk charcoal medium (1: 1 v/v) can increase plant growth [21] .

3.2. Nutrient solution pH value

Water quality is critical to the successful plants production in hydroponics [22]. Adjustment pH value and EC value of the nutrient solution affect the nutrient uptake by plants. The pH value is an indicator of the acidity or alkalinity of the nutrient solution, the acid nutrient solution causes the plant cannot absorb macro elements maximally otherwise if the nutrient solution alkaline plant can absorb macro elements excessively and uptake of micro elements by the plant is inhibited neither acid or alkaline but close to neutral. The optimum pH value (5.5-6.5) required by the plant to absorb nutrients adequately according to its growth phase [20]. The optimum pH hydroponic paddy plants is in the range of pH values of 5.6 ± 0.2 [18]. The initial pH values (treatments A and B) used in this research were 6.5, after 21 DAP pH value change in the nutrient container (Table 2).

Table 2. Change of pH of Nutritional Solution after 21 DAP

Treatments	Average of pH value*)
A	6.17 a
B	6.71 b
Coefficient of Variance (CV)	2.23 %

Note : *) The numbers followed by the same letters (lowercase letters vertical in column) indicate no significant difference according to Duncan Multiple Range Test at 5% level

The addition of inorganic Si (Na_2O_3Si) significantly affected the change of pH of the solution after 21 DAP. The addition of Na_2O_3Si compounds is suspected cause a higher uptake of nutrient-enriched nutrient absorption than without Na_2O_3Si, so that the OH concentration increases.

3.3. Electrical Conductivity (EC)

EC values are widely used as parameters in hydroponic plant nutrition. The increase in EC values is affected by the concentration of ions in the concentrated solution. The concentration of nutrient solution of EC value increases and vice versa [23]. Fluctuations in EC values other than temperature-induced are also affected by decreased bicarbonate (HCO_3^-) during the process of photosynthesis (in Furtbach condition becomes evident uptake CO_2 for primary biomass during photosynthesis that reduction bicarbonate [24].

The initial EC value assigned to each treatment was 2 mS/cm (EC = 2). EC value after 21 DAP has increased (Table 3). The increase in EC values in this research was thought to be influenced by an increase in the temperature of the nutrient solution, the condition confirming the statement [25] that the value of EC will increase along with the increase in temperature. The EC's value of nutritional oceans should theoretically decrease because the ions present in the nutrient solution are absorbed by the plant. Increasing the value of EC compared to the initial EC value is a sign that plants experience barriers in absorbing nutrients. This condition is visually reinforced by the plants undergoing chlorosis and the addition of inhibited height (Figure 3). The variation of EC values on mustard plants (dicotyl) have an effect on plant height but in the rice plant (monocot) various EC value un-significant affect vegetative phase.

Table 3. Change of EC value after 21 DAP

Treatments	Average
A	2.758 b
B	2.442 a
Coefficient of Variance (CV)	5.20 %

The numbers followed by the same letters (lowercase letters vertical in column) indicate no significant difference according to Duncan Multiple Range Test at 5% level

4. Conclusions

The application of 20 ml/L rice husk silica extract can be used as an alternative source of Si for paddy plants. The role of Si element in vegetative phase do not affect paddy plant growth especially plant height.

Acknowledgements

We would like thank to Rector of UIN Sunan Gunung Djati and LP2M UIN Sunan Gunung Djati who has provided research grant (BOPTN 2018).

References

[1] Pusat Data dan Sistem Informasi Pertanian 2015 Outlook Komoditas Pertanian Subsektor Tanaman Pangan Jakarta : *Pusat Data dan Sistem Informasi Pertanian Kementerian Pertanian.*

[2] Yamori W, Zhang G 2014 Feasibility Study of Rice Growth in Plant Factories *Rice Res. Open Access* **2:** 1–6.

[3] Godfray H C J 2014 Food Security: The Challenge of Science **2(1):** 812–818.

[4] Andani A 2008 Analysis rice production and rice consumption in Indonesia *Agrisep* **8:** 1–20.

[5] Husnain, Rochayati S, Adamy I 2012 Pengelolaan Hara Silika pada Tanah Pertanian di Indonesia *Pros. Semin. Nas. Teknol. Pemupukan dan Pemulihan Lahan Terdegradasi* **12:** 237–246.

[6] Qurrohman B F T 2017 Formulasi Nutrisi Hidroponik AB Mix dengan Aplikasi MS Excel dan Hydrobuddy Yogyakarta: *Plantaxia.*

[7] Cuong T X, Ullah H, Datta A, Hanh T C 2017 Effects of Silicon-Based Fertilizer on Growth, Yield and Nutrient Uptake of Rice in Tropical Zone of Vietnam *Rice Sci* **24:** 283–290.

[8] Amrullah, Sopandie D, Sugianta, Junaedi A 2014 Peningkatan Produktivitas Tanaman Padi (

Oryza sativa L.) melalui Pemberian Nano Silika Increased Productivity of Rice Plants (*Oryza sativa* L .) through The Application of Nano Silica *PANGAN* **23:** 17–32.
[9] Chandra A, Miryanti Y I P A, Widjaja L B, Pramudita A 2012 Isolasi dan karakterisasi silika dari sekam padi Bandung.
[10] Shen Y 2017 Rice husk silica derived nanomaterials for sustainable applications *Renew. Sustain. Energy Rev* **80:** 453–466.
[11] Laboratorium Kesuburan Tanah dan Nutrisi Tanaman 2018 Hasil Analisis Pupuk Cair Organik Jatinangor.
[12] Suka I G, Simanjuntak W, Sembiring S, Trisnawati W 2008 Karakteristik silika sekam padi dari provinsi lampung yang diperoleh dengan metode ekstraksi *MIPA* **37:** 47–52.
[13] Peckenpaugh D 2004 Hydroponic Solutions: Volume 1: Hydroponic Growing Tips Orlando: *Tom Alexander*.
[14] Domingues D S, Takahashi H W, Camara C A P, Nixdorf S L 2012 Automated system developed to control pH and concentration of nutrient solution evaluated in hydroponic lettuce production. *Comput. Electron. Agric* **84:** 53–61.
[15] Hidayat C, Pahlevi M R, Taufiqqurahman B F, Ramdhani M A 2018 Growth and yield of chili in nutrient film technique at different electrical conductivity *IOP Conf. Ser. Mater. Sci. Eng.* **288:** 1 - 4
[16] Köhl K 2015 Growing rice in controlled environments *Ann. Appl. Biol.* **167:** 157–177.
[17] Dobermann A, Fairhurst T 2000 Rice: nutrient disorders & nutrient management.
[18] Zanão J L A, Fontes R L F, Neves J C L, Korndörfer G H, Ávila V T 2010 Rice Grown in Nutrient Solution With Doses of Manganese and Silicon Rice Grown in Nutrient Solution With Doses of Manganese and Silicon *Rev. Bras. Ciência do Solo* **34(1):** 1629–1639.
[19] Ying-Hua D, Zhang Y L, Qi-Rong S, Song-Wei W 2006 Nitrate Effect on Rice Growth and Nitrogen Absorption and Assimilation at Different Growth Stages *Pedosphem* **16:** 707–717.
[20] Kulkarni S, Abraham P S, Mohanty N 2018 Techno-Societal 2016: 117–125.
[21] Embarsari R P, Taofik A, Frasetya B 2015 Pertumbuhan dan Hasil Seledri (*Apium Graveolens* L.) pada Sistem Hidroponik Sumbu dengan Jenis Sumbu dan Media Tanam Berbeda *J. Agro* **2:** 41–48.
[22] Van Os E, Blok C, Voogt W, Waked L 2016 Water quality and salinity aspects in hydroponic cultivation 1–7.
[23] Schmidt C, Musolff A, Trauth N, Vieweg M, Fleckenstein J H 2012 Transient analysis of fluctuations of electrical conductivity as tracer in the stream bed *Hydrol. Earth Syst. Sci* **16:** 3689–3697.
[24] Ort C, Siegrist H 2009 Assessing wastewater dilution in small rivers with high resolution conductivity probes *Water Sci. Technol* **59:** 1593–1601.
[25] Bai W, Kong L, Guo A 2013 Effects of physical properties on electrical conductivity of compacted lateritic soil *J. Rock Mech. Geotech. Eng* **5:** 406–411.

Utilize Open Space of Oil Palm (*Elaeis Guineensis* Jacq.) Plantation at Immature Stages for Growing Maize (*Zea Mays* L.) which is Applied Biofertilizer

V Z C Suherman[1], A Nuraini[1], Yuditri R E[1], L Batu[1] and M A Soleh[1]

[1]Departement of Agronomy Agricultural Faculty, Universitas Padjadjaran

Email: cucu.suherman@unpad.ac.id

Abstract. Land optimizing in oil palm plantation could be conducted by adopting polyculture cropping system of maize grown under oil palm trees during immature stage (IS) due to wider plant distance. The maize growth depends on genetic and environmental condition such as soil fertility, in fact oil palm plantation is often located in marginal area which is lack of fertility like Inceptisol soil type in Jatinangor, so that it needs supply of biofertilizer to improve soil fertility. This study was to get interaction effect between biofertilizer application and kind of maize varieties that is giving best effect on growth and yield of maize under polyculture system of immature oil palm trees. The experiment was conducted from November 2017 to February 2018. The experimental design was using split plot which was biofertilizer as a main plot and biofertilizer application as a subplot. The result showed all varieties of maize gave significant effect on plant height, leaves number, and cob size. Independently, maize variety of DR6 gave the best effect on plant height at 4 and 6 WAP (weeks after planting), DR1 gave the best effect on leaves number at 4, 6, 8 WAP, and DR5 gave the best effect on cob size.

1. Introduction

Oil palm tree is one of important plantation crop due to its contribution to Indonesian national revenue in vegetable oil production. [1] reported oil palm is popular product in many industries, this is followed by increase in land use of oil palm plantation. For an example in 2016 the oil palm plantation field in Indonesia reached to 11,300,307 ha consisted of 22.74% of immature trees, 75.2% produced trees, and 2.06% unproduced trees [2].

Wider plant distance of palm oil i.e. 9 m x 9 m, there is open spaces between trees which could be grown another crop such as maize. [3] reported immature oil palm trees (0-12 months) there is 75% of open spaces among the trees, whereas at the later stage of immature trees (13-24 months) there is 60% of open spaces which is potential to adopt polyculture system in the oil palm plantation. Generally, under monoculture system of oil palm, the open space uses for growing soil cover crop such as legume, however under polyculture system, soil cover crop could be used edible crop so that the beneficial may increase rather than under monoculture system. Polyculture system gives some beneficial in economic and ecological sector [4]. There are some conditions that required for beneficial value under polyculture system namely the crops should be complements and or supplements each other's, and the crops have economical value [5].

Maize is popular crop grown under polyculture system due to its beneficial both are economic and ecology, additionally, in Indonesia maize is main staple food besides rice [6]. National consumption of maize tends to be increases by increases people population and industries of food and feed. The consumption has reached to 22.08 million ton, however, the national production just reached to 20.31 million ton or less than its consumption. In other situation, the maize productivity of farmers is quite low reached 3-4 ton/ha, so that the polyculture system between oil palm tree and maize will be the best alternative for increase maize production in the future. To do best practice in polyculture cropping system, selection of crop e.g. variety and application of fertilizer which is enriching soil fertility will be needed to get proper growth and development of crop under such condition. One of maize varieties that is developed by University Padjadjaran namely: DR Unpad known as improved variety with advantageous traits of shading tolerant will be proper to grown under polyculture system [7]. This study was to evaluate maize growth performance under oil palm tree in polyculture cropping system.

2. Materials and methods

The experiment was conducted at the experimental station Faculty of Agriculture Universitas Padjadjaran-Bandung from October 2017 to March 2018. Soil type was Inceptisol and the land location at the 752 m above sea level Oil palm tree varieties of SEU Supreme were grown with plant distance of 8 m x 8 m at 1 year before growing maize in polyculture system in the land of 1 hectar. Seven maize varieties were used in this experiment consisted of DR-1, DR-2, DR-3, DR-4, DR-5, DR-6, and DR-7. These varieties are developed by Faculty of Agriculture Universitas Padjadjaran-Bandung. Maize was grown in between oil palm trees row with plant distance of 75 cm x 20 cm, so that it was 2 m distance to oil palm trees. Manure and inorganic fertilizer of NPK were added to the soil as a basic fertilizer in amount of 10 kg per plant of oil palm trees and 0.5 kg per plant of maize. All treatments in the experiments were designed to ensure that plants are not lack of nutrients at any point.

2.1. Biofertilizer application

Biofertilizer of Bion-up BION-UP (Pupuk Kujang Corp. INA), which contains living microbes, was applied to the soil in around root zone at two doses of application i.e. 0 and 150 ml per plant at 14, 21, and 28 days after sowing (DAS).

2.2. Measurements

Maize growth performance was measured for plant height, leaves number, yield and yield component. Another measurement was conducted for stomatal conductance (g_s) by using porometer (Decagon device Inc. US).

Stomatal conductance were measured at the 6 weeks after sow (WAS) to 2nd leaf from upper of maize. All measurement of gs were conducted on 4 leaves for each maize variety.

2.3. Yield and yield component measurement

Harvest of maize was conducted at the 13 WAS by collected cob then it was separated from grains to measure 100 grains weight.

2.4. Experimental design and statistical measurements

Experimental design was using split plot which was biofertilizer as a main plot and varieties as a subplot. There were 2 level of biofertilizer application i.e. b_1: application biofertilizer of 150 ml per plant; b2: 0 application of biofertilizer. Varieties contained of 7 level i.e. c_1: DR-1; c_2: DR-2; c_3 : DR-3; c_4 : DR-4; c_5 : DR-5; c_6 : DR-6 and c_7 : DR-7. In total there was 7 x 2 treatments and 3 replications. Significantly difference of data was analysed by LSD-Tukey tests, using SAS program.

3. Results and discussion

All measurement on plant such as plant height, leaf numbers, stomatal conductance and yield had shown no interaction between biofertilizer and variety of maize, even if there were significantly difference on the independent response of varieties and or biofertilizer application according to statistical analyses.

3.1. Plant height

At the 8 weeks after sowing, plant height of c_6 was significantly higher than c_2 and c_3, even if there were no significantly difference to DR 1, DR 4, DR 5, and DR 7 (Table 1).

Table 1. The effect of biofertilizer and varieties of maize on plant height at 4, 6, and 8 WAS

Treatments	Average of Plant Height (cm)		
	4 WAS	6 WAS	8 WAS
Biofertilizer			
b_1 = 150 ml/plant	63.03 a	115.89 a	164.49 a
b_2 = 0 ml/plant	59.81 a	110.23 a	163.98 a
Variety of maize			
c_1 = DR 1	62.91 a	114.56 a	169.01 ab
c_2 = DR 2	66.22 a	114.88 a	157.72 b
c_3 = DR 3	59.59 a	111.89 a	153.65 b
c_4 = DR 4	59.75 a	110.38 a	163.59 ab
c_5 = DR 5	58.22 a	111.41 a	170.20 ab
c_6 = DR 6	64.83 a	119.11 a	176.09 a
c_7 = DR 7	58.44 a	109.19 a	159.38 ab

Same letters in same column indicates no significantly difference according to LSD Tukey test 5%.

There was no significance interaction between biofertilizer and varieties of maize, it might be due to several reasons e.g. improper dose, idle of microbes, and or soil leaching during heavy rain. In the experiment, rainfall average had been recorded up to 275.8 mm/month, it is higher than an ideal rainfall for maize cultivation namely 85-200 mm/month [8]. Maize variety itself had shown significantly difference on plant height due to difference on the traits which responded to growth environment.

There was significantly difference on leaf number of maize at 4 and 6 WAS (Table 2), variety of c_1 was higher than c_4, however it was not significantly difference with other varieties. Plant height at 8 WAS showed no significantly difference among the variety.

Table 2. The effect of biofertilizer and varieties of maize on leaf number at 4, 6, and 8 WAS

Treatments	Average of leaf number		
	4 WAS	6 MST	4 WAS
Biofertilizer			
b_1 = 150 ml/plant	7.02 a	9.29 a	10.31 a
b_2 = 0 ml/plant	6.93 a	9.25 a	10.19 a
Variety of maize			
c_1 = DR 1	7.44 a	9.88 a	10.58 a
c_2 = DR 2	6.94 ab	9.12 ab	10.00 a
c_3 = DR 3	7.00 ab	8.88 ab	10.04 a
c_4 = DR 4	6.50 b	8.81 b	10.08 a
c_5 = DR 5	7.06 ab	9.44 ab	10.25 a
c_6 = DR 6	7.19 ab	9.25 ab	10.54 a
c_7 = DR 7	6.69 ab	9.50 ab	10.25 a

*Same letters in same column indicates no significantly difference according to LSD Tukey test 5%.

3.2. Leaf number

DR 4 has lowest leaf number, it is possible due to its traits of higher in yield potential compared to DR 1 (Table 4), too many leaves will make inefficient on photosynthesis distribution so that will affect on yield [9]. High in leaf number is usually followed by high in yield, however, there is ideal number of leaf which could be increase efficiency of photosynthesis. This result is needed to be clarified furthermore how many leaf ideally could improve high efficiency of yield. Application of Bion up did not improve leaf number due to other study on biofertilizer of Bion up showed the application of biofertilizer could improve leaf canopy, plant height and yield [10].

Table 3. The effect of biofertilizer and varieties of maize on stomatal conductance at 8 WAS

Treatments	Average Stomatal Conductance (mmol/m^2/s)
Biofertilizer	
b_1 = 150 ml/plant	394.58 a
b_2 = 0 ml/plant	354.00 a
Variety	
c_1 = DR 1	307.23 a
c_2 = DR 2	435.14 a
c_3 = DR 3	289.38 a
c_4 = DR 4	339.43 a
c_5 = DR 5	382.56 a
c_6 = DR 6	449.18 a
c_7 = DR 7	417.14 a

Same letters in same column indicates no significantly difference according to Tukey test 5%.

3.3. Yield and yield component

Cob length of DR 5 was a significantly longer than DR 1, DR 2, and DR 4, however, it was not significant difference to DR 3, DR 6, and DR 7 (Table 5).

Table 4. Yield potential of maize variety of DR

Yield potential of maize (ton/ha)	
DR 1	3.16
DR 2	2.47
DR 3	1.45
DR 4	4.63
DR 5	2.76
DR 6	6.45
DR 7	4.91

Source: Description of DR maize [7]

3.4. Stomatal conductance (g_s)

Application biofertilizer did not give significantly difference on stomatal conductance among maize varieties (Table 3), even though variety of DR6 showed a tendency of higher in gs compared to other, it might be due to DR6 has a highest yield potential among six varieties, conversely, DR 3 has tendency a lowest in gs among the varieties due to lowest in yield potential among six varieties (Table 4), even if this result was not significant difference. Stomatal conductance could be explains plant photosynthetic response due to its role on CO_2 diffuse on leaf. High in gs means high in CO_2 to diffuse in the leaf so that it will high in photosynthesis activity as well. Value of g_s was more than 250 mmol m^{-2} s^{-1} indicates there were no stress on plants.

Table 5. The effect of biofertilizer and varieties of maize on cob length and diameter at 13 WAS

Treatments	Average (cm)	
	Cob length	Cob diameter
Biofertilizer		
b_1 = 150 ml/plant	18.44 a	4.22 a
b_2 = 0 ml/plant	18.86 a	4.23 a
Variety of maize		
c_1 = DR 1	17.91 b	4.21 a
c_2 = DR 2	17.92 b	4.16 a
c_3 = DR 3	19.15 ab	4.12 a
c_4 = DR 4	18.16 b	4.34 a
c_5 = DR 5	20.18 a	4.19 a
c_6 = DR 6	19.03 ab	4.30 a
c_7 = DR 7	18.30 ab	4.27 a

Same letters in same column indicates no significantly difference according to LSD Tukey test 5%.

Cob length effects on seed number, consequently, plant needs higher energy for producing more grain, it is well known balance of sink and source could be result in yield. Cob length if it does not followed by high in source i.e. photosynthetic leaf will show in low yield [11], this could be seen at the longer cob of DR 3 and DR 5 did not correlated with high yield such as in DR 6 (Table 4). In facts, genetic diversity of maize varieties and their interaction effect on yield [12].

3.5. A hundred grains weight

Table 6. The effect of biofertilizer and varieties of maize on 100 grains per cob (g) at 13 WAS

Treatments	Average
	100 grains
Biofertilizer	
b_1 = 150 ml/plant	17.24 a
b_2 = 0 ml/plant	16.46 a
Variety	
c_1 = DR 1	15.32 a
c_2 = DR 2	16.77 a
c_3 = DR 3	16.24 a
c_4 = DR 4	17.27 a
c_5 = DR 5	17.86 a
c_6 = DR 6	16.93 a
c_7 = DR 7	17.54 a

Same letters in same column indicates no significantly difference according to Tukey test 5%.

There were no significantly differences on a hundred grains among DR 1 – DR 7. A hundred grain was ranged from 15.32 g of DR 1 to 17.86 g of DR 5 (Table 6). Grain weight of maize is affected by temperature, proper temperature for maize growth and development ranged from 18 °C to 27 °C. In this experiment temperature average was 23.3 °C.

Further research is needed to clarify best dose of biofertilizer at some levels to get clearer effect on maize growth and development under oil palm polyculture cropping system. Maize could be grew at the immature stage of oil palm with a productivity without reduce its productivity.

4. Conclusion

Maize cultivation under polyculture system at the immature stage of oil palm tree could be done to increase land utilities without any disadvantages of plant growth and yield.

References

[1] Setyamidjaja D 2006 Teknik Budidaya, Panen dan Pengolahan Kelapa Sawit *Kanisius* Yogyakarta.

[2] Direktorat Jenderal Perkebunan 2015 Statistik Perkebunan. Departemen Pertanian. Direktorat *Jenderal Perkebunan Indonesia* Jakarta pp. 4.

[3] Mawarni L 2011 Kajian Awal Varietas Kedelai Tahan Naungan untuk Tanaman Sela pada Perkebunan Kelapa Sawit *Jurnal Ilmu Pertanian Kultivar* 5(2): 54 – 59.

[4] Armaini, Ariani E, Yoseva S, Anom E 2012 Optimalisasi Produksi Kedelai [*Glycine max* (L) Merril] pada Kebun Kelapa Sawit di Lahan Gambut dengan Aplikasi Beberapa Komposisi Pupuk dan Pembenahan Tanah *J. Agroteknologi Tropikal* 1(2): 11-15.

[5] Yuwariah Y 2011 Dasar-Dasar Sistim Tanam Ganda *Universitas Padjadjaran* Bandung.

[6] Riwandi, Merkati H 2014 Teknik Budidaya Jagung Dengan Sistem Organik Di Lahan Marjinal.

[7] Ruswandi D, Saraswati M, Elia 2017 Deskripsi Varietas Jagung DR *Universitas Padjadjaran*.

[8] Zulkarnain M F, Karamoy T L, Jody M W 2017Analisis Ketersediaan Air untuk Tanaman Tomat (*Lycopersicum esculentum* Mill) dan Jagung (*Zea mays* L.) di Tonsewer *Jurnal Cocos* 1(5) : 1-20.

[9] Gardner 1991 Fisiologi Tanaman Budidaya *Universitas Indonesia Press* Jakarta

[10] Suherman C, Soleh M A, Nuraini A, Annisa N F 2018 Pertumbuhan dan hasil tanaman cabai (*Capsicum* Sp.) yang diberi pupuk hayati pada pertanaman kelapa sawit (*Elaeis guineensis* Jacq.) TBM I *Jurnal Kultivasi* 17(2): 648-655.

[11] Zhang H, Turner C N, Poole L M 2010 Source–sink balance and manipulating sink–source relations of wheat indicate that the yield potential of wheat is sink-limited in high-rainfall zones *Crop and Pasture Science* 61(10): 852-861.

[12] Wahyudin A, Ruminta, Bachtiar D C 2015 Pengaruh Jarak Tanam Berbeda pada Berbagai Dosis Pupuk Organik terhadap Pertumbuhan dan Hasil Jagung Hibrida P-12 di Jatinangor *Jurnal Kultivasi* 14(1): 1-8.

IOP Conf. Series: Earth and Environmental Science **334** (2019) 012028 doi:10.1088/1755-1315/334/1/012028

Coal Utilization as a Growth Medium of Microbial Consortium from Dairy Cow Feces

E Harlia[1], E T Marlina[1], Y A Hidayati[1], T B A Kurnani[1], K N Rahmah[1], I M Joni[2] and R Ridwan[3]

[1]Faculty of Animal Husbandry, Universitas Padjadjaran, Jl. Raya Bandung Sumedang km 21, Jatinangor, West Java 45363, Indonesia
[2]Dept.of Physics, Faculty of Mathematics and Natural Sciences, Universitas Padjadjaran, Jl. Raya Bandung Sumedang km 21, Jatinangor, West Java 45363, Indonesia
[3]Research center for Biotechnology, Lembaga Ilmu Pengetahuan Indonesia, Jl. Raya Bogor Km 46, Cibinong Bogor, West Java 16911, Indonesia

E-mail: ellin.harlia@unpad.ac.id

Abstract. Many uneconomical coal mining areas which have low calorific value can be found in Kalimantan, Indonesia. In those areas, coalbed methane (CBM) mainly methane (CH_4) is formed, trapped and accumulated in the pores or cleats during the lifetime coal formation. This condition can potentially be utilized as an alternative energy source. Adding a source of methanogenic bacteria and organic matter is an option to maximize methane in CBM. High carbon content causes young coal to become a source of nutrients for methane-forming microorganisms. Methanogen type microorganisms play an important role in the manufacture of CBM and they can only work on anaerobic conditions. The purpose of this research is to know the relationship between coal characteristic and coal function as microbial growth media to gas formation which can be utilized as environmentally friendly alternative energy. The research used explorative method with descriptive analysis. The research parameters were three coal characteristics, methane production, and Volatile Fatty Acids. The results indicated that coal characteristics support symbiosis and microbial growth with total bacterial counted > 10^{10} CFU/ml, while the amount of acetic acid larger than butyric acid and propionic acid which support sustainable methane gas formation.

1. Introduction

Coal is a complex heterogeneous stone with various physics and chemical characteristics. It composes of organic and inorganic materials. The organic content of coal consists of carbon, oxygen, hydrogen, and sulfur. The carbon content is around 65%-95% of coal and increases as the coal-forming process. During coal formation process, the level of oxygen and hydrogen decrease approximately about 2%-30% and 2%-7% respectively while nitrogen and sulfur decrease around 1% -4% [1]. Microorganisms can grow on coal utilizing organic material contained in coal as a substrate. Complex microbes degrade C_3, n-C_4, hydrocarbon liquids, and solid organic mixtures contained in coal, to the stage of biogas formation [2]. Coalbed Methane (CBM) can be offer as a new and renewable energy source. Based on potential study conducted on 10 coal basins, Indonesia has a potential CBM of 337 tcf [3]. This is a potential to be developed, considering Indonesia is also one of the countries with the largest coal reserves in the world. CBM is methane contained in coal and trapped in micro pores of coal through biogenic or thermogenic process during coal formation at 600-1000 m depth. CBM is a clean and renewable energy source, as it will continue to be produced as long as coal seams are available. The yield of coal excavation will produce the remains of coal that is left piled up. This coal can be utilized as a medium of microbial growth in biogas digester through bio augmentation and bio stimulation process.

Bioaugmentation is carried out by using microbial feces from livestock waste with addition of coal (lignite, bituminous, and subbituminous) media. Characteristics of coal include cleats, pores and permeability. The content of organic matter in coal can be used as nutrient source to support microbial growth so that methane production increases with a longer time. Similar process on anaerobic fermentation such as hydrolysis, acidogenesis, acetogenesis and metanogenesis occurs in coal. The similarity of process with anaerobic digestion becomes the basis of bioaugmentation using microbial from livestock waste. The purpose of this study is to determine the characteristics of lignite, bituminous and subbituminous coal, and to what extent coal can be utilized as a medium for microbial growth from dairy cattle waste.

2. Materials and methods

2.1. Materials

The research materials were three types of coal namely lignite, bituminous and subbituminous. The data of coal characteristic obtained through scan process on each coal. Other research materials used were dairy cow feces, medium 985 for anaerobic microbes, and diluent solution no. 14 [4].

2.2. Research procedure

The research procedures were divided into several steps. First, preparation of dairy cow feces sample. Sample of feces was taken from dairy cow farm in Faculty of Animal Husbandry, Universitas Padjadjaran. Fresh feces was taken from the middle part of feces pile which contained relatively low oxygen contamination. After that, sample of feces was squeezed using gauze so that the liquid part separated and put into closed flask. Second, activation of microbes from dairy cow feces. Third, adaptation of microbes on enriched medium. Fourth, preparing of dairy cow feces microbial inoculum. Fifth, addition of microbial consortium from dairy cow feces into coal medium. As much as 70 ml of medium 985 were put into a 100 ml bottle, followed with coal about 10% of media volume. Microbial consortium was added using syringe. The study was conducted with six treatments of microbial consortium addition, namely 0%, 2%, 4%, 6%, 8%, and 10% of medium volume with triplicates. Each sample was observed on day 2, 5, 10, and 15. CO_2 was passed during the process to create anaerobic condition.

2.3. Research methods

This research was used explorative methods with descriptive explanation. Parameter observe were coal characteristics (porosity and permeability), total number of microbes, volatile fatty acids (VFA), and methane content.

3. Result and discussions

3.1. Coal characteristics

Coals were obtained from different location. Bituminous coals (Tanjung formation) were obtained from Muser Village, Paser District, East Kalimantan. Subbituminous coals (Pamaluan formation) were obtained from Liburdinding Village, Paser District, East Kalimantan. Lignite coals (Sajau formation) were obtained from Kasai Village, Berau District, East Kalimantan. The analysis results of coal characteristics can be seen in Table 1.

Table 1 described the higher rank of coal, the lower the coal's ability to store and passed fluids rich in volatile substances, including methane. Methane in coal was stored through absorption. If porosity and permeability were low, the liquid would not be stored properly. Lack of absorbed fluid would also reduce methane production by coal bed [5].

Lignite had an average distance between clit (b) 0.55 mm with standard deviation 0.0079 mm; opening clit 0.12 mm. While subbituminous had average distance between clit 0.32 mm with standard deviation 0.0336 mm; clit openings ranged from 0.05 to 0.12 mm with an average of 0.0247 mm. Bituminous coal (Tanjung formation) had a distance between clit range of 0.29 to 0.39 mm with an average of 0.34 mm; with the average value of clit opening was 0.07 mm with standard deviation 0.0317 mm. The presence of clit would support the escape of liquid from the coal bed. The higher the maturity level of coal, the less clit visible in coal was. Lignite had the largest average distance between clit compared to bituminous and subbituminous coal.

The gas content in coal increased with the additional level of coal hardness associated with the maturity level of coal. The permeability value would decrease along with the addition of coal hardness level. The pores were part of the total volume of coal that could be filled by the fluid. Coal pores were classified by macropores (> 500Å), mesopores (20-500Å), and micropores (8-20Å). The volume and size of the pores would shrink as the maturity level of coal increases. Most macropores were filled with water, free gas, and water-soluble gases. In the micropores structure the water rate and permeability capacities were low, whereas in the clusters of water rate and permeability capacity were larger therefore coal was a reservoir having a double porosity system [6]. The observation of coal characteristics such as porosity, permeability and clit appearance could be a basis for utilizing coal as a growth medium for microbial consortium from dairy cow feces.

Tabel 1. Sampling site and coal characteristics

	Type of coals		
	Bituminous	Subbituminous	Lignit
Coal characteristics	Black in color, black scratched, glossy glass, containing many fractures, hard and solid, conchoidal shards	Black colored, black scratch, glossy glass, contains many fractures, compact, conchoidal shards.	Black brownish colored, black scratch, glossy earth-glass, containing fractures, soft, conchoidal shards.
Permeability	30.2 md[a]	7.05 md[a]	37.21 md[a]
Porosity	9.2 %	5.20 %	9.20 %

[a] md: millidarcy

3.2. The number of anaerobic bacteria

Microorganisms can grow on coal using the organic material contained as a substrate. Complex microbes degrade C_3, n-C_4, hydrocarbon liquids, and solid organic mixtures contained in coal until the stage of biogas formation [7]. The proximate analysis by Applied Microbiology Laboratory-Biotechnology LIPI, Cibinong, showed that sample of coal contained 88.53% dry matter component, 11.47% water content, 58.05% ash and 3.54% total nitrogen. These components support the growth of microbes. Data on the number of anaerobic bacteria is shown in Figure 1.

The highest amount of anaerobic bacteria on the lignite medium was achieved at a concentration of 4%, while the highest anaerobic bacteria on subbituminous and bituminous coal media were achieved at a concentration of 6%. Increase in the number of bacteria and the presence of gas formation showed that the microbes from dairy cattle feces can adapt in coal media and the existence of microbial symbiosis between dairy cow feces microbes and coal microbe. Basically methanogens are microorganisms that can only utilize simple carbon components as substrates. During the process of biogas formation of coal, methanogens need to collaborate with other bacterial species that can degrade complex organic materials in coal to a simpler carbon component [8].

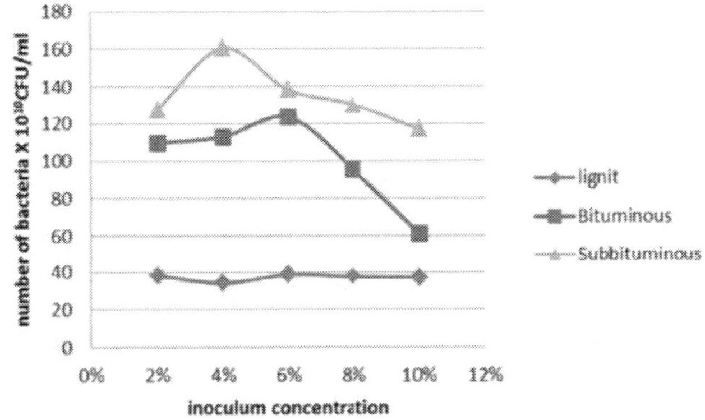

Figure 1. The number of anaerobic bacteria

3.3. VFA production

The formation of VFA is started by a consecutive processes namely hydrolysis, acidogenesis and acetogenesis. VFA, apart from acetate, specifically propionate and butyrate is converted into acetate and hydrogen by acetogenic bacteria such as *syntrophomonas* or *syntrophobacter*. Acetate and hydrogen are used by methanogens to form methane [9]. Generally, the accumulation of VFA in the biogas formation process can decrease the pH value, but this depends on the buffer content formed, and the type of biomass used. Methane formation can be inhibited by high VFA concentration [10]. VFA production can be seen in Figure 2 to Figure 4.

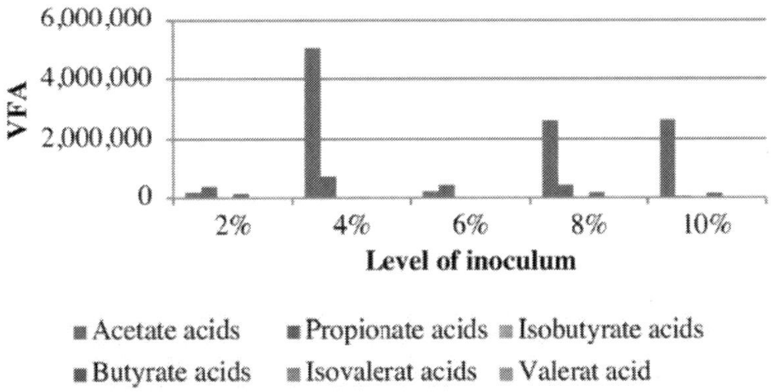

Figure 2. VFA content in lignite coal media

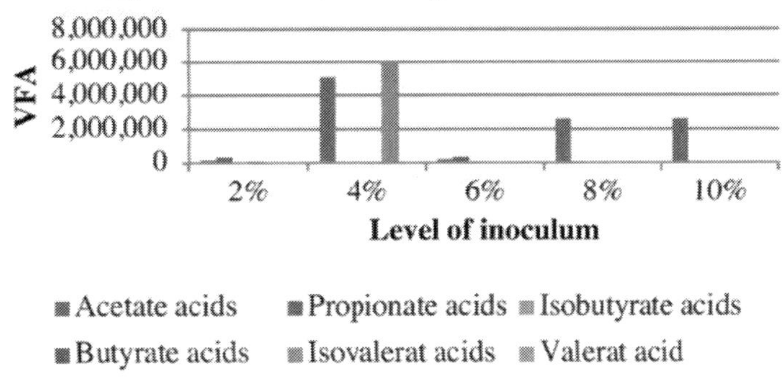

Figure 3. VFA content in bituminous coal media

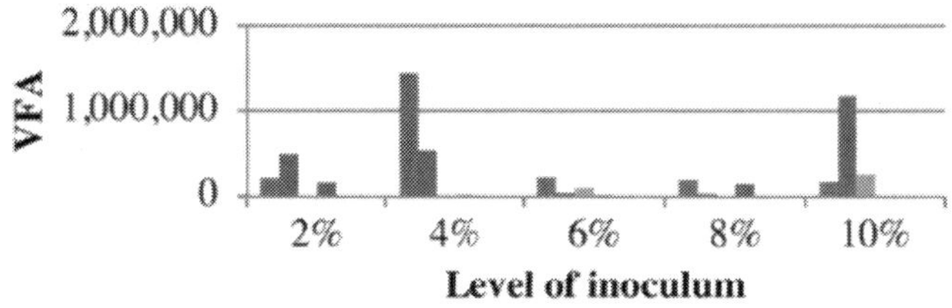

Figure 4. VFA content in subbituminous coal media

The highest VFA produced on the lignite medium was acetic acid. In acetogenetic phases about 25% acetate and 11% hydrogen was formed [11]. Acetate, H_2, CO_2, formate, methanol, methylamine, and dimethyl sulfide are substrate in methane formation. The process of converting polymers such as polysaccharides, proteins, amino acids, and fats to CO_2 and CH_4 was called methanogenesis that involves interaction of various species of microorganisms. Methane was formed through conversion of acetate, formate, and hydrogen into methane and CO_2 with the help of methanogenic microbes [11]. The process of acetate formation represented the effectiveness level of biogas production, because about 70% of methane was formed from the conversion of acetate to methane. Only a few types of bacteria were capable of producing methane from acetate, but the majority of CH_4 was formed from the process of digestion of acetate by heterotrophic methane bacteria [12]. Only about 30% of methane was formed from CO_2 and H_2 reduction by methane autothropic bacteria [13,14].

The concentration of VFA was aligned with the activity of microorganisms in substrate degradation [15]. The addition of 4% microbial inoculum in lignite and bituminous coal medium and addition of 6% microbial inoculum on subbituminous coal media showed optimal results in producing VFA. This related to the population and the ability of microbes to degrade organic matter contained in coal. Too dense microbial population caused competition in obtaining substrate, otherwise a small microbial population could cause non optimal substrate degradation. Competition between microorganisms because of population density might cause dominance in one species [16]. According to [17], if two species of microorganisms competed for a limited source of nutrients they could not co-exist with a constant population and a superior species would dominate.

Such dominance induced the production of VFA during incubation to be not maximal. VFA was formed by the activity and symbiosis between microbial species through 2 stages of formation. Stage one was the degradation of organic matter into simple sugars by microbial enzymes. The second stage was the conversion of pyruvic acid into VFA through microbial intracellular metabolism process. VFA results were influenced by the types of microorganisms that played role in degrading pyruvic acid into acetate, butyrate, or propionate [15]. There were at least 4 genera of bacteria consisting of *Bacteroides, Geobacter, Clostridium*, and *Sporochaeta* which were known to have a role in the hydrolysis geopolymer process in coal [8].

Coal organic content was converted into fatty acids with long chains and acetic acid through the process of bacterial respiration, while the single ring aromatics component of coal was fermented into phenol and benzoate compounds by fermentative bacteria. Long-chain fatty acids as well as phenol and benzoate compounds were then converted to propionate and butyrate through the activity of syntrophic bacteria [18].

3.4. Methane production

CH_4 is formed in two ways: acetic acid fermentation and CO_2 reduction using formic acid or hydrogen [19]. Acetate is formed during the acidification process of VFA formation and VFA conversion of long carbon chains through the process of acetogenesis [10]. In addition, [10] states that if VFA concentration too high, it may inhibit gas formation. High VFA concentrations should be accompanied by high methanogen populations so that VFA conversion process becomes balanced with VFA concentration levels. VFA that accumulates in the medium can decrease the pH value. pH is one of the environmental indicators that affect the growth of methanogens. Most methanogens grow optimally at neutral pH, but can still survive at a pH between 3-10 [10,20].

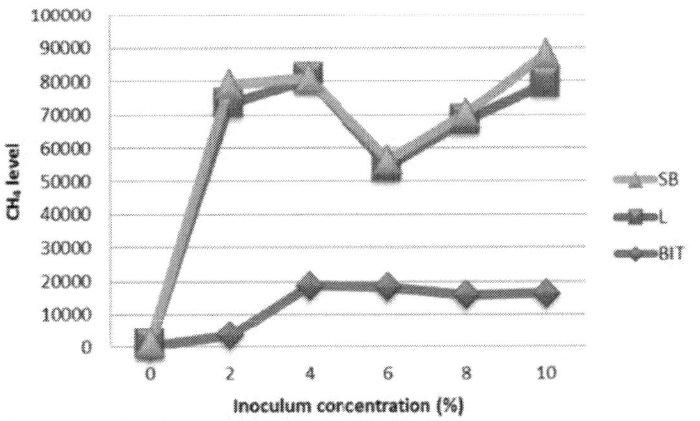

Figure 5. Methane content on various inoculum concentration

The addition of 4% inoculum treatment might increase methane gas production in lignite, bituminous and sub bituminous coal media. This was an indication that bioaugmentation could accelerate the biogenic process of biogas formation on the coal substrate. The addition of selected bacterial and methanogen consortiums could increase methane production by 0.1 μmol methane/day/g coal in the first 50 days of incubation period, and 3 μmol methane/day/g coal in the 50-70 day of incubation period [18]. Research conducted by [21], states that the addition of Na-Acetic nutrients can increase methane production to 5,034 mmol/g of coal on day 24.

4. Conclusion

Coal can be utilized as a growth medium for microbial consortium from dairy cow feces because it has porosity, permeability and clit that can accommodate liquid and moisture. Bioaugmentation with microbes from dairy cattle feces on the three types of coal media can produce VFA with the highest acetic acid to support the production of methane gas and the growth of sustainable anaerobic bacteria.

Acknowledgments

The authors thank to the Rector of Universitas Padjadjaran for the financial support under the Academic Leadership Grant (ALG)

References

[1] Takanohashi T 2009 Coal Structure and Properties *Coal, Oil, Shale, Natural Bitumen, Heavy Oil and Peat* **1**.

[2] Strapoc D, Aderhold J, Davydov V Yu, Fedler F, Klausing H, Mistele D, Rotter T, Semchinova O, Stemmer J, Graul J 2001 *J. Cryst. Growth* **222** 701 Metanogenic Microbial Degradation of Organic Matter in Indiana Coal Beds-Methane Producing Microbial Community in a Coal Bed of the Lillinois Basin *Journal of Applied and Environmental Microbiology* **74**.

[3] Arista M, Iskandar E 2006 Penentuan Peta Sebaran Potensi GMB (Sweet Spot Area) di Daerah Bukit Asam, Sumatra Selatan *Jurn. Mindagi* **10(1)**.

[4] Ogimoto K, Imai S 1980 Atlas of Rumen Microbiology Tokyo *Japan Scientific Societies Press.*

[5] Zhao Y, Cao S, Li Y, Zhang Z, Guo P, Yang H, Zhang S, Pan R 2018 RSC Adv **8** 5420-5432

[6] Sarsono A 2008 Kajian keekonomian pengembangan lapangan gas metana-B dengan mempertimbangkan manajemen pengelolaan air terproduksi *Universitas Indonesia.*

[7] Strapoc D 2008 Metanogenic Microbial Degradation of Organic Matter in Indiana Coal Beds-Methane Producing Microbial Community in a Coal Bed of the Lillinois Basin *Journal of Applied and Environmental Microbiology* **74**.

[8] Susilawati S S R 2015 Biogenic Methane in Indonesian Coal: Assessing The Potential Using Culture Enrichment, Molecular Phylogenetic and Isotopic Analysis Disertation *School of Earth Science* The University of Queensland

[9] Schink B 1997 Energetics of Syntrophic Cooperation in Metanogenic Degradation *Microb. Mol. Biol.* Rev **61**: 262-280

[10] Seadi T A 2008 Handbook of Biogas University of Southern Denmark Esbjerg *Niels Bohrs Vej* 9-10 DK-6700 Esbjerg Denmark (http://lemvigbiogas.com/ accessed at 25 Mei 2017)

[11] Krzysztof Z, Magdalena F 2012 Methane Fermentation Process as Anaerobic Digestion of Biomass: Transformations, Stages, and Microorganism *African Journal of Biotechnology* **11(18)**: 4127-4139

[12] Demirel B, Scherer P 2008 The roles of acetotrophic and hydrogenotrophic metanogens Turing anaerobic conversion of biomass to methane: a review Rev. *Environ. Sci. Biotechnol.* **7**: 173-190

[13] Griffin M E, McMahon K D, Mackie R I, Raskin L 2000 Metanogenic population dynamics during start-up of anaerobic digesters treating municipal solid waste and biosolids *Biotechnol. Eng.* **57**:342-355

[14] Karakashev D, Batson D J, Angelidaki I 2005 Influence of environmental conditions on methanogenic compositions in anaerobic biogas reactors *Appl Environ Microbiol.* **71(1)**: 331-338.

[15] McDonald P, Edwards R A, Greenhalgh J F D, Morgan C A, Sinclair L A, Wilkinson R G 2010 Animal Nutrition. 7th *Pearson Press.*

[16] Yoon H, Klinzing G, Blanch H W 1977 Competition for Mixed Substrates by Microbial Populations *Biotechnology and Bioengineering* **19**: 1193-1210

[17] Gauss G F 1934 The Struggle for Existence *Williams and Wilkins* Baltimore.

[18] Elizabeth J P Jones, Mary A Voytek, Margo D Corum, William H O 2010 Stimulation of Methane Generation from Nonproductive Coal by Addition of Nutrients or a Microbial Consortium *Appliedand Environmental Microbioogy* p. 7013–7022

[19] Campbell J 1983 Biomass Catalysts and Liquid Fuals *Holt Rainheart and Winston ltd* Pensylvania.

[20] Alperin M, Hoehler T 2010 The ongoing mystery of sea-floor methane *Science* **329(5989)**: 288-289 DOI: 10.1126/science.1189966

[21] Pingkan A, Agus P, Indra R, Harry R 2013 Effect of Stimulants on Biogenic Methane Formation and Dynamics of Bacteria Population *J. Math. Fund. Sci.* **45(3)**: 274-285.

The Effect of Green Tea (*Camellia Sinensis*) and Turmeric (*Curcuma Longa*) as Feed Additives on Performance of Local Sheep

D Heriyadi[1], H Yurmiati[1] and D Ramdani[1]

[1]Department of Animal Production, The Faculty of Animal Husbandry, Universitas Padjadjaran, Jatinangor Campus, Sumedang 45363, Indonesia

email: diky.ramdani@unpad.ac.id

Abstract. Studies on green tea and turmeric utilizations as natural additives in ruminant diets are still limited. In this experiment, randomized completely design was used to compare the effects of 4 experimental diets: T0 = control diet without green tea and turmeric powders; T1 = diet with 1% turmeric powder; T2 = diet with 2% green tea powder; and T3 = diet with 1% turmeric acid and 2% green tea powders on performance of local lambs. Based on the results, adding 2% green tea or 1% turmeric powders or their combination had no a significant effect (P>0.05) on feed intake (g DM/head/day) of local lambs confirming that their uses did not reduce diet's palatability. The two herbal inclusions had also no significant effect (P>0.05) on average daily gain (ADG, g/head/day) and carcass percentage (%).

1. Introduction

Tea drink is popular among people as a healthy consumption. Tea leaves is rich in antioxidants to inactivate free radical oxygen in the human body, reducing the risk of obesity, and preventing *cardiovascular* diseases [1,2,3,4]. On the other hand, turmeric is well-known for its uses as herbs and herbal medicine for human. However, study on green tea, turmeric, and their mixed utilizations as feed additives for ruminants have not been done intensively. Ramdani et al. [5] reported that green tea leaves contain considerable amounts of protein, fiber, minerals, and bioactive compounds especially tannins and saponins. Meanwhile, turmeric is rich in phenolic curcumin that has antioxidant and antibacterial properties. In broiler chicken, adding turmeric powder into a diet increased erythrocyte and leucocyte of blood cells of broiler chickens leading to improve their performances [6]. Therefore, this research aimed to test the effect of green tea (*Camellia Sinensis*) and turmeric (*Curcuma Longa*) powders as additives in a sheep diet on feed intake, ADG, and carcass percentage of local sheep.

2. Material and methods

2.1. Animals

About 24 heads of West Java local lambs were used in this experiment aged 8-10 months at average live weight of 20.2 ± 1.85 kg. Each lamb was caged in an individual pen. Before experiment, lambs were adapted to experimental situation for about 2 weeks and fed ad libitum field grass with about 200 g concentrate. During experiment, each lamb was fed experimental diets for 3 months.

2.2. Experimental diets

All diets consisted of 60% concentrate and 40% field grass that were given as feed while green tea and turmeric powders were added into concentrate. There were 4 experimental diets: T0 = control diet without green tea and turmeric powders; T1 = diet with 1% turmeric powder; T2 = diet with 2% green tea powder; and T3 = diet with 1% turmeric acid and 2% green tea powders. Diets were formulated to meet lambs nutrient requirement to reach ADG of about 100 g/head/day at crude protein 9.5%, TDN 55%, and dry matter intake minimum 3% live weight [7].

2.3. Measured Parameters
ADG was measured by calculating the difference between initial and final weights divided by the number of days of experiment and expressed as g/head/day. Meanwhile, feed intake was measured by calculating the difference between the amounts of given and remaining feeds on feed bunk and expressed in gram DM/head/day. Carcass percentage was measured by calculating the percentage of carcass from the live weight before slaughtering.

2.4. Statistical Analysis
Randomized experimental design was used to compare 4 different experimental diets as explained previously whilst One-way Analysis of Variance in MINITAB 16 was used to see the different among 4 experimental diets at P<0.05.

3. Results and discussion

Table 1. shows the effect of diets containing green tea and turmeric powders on grass, concentrate, and total feed intakes (g DM/head/day) of local lambs. It can be seen that grass intakes of local lambs ranging from 290.50 ± 23.40 to 294.40 ± 13.89 g DM/head/day while concentrate intake ranging from 566.20 ± 3.59 to 569,50± 2,09 g DM/head/day. In combination, lambs consumed about 858.70 ± 25.77 to 863.90 ± 12.88 g DM/head/day.

Table 1. Means (± SD) grass, concentrate, and total feed intakes (g DM/head/day) of local lambs fed diets containing green tea and turmeric powders.

Treatments	Grass (g DM/head/day)	Concentrate (g DM/head/day)	Total (g DM/head/day)
T0	294.40± 13.89	569.50± 2.09	863.90± 12.88
T1	294.10± 8.67	566.20± 3.59	860.10± 11.28
T2	291.90± 17,40	568.80± 2.70	860.80± 18.06
T3	290.50± 23.40	568.20 ± 2.72	858.70± 25.77
Significant	P = 0.780	P = 0.242	P = 0.741

T0 = control diet without green tea and turmeric powders; T1 = diet with 1% turmeric powder; T2 = diet with 2% green tea powder; and T3 = diet with 1% turmeric acid and 2% green tea powders.

Table 2. Means of initial weight (kg/head), final weight (kg/head), ADG (g/head/day) and carcass percentage (%) of local lambs fed diets containing green tea and turmeric powders.

Treatments	Initial weight (kg/head)	Final weight (kg/head)	ADG (g/head/day)	Carcass percentage (%)
T0	20.10 ± 1.79	27.30 ± 1.95	86.60± 24.4	46.60± 1.71
T1	19.90± 2.34	27.50 ± 3.25	92.40 ± 19.8	44.80± 1.95
T2	20.30± 1.62	27.60± 2.59	88.20± 21.1	45.50± 1.41
T3	20.40 ± 1.66	28.40 ± 1.39	96.00± 12.3	45.30± 1.93
Significant	P = 0.965	P = 0.925	P = 0.253	P = 0.394

T0 = control diet without green tea and turmeric powders; T1 = diet with 1% turmeric powder; T2 = diet with 2% green tea powder; and T3 = diet with 1% turmeric acid and 2% green tea powders.

Statistically, adding 2% green tea or 1% turmeric powders or their combination had no a significant effect (P>0.05) on grass, concentrate, and total feed intakes. Mueller-Harvey [8] and Po *et al.* [9] reported that adding herbals containing high amounts of plant secondary metabolites such as tannins into ruminant's diets could reduce the palatability of the diets because of bitter taste of tannins. However, adding green tea and turmeric into sheep diets in the current study did not reduce feed intake or the palatability of the diets. This might be due to the green tea and turmeric powders were mixed with high palatable concentrate before feeding so that it did not affect the palatability significantly.

Table 2 describes the effects of adding green tea and turmeric powders into a sheep diet on ADG (g/head/day) and carcass percentage (%) of local lambs. As can be seen that ADG of local lambs fed experimental diets in the current study ranging from 86.60 ± 24.4 to 96.00 ± 12.3 g/head/day. However, adding green tea and turmeric powders into a sheep diet in the current experiment had no significant effect ($P>0.05$) on performance of local lambs.

4. Conclusion

Plant secondary metabolites such as tannins have widely reported to taste bitter leading to have poor palatability. However, adding green tea and turmeric into sheep diets in the current study did not reduce feed intake or the palatability of the diets. Adding 2% green tea, 1% turmeric, and their combination into a sheep diet has no significant effect on lamb's performances.

Acknowledgement

This research was funded by Universitas Padjadjaran through Academic Leadership Grant 2016 under supervision of Prof. Husmy Yurmiati.

References

[1] Andlauer W, Heritier J 2011 Rapid electrochemical screening of antioxidant capacity (RESAC) of selected tea samples *Food Chem.* **125**:1517–1520

[2] Chen C, Yu R, Owour E D, Kong A N T 2000 Activation of antioxidant-response element (ARE), mitogen-activated protein kinases (MAPKs) and caspases by major green tea polyphenol components during cell survival and death *Arch. Pharmacal Res.* **23**: 605-612.

[3] Leung L K, Su Y, Chen R, Zhang Z, Huang Y, Chen Z Y 2001 Theaflavins in black tea and catechins in green tea are equally effective antioxidants *J. Nutr.* **131**: 2248-2251.

[4] Higdon J V, Frei B 2003 Tea catechins and polyphenols: health effects, metabolism, and antioxidant functions *Crit. Rev. in Food Sci. Nutr.* **43**: 89–143

[5] Ramdani D, Chaudhry A S, and Seal C J 2013 Chemical Composition, Plant Secondary Metabolites, and Minerals of Green and Black Teas and the Effect of Different Tea-to-Water Ratios during Their Extraction on the Composition of Their Spent Leaves as Potential Additives for Ruminants *J. Agric. Food Chem.* **61**: 4961-4967.

[6] Al-Sultan S I 2003 The effect of Curcuma Longa (turmeric) on overall performance of broiler chicken *Int. J. Poult. Sci.* **2**: 351-353.

[7] National Research Council 1985 Nutrient requirements of sheep 6th revised ed. *National Academic Press Washington DC.*

[8] Mueller-Harvey I 2006 Review unravelling the conundrum of tannins in animal nutrition and health *J. Sci. Food Agric.* **86:** 2010-2037.

[9] Po E, Xo Z, Celi P 2012 The effect of Yerna Mate (*Ilex paraguarensis*) supplementation on the productive performance of Dorper ewes and their progeny *Asian-Australas. J Anim Sci.* **25**: 945-949.

Duckweed Quality Improvement Through Fermentation Using *Trichoderma harzianum* and *Saccharomyces cerevisiae* on Dry Matter, Ash and Crude Fat

H Setiyatwan[1], E Harlia[1] and D Rusmana[1]

[1]Faculty of Animal Husbandry, University of Padjadjaran, Bandung-Sumedang Street KM. 21, 45363, Sumedang, Indonesia, Phone:+6222 7798241, Fax: +6222 7798212

E-mail: hendisetiyatwan@gmail.com

Abstract. Duckweed has been used as animal feed which contains high nutrients, ie proteins and amino acids, but the water content is high. This is a limitation in its use Therefore, research was conducted to improve dry matter, mineral content and lower duckweed crude fat content. Duckweed fermentation is divided into two stages, firstly, *Trichoderma harzianum* (Th) ($3x10^7$ spores/100 grams substrate) with added $ZnCO_3$ (186 ppm) and dl-methionine (286 ppm), secondly, *Saccharomyces cerevisiae* (Sc) ($3x10^7$ spores /100 grams of a substrate). Completed randomized design (CRD) was used in experimental design with 20 experimental units, the treatments consist of P1 (Fermentation using Th for 1day continued with Sc for 9 days), P2 (Th for 3 days continued with Sc for 7 days), P3 (Th 5 days continued with Sc 5 days), P4 (Th 7 days continued with Sc 3 days), P5 (Th 9 days continued with Sc 1 day) with four replications. Analysis of variance was conducted to know the treatment effect, then followed by the Duncan test to know the difference between treatments. Fermentation using Thand Sc has a significant effect ($P<0.05$) on dry matter, ash, and crude fat in the duckweed. The best duration of fermentation is Th for 3 days and Sc for 7 days (P2) that increasing dry matter (92.95%) and ash (21.35%), and lower crude fat (2.37%). Duckweed fermentation with the combination of Th for 3 days and continued Sc for 7 days with added dl-methionine and Zn yield the highest dry matter and ash content, and lowest crude fat contents.

1. Introduction

Duckweeds are small floating aquatic plants [1, 2]. The productivity of *Lemna sp.* planted in an effective planting system reach to 12−38 tons of dry weight/ha/year [3]. *Lemna sp.* has high protein content (38%), however it also rich in crude fiber content. Duckweed has a 6-7% lower ash content. Duckweed has been used as animal feed. The use of duckweed in poultry nutrition is limited to 5% level due to high water content, low ash and fat content.

Fermentation is considered as one way to solve the limited use of duckweed. This is because fermentation can increase nutrient availability and reduce anti-nutrition [4]. Fermentation process raised the content of minerals. Fermentation can raise the content of crude fat in fermented Lemna meal increased up to 5.76% after fermentation [5].

The use of *Trichoderma harzianum* for fermentation has been reported because of its positive effect on the increase of dry matter and ash due to the decrease in crude fiber. The duckweed crude fiber content decreased by 76.06% after fermentation by *Trichoderma harzianum* for 6 - 9 days [6].

Saccharomyces cerevisiae is a yeast that can produce heterologous proteins. The fermentation

IOP Conf. Series: Earth and Environmental Science **334** (2019) 012030 doi:10.1088/1755-1315/334/1/012030

using *Saccharomyces cerevisiae* yeast can enrich the amino acid dl-methionine and inorganic Zinc (Zn) mineral because processing of vegetable-based feed usually results in a limiting amount of amino acids and essential minerals, in turn such a process will increase the content of dry matter and duckweed ash.

Duckweed quality can be improved by fermentation process using *Trichoderma harzianum* and *Saccharomyces cerevisiae*. The secondary metabolic pathway of *Trichoderma harzianum* and *Saccharomyces cerevisiae* are synergistic. *Trichoderma harzianum* can provide nitrogen for the growth of *Saccharomyces cerevisiae* and subsequently used as a raw material for the formation of metallic bonds and amino acids. The activity of both microbes produces complementary enzymes that will result in better results. That way can be used to imitate a natural microbial growth place that is coexisting in a complex microbial community. Duckweed fermentation process using two microbes (co-culture) is expected to produce functional complementary enzymes. This is due to the mutual expression of the metabolic pathways in the substrate utilization. However, the research on the quality improvement of duckweed through fermentation using *Trichoderma harzianum* and *Saccharomyces cerevisiae* has not been carried out. Therefore, the purpose of this research is to determine the best nutrient content of duckweed from the fermentation process based on two types of microbes.

2. Materials and Methods

2.1. Materials and Research Tools

The research materials consisted of duckweed, potato dextrose, PDA agar, Aquadest, $ZnCO_3$, dl-mehionin, rice, tripton, cotton, *Trichoderma harzianum*, and *Saccharomyces cerevisiae*. The equipment consisted of analytical scales, technical scales, measuring flask 1000 ml, Erlenmeyer 1000 ml, stirrer mugs, reaction tubes, Oase needles, spirtus lights, Erlenmeyer glasses, sterilisator, beaker glass, etiquette paper, electric oven, refrigerator, autoclave, and fermentor.

2.2. Research Procedure

Substrate consisted of a mixture of duckweed and selective medium. The mixtures were boiled in the water for 60 minutes at a temperature of 115°C and a pressure of 1.1 atmospheres. *Trichoderma harzianum* microbes were incubated in each treatment of 3×10^7 spores / 100 grams of the substrate for 1, 3, 5, 7, and 9 days at room temperature. After first stage fermentation, $ZnCO_3$ (186 ppm) and dl-methionine (286 ppm) were added. Next, each substrate was fermented using *Saccharomyces cerevisiae* at 3×10^7 spores / 100 grams substrate for 9, 7, 5, 3, and 1 day. After fermentation, the product was dried in the oven.

2.3. Experimental Design

The experimental design used Completely Randomized Design (CRD) for five treatments with a total of 20 experimental units. The treatments consisted of P1 (Fermentation using *Trichoderma harzianum* for 1 day continued with *Saccharomyces cerevisiae* for 9 days), P2 (*Trichoderma harzianum* for 3 days continued with *Saccharomyces cerevisiae* for 7 days), P3 (*Trichoderma harzianum* for 5 days continued with *Saccharomyces cerevisiae* for 5 days), P4 (*Trichoderma harzianum* for 7 days continued with *Saccharomyces cerevisiae* for 3 days), P5 (*Trichoderma harzianum* for 9 days continued with *Saccharomyces cerevisiae* for 1 day) with four replications.

2.4. Measurement Variables

2.4.1. Dry Matter

Dry matter refers to material remaining after removal of water. Dry matter content was determined by Wendee method. Proximate analysis was done by drying method of the material at temperature 105°C.

2.4.2. Ash Content
The analytical method used to determine ash content is the proximate analysis of the Wendee method. The sample is added to the completely dry crucible and lid and together they are weighed to determine the mass of the sample by difference. The sample is placed in the hot furnace long enough so that complete combustion of the sample occurs. The crucible, lid and ash then are re-weighed.

2.4.3. Fat Content
Methods currently used for quantitating total fat and fatty acid composition in feedstuffs require solvent extraction, purification, and esterification followed by gas Chromatographic analysis.

2.5. Statistical Analysis
Data collected were subjected to one-way analysis of variance (ANOVA) as per Steel and Torrie. If there are mean differences, they were compared using Duncan's multiple range test with 5% significant level.

3. Results and Discussion
3.1. Fermented Duckweed Dry Matter
The fermentation using *Trichoderma harzianum* followed by *Saccharomyces cerevisiae* has a significant effect (P<0.05) to increase the duckweed dry Matter content. The rank of treatments effect are (higher to lower) P2 (92.71%), P3 (92.55%), P1 (92.50 %), P4 (92.49), and P5 (91.53 %) (Table 1).

Table 1. Dry Matter, Ash Content, Fat Content of Duckweed fermented with *Trichoderma harzianum* followed by *Saccharomyces cerevisiae*

Parameters	Treatments				
	P1	P2	P3	P4	P5
Dry Matter (%)	92.50±0.57^{a}	92.71±0.36^{b}	92.55±0.38^{a}	92.49±0.46^{a}	91.53±0.45^{ab}
Ash Content (%)	21.73±0.41^{c}	21.35±0.38^{c}	18.42±0.38^{a}	19.65±0.73^{b}	22.76±0.37^{d}
Fat Content (%)	2.23±0.13^{a}	2.57±0.37^{c}	2.37±0.24^{b}	1.82±0.15^{ab}	1.66±0.19^{ab}

Mean in the same column with different superscript differs significantly (P <0.05),
P1: Fermentation using *Trichoderma harzianum* (Th) for 1 day continued with *Saccharomyces cerevisiae* (Sc) for 9 days, P2: Th 3 days continued with Sc 7 days, P3: Th 5 days continued with Sc 5 days, P4: Th 7 days continued with Sc 3 days, P5: Th 9 days continued with Sc 1 day.

The results of the present study are in-line with other studies. In the present fermentation process, dry matter weight loss was highly correlated with the biomass [7]

The Duncan's multiple-range test showed that the fermentation using *Trichoderma harzianum* for one day followed by *Saccharomyces cerevisiae* for nine days (P1) on dry matter content was not significantly different from fermented duckweed using *Trichoderma harzianum* for five days followed by *Saccharomyces cerevisiae* for five days (P3). Similarly, the dry matter content of fermented duckweed using *Trichoderma harzianum* for seven days followed by *Saccharomyces cerevisiae* for three days (P4) was not significantly different.

One day fermentation using *Trichoderma harzianum* was expected to increase the availability of dry matter. However, it could not break crude fiber of duckweed. Therefore, many complex cellulose bonding could be found resulting in low dry matter content in P1 (92.50%). On the other hand, the dry matter content has increased to 92.55% (P3) because five days fermentation were sufficient to form a dry matter.

The dry matter content in P4 (92.49%) were influenced by the fermentation duration of *Saccharomyces cerevisiae* (3 days). In P5, nine days fermentation of *Trichoderma harzianum* and one days of *Saccharomyces cerevisiae* are not balanced causing declines dry matter content at 91.53%. This is in line with the study by [7] found that the loss of dry matter weight was caused by chemical composition changes (in fibres, starch and free-sugar content) of solid state fermentation

Saccharomyces cerevisiae for seven days (P2) has significantly ($P<0.05$) increased the dry matter content (92.71%) compared with other treatments (P1, P3, P4, and P5). It showed that the addition of *Saccharomyces cerevisiae* can be used as a duckweed modification agent to increase dry matter content.

3.2. Fermented Duckweed Ash Content

The highest ash content of fermentation to the lowest in sequence is P5 (22.76%), P1 (21.73%), P2 (21.35%), P4 (19.65%), and P3 (18.42%) as shown in Table 1. The crude fiber content in the Fermented duckweed reduced 15.27% as a result of the activity of the cellulose enzyme contained in *Saccharomyces cerevisiae*.

The fermentation by using *Trichoderma harzianum* followed by *Saccharomyces cerevisiae* was significantly ($P<0.05$) decreasing the content ash of duckweed. In terms of the differences of the treatments, the results showed that fermented ash content by using *Trichoderma harzianum* for one day continued by *Saccharomyces cerevisiae* for nine days (P1) was not significantly different from the fermentation using *Trichoderma harzianum* for three days continued by *Saccharomyces cerevisiae* for seven days(P2).

Trichoderma harzianum has higher cellulolytic activity than other *Trichoderma sp* groups. The longer fermentation time of the crude fiber content the higher substrate will be gained. This is due to the decrease in water content on the substrate that result in the more concentrated crude fiber. In addition, increasing the amount of *Trichoderma harzianum* raise the crude fiber from the cell wall. The amount of inoculum (3 x 10^7 spores/100 grams of substrate) used as the exact dose, and the minimum fermentation time by *Trichoderma harzianum* are the determinant of the crude fiber content of fermented duckweed.

The average duckweed content of fermented ash using *Trichoderma harzianum* for one day followed by *Saccharomyces cerevisiae* for nine days (P1) was significantly higher than other treatments (P2, P3, and P4). Running the fermentation for one day has not successfully broken the crude fiber in the duckweed. Fermentation using *Trichoderma harzianum* for three days followed by *Saccharomyces cerevisiae* for seven days (P2) has significantly decreased ash content originally from 31.36% to 21.35%.

3.3. Fermented Duckweed Fat Content

Duckweed fermented in *Trichoderma harzianum* followed by *Saccharomyces cerevisiae* was significant ($P<0.05$) to increase the content of fat in the duckweed. The highest to the lowest mean scores in sequence are P2 (2.57%), P3 (2.37%), P1 (2.23 %), P4 (1.82%) and P5 (1.66%). Crude fat in fermented lemna minor increased up to 5.76% after fermentation. The increase of fat content was caused by the microorganism's capability to produce lipid / fat during fermentation process. Microorganisms, as other life cell systems, produce lipid or fat [5].

Fermentation by using *Trichoderma harzianum* for three days followed by *Saccharomyces cerevisiae* for seven days (P2) has significantly ($P<0.05$) increased the content of fat in the duckweed (2.57%) compared with the other treatments (P1, P3, P4, and P5). It showed that the addition of *Trichoderma harzianum* and *Saccharomyces cerevisiae* can be used as a duckweed modification agent to increase the content of fat in the duckweed.

4. Conclusion

Fermentation with the combination of *Trichoderma harzianum* for three days and continued with *Saccharomyces cerevisiae* for seven days that are supplemented with amino acid dl-methionine and Zn is the best treatment to improve duckweed quality. It can increase the content of dry matter content (92.71%) and fat (2.57%), and decreased ash content (21.35%).

Acknowledgements

The authors would like to thank the honorable Rector of Universitas Padjadjaran and Ministry of Research, Technology and Higher Education of Universitas Padjadjaran that have supported this research. The research can be carried out with the financial support from DIPA Fund of Universitas Padjadjaran no. 855/UN6.3.1/PL/2017 dated on March 8th, 2017.

References

[1] Talukdar M Z H, Shahjahan M, Rahman M S 2012 Suitability of duckweed Lemna minor as feed for fish in polyculture system *International Journal of Agricultural Research, Innovation and Technology* **2**: 42–46

[2] Uddin M N, Rahman M S, Shahjahan M 2007 Effects of duckweed Lemna minor as supplementary feed on monoculture of gift strain of tilapia Oreochromis niloticus *Progressive Agriculture* **18**: 183–188

[3] Ansal M D, Dhawan A, Kaur V I 2010 Duckweed based bio-remediation of village ponds: An ecologically and economically viable integrated approach for rural development through aquaculture *Livestock Research for Rural Development* **22**: 130

[4] Haustetn A T, Gilman R H, Skillicorn P W, Vergara V, Guevara V, Gastanaduy A 1990 Duckweed, a useful strategy for feeding chickens: performance of layers fed with sewage-grown Lemnacea species *Poultry Science.* **69(11)**: 1835-1844. DOI: https://doi.org/10.3382/ps.0691835

[5] Rostika R, Andriani Y, Abram A H, Vinasyiam A 2017 The growth rate of Nile tilapia Oreochromis niloticus fry fed on fermented *Lemna sp.* Meal. *Jurnal Akuakultur Indonesia* **16(1)**: 101–106

[6] Setiyatwan H 2007 Quality Improvement of Duckweed Nutrition through Fermentation using Trichoderma harzianum *J. Animal. Sci*. Padjadjaran University **7**.

[7] Smits J P, Rinzema A, Tramper J, Van Sonsbeek H M, Knol W 1996 Appl. Microbiol. *Biotechnol.* **46**: 489–496.

Relationship of Free Fatty Acid and Natural Autoantibodies 2 Weeks Postpartum

N Mayasari[1,2], J Chen[2], B Kemp[1], H Parmentier[1] and A Van Knegsel[1]

[1]Department of Animal Nutrition and Feed Technology, The Faculty of Animal Husbandry, Universitas Padjadjaran, Jatinangor Campus, Sumedang 45363, Indonesia
[2] Adaptation Physiology Group, Department of Animal Sciences, Wageningen University, P.O. Box 338, 6700 AH Wageningen, The Netherlands

E-mail: novi.mayasari@unpad.ac.id

Abstract. Natural autoantibodies (NAAb) have been found in plasma and milk of dairy cows. Natural autoantibodies maintain homeostasis and may prevent infections. The objective of this study was to evaluate the relationship between concentration of plasma free fatty acid (FFA) and the levels of NAAb in plasma binding glutamate dehydrogenase (GD) and carbonic anhydrase (CA) in cows 2 weeks after calving. In total, 55 Holstein-Friesian dairy with 60-d dry period lengths and fed 2 early lactation diets (glucogenic or lipogenic). Blood was sampled at week 1 and 2 after calving. Result showed that from week 1 and 2 relative to calving, levels of IgG and IgM binding CA were positively related with concentration of FFA in plasma. Levels of IgM binding GD were negatively related with concentration of FFA in plasma. It is concluded that IgG and IgM binding CA in plasma might be an indicator for energy status and reflect metabolic status in dairy cows 2 weeks after calving.

1. Introduction

In early lactation, high yielding dairy cows experience a negative energy balance (NEB) caused by a sudden increase of energy requirement for milk production and maintenance while the dietary energy intake in this period is limited. The NEB in early lactation has been related to metabolic disorders[1], infectious diseases[2] and immunosuppression[3]. An elevated level of serum NEFA is one of the indicators of negative energy balance (NEB) in postpartum dairy cows. Previous studies showed that cows with NEB had elevated plasma free fatty acid (FFA) and β-hydroxybutyrate (BHB) concentrations which are related to higher risk of health problems [4-6]. Circulating FFA from lipolysis can be partially oxidized to form ketone bodies or esterified to form triacylglycerol (TAG) in liver because hepatic capacity of complete oxidation is not sufficient[1]. Parts of the formed TAG are exported as very-low-density-lipoproteins (VLDL). It was suggested that FFA and VLDL may be used for milk fat synthesis [7].

Cows with NEB have been attributed to suboptimal immune function like natural antibodies (NAb) in early lactation [3]. In the study of Van Knegsel [12] cows with severe NEB had low NAb. Natural antibodies act as a first line of defence against infection [8, 9] but also to maintain homeostasis, to promote phagocytes of dead cells, and to prevent infectious and autoimmune diseases [10]. Natural antibodies have been divided into two classes: overt and cryptic NAb. Overt NAb bind antigens that the individual has never encountered before. Cryptic NAb or so-called natural autoantibodies (NAAb) are antibodies that bind to self-antigens or slightly changed self-antigens (neo-epitopes). Natural autoantibodies are thought to be involved in inactivation of cytokines, prevention of inflammation, clearance of metabolic waste, perform various homeostatic roles within the immune response [10], and have a role in the prevention of autoimmunity [10].

Earlier study in the present work, cows with higher somatic cell count (SCC) had higher natural antibodies (NAb) titers for isotypes IgG and IgM binding keyhole limpet hemocyanin (KLH) and lipopolysaccharides (LPS) in milk, and higher titer of IgG binding LPS in plasma [11]. In an earlier study, high NAAb binding myosin in milk of cows were associated with severe NEB (β = -0.001, $P <$ 0.10), low dry matter intake (DMI) (β = -0.049, $P < 0.05$) and occurrence of clinical mastitis [12]. In the study of Van Knegsel [12], NAAb binding myosin was measured in milk, but not in plasma. Little is known of the relationship between plasma FFA concentration and NAAb in early lactation. In this study, there are two autoantigens will be used glutamate dehydrogenase (GD) and carbonic anhydrase (CA).

2. Material and methods

2.1. Animals and Experimental Design

All the data including animals and experimental design of this study were obtained from the project of WHY DRY at Adaptation Physiology Group Wageningen University, the Netherlands. The Institutional Animal Care and Use Committee of Wageningen University approved the experimental protocol. Holstein-Friesian dairy cows (n = 167) were selected from the Dairy Campus Research dairy herd (WUR Livestock Research, Lelystad, The Netherlands). The experimental design, DP length, and dietary contrasts were described by van Knegsel *et al.* [13]. In summary, cows were blocked for parity (primiparous or multiparous), expected calving date, milk yield in the previous lactation, and body condition score (BCS). Within blocks, each cow was assigned randomly to a 3 × 2 factorial design with 3 DP lengths (0, 30, or 60 d) and 2 early lactation diets (glucogenic or lipogenic). For this study, we used only the data from cows with 60-d DP.

Precalving period, lactating cows were fed a lactation diet supporting 25 kg of milk, and dry cows were fed a dry cow ration. From d 10 before the expected calving date and onward, cows of all treatments were fed 1 kg/d of glucogenic or lipogenic concentrate and increased postcalving stepwise with 0.5 kg/d until the concentrate supply reached 8.5 kg/d. The main ingredient for the glucogenic concentrate was corn and the main ingredients for the lipogenic concentrate were sugar beet pulp, palm kernel, and rumen protected palm oil. Forage was supplied ad libitum and was composed of grass silage, corn silage, wheat straw, and rapeseed meal or soybean meal (51:34:2:13, DM basis). Diets were formulated to be isocaloric (Net energy basis; VEM system; [14] and equal in intestinal digestible protein and degraded protein balance (DVE/OEB system; [15]. Blood samples were collected from 56 of the 167 cows, Cows were housed in a free stall with a slatted floor and cubicles and were milked twice daily (0500 and 1630 h).

2.2. Blood Sampling and Analysis

Blood samples were taken from the coccygeal vein at week 1 and 2 after calving at 3 h before the morning feeding. Blood was collected in evacuated tubes (Vacuette, Greiner BioOne, Kremsmunster, Austria) containing EDTA for FFA analysis; or heparin for NAAb. Blood samples were centrifuged at 3000 x g at 4°C, for 15 min within two hours after collection. Plasma was transferred and stored at -20°C until analysis.

2.3. Analysis of NAAb in plasma of cows

Natural autoantibody titers to two self-antigens GD and CA in plasma of cows were measured by an indirect enzyme-linked immunosorbent assay (ELISA) technique as outlined by Mayasari *et al.* [24]. In brief, plates were coated with 4 µg/ml of GD (G7882, Sigma) and CA (C3934, Sigma), respectively in 100 µl/well. Natural autoantibodies of the IgG isotype were detected using 1:20,000 (cow) diluted sheep polyclonal anti-bovine IgG-heavy chain conjugated to horseradish peroxidase (PO) (Cat. No. E10-118P, Bethyl). Natural autoantibodies of the IgM isotype in plasma of cows were detected using 1:20,000 diluted rabbit polyclonal anti-bovine IgM-whole molecule conjugated to PO (Cat. No. A10-

100P, Bethyl). Four step serial dilutions for both IgG and IgM in plasma samples of cows started at 1:40. After washing, a substrate containing tetra methyl benzidine (Sigma Aldrich Chemie, Steinheim, Germany) and 0.05 percent hydrogen peroxide was added, and incubated for 10 minutes at room temperature. The reaction was stopped by adding 1.25 M sulfuric acid. Extinctions were measured with a Multiskan reader (Lab Systems, Helsinki, Finland) at a wavelength of 450 nm. Titers were expressed as log2 values of the dilutions that gave an extinction closest to 50 percent of Emax, where Emax represents the highest mean extinction of a standard positive (pooled) serum present on every microtiter plate[16]

2.4. Statistical analyses

All data of NAAb titers and plasma FFA in cows approximated normality of residuals by examining whether skewness and kurtosis were in a range of -2 until 2. To assess the relationship between NAAb titers in plasma of cows with FFA concentration in plasma of cows, the Pearson Correlation were used. The average, minimum and maximum data from the statistical model and the P-value corresponding to the r are displayed.

3. Results and discussion

As described in the result of plasma FFA concentrations were different over time [29]. The highest level of plasma FFA concentrations of cows with a 60 d dry period (DP) is at week 1 and 2 after calving (Figure 1). Additionally, diet did not affect plasma FFA. Immediately after calving, high milk production coupled while with inadequate DMI require the use of energy from body fat stores, which are mobilized as FFA. Most FFA are metabolized in the liver completely into acetyl-coA, incompletely into ketone bodies such as BHBA, or reesterified into triglycerides.

In early lactation, increased mobilization of adipose tissue is associated with high plasma FFA concentration [17]. The increased rate of lipolysis results in increased plasma FFA concentration [18]. Thus, high plasma FFA concentration in cows at week 1 and 2 after calving reflects increased fat mobilization, Free fatty acids can be esterified and stored as TAG in liver. Cows with a moderate (50 to 100 mg/g liver TAG on a wet weight basis) or severe fatty liver (> 100 mg/g liver TAG) were at high risk for metabolic and reproductive problems [18]. Although FFA can be used for ketogenesis through mitochondrial oxidation, we did not found many cows with ketosis in this study. Plasma FFA concentration during early lactation was suggested does not determine the extent of β-oxidation and plasma BHB concentration [19]. Plasma BHB could come from the partial β-oxidation of FFA in liver or from conversion of butyrate in rumen [20]. However, plasma BHB could be used for synthesis of milk fat in the mammary gland [21] or as an alternative energy source for glucose in brain and muscles [22]. Elevated precalving FFA and decreased DMI were associated with impaired neutrophil function [23], which decreases the first line of defence against infectious agents.

In line with previous study that IgM and IgG antibodies binding two self-antigens: GD and CA were detected in plasma of cows [24]. NAAb are present in young and old dairy cows. Previous studies showed that antibodies reacting to self-antigens can be found in plasma and milk from healthy non-immunized dairy cows [12] and in plasma from calves of 20-26 of weeks old [25].

In present study, high plasma FFA concentration was associated with high NAAb both IgG and IgM binding CA at week 1 and 2 after calving. In addition, plasma FFA concentration tended to be associated with IgM binding GD. Previous study showed that cows with subclinical ketosis had higher free fatty acid concentration and GD activation after calving [26]. Carbonic anhydrase also functions in regulation of pH and fluid balance [27] as observed during inflammation [28]. Another study reported NAAb binding myosin and transferrin in milk of cows were associated with sensitivity for mastitis [12]. It is suggested that high plasma FFA concentration was related to regulation of pH and fluid balance during inflammation such as clinical mastitis.

Table 1. Mean, standard deviation, sum, minimum and maximum for the main effects of free fatty acid, natural antibodies (NAAb) isotipe IgG and IgM binding glutamate dehydrogenase and carbonic anhydrase.

Variable	N	Mean	Std Dev	Sum	Minimum	Maximum
Non-esterified fatty acid	54	0.56	0.35	30.24	0.07	1.94
IgG binding Glutamate dehidrogenase	56	7.63	1.54	427.2	3.8	10.8
IgM binding Glutamate dehidrogenase	56	5.94	0.97	332.7	3.8	7.5
IgG binding Carbonic anhydrase	56	5.51	1.09	308.7	2.9	8.7
IgM binding Carbonic anhydrase	56	5.48	1.02	306.6	3.2	9.5

Table 2. Pearson Correlation Coefficients between free fatty acid (FFA) and natural antibodies (NAAb) isotipe IgG and IgM binding glutamate dehydrogenase dan carbonic anhydrase.

	Natural autoantibodies			
	IgG_GD	IgM_GD	IgG_CA	IgM_CA
Cows (n)	54	54	54	54
Free Fatty Acid	0.17	-0.25	0.52	0.24
P-value	0.17	0.07	<0.01	0.08

n, number of replicates; GD, *glutamate dehidrogenase*; CA, *carbonic anhydrase*

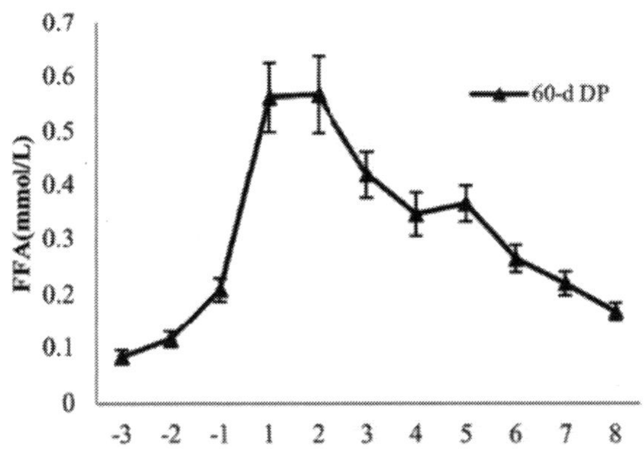

Figure 1. Plasma Free fatty Acid (FFA) (mmol/L) at week 1 and 2 after calving. Values represent means (± SEM) per week [29]

4. Conclusion

It is concluded that IgG and IgM binding carbonic anhydrase in plasma might be an indicator for energy status and reflect metabolic status indicated with high plasma free fatty acid concentration in dairy cows 2 weeks after calving. Carbonic anhydrase involved on the regulation of pH and fluid balance during inflammation which occur mostly in early lactation.

Acknowledgement

The authors thank the Dutch Dairy Board (PZ, the Netherlands), Product Board Animal Feed (PDV; the Netherlands), and CRV (the Netherlands) for financing the experiment. The first author thanks the DIKTI, Indonesian scholarship for financial support. Additionally, the authors thank the staffs of the Dairy Campus (Lelystad, the Netherlands) and the staffs from the Veterinary Physiology Group of the University of Bern for their technical support and laboratory analyses during the experiment.

References

[1] Grummer R R 1993 Etiology of lipid-related metabolic disorders in periparturient dairy cows *J. Dairy Sci.* **76**: 3882-3896.

[2] Collard B L, Boettcher P J, Dekkers J C M 2000 Relationships between energy balance and health traits of dairy cattle in early lactation *Journal of Dairy Science* **83**: 2683-2690.

[3] Mallard B A, Dekkers J C, Ireland M J 1998 Alteration in immune responsiveness during the peripartum period and its ramification on dairy cow and calf health *Journal of Dairy Science* **81**: 585-595.

[4] Ospina P A, Nydam D V, Stokol 2010 Associations of elevated nonesterified fatty acids and β-hydroxybutyrate concentrations with early lactation reproductive performance and milk production in transition dairy cattle in the northeastern United States *Journal of Dairy Science* **93**: 1596-1603.

[5] Seifi H A, LeBlanc S J, Leslie K E 2011 Metabolic predictors of post-partum disease and culling risk in dairy cattle *The Veterinary Journal* **188**: 216-220.

[6] Chapinal N, Carson M, Duffield T F 2011 The association of serum metabolites with clinical disease during the transition period *Journal of Dairy Science* **94**: 4897-4903.

[7] Bell A W 1995 Regulation of organic nutrient metabolism during transition from late pregnancy to early lactation *J. Animal Sci* **73**: 2804-2819.

[8] Ochsenbein A F, Fehr T, Lutz C 1999 Control of early viral and bacterial distribution and disease by natural antibodies *Science* **286**: 2156-2159.

[9] Zinkernagel R M 2012 Immunological memory≠ protective immunity *Cell Mol. Life Sci.* **69**: 1635-1640.

[10] Cojocaru M, Cojocaru I M, Silosi I 2009 The significance of natural autoantibodies *Maedica (Buchar)* **4**: 22-25.

[11] Mayasari N, Rijks W, de Vries R G 2016 The effects of dry period length and dietary energy source on natural antibody titers and mammary health in dairy cows *Preventive Veterinary Medicine* **127**: 1-9.

[12] Van Knegsel A T M, Hostens M, de Vries Reilingh G 2012 Natural antibodies related to metabolic and mammary health in dairy cows *Preventive Veterinary Medicine* **103**: 287-297.

[13] Van K A, Remmelink G, Jorjong S 2014 Effect of dry period length and dietary energy source on energy balance, milk yield, and milk composition of dairy cows *J. Dairy Sci.* **97**: 1499-1512.

[14] Van E A 1975 Feed evaluation for dairy cows *Livest Prod. Sci.* **2**: 95-107.

[15] Tamminga S, Van S W, Subnel A 1994 The Dutch protein evaluation system: the DVE/OEB-system *Livest Prod. Sci.* **40**: 139-155.

[16] Ploegaert T C W, Tijhaar E, Lam T J G M 2011 Natural antibodies in bovine milk and blood plasma: Variability among cows, repeatability within cows, and relation between milk and plasma titers *Veterinary Immunology and Immunopathology* **144**: 88-94.

[17] Bell A W 1995 Regulation of organic nutrient metabolism during transition from late pregnancy to early lactation *J. Anim. Sci.* **73**: 2804-2819.

[18] Locher L, Meyer N, Weber E M 2011 Hormone-sensitive lipase protein expression and extent of phosphorylation in subcutaneous and retroperitoneal adipose tissues in the periparturient dairy cow *J. Dairy Sci.* **94**: 4514-4523.

[19] Van K A, Van den B H, Graat E 2007 Dietary energy source in dairy cows in early lactation: metabolites and metabolic hormones *J. Dairy Sci.* **90**: 1477-1485.

[20] Bergman E 1990 Energy contributions of volatile fatty acids from the gastrointestinal tract in various species *Physiol. Rev.* **70**: 567-590.

[21] Bauman D E, Griinari J M 2003 Nutritional regulation of milk fat synthesis *Annu. Rev. Nutr.* **23**: 203-227.

[22] Veech R L 2004 The therapeutic implications of ketone bodies: the effects of ketone bodies in pathological conditions: ketosis, ketogenic diet, redox states, insulin resistance, and mitochondrial metabolism *Prostaglandins Leukot Essent Fatty Acids* **70**: 309-319.
[23] Hammon D S, Evjen I M, Dhiman T R 2006 Neutrophil function and energy status in Holstein cows with uterine health disorders *Veterinary Immunology and Immunopathology* **113**: 21-29.
[24] Mayasari N, Van Knegsel A, de Vries R G 2016 Natural autoantibodies in Bos taurus calves during the first twelve weeks of life *Veterinary Immunology and Immunopathology* **178**: 70-78.
[25] Khobondo J O, Nieuwland M G B, Webb L E 2015 Natural (auto) antibodies in calves are affected by age and diet *Veterinary Quarterly* **35**: 64-75.
[26] Schulz K, Frahm J, Meyer U 2014 Effects of prepartal body condition score and peripartal energy supply of dairy cows on postpartal lipolysis, energy balance and ketogenesis: an animal model to investigate subclinical ketosis *Journal of Dairy Research* **81**: 257-266.
[27] Badger M R, Price G D 1994 The role of carbonic-anhydrase in photosynthesis *Annual Review of Plant Physiology and Plant Molecular Biology* **45**: 369-392.
[28] Mihaylova N M, Dimitrov J D, Djoumerska-Alexieva I K 2008 Inflammation-induced enhancement of IgG immunoreactivity *Inflammation Research* **57**:1-3.
[29] Chen J, Remmelink G J, Gross J J, Bruckmaier R M, Kemp B, van Knegsel A T M 2016 Effects of dry period length and dietary energy source on milk yield, energy balance, and metabolic status of dairy cows over 2 consecutive years: Effects in the second year *J. Dairy Sci.* **99**: 4826–4838.

Effects of Addition of Noni Fruit Juice (*Morinda Citrifolia*) Into Drinking Water on Final Body Weight, Carcass Percentage, Abdominal Fat and Meat Cholesterol of *Sentul Debu* Chicken Phase Growth

T Widjastuti[1], Abun[1], I Setiawan[1] and Indrawati [1]

[1] Faculty of Animal Husbandry, Universitas Padjadjaran, Sumedang-West Java, Indonesia

Email : tuti_widjastuti@yahoo.com

Abstract. The purpose of research to determine the effect of addition Noni fruit juice into drinking water on body weight, percentage of carcass, abdominal fat and meat cholesterol on *sentul debu* chicken phase growth. The research used 120 day old chicks of Sentul chicken and were raised until 12 weeks old. The research were conducted using a completely randomized design with six levels of noni fruit juice into the drinking water, P0 (without the addition of noni juice into drinking water), P1 (1 ml of noni juice/1 liter of drinking water), P2 (2 ml of noni juice/1 liter of drinking water), P3 (3 ml of noni juice/1 liter of drinking water , P4 (4 ml of noni juice /1 liter of drinking water and P5 (5 ml of noni juice/1 liter of drinking water) and four replications, each replication consisted of five sentul chickens. The observed variables were final body weight, carcass percentage, abdominal fat and meat cholesterol on sentul chicken. Data were statistically analyzed for variance using a Duncan's Multiple Range test. The results showed that the addition of noni juice until 3 ml/1liter into drinking water gave no significant effect on the final body weight, percentage of carcass and abdominal fat and meat cholesterol ,but the addition of noni juice 4 -5 ml into drinking water was decreases the final body weight, carcass percentage and meat cholesterol of sentul chicken. It can be concluded that the addition of noni juice into drinking until 3 ml/1 liter the best result on final body weight, carcass percentage, abdominal fat and meat cholesterol Sentul Debu chicken phase growth.

1. Introduction

Sentul chicken is one of 32 local chicken clumps that have been identified in Indonesia. The original habitat of Sentul chicken is from Ciamis Regency, West Java [1]. Sentul chicken has the advantage that is as a producer of meat and eggs (the type of dual purpose), body weight of adult male Sentul chicken 1.3 – 3.5 kg and female 0.8 – 2.2 kg, 118 eggs of eggs per year [2]. Sentul chicken as meat producer can be cut at various age of maintenance depending on market demand, are usually cut at the age of about 8 to 12 weeks or after reaching body weight between 700 g to 1200 g. The percentage of carcass is affected by cutting age. The longer the age of the cut will be the greater the percentage of carcasses because the carcass is part of the production, so grows larger until the age of 20 weeks. The results of the percentage of carcasses vary in various studies because influenced by various factors such as genetic differences, maintenance management, rations, chicken age and others. The cut weights and percentage of chicken carcasses at 12 weeks old reached 713.70 g and 60.05% [3]. Ration is the environmental factor that can affect poultry business success, and cause to production cost more or less 60-70 percent. The increasing of production cost can be overcome by finding other alternative feed ingredients which have good quality [4]. The alternative food ingredients which has good quality, is expected to be able to reduce the production cost, is usually also added with additional feed ingredients (feed additive). Feed additive is an additional feed ingredient given to livestock through mixing feed or through drinking water. Feed additive there are two types of natural and synthetic additive feed [5]. One of which is the fruit Noni (*Morinda citrifolia*) is useful to maintain endurance and help metabolism of food substances in the body. Noni fruit produces antioxidant such as: scopoletin, nitric oxide, vitamin C and vitamin A, and has the efficacy to increase the secretion of bile and substance NO (Nitrit Oxide) that can stimulate the excretion of cholesterol through feces. According [6] Noni fruit contains xeronine active substances that can reduce fat and cholesterol levels. Xeronine has a high activity in the formation of proteins for hormones, one of which is the hormone insulin that can increase the number of LDL receptors (low density lipoprotein) hepatic and extra hepatic [7, 8]. If a lot of cholesterol is wasted through the feces, cholesterol in the blood and tissues will decrease, it is good for abdominal fat reduction or waste in chicken carcass which should be reduced to make the product of animal origin more healthy and reduce production costs [9]. The main active substances contained in the noni fruit include: polysaccharide, scopoletin, ascorbic acid, β-carotene, L-arginine, proxeronine and proxeroninase [10]. The role of these active substances is closely related to the metabolic activity that supports the addition of weight and will eventually result in a high percentage of carcass. Needs to drink humans per day 1500 ml, in 1 day the use of noni fruit in humans ± 2% of the needs of drinking/day. 4-week-old broiler for 250 ml/day, so the juice noni given to broiler ± 5 ml. Comparison of noni juice for dose of 1 ml/liter and 3 ml/liter is no significant than 2 ml/liter noni juice in drinking water that gives higher benefit and low feed conversion value The purpose of research to determine the effect of addition Noni fruit juice into drinking water on body weight, percentage of carcass, abdominal fat and meat cholesterol of sentul chicken phase growth.

2. Materials And Methods

2.1. Livestock Experiments
One hundred and twenty day old chick of Sentul Debu chicken were raised in cages until 12 weeks old. The cage is use as 24 cages and measuring 90 cm long, 90 cm wide and 60 cm high, each cage consisted of 5 chickens.

2.2. Trial Rations
The feed ingredients of ration comprised of yellow corn meal (56.00%), soy-bean meal (12.00%), rice bran (21.50%), fish meal (9.25%), CaCo3 (0.50%) and bone meal (0.75%). Rations were prepared based on protein and metabolic energy requirement for Sentul Debu chicken growth phase, ie.17 percent protein and metabolic energy 2750 kcal/kg [11]. The treatment consisted of the use noni fruit juice into the drinking water, P0 (without the addition of noni juice into drinking water), P1 (1 ml of noni juice/1 liter of drinking water), P2 (2 ml of noni juice/1 liter of drinking water), P3 (3 ml of noni juice/1 liter of drinking water, P4 (4 ml of noni juice /1 liter of drinking water and P5 (5 ml of noni juice /1 liter of drinking water.

2.3. Experimental design
The research were conducted using a completely randomized design with six levels of noni fruit juice into the drinking water, and four replications, each replication consisted of five sentul chickens. The data was analyzed with random simple test, and with Duncan's Multiple range Test among the treatment the research, chickens were given only water addition noni juice without antibiotic and no vaccinations. The parameters were final body weight, carcass percentage, gizzard weight and the meat cholesterol content of Sentul chicken.

3. Results And Discussion
The effect of noni fruit juice into the drinking water on final body weight, carcass percentage, gizzard weight and the cholesterol content of Sentul Debu chicken, is shown in Table 1 and Figure 1.

Table 1. The final body weight, carcass percentage, gizzard weight and the meat cholesterol content of Sentul Debu Chicken.

Variables	P0	P1	P2	P3	P4	P5
Final body weight (g)	721.83^a	761.28^a	787.23^a	794.90^a	650.66^b	619.25^b
Carcass weight (g)	410.43^a	458.75^a	467.00^a	468.25^a	350.45^b	329.50^b
Carcass percentage (%)	56.86^a	57.32^a	59.32^a	58.90^a	53.87^b	53.21^b
Abdominal fat (g)	3.11^a	2.86^a	2.76^a	2.60^a	2.01^b	1.35^b
Cholesterol mg/100 ml	85.50^a	76.55^b	74.33^b	72.55^b	71.50^b	68.67^b

Note: The similar superscript in the same row no significant difference (P >0.05)

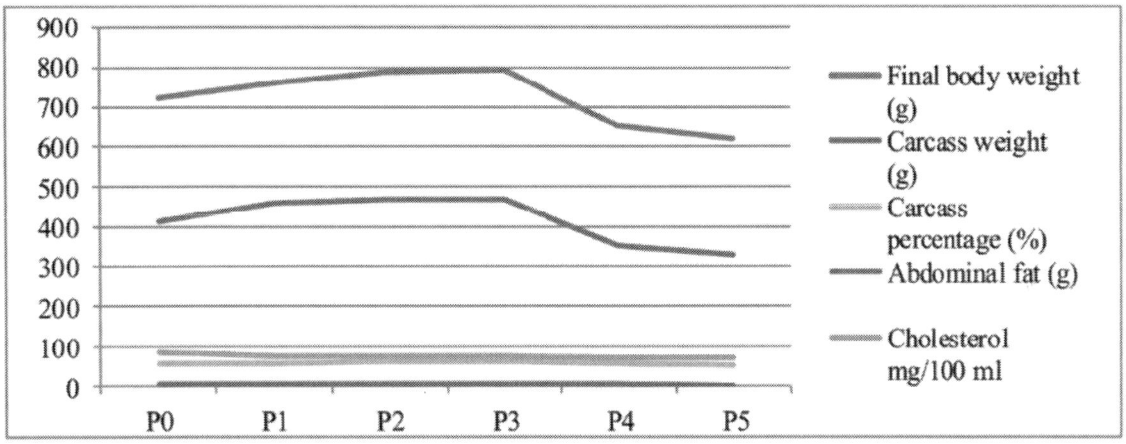

Figure 1 The final body weight, carcass percentage, gizzard weight and the meat cholesterol content of Sentul Debu Chicken

3.1. *Final Body Weight*

The final body weight were variation from the lowest P5= 619.25 gram to the highest P3=794.90 gram. Analysis of variance showed (Table 1) that addition noni juice into drinking water until 3 ml did not significant influence (P0,P1,P2,P3) on final body weight of sentul chicken, but has significant effect (P <0.05) when using 4 ml (P4) and 5 ml in drinking water (P5). Adding noni juice until 3 ml/1 liter of drinking water still gave a good result, because the noni fruit contains bioactive polyphenol compounds. Polyphenolic compounds such as anthraquinone compounds ranges from 5 - 36 g / 100 g of dry matter, that can increase the body's metabolism and have the ability to fight infectious bacteria. These compounds are useful for inhibiting the growth of gram-positive and negative bacteria that can eradicate pathogenic bacteria in the gastrointestinal tract and also make the pH of the gastrointestinal tract into an acid that enables the protein-breaking enzyme to work optimally [12]. While saponin compounds are also found in noni fruit which can increase the permeability of the cell wall of the intestine so as to increase the absorption of food substances. Then [13] added that the noni fruit contains nutrients comprising proteins, minerals and vitamin C that can give a good effect on final body weight. A decrease in final body weight to increased level noni juice 4 - 5 ml (R5 and R6), because in the treatment of P4 and P5 the content of crude fiber in noni juice too high so that the performance of the less optimum proxeronine enzyme that causes little growth decreased. This is in line with research conducted [14] that the use of noni fruit juice decrease body weight if used of 5% through drinking water, which caused increased crude fiber content proxeronin compounds contained in noni fruit does not work optimally.

3.2. Carcass Weight and Carcass Percentage
The carcass weight and carcass percentage were variation, from analysis of variance showed that by addition noni fruit juice into drinking water until 3 ml did not significant influence (P0,P1,P2,P3) on carcass weight and carcass percentage, but has significant effect ($P < 0.05$) when using 4 ml (P4) and 5 ml in drinking water (P5). These results indicate that using noni juice during maintenance provides and increased response to carcass weight and carcass percentage. This is because the active substances of xeronine in noni juice will reactivate the dead cells so that the respiration process from the cell back to run, the nutrients consumed by the chicken will be absorbed perfectly. The presence of xeronine will balance or normalize body [8]. Noni juice also contains Scopoletine and L-arginine, both active substances can work synergistically in optimizing the absorption of nutrients in the small intestine villi, such as protein. The function of protein is primarily to build muscle or meat. Carcass weight a part of chicken that contains muscle or meat and is proportional to the final body weight, so when calculated on a percentage of final body weight result is relatively the same percentage [15]. Treatment P4 and P5 was the lowest produced on carcass weight and carcass percentage, because proportion of noni juice in P4 and P5 was higher than those in P1,P2 and P3. The limiting factor for the use of noni juice is the content of crude fiber which is consumed poultry through Noni juice. At the treatment P4 and P5 crude fiber content in noni juice too high so the performance of the less optimum proxeronin enzyme that causes final body weight decreased. This is in line with research conducted [14] that the use of fruit juice noni decrease in body weight use of 5% juice noni fruit through drinking water, which caused increased crude fiber content so proxeronine compounds contained in noni fruit does not work optimally.

3.3. Abdominal Fat
The average levels of abdominal fat content of the treatments showed in Table 1. The weight of abdominal fat were variation, where P6 by giving 5 ml of noni juice /1 liter of drinking water is most lowest (1.15 g) and those in P0 without noni juice in drinking water was the highest (3.11 g). Analysis of variance showed by adding 1–3 ml of noni juice /1 liter of drinking water have no significantly effect ($P>0.05$) on the abdominal fat of sentul chicken, but has significant effect $P < 0.05$) when using 4 ml (P4) and 5 ml in drinking water (P5). By adding 4 - 5 ml of noni juice /1 liter of drinking water, these was a tendency levels abdominal fat content going to decreased ($P<.05$). The decrease in abdominal fat because the bioactive substance as xeronine would improved the metabolism of carbohydrate and fat in the carcass. In line with the opinion [6] states that noni fruit contains xeronine active substances that can reduce fat and cholesterol levels.

3.4. Meat Cholesterol
The average meat cholesterol levels were showed at Table 1. Analysis of variance showed that by addition of noni juice into drinking water has significantly effect ($P< 0.05$) on the meat cholesterol of sentul chicken. The result indicated that by treatment adding until 4 ml of noni juice /1 liter of drinking water gave the best results of meat cholesterol. Noni fruit produces antioxidant such as: scopoletin, nitric oxide, vitamin C and vitamin A, and has the efficacy to increase the secretion of bile and substance NO (Nitrit Oxide) that can stimulate the excretion of cholesterol through feces. According [6] Noni fruit contains xeronine active substances that can reduce fat and cholesterol levels. Xeronine has a high activity in the formation of proteins for hormones, one of which is the hormone insulin that can increase the number of LDL receptors (low density lipoprotein) hepatic and extra hepatic [7, 8].

4. Conclusion
It can be concluded that the addition of noni juice into drinking until 3 ml/1 liter the best result on final body weight, carcass percentage, abdominal fat and meat cholesterol Sentul Debu Chicken phase growth.

Acknowledgements
The research work have been conducted in the Grand Research Academic Leadership project. Sources of funds from Padjadjaran University, through the Directorate of Research, Community Service and Innovation Padjadjaran University, Indonesia.

References
[1] Nataamijaya A G, Setioko A R, Brahmantyodan B, Diwyanto K 2003 Performans dan karakteristik tiga galur ayam local (Pelung, Arab, Sentul) *Pros. Seminar Nasional Teknologi Peternakan dan Veteriner.* Puslitbang Peternakan Bogor pp. 353–359
[2] Diwyanto K, Zainuddin D, Sartika T, Rahayu S, Djufri, Arifin C, Cholil 2011 Model pengembangan peternakan rakyat terpadu berorientasi agribisnis: komoditas ayam lokal *Laporan Kerjasama Direktorat Jenderal dengan Balitnak* Ciawi Bogor.
[3] Muryanto, Hardjosworo P S, Herman R, Setijanto H 2002 Evaluasi Karkas Hasil Persilangan Antara Ayam Kampung Jantan dengan Ayam Ras Petelur Betina *J. Anim. Prod.* **4 (2)**: 71−76
[4] Dutta Mrigen 2009 Feed cost savings via the phytase matrix *Poultry Internasional* **48 (2)**: 136 - 137
[5] Wahyu, Juju 2004 Ilmu Nutrien Unggas Cetakan III *Gadjah Mada University Press* Yogyakarta
[6] Salleh, Runnie I, Roach 2002 Inhibitor Of Low Density Lipoprotein Oxidation And Up Regulation Of Low Density Lipoprotein Recept Or In Helpg2 Cell By Tropical Plant Extracts **19(50)**
[7] Linder M C 1992 Biokimia Nutrisi dan Metabolisme dengan Pemakaian secara Klinik Edisi I. Alih bahasa Aminudin Parrakasi *Penerbit Universitas Indonesia (UI-Press)* Jakarta
[8] Heinicke R 2001 Tahitian Noni: What Is Xeronine? *Tahitian Noni Home Page* Http://Www.Hmt.Com/Noni/Active.Html
[9] Maslachah L 2005 Potensi ekstrak mengkudu (*Morinda citrifolia*) terhadap sekresi nitric oxide (NO) dan endothel pembuluh darah hiperk olesterolemia. *Lembaga Penelitian, Unair*
[10] Sjabana D, Bahalwan R R 2002 Seri Referensi Herbal Pesona Tradisional dan Ilmiah *Salemba Medika* Jakarta
[11] Widjastuti T 1996 Penentuan Efisiensi Penggunaan Protein, Kebutuhan Protein dan Energi untuk Pertumbuhan dan Produksi Telur Ayam Sentul pada Kandang Sistem Cage dan Sistem Litter *Disertasi* Program Pascasarjana Unpad Bandung
[12] Bintang I A K, Sinurat A P, Purwadaria T 2008 Penambahan antibiotika dan ampas mengkudu sebagai sumber senyawa bioaktif terhadap performans ayam broiler JITV **13(1)**: 7-12
[13] Amar A, Sumarmo L, Makosim S, Magdalena M, Yulianto D T 2004 Analisis Mikroorganisme, Kandungan Alkohol Dan Asam Lemak Sari Buah Mengkudu Dengan Gas Chromatography *Proseding Seminar Nasional Dan Konggres Perhim - Punan Ahli Teknologi Pangan Indonesia (Patpi)* Jakarta
[14] Nurhayati 2008 Pengaruh Pemberian Jus Buah Mengkudu (*Morinda citrifolia*) dalam Air Minum Terhadap Penampilan Ayam Broiler Jantan. *Fakultas Peternakan Universitas Jambi* Jambi
[15] Widjastuti, Sujana E, Darana S 2012 The Yielding Characteristic of Sentul Chickens Feed Diet Containing Papaya Leaves Meal (Carica papaya L.Less) *Scentific Papers Animal Science Series D* **55**: 99-103

IOP Conf. Series: Earth and Environmental Science **334** (2019) 012033 doi:10.1088/1755-1315/334/1/012033

The Effect of Liquid Organic Fertilizer of Vegetable Waste and Planting Media on Growth and Yield of Strawberry (*Fragaria* spp.) Earlibrite Cultivar

S Ginandjar[1], B Frasetya[1], W Nugraha[1] and M Subandi[1]

[1]UIN Sunan Gunung Djati Bandung, Jl. A.H. Nasution No. 105 Bandung 40614 Indonesia

E-mail : salametginandjar@uinsgd.ac.id

Abstract. Strawberry becomes one of the popular plants in agro-tourism by combining strawberry plantation with culinary tourism. Increasing growth and yield of strawberry plant can be achieved by applying liquid organic fertilizer (LOF) of vegetable waste and planting media. This research was conducted to study interaction between LOF and planting medium (PM) and to inquire about the best dosage of LOF and planting medium for the growth and yield of strawberry. This research was conducted in the sub-district of Cibiru, Bandung City, West Java. The research design was Randomized Block Design with two factors and three replications. The first factor was vegetable waste of LOF with four treatments level respectively: p0: without LOF application, p_1:250 ml, p_2: 500 ml, and p_3: 750 ml. The second factor was planting media with three treatments level respectively : m_0: without husk charcoal, m_1: (1:1) soil and husk charcoal, m_2: (1:2) soil and husk charcoal. The research finding came up by the existence of interaction in leaf width with the highest number by 835.42 cm^2 in combination of p_2m_2 treatment.

1. Introduction

The strawberry plants are herbaceous fruit plants found first in Chile, South America. One species of strawberry plants (*Fragaria choiloensis* L). spread to various countries of America, Europe and Asia. Furthermore, another species, namely *Fragaria vesca* L is more widespread than other species. This strawberry is the first introduce to Indonesia. The strawberries we find in supermarkets are hybrids produced from the North American *Fragaria virgiana* L. var Duchesne cross with *Fragaria Chiloensis* L. var Duchesne from Chile. The crossing resulted in a hybrid of modern strawberries (commercial) *Fragaria* x *annanassa* var Duchesne [1].

Today we encounter many of the restaurants that are integrated with the agro-tourism of strawberry plants as an appeal to consumers. The number of strawberry-based agro-tourism restaurants indicates the large market demand for agro-tourism of strawberry plants. Strawberry plants can be used as an ornamental plant in the yard of the house in addition to beautifying the home page can also be used as an additional nutritional family. Selection of appropriate and appropriate strawberry plant media is expected to make the media easy to carry and have a long-lasting production period and good for the growth of strawberry plants during the production period. Strawberry planting medium is a good planting medium in accordance with the needs of the production of strawberry plants that have a water holding, light, filled with oxygen and has a good soil structure and texture. Strawberry planting media usually use loose soil because strawberry plants can grow well on loose soil.

A lot of rubbish is scattered in landfills, the composition of Bandung's highest garbage is organic waste 63.56%, 10.42% paper, 9.76% Metal, 1.7% Glass, 0.95% Textile Plastic or Rubber 1.45% and others 12.6%. Judging from its characteristics, waste can usually still be processed into a more useful product that has economic value. Even the waste can be used as a new product (Waste to Product) and become energy (Waste to Energy). For example, organic solid waste can still be used for raw materials of organic fertilizer, animal feed, and biogas [2].

Based on the shape, organic fertilizer is divided into two, namely liquid organic fertilizer and solid. Liquid fertilizer is a soluble solution containing one or more nutrient carriers that the plant needs. The advantages of liquid organic fertilizer are to provide nutrients in accordance with the needs of plants. other than that the application can be more evenly and concentration can be arranged according to the plants requirement [3]. Organic material is now promoted by farmers to support the application of organic farming. The number of microorganism population at soil used as an indicator for soil degradation [4]. The organic farming is not allowed the use of artificial chemical substances even the pesticides in the control of pests and diseases. The natural pesticide was proven to be better and more profitable, biological control of pest will avoid ecology contamination with chemical pesticides, and be compatible with other methods of pest control [5].

2. Materials and Methods

This research was conducted at the field laboratory of UIN Sunan Gunung Djati Bandung, West Java. The method used is the experimental (experimental) method, the experimental design used in this study was the Factorial Randomized Block Design (RBD) of two treatment factors with three replications. The first factor is liquid organic fertilizer. Mite waste that is p_0 = without liquid organic fertilizer. Waste, p_1 = 250 ml, p_2 = 500 ml, p_3 = 750 ml is applicated 2 times each stages at each dose. The second factor is Planting medium, respectively; m_0 = Without Charcoal Skin, m_1 = 1: 1 (Soil: Husk Charcoal), m_2 = 1: 2 (Soil: rice husk charcoal). There are 12 treatment combinations were obtained and each treatments was repeated three times.

Research Procedures include the manufacture of vegetable waste LOF (Liquid Organic Fertilizer), media preparation, planting, fertilizing, maintenance, harvesting, and pests and diseases.

a. LOF vegetable waste preparation of liquid organic fertilizer for vegetable waste is done for 2 weeks according to the method of making LOF for vegetable waste.
b. Media plant preparation by mix the soil until soil mix uniformly. The soil than mixed with rice husk charcoal to fulfill the treatments requirement that is m0 = without husk charcoal, m1 = 1: 1 (Soil: Husk Charcoal), m2 = 1: 2 (Land: Husk Charcoal)
c. Planting strawberry that have been 10 cm in size and have been propagated in good condition, fresh, upright, and free from pests and diseases. Then the seedlings are planted into the planting medium by first making the planting hole, with the depth of the root length and then the planting medium is compacted around the root.
d. Planting Seeds of strawberries vegetative propagation that has been measuring 10 cm and has been cultivated in a good state, fresh, upright, and free from pest and disease attacks. Then the seedlings are planted into the planting medium by first making a planting hole, with the depth of the root length and then the planting medium is compacted around the roots.
e. Fertilization Application of fertilization is done on the leaves and soil in the form of liquid organic fertilizer of vegetable waste in accordance with the concentration on the treatment by way of the pour. Application of liquid organic fertilizer vegetable waste done on 7 days before planting on planting medium and 21 DAP (day after planting). Application of artificial fertilizer application was 7.30 and 45 DAP with 75% dose as much as 2.80 gr N cropping, 3.75 g P cropping, and 3.5 g K planting by means of buried on medium parallel to strawberry leaf tip canopy.
f. Cultivation Plant maintenance includes weed weeding by weeding that grows around the plant and applied once every week, pruning is done by pruning old leaves and the disease because it can transmit disease to other plants, stolon pruning is intended that all the energy obtained by plants can be centered on growth and the development of the main crown so that it can bear fruit well.
g. Harvest Strawberry plants bear fruit when aged 75 DAP. Harvesting is done by way of picking by using scissors after matching the characteristics of ready-to-harvest with magenta red ripe traits.
h. Pests and Diseases Control of pests and diseases by sanitation, and physically and chemically. Pests and diseases contained in the study will be physically controlled or the disposal of pests or parts of plants affected by the disease, and if pests and diseases exceed the economic threshold it will be overcome by chemical means that is using pesticides and insecticides.

3. Results and Discussion

3.1. Plant Height (cm)

Plant height is one of the growth observation parameters to know the rate of photosynthate quantitatively, and in this study, the height measurement of the strawberry plant is done at age 30 DAP, 45 DAP and 60 DAP with unit cm. The average Table of observation of plant height can be seen in Table 1. The effect of liquid organic fertilizer on vegetable waste and media type on the growth of plant height gives effect independently. At the age of 30 DAP and 45 DAP treatment p_2 with dose 500 ml and p_3 with dose 750 ml gave significant effect on p_0 without LOF, and did not give significant effect to p_1 with dose 250 ml, while at age 60 DAP, p_3 give significant effect to p_1 and p_2, and p_3 have a very significant effect on p_0, whereas at p_2 it has no significant effect on p_1 and gives a significant effect on p_0.

The effect of plant media use on plant height at age 30 DAP did not give any significant effect on each treatment, and at age 45 DAP gave a significant effect of giving a significant effect on m_2 with comparison of rice husk charcoal (1: 2) compared with m1 with soil and rice husk charcoal (1: 1) and m_0 without charcoal husk, whereas at 60 DAP m_2 gives a significant effect on m_0 and does not give significant to m_1.

The application of liquid organic fertilizer is very good against the growth of plant height in each observation is indicated by there is a significant influence on the dosing of 500 ml and 750 ml but the use of 500 ml doses more efficient at 30 DST and 45 DAP observation age because the dose of 750 ml is not gives a significant effect on the dosing of 500 ml this is because the vegetable waste LOF has enough nitrogen as the protein and plant chlorophyll material [6]. The highest rate of plant growth was demonstrated at 60 DAP at treatment of p3 with dose of 750 ml with a very significant effect on this matter.

Table 1. Effect of Liquid Fertilizer Dosage and Type of Planting Media to Average Plant Height at 30 DAP to 60 DAP

Treatment	Mean of plant height (cm)		
	30 DAP	45 DAP	60 DAP
p_0	3.36 a	4.17 a	4.91 a
p_1	3.52 ab	4.47 ab	5.56 ab
p_2	3.70 b	4.68 b	5.79 b
p_3	3.93 b	4.74 b	5.87 c
Planting Medium			
m_0	3.40 a	4.30 a	5.28 a
m_1	3.80 a	4.61 a	5.52 ab
m_2	3.68 a	4.64 b	5.80 b

The average value of each column marked with the same letter shows a significant difference based on Duncan Multiple Range Test (α= 5%)

Plant media has a significant effect on the treatment of m2 with the ratio of soil medium to charcoal husk (1: 2) compared with m1 and m0, this is because the soil media and charcoal husk can improve the porosity of the media so good for root respiration, maintaining moisture that the growing media need to be considered to support for plants to grow more fertile. Sorts of planting medium that is often used is sand, charcoal husk, and fern, the three kinds of media have a good porosity so that the plant propagation easy to penetrate.

3.2. Leaf Area

There is an interaction on leaf area between liquid organic fertilizer of vegetable waste (LOF) with treatment of p_0 without LOF; p_1 250 ml dose; p_2 500 ml; and p_3 750 ml with medium m_0 without rice husk charcoal; m_1 comparison of soil and charcoal husk (1: 1) and m_2 of husk charcoal (1: 2) ratio. The highest rate was in the treatment of p_2m_2 with the number 835.42 compared with the combination of other treatments and showed a significant effect, while the lowest number in the treatment combination p_0m_0 with the number 208.33 and did not give a significant effect.

Table 2. Effect of Liquid Organic Fertilizer Dosage and Type of Planting Media on Average Area Leaves at Age 75 (DAP)

LOF	Planting Medium		
	m_0	m_1	m_2
p_0	208.33a A	285.42b A	372.92c A
p_1	452.08a C	543.75b B	622.92c C
p_2	385.42a B	52.92b B	835.42c D
p_3	437.50a C	541.67b B	564.58b B

The average number followed by the same letter (capital letters horizontal and vertical italic letters) shows no significant difference according to Duncan Advanced Test at a significant level of 5% p = liquid organic fertilizer of vegetable waste; m = planting medium

Strawberry plants absorb nutrients in the form of ions and cations that have been dissolved by water, liquid organic fertilizer can easily be directly absorbed by the plant because there is no dissolution process, charcoal granting can provide porosity also provide enough oxygen for plants so that nutrients are available for plants, this makes the interaction between LOF and planting media.

The average strawberry leaf area (cm^2) in Table 7 shows p2m2 with a 500 ml LOF dose and media ratio (1: 2) had a significant effect and was the highest number compared with other treatments. This is because with the addition of organic liquid fertilizer dosage of vegetable waste, nutrient availability will be fulfilled, so that if enough nutrients then the leaves will be more extensive and will affect the results of photosynthate to be produced. The fulfilment of plant nutrients, the result of photosynthate is increasing. This is inseparable from the role of organic fertilizer application of liquid vegetable waste, in accordance which states that plants that do not get additional N elements as needed will have a small leaf formed, otherwise plat will get N elements as needed wide leaves [7]. Nitrogen has a significant effect on leaf expansion, especially on the width and extent of the leaves. The element of nitrogen as an element that plays an important role in the growth of plant leaves [8].

Strawberry plants not only require good fertilization but strawberry plants also require planting media that has a good porosity, in Table 7 average leaf area (cm^2) p2m2 as the best treatment with the ratio of land charcoal husk (1: 2), This shows that strawberry plants require a good porosity because charcoal husk is one of the mixture of planting medium that can provide good porosity of soil and is a natural nutrient material that can fertilize the plant because of its crumb and easy structure to store oxygen. Root plants will grow perfectly when sufficient water and oxygen in the soil is available [9].

3.3. Fresh Treaded Weights (g)

The effect of organic fertilizer application of liquid vegetable waste to fresh weight of strawberry berries made a very significant effect on the treatment of p_3 with dose of 750 ml compared to p_0 without vegetable LOF, and significantly more significant compared to p_2 with 500 ml dose and p_1 with 250 ml dose .

The effect of using planting media on fresh weight of strawberry berries has a significant effect on m_1 with soil composition: husk charcoal (1: 1) and m_2 with soil composition: husk charcoal (1: 2) compared with m_0 without charcoal husk, treatment m_1 is most efficient than m_2 because m_2 does not have a significant effect on m1. The average fresh weight of trimmed can be seen in Table 3.

The fresh weight of the tangle is measured by weighing the whole plant part. This means that the fresh, tangled weight is supported by the height of the plant and the number of leaves. The large number of leaves and plant height is directly proportional to the size of the fresh weight of trimming plants. The amount of leaf and plant height depends on the availability of nutrients absorbed. Liquid organic fertilizer vegetable waste plays a role in the provision of nutrients and nutrients for plants.

Table 3. Effect of Liquid Organic Fertilizer Dosage and Type of Planting Media Against Average Freshly Weeded Weights (g)

Treatment	Average of Fresh Weight (g) Age of Plant 75 DAP
LOF	
p_0	30.19 a
p_1	46.09 b
p_2	46.45 b
p_3	56.92 c
Growing Media	
m_0	40.04 a
m_1	42.52 b
m_2	42.18 b

Note: The average value of each column marked with the same letter shows a significant difference based on Duncan Advanced Test at 5%

This is consistent with statement, that plant height and number of leaves affect the fresh weight of plant canopy. The larger the plant height and the greater the number of leaves, the fresh weight of the canopy will increase. Liquid organic fertilizer waste can increase the availability of nutrients Ca, Mg and K surrounding soil and the presence of growth regulators such as auxin, spur leaf formation [10].

Organic fertilizers can provide ingredients of amino acids, proteins and contain substances that grow the auxin that plays a role in spurring root growth and ready to build a network of plant growth. Increased root growth and the formation of tissues on and leaves in large and large quantities can support the increase in fresh weight of plants [11]. Good roots required porous planting medium so that can hope well with good planting media in this study can be seen on m2 which have significant effect on m0 without charcoal husk. Charcoal husk provides porosity to the soil pore space (empty space) contained in a volume of soil that can be occupied by water and air, so it is an indicator of drainage conditions and soil aeration. The porous soil means enough tanh to have a pore space for movement of water and air in and out of the ground freely [12].

3.4. *Plant Dry Weight (g)*

The dry weight of the plant was weighed with analytical scales after the oven at 80 ^{0}C for 3 days. The influence of the use of liquid organic fertilizer of vegetable waste to the dry weight of strawberry plants had a very significant effect on the treatment of p_3 with a dose of 750 ml compared with p_0 without liquid organic fertilizer of vegetable waste, and p_3 influenced compared with p_2 with dose 500 ml and p_1 250 ml.

The effect of plant media use gave a significant effect on the treatment of m_2 with the ratio of soil and charcoal husk (1: 2) to m1 with the ratio of soil and husk charcoal media (1: 1) and m0 without using husk charcoal (Table 4).

Growth and development of plant tissue will cause the increase in the number of leaves, the more widely formed leaves, stems and roots are getting bigger so that the weight of plat and dry weight will increase. Provision of liquid organic fertilizer vegetable waste with a dose of 750 ml produces plant height, leaf area, fresh weight of plat and also the highest dry weight of strawberry plants.

Table 4. Effect of liquid organic fertilizer vegetable waste and planting media on average dry weight of strawberry plants (g)

Treatment	Average of Plant Dry Weight (g)
LOF	
p_0	6.55 a
p_1	8.58 b
p_2	9.05 b
p3	11.62 c
Growing Media	
m_0	8.05 a
m_1	8.90 ab
m_2	9.90 b

Note: The average value of each column marked with the same letter shows a significant difference based on Duncan Multiple Range Test at 5%

This is certainly related to the increasing number of nutrients that can be supplied and absorbed by strawberry plants as a result of increasing doses of liquid organic fertilizer vegetable waste is given. This result confirmed previous research that the dosage of nitrogen fertilizer used can change the plant physiology variables including leaf chlorophyll, leaf specific weight, photosynthesis rate, light efficiency reduces CO_2, and leaf maintenance respiration coefficient.

The average dry weight of the plant shows the highest number at m_2 level with the number 9.90 and has a significant effect on m1 and m0, because if the charcoal husk is added to the soil and the plant will be able to give good porosity, then released to the micro pore to be absorbed by the plant and promotes the growth of useful microorganisms for soil and plants. This is confirmed that charcoal husk is not easily decayed and become the source of potassium needed by the plant, besides that this material does not quickly coagulate or solidify so that plant roots can grow perfectly.

Table 5. Effect of liquid organic fertilizer of vegetable waste and planting media on average fruit crop weight (g)

Treatment	Average of Weight per Plant (g)
LOF	
p_0	4.59 a
p_1	6.52 b
p_2	8.24 c
p_3	8.61 d
Planting Media	
m_0	6.51 a
m_1	6.72 b
m_2	7.75 c

The average value of each column marked with the same letter shows a significant difference based on Duncan Multiple Range Test at 5%

3.5. *Fruit Weight (g)*

Measurement of fruit weight by weighing fruit one by one in one sample of treatment, performed simultaneously at 75 DAP. The effect of organic fertilizer application of liquid vegetable waste to fruit weight of planting showed that p_3 with doses of 750 ml had a very significant effect on p_1 with a dose of 250 ml and a significant effect on p_2 with a dose of 500 ml.

The effect of planting media on strawberry crop production has a very significant effect on m_2 level with the proportion of soil and 1: 2 husk to m_0 without charcoal husk, and m_2 has significant effect on m_1 with ratio of 1: 1 husk charcoal soil (Table 5).

The application of adequate nutrients can increase the production of strawberry crops so well, by providing liquid organic fertilizer at several treatment doses show a significant difference. This is consistent which states that plant nutrients and water availability affect the growth or expansion of cells such as vegetative or plant-fertilizing organs [13]. The increase of fruit weight based on planting medium in each comparison gives a significant effect because charcoal husk as a mixture of planting medium can provide nutrients for this plant because in charcoal husk has potassium and phosphor. The husk charcoal is the residual solid of combustion of non-volatile organic matter, may increase pH and liberate or increase essential soil such as potassium, magnesium, potassium and phospor, so as to increase crop yield.

3.6. *Quality of Fruit*

The quality of fruit is sorted by its classification of A grade> 20 g, grade B 12-20 g, grade C 8-12 g, and grade D 7-8 g, fruit yield average yield on p_2m_1, p_3m_0, and p_3m_2 with average on each replication obtained Grade C (8-12 g) (Table 6). Not all treatments get grading and a decrease The quality of fruit in strawberry plants occurs due to disturbed metabolism in plants strawberries as a result of an incompatible growth requirement.

Table 6. Effect of liquid organic fertilizer vegetable waste and planting medium on fruit quality

Treat ment	Replication			Grade	
	I	II	III	C(8-12 g)	D(7-8 g)
p_0m_0	3.35	4.78	3.55		
p_0m_1	3.90	3.83	5.20		
p_0m_2	4.40	5.15	7.20		1
p_1m_0	5.78	5.78	7.58		1
p_1m_1	6.60	6.33	6.00		
p_1m_2	7.15	6.83	6.63		1
p_2m_0	6.88	7.40	7.78		2
p_2m_1	5.98	8.00	9.83	1	1
p_2m_2	8.35	9.40	10.58	3	
p_3m_0	8.38	8.60	8.25	3	
p_3m_1	7.93	9.55	7.55	1	2
p_3m_2	9.55	9.18	8.55	3	
Total				11	8

This is because strawberries can grow optimally at an altitude of 1,000-1,200 m above sea level. The Strawberry plants can grow well in areas with rainfall 600-700 mm per year [14]. The duration of sunlight exposure required in growth is 8 to 10 hours per day. Strawberries are subtropical plants that can adapt well in the tropical highlands that have a temperature of 17-20 ^{0}C. Good air humidity for the growth of strawberry plants between 80-90%.

There is the fruit crop the so called strawberry which is famous as refreshing fruit and served as juice and drink after or before lunch. It is a value of domestic plant and there are many natural plants producing valuable usage for human both food and medicine. Reserpine is the main component of serpentine which accumulated in roots and is known to have antihypertensive properties [15]. The active compounds in the root of serpentine can lower the high blood pressure than other alkaloids. And another plant of dual benefits as refreshing and medicine is the dragon fruit plant. Dragon fruit is delicious to serve as desert and it also contains medicinal effect. The prebiotic properties of dragon fruit performed because of carbohydrate in dragon fruit is white and red are glucose, fructose and some classes of oligosaccharides with a total concentration of between 86.2 g/kg and 89.6 g/kg which has a function as a prebiotic [16].

4. Conclusions

There is an interaction of the observation of leaf area between the provision of liquid organic fertilizer (LOF) of vegetable waste and media type in earlbrite variety strawberry plant, with the best treatment combination on p_2m_2 treatment combination with the highest number of 835.42 cm^2. The best LOF dosage was found in treatment of p_3 with dosage of 750 ml and the best media type was m_2 with 1: 2.

References

[1] Darwis V 2007 Cultivation, Analysis of Farming, and Partnership of Bali Tabanan Strawberries Jakarta: *Center for Socio-Economic and Agricultural Policy Analysis.*

[2] Wahyono S, Firman L S, Feddy S 2011 Making Organic Granul Fertilizer of Various Waste Jakarta: *Agromedia.*

[3] Sukamto H 2012 Making Liquid Compost Fertilizer South Jakarta: *PT Agromedia Pustaka.*

[4] Qurrohman B F T, Suriadikusuma A, Haryanto R 2014 Analisis Potensi Kerusakan Tanah Untuk Produksi Ubi Kayu (Manihot utilisima) Pada Lahan Kering Di Kecamatan Tanjungsiang, Kabupaten Subang *Jurnal Agro* **1(1)** Retrieved from http://repository.unpad.ac.id/20083/1/Analisis-Potensi-Kerusakan-Tanah-Untuk-Produksi-Ubi-Kayu.pdf

[5] Subandi M, Setiati Y, Mutmainah N H 2017 Suitability of Corcyra cephalonica eggs parasitized with Trichogramma japonicum as intermediate host against sugarcane borer Chilo auricilius *Bulgarian Journal of Agricultural Science* **23(5)**: 779-786.

[6] Salisbury F B, Ross C W 1995 Plant Physiology III Development of Plant and Environmental Physiology Translation Lukman D R, Sumaryono Bandung: *ITB Publisher.*

[7] Lakitan B 2001 Basics of Plant Physiology Jakarta: *Raja Grafindo Persada.*

[8] Bojovic B, Markovic A 2009 Correlation between nitrogen content and chlorophyll in wheat (*Tritikum aestivum* L.) *Kragujevac. J. Sci.* **57(1):** 69-74.

[9] Septiani D 2012 The Influence of Charcoal Husk Giving on the Growth and Results of Rawit Chili Plants Lampung State Polytechnic: *Seminar on Holticulture Studies Program.*

[10] Fahrudin F 2009 Caisim Cultivation Using Tea Extract and Kascing Fertilizer Surakarta: *Sebelas Maret University.*

[11] Kariada I K, Sukadana I M 2000 Organic Vegetables *Agricultural Science* **11(2)**: 35-42. Via http: // Pustaka_ Deptan. Go.Id/Agritek/Bali0208. [19/12/2014].

[12] Judge N M, Nyakpa Y, Lubis A M, Nugroho S G, Soul M R, Dhina M A, Hong G B, Bailey H H 1986 Fundamentals of Soil Science *Faculty of Agriculture* University of Lampung.

[13] Gardner P F, Pearce R B Mitchell R L 1991 Physiology of Crop Plant Cultivation Plant Physiology translation Jakarta: *University of Indonesia.*

[14] BPP Technology 2000 About Strawberry Cultivation Jakarta: *Field of Utilization and Socialization of Science and Technology (BPPT).*

[15] Subandi M, Dikayani, Firmansyah E 2018 Production of reserpine of Rauwolfia serpentina (L) kurz ex benth through in vitro culture enriched with plant growth regulators of NAA and kinetin *International Journal of Engineering & Technology* **7(2.29)**: 274-278.

[16] Subandi M, Mustari E, Setian A 2018 The Crossing Effect of Dragon Fruit Plant Caltivars (*Hylocereus* sp.) On Yield *International Journal of Engineering & Technology* **7**: 762-765.

Allelopathic Screening of Several Weed Species as Potential Bioherbicides

G Erida[1,2], N Saidi[3], Hasanuddin[2] and Syafruddin[2]

[1]Doctoral Program in Agricultural Sciences, Syiah Kuala University, Darussalam, Banda Aceh 23111, Indonesia;
[2]Department of Agrotechnology, Faculty of Agriculture, University of Syiah Kuala, Darussalam, Banda Aceh 23111, Indonesia
[3]Department of Chemistry, Faculty of Mathematics and Natural Sciences, University of Syiah Kuala, Darussalam, Banda Aceh 23111, Indonesia

E-mail: ginaerida@gmail.com

Abstract Many plants that are considered to be weeds have the potential to exhibit allelopathy; they might therefore be used as a herbicides to control other weeds. The aim of this study was to identify the best potential bioherbicide against spiny *Amaranthus spinosus* growth at various concentrations from among five weed species: *Imperata cylindrica*, *Cyperus rotundus*, *Chromolaena odorata*, *Ageratum conyzoides*, and *Axonopus compressus*. The study followed a non-factorial, completely randomized design, examining 17 treatments with three replicates. Extracts of the leaves, tuber, and/or rhizome of the five sample plants were obtained using maceration and were applied to the indicator weed *A. spinosus* at concentrations of 10%, 20% and 30%. A synthetic herbicide (2,4-dichlorophenoxyacetic acid at 0.686 kg/ha) was used as a positive control and distilled water as a negative control. The application of allelopathic extracts of the weeds at various concentrations significantly affected the height growth rate, the percentage of weed control, leaf area, root length, and shoot and root dry weight growth of *A. spinosus*. The greatest inhibition on *A. spinosus* growth at 7 days after application was observed with a 20% concentration of *A. conyzoides*, followed by 20% concentrations of *C. rotundus*, *C. odorata*.

1. Introduction

Weeds growing in crop fields compete with the planted crops for light, moisture, and other essential nutrients [1]. In addition, some weeds interfere with the crop plants by producing allelochemicals, which inhibit the growth and development of the crops [2-4], and some act as hosts for pests and diseases [5]. As a result, weeds can be serious pests that can damage crops, reducing the quality and yield and increasing the cost of production.

Weeds are generally controlled either through conventional methods or by using herbicides. Although chemical herbicides can be effective for controlling weeds, they are associated with risks such as a negative impact on soil [6], the development of resistance to herbicides [7], harm to non-targeted organisms, the disturbance of the ecology as a whole, and remnant chemical residues in the environment [8]. An increased awareness of risks involved in the use of herbicides has resulted in greater emphasis on searching for alternative methods of weed control that are safe for the environment. One such approach is allelopathy, which is recognized as a natural method of weed control [3,9-12].

Allelopathy can be defined as the effect of one plant on another plant through the release of chemical compounds into the environment [13]. Allelochemicals can be advantageous, stimulating the target organism, or they may harm the other plant [13,14]. A great variety of natural compounds are known to cause allelopathy, with secondary metabolites constituting the most important group of allelochemicals [13,15]. Known allelochemicals include simple water-soluble organic acids and unsaturated lactones, long-chain fatty acids and polyacetylenes, naphthoquinone, anthraquinones and complex quinones, simple phenols, benzoic acid and derivatives, cinnamic acid and derivatives, flavonoids, tannins, terpenoids, steroids, amino acids and polypeptides, alkaloids and cyanohydrins, sulfides and glucosides, purines and nucleotides, coumarins, thiocyanates, lactones, and acetogenins [16]. It has been reported that allelochemicals are present in almost all plant tissues, stems, roots, leaves, and fruits. Allelochemicals are released from the plant residues via various processes such as root exudation, leaching, decomposition, and volatilization [13,15].

Perennial weeds have allelopathic potential that can severely affect crop survival and productivity. Such weeds were used in the present study. Cogon grass (*Imperata cylindrica*) extracts have inhibitory effects on the seed germination and seedling growth of butterfly pea (*Centrosema pubescens*) [17]. In addition, extracts of ≥10% of concentrations of *I. cylindrica* significantly reduced the early seedling growth of the weed *Parthenium hysterophorus* [18]. A previous investigation showed that 15% concentrations of purple nutsedge (*Cyperus rotundus*) extracts reduced the germination percentage of finger millet (*Eleusine coracana* Gaertn.) by 48% and 60% [19]. This extract has also been shown to inhibit the germination and growth of several broadleaf weed species [20] and of maize [21]. Conversely, blanket grass (*Axonopus compressus*) may increase the yield of oil palm crops when used as a soil cover, when compared to beans [22]. Siam weed (*Cromolaena odorata*) leaf and root extracts have been shown to inhibit the germination and growth of 10 herbaceous plants, with the leaf extracts providing the strongest inhibition [23]; at 15% concentration, they inhibit the germination and growth of *C. pubescens* seedlings [17], the germination of chickpea (*Cicer arietinum*) seeds in the laboratory [24], and the germination of mung beans (*Vigna radiata*) and the shrub *Mimosa invisa* [25]. Intercropping annual billygoat weed (*Ageratum conyzoides*) in citrus orchards may effectively suppress weeds and control insect pests and diseases [26,27,28]. It has the potential to be used as a natural herbicide for weed control in paddy fields [29]; however, leaf debris of *A. conyzoides* deleteriously affect the easy growth of rice by releasing water-soluble phenolic acids into to the soil [12].

The essential aspect of a bioassay test for identifying allelopathy is to detect allelopathic action in the selected species [30,31]. The most common target species are plants with fast germination, uniformity, and sensitivity. For this reason, spiny amaranth (*Amaranthus spinosus*) was selected as the target plant for the present study because it is known to exhibit rapid germination, uniformity, and sensitivity [32], and it can behave as an active competitor [33].

The aim of this study was to identify the best potential bioherbicide against the growth of *A. spinosus* from among five weed species, described above, that are known to have an allelopathic influence on other plants, by investigating extracts at various concentrations.

2. Materials and methods

2.1. Design and setting

This study followed a non-factorial, completely randomized design consisting of 17 treatments with three replicates per treatment and two plants per treatment. The negative control was distilled water, and the positive control was a conventional herbicide, 2,4-dichlorophenoxyacetic acid (2,4-D), at 0.686 kg/ha. The treatments comprised various concentrations (10%, 20%, and 30%) of extracts from all parts of five plants: *I. cylindrica* (rhizome), *C. rotundus* (tuber), *C. odorata* (leaves), *A. conyzoides* (leaves), and *A. compressus*. The indicator weed *A. spinosus* was grown from seeds.

The study was conducted from May to October 2015 at the Science of Weed Laboratory, Faculty of Agriculture, the Laboratory of Biology (Herbarium) and the Laboratory of Chemistry, Faculty of Mathematics and Natural Sciences, and the Experimental Farm, Faculty of Agriculture, Syiah Kuala University, Indonesia (Unsyiah).

2.2. Sample preparation of plant extracts and the indicator plant

Samples of *I. cylindrica*, *C. rotundus*, and *C. odorata* were obtained from Meunasah Jambe, Sigli, Pidie, *A. conyzoides* from Indrapuri, Aceh Besar, and *A. compressus* from the gardens of the Faculty of

Agriculture, Unsyiah, Banda Aceh. The *A. spinosus* samples were obtained from Meunasah Gle, Sigli, Pidie. The collected samples were identified by a botanist at the Faculty of Mathematics and Natural Sciences Herbarium, Biology, Unsyiah.

Optimally grown leaves of *A. conyzoides* and *C. odorata*, tubers of the same color and shape of *C. rotundus*, rhizomes of *I. cylindrica*, and all parts of *A. compressus* were dried for 2 weeks at room temperature, cut into pieces, and then crushed with a mortar and pestle. Approximately 1 kg (dry weight) of material was obtained from each plant species. The mashed material was transferred into a can and was extracted using methanol for 24 h by maceration. The resulting filtrate was then evaporated using a rotary evaporator (R-210/R-215, Buchi Corp.) at a temperature of 40°C to yield a concentrated extract.

2.3. *Phytochemical testing*

Phytochemical testing for alkaloids, steroids, terpenoids, saponins, and flavonoids was performed in the organic chemistry laboratory of the Natural Materials Division at Unsyiah. For the alkaloid test, 0.5 g samples extracted from the source plant were basified using ammonia and were then added to 1 mL of chloroform and crushed. The filtrate was added to 10 mL 5% hydrochloric acid, shaken vigorously, and kept still until the hydrochloric acid and chloroform formed two separate layers. The hydrochloric acid layer was divided among three tubes, each of which was used for testing the presence of alkaloids by adding Mayer's, Dragendorff's, and Wagner's reagents. The presence of alkaloids was also tested using thin-layer chromatography, in which a small amount of concentrated ammonia was added to a fresh 0.5 g sample of the plant extract. This was stirred until mixed, left for an hour, and the same amount of dichloromethane was added and left for 30 min until separation occurred. The dichloromethane fraction was speckled on a chromatography plate and eluted using ethyl acetate and hexane (8:2) and inserted into the chamber. After removing it from the chamber, it was inserted into an acid cabinet and evenly sprayed with Dragendorff's reagent and then heated on a hotplate until the color changed to brown.

Terpenoid and saponin tests were performed to identify steroids. In these tests, 0.5-g samples from the source plant were finely crushed and extracted with hot methanol. The filtrate obtained was concentrated to obtain the methanol extract, which was extracted again with dichloromethane. The dichloromethane layer was tested for steroids and terpenoids using the Liebermann–Burchard reagent. The insoluble residue in dichloromethane layer was vigorously shaken. The presence of terpenoids and steroids was also tested using thin-layer chromatography. A 0.5-g sample was diluted with 1 mL of ethyl acetate and then speckled on a thin-layer chromatography plate and eluted using the ethyl acetate eluent system described above for approximately a minute or until the boundary mark was reached. The plate was then left to dry, sprayed with vanillin sulfate and heated on a hotplate until a color change was noted, which indicated steroids, terpenoids, or if a stable form was noted after approximately 30 min, which indicated the presence of saponins. This was then hydrolyzed with up to 4 mL of 2 N hydrochloric acid and filtered and tested for triterpene and steroid saponins using the Liebermann–Burchard reagent.

For the flavonoid test, 0.5 g leaf samples were extracted using methanol and then concentrated; the concentrated sample was then extracted again using n-hexane. The residue was extracted with 10 mL of 80% ethanol and added to 0.5 mg magnesium metal and 0.5 M hydrochloric acid. A pink or purple color indicated the presence of a flavonoid. The presence of phenolic compounds was tested by treatment with FCl_3 solution [34].

2.4. *Preparation of growth media and planting spiny amaranth (A. spinosus)*

The topsoil layer was obtained from Lampakuk Village, Aceh Besar, taken to a depth of 20 cm. The soil was dried for 7 days, separated from the plant remains, sieved, and put into 1 kg plastic pots. Selected *A. spinosus* seeds were soaked in water for 2 h and grown in the pots (with up to five seeds per pot). At 7 days after planting, selective pruning reduced the number of plants to one per pot. The selected plants were those that exhibited good growth and showed uniformity. The plants were watered (200 mL/pot of water purified by reverse osmosis) twice a day at regular intervals. Any water that flowed out through the pot holes was returned to the pot.

2.5. *Application of plant extracts*

Plant extracts were applied to each experimental treatment apart from the control. Water was calibrated before the application to determine the amount of solution required (8 mL for each plant). The liquid extract was separately sprayed on all parts of the plant for each treatment and replicated.

2.6. *Measures*

Percentage control of *A. spinosus* observed at 7, 14, and 21 days after the application (DAA) of each extract were recorded. The level of inhibition was assessed by five observers using a 0–100 rating system of weed control (Table 1).

Table 1. The rating system used to assess weed control

Rating (%)	Description of main categories	Detailed description
0	No effect	No weed control
		No crop reduction or injury
		Very poor weed control
10		Slight crop discoloration or stunting
		Poor weed control
20	Slight effect	Some crop discoloration, stunting, or stand loss
		Poor to deficient weed control
30		Crop injury more pronounced, but not sustained
40		Deficient weed control
		Moderate injury, crop usually recovers
50	Moderate effect	Deficient to moderate weed control
		Crop injury lasts longer, recovery doubtful
60		Moderate weed control
		Lasting crop injury without recovery
70		Weed control somewhat less than satisfactory
		Heavy crop injury and stand loss
80	Severe effect	Satisfactory to good weed control
		Crop nearly destroyed, with only a few surviving plants
90		Very good to excellent weed control
		Only occasional crop plants left alive
100	Complete effect	Complete weed destruction
		Complete crop destruction

Source: Frans and Talbert, 1977

Weed height rate was measured with a ruler from the bottom of the stem to the top of the plant. Measurements were conducted at 0–7, 7–14, and 14–21 DAA. The formula used to find the height growth rate was as follows:

$$HGR = \frac{H2-H1}{T2-T1} \quad (1)$$

Where HGR is the height growth rate; H1 and H2 are the starting and ending heights, respectively; and T1 and T2 are the start and end times.

At 21 DAA, the leaf area was measured using a leaf area meter, and the root length was measured by pulling the plant from the pot and measuring the roots using a ruler. Dry weights of shoots and roots were measured by placing them in an oven for 48 h at 60°C and then weighing them using an analytical scale.

2.7. *Data analysis*
All the observed parameters were analyzed using analysis of variance. If an effect of treatment was observed, post hoc testing was performed using Duncan's new multiple range test at a 5% probability level. The analyses were performed using the SPSS version 16 (SPSS Inc., Chicago, IL).

3. Results and discussion

3.1. *Phytochemical testing*
The results of the phytochemical tests are shown in Table 2.

Table 2. Results of phytochemical testing of plant extracts

Plant Species	Phytochemical Content					
	Alkaloid	Terpenoid	Saponin	Steroid	Flavonoid	Phenolic
I. cylindrica	−	+	−	+	+	+
C. rotundus	−	+	−	−	+	+
C. odorata	−	+	−	+	+	+
A. conyzoides	−	+	−	+	−	+
A. compressus	−	+	−	+	+	+

+ Present; − Not present.

All five plant extracts contained terpenoids and phenolic compounds. All except *C. rotundus* contained steroids, and all except *A. conyzoides* contained flavonoids. None of the extracts contained saponins. A previous study found that *I. cylindrica* contained saponins, tannins, arundoin, femenol, isoarborneol, cylindrin, stigmasterol, simiarenol, campesterol, β-sitosterol, scopoletin, scopolin, p-hydroxybenzaldehyde, catechol, chlorogenic acid, oxalate acid, d-malic acid, citric acid, potassium, and 5-hydroxytryptamine and its leaves contained polyphenolic compounds [35]. *C. rotundus* has been found to contain cyperene, flavonoids, sitosterol, and ascorbic acid [36], whereas *C. odorata* has been found in various studies to contain palmitic acid, flavonoids, alkaloids/terpenoids, tannin/polyphenols, triterpenoids, steroids, and saponin [37]; and tannins, phenols, alkaloids, coumarin, and flavonoids [38]. Leaves of *A. conyzoides* contain alkaloids, flavonoids, tannins, saponins, cardiac glycosides, anthraquinones, and terpenoids along with minerals, vitamins, and other compounds of known pharmacological activity [39]. It has been reported that *A. conyzoides* contains mono- and sesquiterpenes, chromene, chromone, benzofuran, and coumarin as well as flavonoids, triterpenes, and steroids [40]. In addition, another researcher found that *A. conyzoides* L contained ageratochromene and its derivatives, monoterpenes, and sesquiterpenes as major volatile components as well as flavones under adverse conditions [26,27].

3.2. *Height growth rate of A. spinosus*
The results of analysis of variance applied to the plant extracts at various concentrations showed a highly significant effect on the height growth rates at 0–7, 7–14, and 14–21 DAA. Mean values for height growth rates of *A. spinosus* are presented in Table 3.

Table 3 shows that 20% of *A. conyzoides* extract was the most effective for inhibiting the height growth rate of *A. spinosus* at 7 DAA. This concentration may kill 100% of *A. spinosus* as a target plant. The inhibition of height growth rate can be prevented using allelochemical compounds contained in the extracts, which can inhibit the hormonal activity facilitating cell division and elongation in the apical shoot and root areas. Results of phytochemical test showed that *A. conyzoides* contains allelochemical compounds such as phenols, terpenoids, and steroids (Table 2). In a previous study, it has been reported that allelochemical compounds inhibit growth by inhibiting auxin activity during cell division and elongation [41].

Table 3. Height growth rate of *A. spinosus* following treatment with plant extracts

Treatment		Average height growth rate (cm)					
		0–7 DAA		7–14 DAA		14–21 DAA	
2,4 D		2.34	def	0.00	a	0.00	a
Distilled water		2.66	def	2.77	de	3.09	ef
	10%	2.16	c-f	2.73	cd	3.18	f
I. cylindrica	20%	2.79	f	3.38	def	3.19	f
	30%	2.90	ef	3.42	ef	3.04	ef
	10%	2.23	c-f	3.01	d-f	2.67	c-f
C. rotundus	20%	2.62	c-e	3.04	d-f	2.68	c-f
	30%	1.83	cd	2.87	d-f	2.33	cd
	10%	2.59	d-f	3.24	d-f	2.79	d-f
C. odorata	20%	2.65	d-f	3.10	d-f	2.79	d-f
	30%	2.21	c	2.24	c	2.13	c
	10%	1.91	cd	2.91	d-f	2.40	c-e
A. conyzoides	20%	0.00	a	0.00	a	0.00	a
	30%	1.81	b	1.45	b	1.35	b
	10%	2.99	ef	3.47	f	3.18	f
A. compressus	20%	2.95	ef	3.26	d-f	2.91	d-f
	30%	3.00	d-f	3.13	d-f	2.73	d-f

Numbers followed by the same letter in the same column were not significantly different at the 5% probability level (Duncan's new multiple range test). Data were transformed by $\sqrt{x} + 0.5$. DAA: Days after application; 2,4 D: 2,4-Dichlcrophenoxyacetic acid.

The inhibition of height growth rate also occurs through the activity of phenolic compounds, which inhibit mitosis in the embryo that results in the inhibition of cell division at the growing point, affecting the height growth rate [13,42]. Phenolic compounds destroyed the spindle threads during metaphase, reducing the mitosis index by producing excessive prophasic condensation and metaphasic chromosomes. This ultimately caused metaphase accumulation and a significant increase in cell percentage through chromosome deviations. This primarily resulted in a change in microtubular axis formation, which subsequently resulted in the formation of some spindle rods and asymmetric convergence from the chromosomes. When cell proliferation is inhibited, cell propagation in plant organs may also be inhibited; growth may therefore decrease or stop, and the cell numbers and their sizes may not increase [43].

Phenolic compounds reduced the synthesis of total carbohydrate, protein, and nucleic acids (both DNA and RNA) owing to the distribution of the phosphorylation channel or inhibition of the activation of Mg^{2+} and ATPase activity; in addition, they inhibited non-absorption and disturbed the biosynthesis process and osmotic inhibition [44]. Allelochemicals especially phenols, can reduce cell membrane permeability [45]. Cell elongation may be inhibited by phenolic compounds that affect the hormone gibberellin, which promotes cell division, cell enlargement, and stem elongation; gibberellin inhibition may disturb cell division in the meristematic tissue, thus inhibiting elongation of stem internodes [46].

Secondary metabolite compounds from the terpenoid group may act as plant growth inhibitors [47]. Neophytadiene, a diterpene terpenoid contained in *Nepeta rtanjensis* (a plant from the Balkans used for essential oils) and catnip (*Nepeta cataria*), inhibits growth in the shoots of ragweed (*Ambrosia artemisiifolia*) [48]. Steroids have the potential for allelopathy and are able to inhibit the germination of the grass *Echinochloa crus-galli* and the legume *Senna obtusifolia* [49].

3.3. Percentage growth control of A. spinosus

The results of analysis of variance applied to the plant extracts at various concentrations showed a highly significant effect on the percentage control of *A. spinosus* at 7, 14, and 21 DAA. Mean values for percentage of control of *A. spinosus* are presented in Table 4.

Table 4. Mean percentage growth control of *A. spinosus* by the plant extracts

Treatment		The Mean percentage control of *A. spinosus* (%)					
		7 DAA		14 DAA		21 DAA	
2,4 D		95.83	j	100.00	j	100.00	i
Distilled water		0.00	a	0.00	a	0.00	a
	10%	16.00	b	45.67	bc	58.00	cd
I. cylindrica	20%	24.17	c	47.17	cd	62.50	de
	30%	34.50	d	52.17	de	65.50	de
	10%	50.17	e	40.00	b	40.83	e
C. rotundus	20%	85.83	h	67.67	f	67.67	ef
	30%	91,50	i	76.67	g	73.33	f
	10%	49.67	c	51.67	dc	35.17	b
C. odorata	20%	78.00	g	74.33	g	62.50	de
	30%	93.83	i	93.33	h	82.33	g
	10%	58.17	f	56.67	e	50.83	c
A. conyzoides	20%	100.00	k	100.00	j	100.00	i
	30%	96.83	j	95.83	i	94.17	h
	10%	13.33	b	41.67	bc	51.67	c
A. compressus	20%	15.50	b	45.19	bc	60.00	de
	30%	28.83	c	51.67	de	66.67	ef

Numbers followed by the same letter in the same column were not significantly different at the 5% probability level (Duncan's new multiple range test). Data were transformed by arcsin. DAA: Days after application; 2,4 D: 2,4-Dichlorophenoxyacetic acid.

Table 4 shows results for the assessed percentage growth control of *A. spinosus*. Notably, the 20% concentration of *A. conyzoides* extract had the highest control percentage of 100% at 7 DAA. These scores were categorized as a complete effect [50]. Aprevious study showed that a 20% extract of *A. conyzoides* had a greater inhibitory effect than that of *Cleome viscosa* on the growth of *Sesamum indicum*. L [51].

The field observations recorded that following the application of the 20% extract of *A. conyzoides*, the leaves of *A. spinosus* had the appearance being burnt and rolling down on the first day. This was followed by a partial loss of leaves on the second day. On the third day, all the leaves fell off and the shoots died, even though the stem remained upright. By the fifth day, all *A. spinosus* plants had died.

Table 4 also shows that at 7 DAA, the application of the *C. odorata* extract had a greater effect on *A. spinosus* than the application of the *I. cylindrical* extract. This is in accordance with the findings in a previous study on the germination and seedling growth of *Centrocema pubescens* [17].

Various responses occur due to higher concentrations and the selective nature of the effects of allelochemicals on target plants [10,32,52,53]. Allelochemicals stimulated or inhibited plant growth depending on their concentration [54,55].

At DAA 7, 30% *C. odorata*, *C rotundus*, and *A. conyzoides* extracts did not result in the death of *A. spinosus*, although many remnants of extracts were observed on the leaves, which gave the leaves a burnt and rolling down appearance. The percentage control of *A. spinosus* following the application of 30% *C. odorata* and 30% *C. rotundus* decreased over the longer observation time period. This is in accordance with the previous findings that reported that the active ingredients of bioherbicides are readily biodegradable [13]. In the present study, the percentage control of *A. spinosus* following the application of *I. cylindrica* extract increased over time. This is in contrast to the other researcher who reported that bioherbicide qualities of *I. cylindrica* were unstable [6]. The concentration of the extract affects the control level and mortality rates of the indicator plant. However, if the concentration is not sufficient to cause death, then natural activities of life still continue [56]. Following the application of *I. cylindrica* and *A. compressus* extracts, the stem of *A. spinosus* appeared abnormally pale compared to its original color 7 DAA. It has been reported that the application of extracts of *I. cylindrica* and *C. rotundus* at concentrations of 30%, 45%, and 60% reduced the growth of *A. spinosus* [57]. In the present study, the application of a synthetic herbicide (2.4 D at a dose of 0.686 kg/ha) caused the death of *A. spinosus* by 14 DAA. Thus, the application of 20% *A. conyzoides* extract resulted in faster death of *A. spinosus* than the application of the synthetic herbicide.

At 7 DAA, the second highest inhibition of *A. spinosus* was observed with the application of 30% *A. conyzoides* extract followed by 30% *C. odorata* extract; this result was not significantly different from that observed with the application of the *C. rotundus* extract. Furthermore, a high level of inhibition of *A. spinosus* was also seen following the application of *I. cylindrica* and *A. compressus* extracts.

3.4. Leaf area, root length, and shoot and root dry weights of A. spinosus

The results of analysis of variance applied to the plant extracts at various concentrations showed a highly significant effect of these extracts on the leaf area, root length, and dry weight of shoots and roots. Mean values for the aforementioned parameters of *A. spinosus* are presented in Table 5.

Table 5. Mean leaf area, root length, and shoot and root dry weights of *A. Spinosus* following the application of the plant extracts.

Treatment		Leaf Area (cm^2)		Root length (cm)		Dry weight of shoot (g)		Dry weight of root (g)	
2,4 D		0.00	a	0.00	a	0.58	a	0.00	a
Distilled water		37.17	bc	33.17	c	8.47	c	2.30	e
	10%	31.50	bc	23.00	ab	31.50	b	0.85	bc
I. cylindrica	20%	36.60	bc	25.08	bc	36.60	b	0.59	b
	30%	36.95	bc	30.50	bc	36.95	b	0.95	bc
	10%	32.37	bc	31.42	bc	32.37	B	0.62	b
C. rotundus	20%	38.12	bc	32.17	bc	38.12	B	0.85	bc
	30%	30.45	bc	35.50	c	30.45	B	0.67	b
	10%	39.88	bc	31.75	bc	39.88	B	0.98	b-d
C. odorata	20%	41.40	bc	26.67	bc	41.40	b	0.65	b
	30%	31.33	bc	22.17	bc	31.33	b	0.92	bc
	10%	67.98	c	27.33	bc	67.98	b	0.98	b-d
A. conyzoides	20%	0.00	a	0.00	a	0.00	a	0.00	a
	30%	26.12	b	14.17	b	26.12	a	0.57	b
	10%	51.12	cd	31.25	bc	51.12	c	1.72	de
A. compressus	20%	53.43	cd	37.92	c	53.43	c	1.63	c-e
	30%	41.28	bc	31.58	bc	41.28	b	0.75	b

Numbers followed by the same letter in the same column were not significantly different at the 5% probability level (Duncan's new multiple range test). Data were transformed by$\sqrt{x}+0.5$. 2,4 D: 2,4-Dichlorophenoxyacetic acid.

Table 5 shows the leaf areas, root lengths, and dry weights of the shoots and roots of *A. spinosus* after the application of the various extracts. The results for the 20% *A. conyzoides* extract were not significantly different from those of the synthetic herbicide (the positive control). By 7 DAA, *A. spinosus* plants were completely destroyed, so further measurements were not possible. We believe this was because the allelopathic compounds in the *A. conyzoides* extract are phenolic compounds, terpenoids, and steroids that inhibit metabolic processes such as cell division, especially in the leaf. In a prior study, it has been reporeted that a *Senecio salignus* extract affected photosynthesis in *Physalis ixocarpa* and *E. crus-galli* and identified two sesquiterpenes responsible for this: β-caryophyllene and caryophyllene oxide. A transient fluorescent chlorophyll indicated that 100 μg/mL β-caryophyllene had a major effect on photosynthesis in *P. ixocarpa* plant *in vivo* by inhibiting photosystem II by transforming the active centers to "heat sinks" or the formation of silent reaction centers unable to reduce QA. β-caryophyllene also induces chlorosis on treated leaves [58].

Gallic acid compounds contained in *A. conyzoides* can inhibit the growth of other plants through denaturing proteins. The reduction in macromolecular components resulted in the inhibition of protein synthesis, which also resulted in the inhibition of protoplasm synthesis [59]. This subsequently inhibited the process of cell division and cell elongation, causing narrowing of *A. spinosus* leaves. Other study reported that the residue obtained from an aqueous acetone shoot extract of *A. conyzoides* inhibited the germination and the growth of roots and shoots of *Lactuca sativa, Digitaria sanginalis*, and *Amaranthus caudatus* [60]. Another one reported that the sesquiterpenes β-caryophyllene and caryophyllene oxide inhibited the dry biomass of *P. ixocarpa* plants, with β-caryophyllene having the greater effect [58]. β-Caryophyllene has been shown to inhibit root elongation in *P. ixocarpa* and *E. crus-galli* seedlings [13]. The morphological and physiological effect of phenolic acids on susceptible plants include reductions in leaf expansion, leaf production, net carbon assimilation rate, and stomatal conductance as well as decreased leaf water potential due to the reduced osmotic potential and turgor pressure and lower nutrient contents in the roots and shoots.

A plant's dry weight reflects the accumulation of organic compounds synthesized by the plant from water and carbon dioxide. If the synthesis process is interrupted, this will result in a decrease in the dry weight [61].

4. Conclusion

The phytochemical testing showed that all five of the tested allelopathic plant extracts, *I. cylindrical*, *C. rotundus*, *C, odorata*, *A. conyzoides*, and *A. compressus*, contained terpenoids and phenolic compounds; all but *C. rotundus* contained steroids, and all but *A. conyzoides* contained flavonoids. None contained saponin. All of the plant extracts at the different concentrations affected the control of the target plant, *A. spinosus*, resulting in changes in the height growth rate, leaf area, root length, and shoot and root dry weights. The 20% *A. conyzoides* extract showed the greatest inhibition of the growth of *A. spinosus* at 7 DAA, resulting in 100% control; this was followed by 20% *C. rotundus* (86% control), *C. odorata* (78%,), *I. cylindrica* (24%), and *A. compressus* (13%).

Acknowledgements

The authors would like to thank the Directorate of Research and Community Service, Directorate General of Research and Development Strengthening Ministry of Research, Technology and Higher Education for research funding

References

[1] Soerjani M, Soendaru M, Anwar C 1996 Present status of weed problems and their control in Indonesia *BIOTROP Special Publication* **24:** 387-395
[2] Batish D R, Kaur S, Singh H P, Kohli R K 2009 Nature of interference potential of leaf debris of *Ageratum conyzoides Plant Growth Regulation* **57:** 137–144.
[3] Mahmood A, Cheema Z A 2004 Influence of sorghum mulch on purple nutsedge (*Cyperus rot undus* L.) *International Journal of Agriculture and Biology* **6:** 86–88.
[4] Nektarios P A, Economou G, Avgoulas C 2005 Allelophathic effects of *Pinus halepensis* needles on turfgrasses and biosensor plants *Journal of Horticultural Science* **40:** 246–250.
[5] Sastroutomo S S 1990 Ekologi gulma *PT Gramedia Pustaka Utama* Jakarta.
[6] Inderjit D, Wardle D A, Karban R, Callaway R M 2011 The Ecosystem and Evolutionary Contexts of Allelopathy *Trends in Ecology and Evolution* **26**: 655–662.

[7] De-Martino L, Mancini E, de Almeida L F R, De Feo V 2010 The antigerminative activity of twenty seven monoterpenes *Molecules* **15**: 6630–6637.
[8] Khan Md S I, Islam A K M M, Kato-Noguchi H 2013 Evaluation of allelopathic activity of three mango (*Mangifera indica*) cultivars *Asian Journal of Plant Sciences* **12:** 252–261.
[9] Akhtar N, Javaid A, Bajwa R 2001 Herbical activity of aqueous extracts of *Crisium arvensea* and *Ageratum conyzoizes* againts weeds of wheat *Pakistan Journal of Biological Sciences* **4:** 1364–1367
[10] El Rokiek K G, El-Din S A S, Sahara F A A 2010 Allelopathic behavior of *Cyperus rotundus* L. on both *Chorchorus olitorius* (broad leaved weed) and *Echinochloa crus-galli* (grassy weed) associated with soybean *Journal of Plant Protection Research* **50:** 274–279.
[11] Skinner E M 2012 Allelopathic effects of sunnhemp (*Crotalaria juncea* L.) on germination of vegetables and weeds *HortScience* **47:** 138–142
[12] Jesudas P A, Kingsley J S, Ignacimuthu S 2014 Sorgoleone from Sorghum Bicolor as a Potent Bioherbicide *Research Journal of Recent Sciences* **3** 32–36
[13] Rice E L 1984 Allelopathy (2nd ed) *Academic Press* New York
[14] Enhellig F A 1994 Allelopathy: Current status and future goals, in *"Allelopathy: Organisms, processes and applications"*. In: Inderjit D K M M, Einhellig F A (Eds.) ACS Symposium Series New York: *American Societies* pp. 1–24
[15] Putnam A R 1985 Weed allelopathy, in *"Weed physiology, reproduction and Echophysiology"* In: Duke S O (Ed.) *CRC Press* pp. 131–155.
[16] Weston L A, Mathesius U Flavonoids: their structure, biosynthesis and role in the rhizosphere, including allelopathy *Journal of Chemical Ecology* **39:** 283–297
[17] Rusdy M, Riadi M, Sari A M, Normal A 2015 Comparative allelopathic effect of *Imperata cylindrica* and *Chromolaena odorata* on germination and seedling growth of *Centrosema pubescens International Journal of Scienific and Research Publications* **5**.
[18] Anjum T, Bajwa R, Javaid A 2005 Biological Control of *Parthenium* I: Effect of *Imperata cylindrica* on Distribution, Germination and Seedling Growth of *Parthenium hysterophorus* L. *International Journal of Agriculture and Biology* **7:** 448–450.
[19] Kavitha D, Prabhakaran J, Arumugam K 2012 Phytotoxic effect of purple nutsedge (*Cyperus rotundus* L.) on germination and growth of finger millet (*Eleusine coracana* Gaertn.) *International Journal of Research in Pharmaceutical and Biomedical Sciences* **3:** 615–619
[20] Setyowati 1998 Allelopathic potential of congograss, yellow nutsedge, and sunflower on selected weed species In: *Providing integrated weed management in managed and natural ecosystem* Biotrop-Bogor.
[21] Kristanto B A 2006 Change in characters of corn (*Zea mays* L.) plant due to alelopathy and text competition (*Cyperus rotundus* L.) The Changing of Corn (*Zea mays* L.) character caused by Allelopathy and Competetion with Purple Nutsedge (*Cyperus rotundus* L.) *Tropical Animal Development Journal* **31:** 189–194
[22] Rika I K, Mendra O M G 1990 New forage species for coconut plantation in bali inproceedings of the workshop on forage for plantation crops. In: Shelto H M, Stur W W (Eds.). Australian Centre for International Agricultural Research pp. 41–43
[23] Hu G, Zhang Z 2013 Allelopathic effect of *Chromolaena odorata* on Native and Non-Native Invasive Herbs *Journal of Food Agriculture and Environment* **2:** 878–882
[24] Devi O I, Dutta B K, Choudhuri P 2013 Allelopathy effect of aqueous extract of *Clerodendrum viscosum, Ageratum conyzoides* and *Parthenium hysterophorus* on the seed germination and seedling vigour of chick pea seeds (*Cicer arietinum* L.) in vitro *Journal of Applied and Natural Science* **5:** 37–40
[25] Frastika D, Suwastika R P I N 2017 The effectiveness test of kirinyuh (*Chromolaena odorata* (L.) R.M King dan H. Rob) leaf extract as natural herbicides to seed germination of mung beans (*Vigna Radiata* (L). R. Wilczek) and karuilei (*Mimosa invisa* Mart. Ex Colla) *Journal of Science and Technology* **6:** 225–238

[26] Kong C, Hu F, Xu X 2002 allelopathic potential and chemical constituents of volatiles from *Ageratum conyzoides* Under Stress *Journal of Chemical Ecology* **28:** 1173–1182
[27] Kong C, Hu F, Xu X, Liang W, Zhanng C 2004 Allelopathic Plant *Ageratum conyzoides* L *Allelopathy Journal* 1**4:** 1–12
[28] Kong C, Hu F, Xu X, Zhang M, Liang W 2005 volatile allochemicals in the *Ageratum conyzoides* intercropped citrus orchard and their effects on mites *Amblyseius newsami* and *Panonychus citri Journal of Chemical Ecology* **31:** 2193–2203
[29] Xuan T D, Shinkichi T, Hong N H, Khanh T D, Min C I 2004 Assessment of Phytotoxic Action of *Ageratum conyzoides* L. (Billy Goat Weed) on Weeds *Crop Protection* **23**: 915–922.
[30] Dayan F E, Duke S O 2006 Clues in the Search for New Herbicides, in *"Allelopathy: A Physiological Process with Ecological Implications"* In: Reigosa M, Pedrol N, Gonzale L (Ed.) *Springer* Amsterdam pp. 63–83.
[31] Tigre R C, Silva N H, Santos M G, Honda N K, Falcão E P S, Pereira E C 2012 Allelopathic and bioherbicidal potential of *Cladonia verticillaris* on the germination and growth of *Lactuca sativa* Ecotoxicology and environmental safety **84:** 125–132
[32] Adler M J, Chase C A 2007 Comparison of the allelopathic potential of leguminous summer cover crops: cowpea, sunn hemp, and velvetbean *Hortscience* **42:** 289–293
[33] Nice G, Johnson B, Jordan T 2011 Weed management in pastures: spiny pigweed *Purdue Extension* **4:** 1–4
[34] Harbone J B 1987 Metode fitokimia: penuntun cara modern menganalisis tumbuhan *Penerbit ITB* Bandung.
[35] Wijaya F 2001 pemanfaatan alelopati pada rimpang alang-alang sebagai herbisida organik pengendali gulma teki (*Cyperus rotundus*) Universitas Sumatra.
[36] Nuryana C T, Suryadi E, Harijadi 2007 The effect of cyperus rotundus root ethanol extract on the epithelization in the healing process on skin excision wound *Berkala Ilmu Kedokteran* **39(4):** 147-153
[37] Munte N, Sartini, Lubis R 2016 Phytochemicals and antimicrobial screening extract kirinyuh leaf (*Chromolaena odorata*) on Bacteria *Staphylococcus aureus* and *Escherichia coli BioLink* **2:** 132–140
[38] Mathew J J, Vazhacharickal P J, Sajesshkumar N K, Joy J M K 2016 Phytochemical analysis and in vitro hemostatic activity of *Mimosa pudica*, *Hemigraphic colorata* and *Chromolaena odorata* leaf extract *Journal of Pharmaceutical Sciences* **5:** 16–34
[39] Agbafor K N, Engwa A G, Obiudu I K 2015 Analysis of chemical composition of leaves and roots of *Ageratum conyzoides* Advances in *Biological Research* **9:** 397–400
[40] Okunade A L 2002 *Ageratum conyzoides* (Asteraceae) *Fitoterapia* **73:** 1–16
[41] Sastroutomo S S 1990 Ekologi gulma *PT Gramedia Pustaka Utama* Jakarta.
[42] Yulifrianti E, Linda R, Lovadi I 2015 Potensi alelopati ekstrak serasah daun mangga (*Mangifera indica* L.) terhadap pertumbuhan gulma rumput grinting (*Cynodon dactylon* L.) *Press. Journal Protobiont* **4:** 46–51
[43] Reigosa M J, Sánchez-Moreiras A, González L 1999 Ecophysiological Approach in Allelopathy *Critical Reviews in Plant Sciences* **18**: 577–608.
[44] Krisna, A, Manjunath G O, Rathod R. 2005 Effect of Casuarina, Mango, Eucalyptus and Acacia Leaf Leachates on Seed Germination of Kasturi Bendi, Sanka Pushpa and Honey Plants. Karnataka *Journal Agriculture Science* **18:** 205–207.
[45] Pebriyani, Linda R, Mukarlina 2013 Potensi ekstrak daun sembung rambat (*Mikania micrantha* H.B.K) sebagai bioherbisida terhadap gulma maman ungu (*Cleome rutidosperma* D.C) dan rumput báhia (*Paspalum notatum* Flugge) *Jurnal Protobiont* **2(2):** 32-38
[46] Ashafa A O T, Ogbe A A, Osinaike T 2012 Inhibitory effect of mango (*Mangifera indica* L.) leaf extracts on the germination of *Cassia occidentalis* seeds *African Journal of Agricultural Research* **7:** 4634–4639
[47] Lenny S 2006 Senyawa Terpenoida dan Steroida *USU Repository*.

[48] Dmitrovic S, Perisic M, Stojic A, Misic D 2015 Essential oils of two Nepeta species inhibit growth and induce oxidative stress in ragweed (*Ambrosia artemisiifolia* L.) shoots in vitro *Acta Physiologiae Plantarum* **37(3):** 64

[49] Khanh T D, Chung I M, Tawata S, Xuan T D 2006 Weed suppression by *Passiflora edulis* and its potential allelochemicals *Weed Research* **46:** 296–303

[50] Fransfried R E, Talbert R E 1977 Design of field experiments and the measurement and analysis of plant responses, in *"Research methods in weed science"*. (2nd ed) In: Truelove B (Ed.) Southern Weed Science Society *Arkansas* p. 19

[51] Natarajan A, Elavazhagan P, Prabhakaran J 2014 Allelopathic potential of billy goat weed *Ageratum conyzoides* L. and *Cleome viscosa* L. on germination and growth of *Sesamum indicum* L *International Journal of Current Biotechnology* **2:** 21–24

[52] Rizvi S J H, Rizvi V 1987 Allelopathy: some new terminological consederations *Current Science* **85:** 191–192

[53] Alsaadawi I S, Salih N M M 2009 Alleopathic potential of *Cyperus rotundus* L. interference with *crops Allelopathy Journal* **23:** 297–304

[54] Tesio F, Vidotto F, Ferrero A 2012 Allelopathic persistence of *Helianthus tuberosus* L. residues in the soil *Scientia Horticulturae* **135:** 98–105

[55] Vidotto F, Tesio F, Ferrero A 2013 Allelopathic effects of *Ambrosia artemisiifolia* L. in the invasive process *Crop Protection* **54:** 161–167

[56] Ekayanti N, Indriyanto, Duryat D 2015 Pengaruh zat alelopati dari pohon *Akasia mangium* dan jati terhadap pertumbuhan semai *A. mangium* dan jati *Jurnal Sylva Lestari* **36:** 78–83

[57] Palapa T M 2009 Senyawa alelopati teki (*Cyperus rotundus*) dan alang-alang (*Imperata cylindrica*) sebagai penghambat pertumbuhan bayam duri (*Amaranthus spinosus*) *Journal Agritek* **17:** 1155–1162

[58] Sánchez-Munoz B A, Aguilar M I, Diaz B K, Rivero J F, Lotina-Hensen B 2012 The sesquiterpenes β-caryophyllene and caryophylene oxide isolate from *Senecio sagignus* act as phytogrowth and photosynthesis inhibitors *Molecules* **17:** 1437–1447

[59] Yuliani 2000 Pengaruh alelopati kamboja (*Plumeria acuminata* W. T. Ait) terhadap perkecambahan biji dan pertumbuhan kecambah *Celosia argentea* L. *perennial* **4:** 15

[60] Kato-Naguchi H 2001 Assessment of the allelopathic potential of *Ageratum conyzoides Biologia Plantar* **1919:** 309–311

[61] Lakitan B 2004 Dasar-dasar Fisiologi Tumbuhan *PT Raja Gravindo Persada* Jakarta.

Problems of Pests and Diseases on "Liberika Tungkal Komposit" Coffee in Peatlands, Jambi Province

A Meilin, S Handoko, R Rubiana, Rustam, Endrizal

[1]Assesment Institute for Agriculture Technology of Jambi, Ministry of Agricultural, Indonesia

Email : araz_meilin@yahoo.com

Abstract. Coffee of *Liberika Tungkal Komposit* (Libtukom) is a superior commodity that adapts well to the peatlands in West Tanjung Jabung Regency, Jambi Province. Libtukom coffee, which has a distinctive flavor, is the main source of income for the community in Tanjung Jabung Barat Regency. The problems in the cultivation of Libtukom coffee include pests and diseases. This study aimed to provide information on the type of pests and diseases that attack Libtukom coffee. The study was conducted in March-September 2017. Data were obtained from direct observation (primary data) and other sources (secondary data). Data were analyzed descriptively. The results showed that some of the pests that attack Libtukom Coffee were coffee fruit borer, white lice, and stem borer, with intensity of mild-moderate attacks. Diseases that can infect leaves Libtukom Coffee were mainly (rust and red rust) and roots (root fungi). White root fungus attack is the main problem of Libtukom coffee in peatland, Jambi Province. Farmers generally recognize the root fungus infection, which was already found in more than 30% of the plant population.

1. Introduction

The peatland area in Jambi Province is 0.717 million ha or 9.95% of the total peatland area on Sumatra Island [1]. The peatlands in Jambi are on the eastern side of Jambi Province, located in Muaro Jambi, Tanjung Jabung Barat (Tanjabbar), and Tanjung Jabung Timur (Tanjabtim) [2].

Libtukom coffee is a commodity of specific location in Tanjung Jabung Baratdistrict besides areca nut and coconut. Libtukom coffee is a coffee plant that has been established as a variety of development through the Decree of the Minister of Agriculture of the Republic of Indonesia No. 4968/Kpts/SR.120/12/2013 dated December 6, 2013. Libtukom coffee has distinctive flavors, its fruits and leaves are different from Robusta or Arabica coffee and is able to adapt well in peatlands, and is also a major source of livelihood for local people. Demand from Malaysia reaches up to Rp 33,000/kg at the farm level, exceeding the price of robusta coffee which is Rp. 16,000,-/kg. Libtukom coffee production is potential if the average is 909 grams of coffee beans/tree or equivalent to 950 kg of seeds for planting with a population of 900-1000 trees/ha [3]. In the certificate of geographical indication (IG) issued by the Ministry of Law and Human Rights on July 23, 2015, informed that the value of taste test results from Indonesian Coffee and Cocoa Research Center (Puslitkoka) that Tungkal Jambi liberica coffee processed as wet coffee (ripened) has flavors of herbal, rubbery, rutter sourish and high acidity.

Libtukom coffee plant area in TanjungJabung Barat regency decreased over the last three years, especially in Bram Itam (from 570 Ha in 2014 to 411 Ha in 2016) and Betara Subdistricts (from 1.540 Ha in 2014 to 1.492 Ha in 2016). The decrease of planted area contributes to the decrease of Libtukom Coffee planting area in TanjungJabung Barat Regency that were from 3.028 Ha in 2014 to 2.873 Ha in 2016 [4,5,6].

The decrease of Libtukom coffee plant area affected by several factors, such as pests and diseases that can cause plant death. Surveys on some farmers revealed that their coffee plants in recent years have experienced of deaths (personal communication). Several pest and disease attacks are commonly reported to result in low productivity of coffee and may even result in death in coffee plants. Some pests in coffee plants are coffee fruit borer (*Hypotenemus hampei*), parasitic nematodes (*Pratylenchus*sp., *Meloidogyne*sp., *Radopholus* sp.), the twig borer (*Xylosandrus* sp.), stem borer (*Zeuzera* sp.), white lice (*Planncoccus*sp.), leaf rust disease (*Hemileia vastatrix*), and root disease (*Rigidophoruslignosus*) [7,8,9,10,11] as well as pests from the primate groups of monkeys and monkeys during long dry seasons [12].

It is important to preserve Libtukom coffee as a source of germplasm in an effort to improve the function of peatlands and the income of Libtukom coffee farmers. Therefore, this study aimed to provide information about pest and disease problems affecting the Libtukom coffee in peatlands of TanjungJabung Barat Regency, Jambi Province.

2. Materials and Methods

2.1. Activity Location

The study was conducted during March-September 2017. The location of the research was the land of coffee cultivation in Betara Sub-district, TanjungJabung Barat Regency. Thise location was selected since it was a coffee plantation development and has the largest planting area (>50%) compared to other districts. Determination of Mandala Jaya Village, Serdang Jaya, and Mekar Jaya (ParitTomo, Parit Lapis, ParitPanglong) as the location of the samples as reported as the location of which many of the coffee crops suffered deaths.

2.2. Sampling Method

The method of determining the location of the plantation was done using Accidental sampling was one of the methods in sampling from a population. Where Accidental sampling was included in the technique of non-probability sampling (samples with unequal probability). For sampling method like this was taken by choosing the land where the farmer happened/found in the research area and the land showed symptoms of pest and disease attack. The number of samples may be representative because it depends only on the sample members at that time. Samples to be taken were coffee plants that show symptoms of pest and disease attacks.

2.3. Observation of pest and diseases

Observations were made on the symptoms of coffee fruit borer attack (PBKo), twig borer and other pests found during observation. The observed plants were 15% of the population. Observation of the intensity of PBKo pest attack was conducted by calculating the percentage of fruits attacked on all fruit of each branch productive branch. Sample plant were consisted of four branches observations. Observations were made of the major diseases that attacked the coffee plant. Observations were made on leaves, fruit, stems, twigs and roots. Symptoms were recorded, documented and signs of an attack taken to the Laboratory for identification.

2.4. Data Analysis

Percentage of disease or attack incidence was calculated by dividing the numbers of diseased or attacked fruit per branch by the total number of fruits per branch observed, multiplied by 100%. The intensity data of pests and diseases were analyzed descriptively.

3. Results And Discussion

3.1. Pest Problems on Liberika Tungkal Komposit

Coffee fruit borer attack caused by *Hypothenemus hampei* were found in all observation locations. Meanwhile, the attack of white lice caused by *Planococcus* sp. was found only in the village of Mandala Jaya, and the stem/twig borer caused by *Xylosandrus* sp. was found only in Mekar Jaya (Table 1). The attack intensity of each pest were presented in the Table 2.

Table 1. Identified problems of pests on LiberikaTungkalKomposit Coffee

No.	Name of Pest / Location found	Symptoms / Signs of Attack
1.	Coffee fruit borer (PBKo /*Hypothenemushampei*)/Mandala Jaya Village, Serdang Jaya, and Mekar Jaya (ParitTomo, Parit Lapis, Parit Panglong)	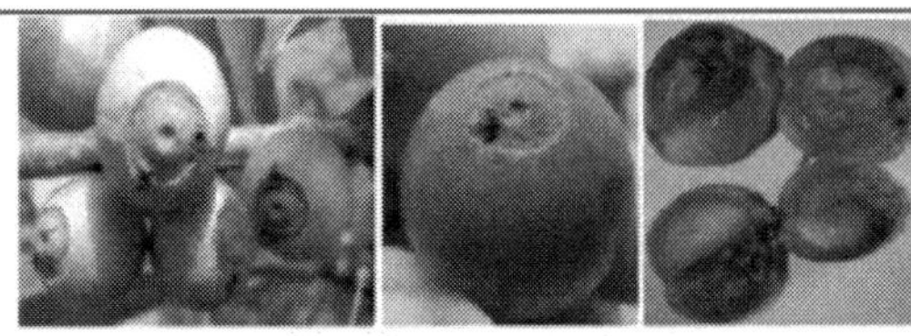
2.	White lice (*Planococcus* sp.)/Mandala Jaya	
3.	Twig borer (*Xylosandrus* sp.) /Mekar Jaya (Parit Tomo)	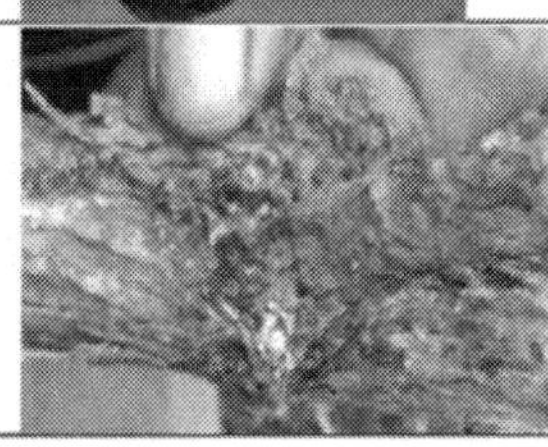

Table 2. The Attack Intensity of pests on LiberikaTungkal Komposit Coffee

No.	Name of Pest	Attack Intensity (%)
1.	Coffee fruit borer	4.17-11.17 (Average 9.03)
2.	White lice (*Planococcus* sp.)	0 – 3.40 /(Average 0.01)
3.	Twig borer (*Xylosandrus* sp.)	0-0.30/(Average 0.07)

The attack intensity of Libtukom coffee borer was obtained from five locations (Table 3). Overall, the average intensity of Libtukom coffee borer attack was below the control threshold (9.03%). The attack intensity above the control threshold was found in ParitTomo (Sri Utomo II and III) and Parit Lapis were 10.74%, 10.24%, and 11.17%,respectively, while in Parit Panglong and Serdang Jaya, the attack intensity of fruit borer attack Libtukom coffee were 8.86 and 4.17%, respectively.

Table 3. Average of Attack Intensity of Libtukom Coffee Fruit Borers (PBKo) on Several Locations in Mekar Jaya and Serdang Jaya Village, Betara Subdistrict

No.	Location of Garden (Village)	Average of Attack Intensity of Coffee Fruit Borer (%)
1	Parit Tomo/Sri Utomo II (Mekar Jaya)	10.74
2	Parit Tomo/Sri Utomo III (Mekar Jaya)	10.24
3	Parit Lapis (Mekar Jaya)	11.17
4	Parit Panglong (Mekar Jaya)	08.86
5	Serdang Jaya	04.17
Average		09.03

The result of research at Parit Panglong showed that percentage intensity of PBKo attack is lower than research conducted by [13], the percentage of PBKo pest attack on Libtukom Coffee in ParitPanglong Village, Betara can reach 17.76%. White lice and twig borer are found with very low average attack intensity. Coffee fruit borer.white fleas and twigs are the most commonly reported pests attack on arabica or robusta coffee plants that live not in peatland [14,15,9,10].

White lice found in the village of Mandala Jaya on coffee fruit that was ripe and clustered on the fruit stalk, removed the honeydew and the soot fungus [16,17,18]. White lice in coffee attack the fruits and flowers when high pest populations can attack the shoots of plants, leaves, and young branches. Attacked flower buds, flowers and young fruit will dry and fall. Fruits that have grown and mature did not fall but experience abnormal growth s resulting in wrinkled fruits and mature before the time. These reduce the quality of coffee beans. Loss of production due to severe attack until 90% [16]. In each coffee fruit found

3 - 19 *Planococcuscitri* that attacks the branches of flowers and young coffee fruit. Coffee flowers were attacked will look brown and fall. Attached young coffee fruit will be blackish green, while the coffee fruit that has been ripe in group who attacked will look red and black and then fall. *Planococcus citri* sucks water and breeds in the underarms of fruits, until adulthood and searches for other branches on the same tree that have not been attacked [17]. *Planococcus citri* produced honey dew which was very preferred by ants. *Xylosandrus*sp. (borer pest) found in Mekar Jaya characterized by one hole and the twig until broken. The *Xylosandruscompactus* characterized by the hole of about 1-2 mm indiameter on the surface of the coffee plant branch [15]. This hole leads until the inside of the twig to reach a length of 20-50 mm.

3.2. Plant Diseases Problems on Liberika Tungkal Komposit Coffee

Symptoms of fungus diseases of root and leaf rust was found in all observation sites of Mandala Jaya, Serdang Jaya, and Mekar Jaya Village (ParitTomo, Parit Lapis, ParitPanglong), whereas the *Cephaleuros* red rust was mainly found in Mekar Jaya (Parit Tomo, Parit Lapis, Parit Panglong) (Table 4). The Average Intensity of disease incidence of root fungus diseases on Liberika Tungkal Komposit Coffee was highly (Table 5).

The dominant disease that occurs was root fungal ~~fungus~~ disease with initial symptoms were the wilting of coffee plants, and wilt symptoms can result in the death of the Libtukom coffee plant. Based on information obtained from farmer groups, farmers, and related agencies, it was stated that the symptoms of the disease that resulted in the death of coffee had started 1-2 years ago. The incidence of the diseases originated from certain spots, then clumped, and further developed into a population with widespread symptoms. At the planting site were also found some partially coffee, withered entirely, withered and partially dead, and died entirely. Observations on the neck roots of coffee plants that showed wilting symptoms found the presence of white mycelium . The fungus is thought to be a white root fungus that attacks Libtukom coffee plants. Morphology of white root fungus resembles a white root fungus that attacks rubber and other plantation crops [7]. The average intensity of disease incidence in Mandala Jaya Village was the highest (35.66%) and the lowest was in Mekar Jaya/ParitTomo/Sri Utomo III (Table 4).

Leaf rust disease symptoms which were encountered on the LiberikaTungkalKomposit coffee in TanjungJabung Barat Regency resembles the symptoms of leaf rust disease proposed by [17] that was characterized by yellow-orange spots such as powder (powder) and can be seen or found on the top and bottom surfaces of the leaves. On the underside of the leaves appears spots that initially light yellow, then turn into dark yellow. The orange or orange flour was a fungus *H. vastatrix* uredospores. In advanced symptoms, the leaveshad brown spots joined together, become larger, then dry and fall off so the plants become bald [19,20,21]. Leaf rust disease in Libtukom coffee is the pathogenic fungus that attacks Libtukom coffee plant in Mekar Jaya Village, Betara Sub-district is *Hemileia* sp. [22].

Table 4. Plant diseases problems on liberika tungkal komposit coffee

No.	Name of Disease/ Location	Symptoms
1.	Root fungal diseases/ Mandala Jaya, Serdang Jaya, and Mekar Jaya (Parit Tomo, Parit Lapis, Parit Panglong)	
2.	Leaf rust/ *Hemileia vastatrix*)/ Mandala Jaya, Serdang Jaya, and Mekar Jaya (Parit Tomo, Parit Lapis, Parit Panglong)	
3.	Alga red rust/ *Cephaleuros* sp./Mekar Jaya (Parit Tomo, Parit Lapis, Parit Panglong)	

Table 5. The Average Intensity of Disease Incidence on LiberikaTungkalKomposit Coffee

No.	Name of Disease / Location	Average Intensity of Disease Incidence (%)
1.	Root fungus diseases/ Mandala Jaya, Serdang Jaya, and Mekar Jaya (Parit Tomo, Parit Lapis, Parit Panglong)	23.41-35.66 (Average: 30.04)
2.	Leaf rust/*Hemileia vastatrix*)/ Mandala Jaya, Serdang Jaya, and Mekar Jaya (Parit Tomo, Parit Lapis, Parit Panglong)	0-12.67 (Average: 0.24)
3.	Alga red rust/ *Cephaleuros*sp./Mekar Jaya (Parit Tomo, Parit Lapis, Parit Panglong)	0-14.26 (Average: 0.67)

Table 6. Intensity Diseases Incidence on Liberika Tungkal Komposit Coffee Root in Betara Subdistrict

No	Location Observation	Average Intensity of Libtukom Coffee Root Diseases Incidence (%)
1.	Mekar Jaya/Parit Tomo/ Sri Utomo II	31.54
2.	Mekar Jaya/Parit Tomo/Sri UtomoIII	23.41
3.	Mekar Jaya /Parit Panglong	26.52
4.	Mekar Jaya /Parit Lapis	30.47
4.	Serdang Jaya	32.64
5.	Mandala Jaya	35.66
	Average	30.04

The symptoms of red rust disease on coffee leaves both on the old leaves and young leaves were the appearance of spots-shaped round, thick, arise, and hairy consisting of colonies of green algae, the colours was grayish green and then turned into the brown orange, finally leaves dry and fall. The result of disease identification was *Cephaleuros* sp. [19, 23].

4. Conclusion

Pests and diseases problems identified on Liberika Tungkal Komposit coffee that grow at peatlands were coffee fruit borer (*Hythenemus hampei)*, white lice (*Planococcus* sp.), twig/stem borer (*Xylosandrus* sp.), root fungus diseases, leaf rust (*Hemileia vastatrix*) and Algae red rust (*Cephaleuros* sp.). The attack intensity of fruit borer and the deseases incident of root fungus diseases on Libtukom coffee was higher than the other pests or diseases.

Acknowledgments

The authors express their sincere thanks and appreciation to Bayu Oktareza for the assistance of data collection and field extension services (Mr. Mardi, SP) and coffee farmers in Betara Subdistrict, Tanjung Jabung Barat District (Hariadi, Sumarno, H. Kadni, Kayan, Rahman, Hanafi, M. Jamrah, Widodo, Jamiludin) for his assistance in conducting the research. Thanks also to the Assessment Institute Agricultural Technology (AIAT) Jambi for funding assistance in this activity.

References

[1] Wahyunto S, Ritung, Suparto H, Subagjo 2005 *Sebaran Gambut dan Kandungan Karbon di Sumatera dan Kalimantan.* Proyek *Climate Change, Forests and Peatlands in Indonesia.* Wetlands International – Indonesia Programme dan Wildlife Habitat Canada. Bogor.

[2] Santoso B 2017 Restorasi Lahan Gambut di Jambi Ditargetkan Capai 151 Ribu Ha.http://regional.liputan6.com/read/2999251/restorasi-lahan-gambut-di-jambi-ditargetkan-capai-151-ribu- ha.

[3] Ditjenbun 2014 Kopi Varietas Liberika Tungkal Komposit-Varietas Kopi Unggulan pelengkap Perbenihan kompositvarietas-kopi-unggulan-pelengkap-perbenihan-kopi-indonesia.html.

[4] BPS Kabupaten Tanjung Jabung Barat 2015 Tanjung Jabung Barat Dalam Angka 2015 Badan Pusat Statistik Kabupaten Tanjung Jabung Barat.

[5] BPS Kabupaten Tanjung Jabung Barat 2016 Tanjung Jabung Barat Dalam Angka 2016. Badan Pusat Statistik Kabupaten Tanjung Jabung Barat.

[6] BPS Kabupaten Tanjung Jabung Barat. 2017. Kabupaten Tanjung Jabung Barat Dalam Angka 2017. Badan Pusat Statistik Kabupaten Tanjung Jabung Barat.

[7] Agrios GN 2005 *Plant pathology* (p. 922) Fifth Edition. USA: Elsevier Academic Press. 922 p.

[8] Vega FE, F Infante, A Castillo & J. Jaramillo 2009 *The coffee berry borer, Hypothenemus hampei*

[9] Wiryadiputra S 2012 Keefektifan insektisida cyantraniliprole terhadap hama penggerek buah kopi (*Hypothenemus hampei*) pada kopi arabika *Pelita Perkebunan no* **28 (2)** Hal 100-110.

[10] Wiryadiputra S 2014 Pola Distribusi Hama Penggerek Buah Kopi (*Hypothenemus Hampei*) pada Kopi Arabika dan Robusta. Pelita Perkebunan 30 (2) : 123-136.

[11] Harni R, E Taufiq dan B Martono 2015 Ketahanan Pohon Induk Kopi Liberika Terhadap Penyakit Karat Daun (*Hemileia Vastatrix* B. Et Br.) Di Kepulauan Meranti *J. TIDP* **2(1)** 35–42.

[12] Rahayu DS & NP Erdiansyah 2015 Serangan Hama Primata pada Perkebunan Kopi. *Warta Pusat Penelitian Kopi dan Kakao, Indonesia* **27 (2)** : 29-32.

[13] Hayata 2016 Hubungan Persentase Serangan Hama Penggerek Buah Kopi (*Hypothenemus hampei* Ferr. (*Coleoptera: Scolytidae*)) dengan Dugaan Kehilangan Hasil Di Kecamatan Betara Tanjung Jabung Barat *Jurnal Media Pertanian* **1 (2)** : 85 – 90.

[14] Kalshoven LGE 1981 The Pest of Crops in Indonesia Revised and Tranlated By PA Van der laan. P.T. Ichtiar Baru-Van Hoeve. Jakarta.

[15] Rahayu S, A Setiawan EA, Husaeni dan S Suyanto 2006 Pengendalian Hama *Xylosandrus*

compactus Pada Agroforestri Kopi Multistrata Secara Hayati: Studi Kasus Dari Kecamatan Sumberjaya, Lampung Barat. *Agrivita* **28 (3)**.

[16] Sulistyowati E 2006 *Pengelolaan Hama Utama Tanaman Kopi.* Pusat Penelitian Kopi dan Kakao Indonesia. Jember.

[17] Rosanti D dan S Purwanto 2009 Pola Distribusi Kutu Dompolan (*Planococcus citri*) Pada Perkebunan Kopi Desa Semidang Alas Kecamatan Dempo Tengah Kota Pagar Alam *Sainmatika* **6 (2)** : 51-57.

[18] Hill DS 1983 *In Agricultural Insect Pests of the Tropics and Their Control* Edition II Cambridge University Press.

[19] Semangun H 2008 *Penyakit-penyakit tanaman perkebunan di Indonesia* Cetakan kelima. Yogyakarta, Gadjah Mada University Press.

[20] Silva MC, Varzea V, Guimaraes LG, Azinheira HG, Fernandez D & Petitot AS, Nicole 2006.Coffee resistance to the main diseases: Leaf rust and coffee berry disease *Braz. J. Physiol* **18(1)** 119–147.

[21] Rodrigues WN, Tomaz MA, Apostólico MA, Colodetti TV, Martins LD, Christo LF, Amaral JFT 2014 Severity of leaf rust and brown eyespot in genotypes of *Coffea arabica* L. cultivated with high plant density *American Journal of Plant Sciences* **5** 3702–3709.

[22] Defitri Y 2016 Pengamatan Beberapa Penyakit Yang Menyerang Tanaman Kopi (*Coffea* sp.) Di Desa Mekar Jaya, Kecamatan Betara, Kabupaten Tanjung Jabung Barat *Jurnal Media Pertanian* **1 (2)** : 78 – 84.

[23] Utami S & Asmaliyah 2007 Pengamatan awal serangan penyakit karat merah pada pulai gading (*Alstonia scholaris* (L.) R.Br) di KHDTK Kemampo, Kabupaten Banyuasin, Sumatera Selatan. Prosiding Seminar Hasil-hasil Penelitian: Optimalisasi Peran IPTEK dalam Mendukung Revitalisasi Kehutanan", 21 Agustus 2007. Pangkalan Balai. Puslitbang Hutan Tanaman.

The Potency of *Anredera cordifolia* as Botanical Pesticide for Sustainable Blast Disease (*Pyricularia oryzae*) Management on Paddy

E Yulia[1], F Widiantini[1], W Kurniawan[1] and I Berliani[2]

[1]Department of Plant Pests and Diseases, Faculty of Agriculture, Universitas Padjadjaran Kampus Jatinangor, Jatinangor 40513, Indonesia
[2]Alumnus Department of Plant Pests and Diseases, Faculty of Agriculture, Universitas Padjadjaran

E-mail : endah.yulia@unpad.ac.id

Abstract. Blast disease (*Pyricularia oryzae*) is one of the most important diseases of rice that can significantly reduce paddy production. This disease causes a potential loss of 3.65 tonnes/ha with loss percentages ranging from 60-100% if the disease is not handled properly. In spite of the issue of the negative effects of synthetic pesticide usage, the blast disease is commonly controlled using synthetic fungicides. Therefore, the exploration of potential botanical pesticides may provide safer and environmentally friendly blast disease management. This study aimed to test the potential of binahong plant (*Anredera cordifolia*) as botanical fungicide to inhibit the growth of *P. oryzae* and the development of blast disease in laboratory testing. The experiments were arranged in a Completely Randomized Design (CRD) consisting of 7 treatments of 5 extract concentrations and 2 controls. All treatments were repeated 5 times. The effectiveness of binahong leaf extract in controlling the pathogen and the disease was tested using poisoned food technique and detached leaf assay. The binahong leaf extract tested in the concentration range of 0.25-2% showed the ability to inhibit the growth of *P. oryzae* with the highest colony growth inhibition of 25.11%. Characteristics of *P. oryzae* mycelium and colony in the binahong leaf extract treatment were different from the control treatment. The treatment of paddy leaves with binahong leaf extract suppressed the blast disease development through the lessening of diseased paddy leaf area percentages. This study indicates the potency of binahong leaf extract for sustainable rice blast disease management.

1. Introduction

The decline in rice production is reported to be caused by the reduction of rice cultivation area in addition to the decrease in rice productivity. Rice production and rice productivity in Java are reduced by 0.83 million tons and 0.17 quintal/ha (0.33%) in 2014 [1]. Several factors contribute to the decline of rice productivity, one of which is the presence of plant pathogens. In some cases, plant pathogen has become a major obstacle in increasing rice production. Pathogenic fungus *Pyricularia oryzae* Cav. is the causal agent of rice blast disease that the infection of this pathogen can result in high rice yield loss or crop failure [2].

Blast disease is a worldwide major disease in rice plants [3,4]. Yield loss in several countries such as Kenya, China, the Philippines, India and Korea can be above 50% or even reach up to 100% [5,6]. In Indonesia, the disease has considerably reduced rice yields in several regions, especially in West Java, Central Java and East Java [7]. Previous research described that *P. oryzae* infection on Ciherang variety in endemic areas (Sukabumi, West Java) can cause yield loss of 3.65 tons/ha, equivalent to 61% yield loss compared to the average production of the variety [8].

Due to its potential to cause high yield loss, it is necessary to control blast disease seriously. Control of blast disease can be done including by applying particular cultivation techniques, the use of resistant varieties and the application of fungicides [7]. However, blast disease control is predominantly carried out chemically using synthetic fungicides. Some active ingredients of synthetic fungicides commonly used in controlling blast disease are reported to be effective in reducing the disease development or inhibiting *P. oryzae* infection [9,10]. Nevertheless, continuous application of synthetic chemicals is believed to have negative effects on the environment and human health. Likewise, in recent time there has been considerable pressure from consumers to farmers to reduce or eliminate synthetic pesticides in agriculture [11], so that there is an urge to develop alternative control methods to synthetic pesticides such as the use of botanical pesticides.

Botanical pesticides are natural chemicals extracted from plants which are mainly secondary metabolites that are used to control pests through inhibition of pathogen growth or suppression of disease development [12,13]. Botanical pesticides can be recommended as an alternative because they can be degraded more quickly and have low toxicity to the environment and humans [11,14,15]. The control of rice plant diseases using plant extracts has been widely reported and provides good and effective results in suppressing targeted pathogens. *Achalypha indica* weed extract had a potential as an antifungal of several rice pathogens such as *P. oryzae*, *Cercospora oryzae*, and *Sclerotium oryzae* in vitro [16]. The use of other plant extracts such as garlic and ginger extracts was found to reduce the pathogen infection of *Pyricularia grisea* in a field trial (Shafaullah and Khan, 2016). In addition, *Azadirachta indica* seed oil extract reduced the growth of *P. oryzae* in vitro and reduced the development of blast disease in rice plants in greenhouse trial [17]. The ethanol extract of *Citrus medica* was also reported to inhibit *S. oryzae* radial growth in vitro with inhibition reached up to 90% [18].

Plants with the potency as botanical pesticides are considered to contain secondary metabolites or phytochemical compounds [18]. These secondary metabolites are the active ingredients of botanical pesticide which are generally obtained through the extraction of plant parts such as seeds and leave [20]. Binahong (*Anredera cordifolia* (Ten.) Steenis) has been widely used as traditional medicinal ingredients and informed to have antifungal and antibacterial properties. In human health, binahong has been reported to suppress the growth of fungal pathogens as well as accelerates wound healing by stimulating more granulation tissue formation [21,22]. Binahong plant extracts showed to have antibacterial substances that can inhibit the growth of *Streptococcus mutans* and *Escherichia coli* bacteria in vitro [23,24]. In some countries such as China, Korea, or Taiwan, binahong plant has been used in healing various diseases and has been consumed for more than hundreds of years [25]. This plant is commonly used to treat wounds, smooth the skin, eliminate body aches, increase body stamina, as an antioxidant, and as an antibacterial drug [23]. Binahong plants are reported to have properties for health and treatment of various diseases because of its antimicrobial compounds [26,27]. Some researchers reported that binahong plants contain alkaloids, polyphenols, flavonoids, saponins, tannins and anthraquinones [25,28,22]. Studies of binahong extracts as antimicrobials of plant pathogens confirmed that binahong leaf water extract inhibited the growth of *Collectrotichum* spp. and *Fusarium oxysporum* with the inhibition percentage of 66.95% and 38.67%, respectively [29,30]. This study aimed to examine the potential of binahong plants as botanical fungicide to inhibit the growth of *P. oryzae* colony and suppress the development of blast disease in vitro.

2. Materials and methods

2.1. Pyricularia oryzae isolation

Pyricularia oryzae was isolated from rice plant leaves that showing symptoms of blast disease. The plant tissue was cut between healthy and diseased appearance. These small pieces were dipped in 70% alcohol solution for 10 seconds then dipped in 1% sodium hypochlorite solution for 1 minute, rinsed with sterile distilled water and dried on sterile filter paper. The leaf tissue pieces are then placed in a Petri dish containing Potato Dextrose Agar (PDA) medium and incubated at room temperature. The obtained isolates were purified and identified as *P. oryzae* using identification book of [31].

2.2. *Binahong leaf extractions*

Binahong leaves were washed and dried using oven for 12 hours at 40ºC to reduce their water content. The oven dried leaves are then prepared in powder form by blending. Binahong leaf powder was then macerated with methanol with a ratio of leaves powder and methanol 1:10 (w/v). The maceration was allowed to stand for two days in a closed container, filtered using filter paper, and evaporated with a rotary evaporator at a temperature of 55-60° C and a pressure of 580-600 mmHg [32]. Obtained crude extract was then prepared in five tested extract concentrations of 0.25, 0.5, 1, 1.5 and 2% w/v using sterile Tween-80 0.01% solution. The extract solution was sterilized by filtering using a 0.2 µm micro-filter.

2.3. *Inhibition test of binahong leaf extract on the growth of P. oryzae colony (Poisoned food technique)*

The inhibition test of binahong leaf extract on the growth of *P. oryzae* colony was performed on PDA medium. Each tested concentration of binahong leaf extract was mixed with sterilized molten PDA (1 ml: 9 ml) aseptically in sterile Petri dishes. The mixtures were then swirled gently to mix content evenly. Mycelial discs (Ø = 5 mm) were taken with a sterile cork-bore from the 7 days old cultures of the *P. oryzae* and were placed centrally on the cooled extract-medium plates and incubated at 28°C. Two control treatments were carried out as described above but with sterile aquadest and Tween-80 0.01% solvent were added to each of the plates. The radial mycelial growth of both test and control plates were then measured. Observations on the diameter of fungal colonies were carried out at 24-hour intervals until the control treatment reached maximum diameter. The percentage of inhibition of the growth of *P. oryzae* colonies was calculated by the following formula:

$$\text{Inhibition} = \frac{(\text{ØC}-\text{ØT})}{\text{ØC}} \times 100\% \ \text{Inhibition} = \frac{(\text{ØC}-\text{ØT})}{\text{ØC}} \times 100\%, \qquad (1)$$

ØC = Colony diameter in control treatment (cm); ØT = Colony diameter in extract treatment (cm)

The study was carried out using Completely Randomized experimental design with 7 treatments that were repeated 5 times. All data were subjected to ANOVA, and significant means were separated by Duncan Multiple Range Test of SPSS Ver. 16.

2.4. *Inhibition test of binahong leaf extract on the blast disease suppression in vitro (Detached leaf method)*

Healthy rice plant leaves were cut and washed using sterile distilled water. In the meantime, plastic boxes were prepared with sterile tissue was placed on the bottom surface of the boxes and straws were then arranged vertically on the tissue layer. The tissue was moistened with sterile distilled water until saturated. To facilitate the infection, the leaf area was wounded using a small needle. Four pieces of leaves that have been cleaned and wounded were arranged in each plastic box horizontally with the both ends of the leaf pieces were inserted under the wet tissue [33] After that, 20 µl of binahong leaf extract at each tested concentration was dripped over the wounded leaf area and left around 60 seconds. The mycelium block (Ø = 5 mm) of *P. oryzae* was then inoculated above the surface of the wounded leaf area and incubated at room temperature. Observations are made up to 7 days after inoculation. Observations were made on the occurrence of symptoms of blast disease, the length and width of the formed lesion, and the characteristics of the lesion.

3. Results and discussion

The results showed that binahong leaf extract was able to inhibit the growth of *P. oryzae* colony at all tested concentrations except the lowest concentration of 0.25%. The diameter of the *P. oryzae* colonies on observation 14 days after incubation at 0.5-2% concentration treatments was significantly shorter than that in the control treatment (Table 1; Figure 1). The result indicates the potential of binahong leaf extract to suppress the growth of *P. oryzae*. The highest inhibition of *P. oryzae* colony growth was shown in the treatment of 2% extract concentration with the inhibition percentage of 25.11% that differed significantly with 0.5% and 1% concentration treatments. The results showed that the higher the extract concentration used, the higher the inhibition of *P. oryzae* growth.

Table 1. Inhibitions of the growth of *P. oryzae* colonies in binahong leaf extract treatments

Treatments	Colony diameter (cm)*	Inhibition (%)
0.25% extract concentration	9.00 a	0
0.5% extract concentration	8.60 a	4.44
1% extract concentration	7.90 b	12.22
1.5% extract concentration	7.40 b	17.78
2% extract concentration	6.74 c	25.11
Tween-80 0.01% solution	9.00 a	0
Sterile aquadest	9.00 a	-

* Means in the same column with the same letter are not significantly different (Duncan Multiple Range Test $\alpha = 0.05$).

In addition to cause *P. oryzae* growth inhibition, the treatment of binahong leaf extract affected the characteristics of *P. oryzae* colony or mycelium. The condition of *P. oryzae* colony or mycelium in the extract treatment was different from the control treatment. The growth of mycelium in the control treatment appeared as normal conditions of lesser and thinner colony or mycelium with dark grey in color. Meanwhile, the colony or mycelium in the binahong leaf extract treatments was seen as thickening (dense) colony or mycelium with brighter colour and even dried (crusty) in some parts. The effect of plant antimicrobial compounds on the characteristics of fungal colonies or mycelium has also been reported. Essential oils of some plants that inhibited the colony growth of *Colletotrichum gloeosporioides* also caused mycelium agglomeration of the fungus and acervulus reduction [35]. Changes in fungal mycelium were also reported in the testing of several fungi such as *Alternaria humicola*, *Rhizoctonia solani* and *Fusarium solani* on the use of *Asarum heterotropoides* essential oil [34]).

The inhibition of fungal colony growth and abnormal growth of *P. oryzae* mycelium suggest that binahong leaf extract has the potential to suppress the growth of *P. oryzae*. The ability to inhibit growth and affect the characteristics of the mycelium of *P. oryzae* is considered to be caused by secondary metabolites contained in binahong leaf extract. It is reported that the ability of plant extracts in suppressing pathogens or disease development is caused by the metabolite compounds contained in the plants. Some bioactive compounds such as flavonoids, saponins, triterpenoids, coumarin and phenolic acids are reported to be present in binahong plants [25,28,36]. Secondary metabolite compounds such as alkaloids, flavonoids, tannins and saponins have properties as antimicrobials. Alkaloids produce allelopathy with characteristics as colourless, toxic, and bitter compounds [37]. According to [23], flavonoid compounds have the function as antibacterial and antivirus for plants. Saponin compounds function as antimicrobials by inhibiting fungal and bacterial growth and protecting from pests [38]. Tannin compounds are mentioned to have a bitter taste and function as an antifungal and antibacterial [39]. In plants, phenol compounds function as plant defence mechanisms against disturbances of plant pests and pathogens [19].

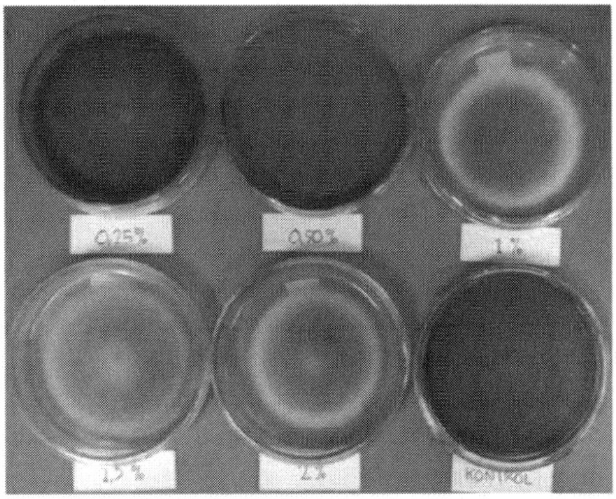

Figure 1. Inhibition test of colony growth of *P. oryzae* using binahong leaf extract at several tested concentrations.

In this test, crude extract of binahong leaves was dissolved using Tween-80 0.01% solution. Addition of Tween-80 0.01% as a carrier was intended to improve the solubility of the extract. However, the inhibition of *P. oryzae* growth with the highest inhibition percentage of 25.11% is still categorized to be low. Several factors might cause this low effectiveness, one of which was the possibility of less dissolved of binahong leaf extract in the PDA medium. In addition, it is reported that the quality of plant extracts was influenced by several factors, including plant materials used, types of solvents and extraction methods [13] Therefore, the effectiveness of the binahong plant extract as a botanical fungicide can still be improved. In this study, crude binahong leaf extract was prepared through the extraction using methanol solvent but other solvents or emulsifiers were then not added in the tests. Thus, the addition of these substances is believed to be able to increase the ability of the extract of binahong plants in suppressing the growth of pathogens.

In detached leaf assay, binahong leaf extract treatments suppressed the development of blast disease symptoms. In the control treatment, typical blast disease lesions were clearly formed and developed. The lesions were longer and wider than in the extract treatments. Some lesions formed in binahong leaf extract treatments were considered not to be disease lesions but stains of leaf wounding effect or due to phytotoxic especially in the use of high extract concentrations (Figure 2). The result of this test indicated that the application of binahong leaf extract on rice plant leaves can prevent the development of blast disease symptoms. It is likely that the presence of extracts on the leaf surface can eliminate *P. oryzae* inoculum or prevent the infection. In general, the results of this study indicate the potential of binahong leaf extract to control blast disease which can be used as an alternative for disease control in environmentally friendly way.

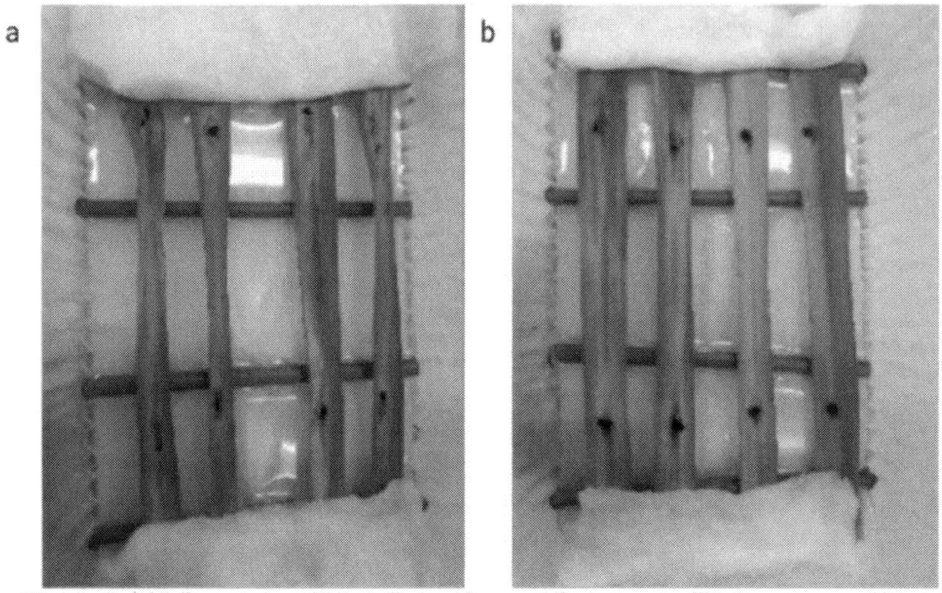

Figure 2. (A) Symptoms of blast disease in control treatment. (B) Symptoms of blast disease in binahong leaf extract treatment.

Acknowledgments
Thanks to the Universitas Padjadjaran for the support of research funding. This research was funded through the Universitas Padjadjaran Internal Grants 2017 and 2018 in the Fundamental Riset Unpad scheme (grant number 021/UN6.E/PL/2018).

References

[1] BPS-Statistics Indonesia: Produksi Padi Tahun 2014 (Angka Sementara) Diperkirakan Turun 0,63 Persen *Badan Pusat Statistik* 2015.

[2] Santoso, Nasution A 2009 Pengendalian Penyakit Blas dan Penyakit Cendawan Lainnya, dalam *"Inovasi Teknologi Produksi Padi"* Buku 2 *Balai Besar Penelitian Tanaman Padi Badan Penelitian dan Pengembangan Pertanian* pp: 531–563.

[3] Dean R A, Talbot N J, Ebbole D J, Farman M L, Mitchell T K, Orbach M J, Thon M, Kulkarni R, Xu J R, Pan H, Read N D, Lee Y H, Carbone I, Brown D, Oh Y Y, Donofrio N, Jeong J S, Soanes D M, Djonovic S, Kolomiets E, Rehmeyer C, Li W, Harding M, Kim S, Lebrun M H, Bohnert, Coughlan H S, Butler J, Calvo S, Ma L J, Nicol R, Purcell S, Nusbaum C, Galagan J E, Birren B W 2005 The Genome Sequence of the Rice Blast Fungus *Maghnaporthe grisea Nature* **434** 980–986.

[4] Kumar A, Sheoran N, Prakash G, Ghosh A, Chikara SK, Rajashekara H, Singh UD, Aggarwal R, Jain RK 2017 Genome Sequence of A Unique *Magnaporthe oryzae* RMg-DI Isolate from India that Causes Blast Disease in Diverse Cereal Crops, Obtained Using PacBio Single-Molecule and Illumina HiSeq2500 Sequencing *Genome Announc* **5(7)** e01570-16.

[5] Kihoro J, Bosco N J, Murage H, Ateka E, Makihara D 2013 Investigating the Impact of Rice Blast Disease on the Livelihood of the Local Farmers in Greater Mwea Region of Kenya *Springerplus* **2** 308.

[6] Wang X, Lee S, Wang J, Ma J, Bianco T, Jia Y 2014 Current Advences on Genetic Resistance to Rice Blast Disease in *"Rice-Germplam, Genetics and Improvement"* In: Yan W, Bao J, Eds).

[7] Sudir AN , Santoso, Nuryanto B 2014 Penyakit Blas *Pyricularia grisea* pada Tanaman Padi dan Strategi Pengendaliannya *Balai Besar Penelitian Tanaman Padi* **9(2)** 85–96.

[8] Suganda T, Yulia E, Widianti F, Hersanti 2016 Intensitas Penyakit Blas (*Pyricularia oryzae* Cav.) pada Padi Varietas Ciherang di Lokasi Endemik dan Pengaruhnya terhadap Kehilangan Hasil

Jurnal Agrikultura **27(3)** 154–159.
[9] Ghazanfar M U, Wakil W, Sahi S T, Yasin S 2009 Influence of Various Fungicides on the Management of Rice Blast Disease *Mycopath* **7(1)** 29–34.
[10] Semangun H 2008 Penyakit-penyakit Tanaman Pangan di Indonesia *Gadjah Mada University Press* Yogyakarta.
[11] Dubey N K, Shukla R, Kumar A, Singh P, Prakash B 2010 Prospects of Botanical Pesticides in Sustainable Agriculture *Current Science* **98(4)** 479–480.
[12] El-Wakeil, N E 2013 Botanical Pesticides and Their Mode of Action *Gesunde Pflanzen* **65(4)** 125–149.
[13] Gurjar M S, Ali S, Ak tar M, Singh K S 2012 Efficacy of Plant Extract in Plant Disease Management *Agriculture Sciences* **3(3)** 425–433.
[14] Isman M B 2006 Botanical Insecticides, Deterrents, and Repellents in Modern Agriculture and an Increasingly Regulated World *Annu. Rev. Entomol* **51** 45–66.
[15] Ndakidemi B. Mtei K, Ndakidemi P A 2016 Impacts of Synthetic and Botanical Pesticides on Beneficial Insect *Agricultural Science* **7** 364–372.
[16] Imrosi A N, Mihardjo P A, Hoesain M 2015 Pemanfaatan Ekstrak Gulma Anting-Anting (*Acalypha indica* L.) sebagai Antifungal Beberapa Patogen Padi secara In Vitro *Berkah Ilmiah Pertanian* **1(1)** 1–4.
[17] Amadioha A C 2000 Controling Rice Blast in Vitro and in Vivo with Extracts of *Azadirachta indica Crop Protection* **19(5)** 287–290.
[18] Shamsi S, Chowdhury P 2016 In Vitro Evaluation of Fungicides and Some Plant Extract against Rice Sheath Rot Pathogen *Sarocladium oryzae Bangladesh Journal of Scientific Research* **29(1)** 47–54.
[19] Cowan M M 1999 Plant Product as Antimicrobial Agents *Clinical Microbiology Reviews* **12(4)** 564–582.
[20] Kabera J N, Semana E, Mussa A R, He X 2014 Plant Secondary Metabolites: Biosynthesis, Classification, Function and Pharmacological Properties *Journal of Pharmacy and Pharmacology* **2(7)** 377–392.
[21] Ariani S, Loho L, Durry M F 2013 Khasiat Daun Binahong (*Andera cordifolia* (Ten.) Steenis) terhadap Pembentukan Jaringan Granulasi dan Repitalisasi Penyembuhan Luka Terbuka Kulit Kelinci *Jurnal e-Biomedik* **1(2)** 914–919.
[22] Kumalasari E, Sulistyani N 2011 Aktivitas Antifungi Ekstrak Etanol Batang Binahong (*Anredera cordifolia* (Tenore) Steen.) terhadap *Candida albicans* Serta Skrining Fitokimia *Jurnal Ilmiah Kefarmasian* **1(2)** 51–62.
[23] Darsana I G O, Besung I N K, Mahatmi H 2012 Potensi Daun Binahong (*Anredera cordifolia* (Tenore) Steenis) dalam Menghambat Pertumbuhan Bakteri *Escherichia coli* secara *In Vitro Indonesia Medicus Veterinus* **1(3)** 337–351.
[24] Rimporok S, Kepel B J, Siagian K V 2015 Uji Efektifitas Ekstrak Daun Binahong (*Anredera cordifolia Steenis*) terhadap Pertumbuhan *Streptococcus mutans* secara *In Vitro Pharmacon* **4(4)** 15–21.
[25] Astuti S M, Mimi S A M, Retno A B M, Risch A 2011 Determination of Saponin Compound from *Anredera cordifolia* (Ten) Steenis Plant (Binahong) to Potensial Treatment for Several Diseases *Journal of Agricultural Sciences* **3(4)** 224–232.
[26] Manoi F 2009 Binahong (*Anredera cordifolia*) Sebagai Obat *Warta Penelitian dan Pengembangan Tanaman Industri* **15(1)** 3–5.
[27] Paju N, Yamelan P V Y, Kojong N 2013 Uji Efektifitas Salep Ekstrak Daun Binahong (*Anredera cordifolia* (Ten.) Steenis) pada Kelinci (*Oryctolagus cuniculus*) yang Terinfeksi Bakteri *Staphylococcus aureus Pharmacon* **2(1)** 51–61.
[28] Djamil R, Wahyudi P S, Wahyono S, Hanafi M 2012 Antioxidant Activity of Flavonoid from *Anredera cordifolia* (Ten) Steenis Leaves *International Research Journal of Pharmacy* **3(9)** 241–243.

[29] Penyebab penyakit layu fusarium (*Fusarium oxysporum* Schlecht. f.sp. *lycopersici* (Sacc.) Synd. et Hans.) pada tomat *Prosiding Seminar Nasional Perkumpulan Agroteknologi/Agroekologi Indonesia (PAGI)* Juli 2016 Surakarta Hlm 503–507.

[30] Yulia E, Widiantini F, Purnama A, Nurhelawati I 2016 Keefektifan Ekstrak Air Daun Binahong (*Anredera cordifolia* (Ten.) Steenis) dalam Menekan Pertumbuhan Koloni Dan Perkecambahan Konidia Jamur *Colletotrichum capsici* Penyebab Penyakit Antraknos Pada Cabai *Jurnal Agrikultura* **27(1)** 16–22.

[31] Barnett H L, Hunter B B 1998 Illustrated Genera of Imperfect Fungi *APS Press St Paul.*

[32] Dono D, Hidayat S, Nasahi C, Anggraini E 2008 Pengaruh Ekstrak Biji *Barringtonia asiatica* l. (Kurz) (Lecythidaceae) terhadap Mortalitas Larva dan Fekunditas *Crocidolomia pavonana* F. (Lepidoptera: Pyralidae) *Jurnal Agrikultura* **19(1)** 5–14

[33] Yulia E, Widiantini F, Suganda T, Mahfudhah H 2016b Kemampuan bakteri endofit dalam meningkatkan ketahanan tanaman padi terhadap penyakit blas (*Pyricularia oryzae* Cav.) *Prosiding Seminar Nasional Pengendalian Penyakit Pada Tanaman Pertanian Ramah Lingkungan II* Agustus 2016 Yogyakarta.

[34] Dan Y, Liu H Y, Gao W W, Chen S L 2010 Activities of Essential Oils from *Asarum heterotropoides* var Mandshuricum against Five Phytopathogens *Crop Protection* **29(3)** 295–299.

[35] Alberida H, Eliza, Lova R N 2014 Pengaruh Minyak Atsiri terhadap Pertumbuhan *Collectrotichum* sp. (Penz.) Sacc. Penyebab Penyakit Antraknosa Buah Papaya (*Carica papaya* L.) secara *In Vitro Jurnal Saintek* **6(1)** 57–64.

[36] Selawa W, Rantuwene M R J, Citraningtyas G 2013 Kandungan Flavanoid dan Kapasitas Antioksidan Total Ekstrak Etanol Daun Binahong (*Anredera cordifolia* (Ten.) Steenis.). *Pharmacon* **2(1)** 18–22.

[37] Enyiukwu D N, Awurum A N, Ononuju C C, Nwaneri J A 2014 Significance of Characterization of Secondary Metabolites from Extracts of Higher Plants in Plant Disease Management International *Journal of Advance Agricultural Research* **2** 8–28.

[38] Saxena M, Saxena J, Nema R, Singh D, Gupta A 2013 Phytochemistry of Medicinal Plants *Journal of Pharmacognosy and Phytochemistry* **1(6)** 168–182.

[39] Ismarani 2012 Potensi Senyawa Tannin dalam Menunjang Produksi Ramah Lingkungan *Jurnal Agribisnis dan Pengembangan Wilayah* **3(2)** 46–55.

Antifungal Activity of Methanol-Extracted Secondary Metabolites of Rhizobacteria Isolated from Rhizosphere of Oil Palm Trees Against *Ganoderma Boninense* Pat.

F Widiantini[1], E Yulia[1] and C Nasahi[1]

[1]Department of Plant Pests and Diseases, Faculty of Agriculture, Universitas Padjadjaran
Jl. Raya Bandung-Sumedang KM 21 Jatinangor West Java 45363, Indonesia

E-mail: fitri.widiantini@unpad.ac.id

Abstract. Several rhizobacteria were isolated from the rhizosphere of healthy oil palm trees showed prospective ability as biocontrol agent against pathogenic fungi *Ganoderma boninense* Pat. The fungus is the causal agent of basal stem rot (BSR) disease can cause yield reduction up to 80%. Therefore, searching for effective compound to control the disease is a continuous process. Rhizosphere bacteria isolates were grown on ISP2 agar media and incubated for 14 days at room temperature. The media with fully grown cultures were cut into cubes and added with methanol. It was shaken and the bacterial cells were then removed through centrifugation and series of filtration. The final filtration was conducted using 0.2 µm to sterilize the filtrate. The filtrates were subjected for antifungal activity against *G. boninense* using agar well diffusion. The result showed that the crude metabolites produced by the rhizosphere bacteria demonstrated the ability to inhibit the growth of *G. boninense* in the agar diffusion method. The highest inhibition reached 51.63%. Furthermore, *G. boninense* mycelia were also experienced malformation by the presence of the secondary metabolites. The isolated rhizosphere bacteria showed promising ability to produce antifungal compounds which were able to inhibit the growth of *G. boninense* as well as causing the morphological changes of its mycelia.

1. Introduction

Ganoderma boninense Pat. is one of important fungal phytopathogens of woody plants especially oil palm trees. The fungal pathogen causing the decay of the oil palm basal stem which in severe infection showing the woody fan-like basidiocarp [1]. The Basal Stem Rot (BSR) disease caused by the fungal is often incurable. Symptoms will only appear when the plant has been severely infected. Young oil palm tree will die within 1 or 2 years following the emergence of the symptoms whilst at older plant, it will die within 3 or so years [2]. At this point no available control measures are effective in controlling the BSR disease [3]. Systemic fungicides are thought to be the immediate short-term solution to control BSR. Together with the application of culture practices methods and the use of biological control agents, systemic fungicides are commonly used to slow down the development of the disease [4]. However, the use of fungicide has not been without problems. One of most serious problems arise due to the use of fungicide is the adverse effects on the environment. Therefore, finding alternative antifungal compounds which are effective to control *G. boninense* as well as save to the environment is necessary to be done.

Antifungal compounds have been found produced by microorganisms. More than 22.000 microbial origin biologically active compounds have been reported [5]. Microbial metabolites have been targeted as source for searching agro-active compounds. This is due to their novel properties including their structure and activities. It has been reported that antifungal produced by microbes cause less pollute and less stress to the environment [6].

Rhizobacteria is one of microbes that reported able to produced antifungal compounds. Bacteria inhabitant soil surrounding the plant roots (rhizosphere) plays important role not only as plant growth promoters but also providing protection against phytopathogens in the form of biocontrol through competition and antibiosis [7]. [8] showed that pyrrole type of compound produced by *Pseudomonas protegens* strain W45 were able to inhibit the growth of Sclerotinia sclerotiorum. Crude bioactive metabolites produced by a plant growth promoting rhizobacteria, *Pseudomonas aeruginosa* PM 105, demonstrated noticeable antimicrobial activities toward several plant pathogens ie Colletotrichum gloeosporioides, Fusarium oxysporum, Pencillium expansum, nad Rhizoctonia solani *[9]*.

Attempts to isolate antagonistic rhizobacteria from oil palm trees have resulted several potential biocontrol isolates namely rhizobacteria BARK4, BARK5, BARK6, BARK7, BARK8, BARK9, BARK15 and BARK16 (reported elsewhere). In vitro testing demonstrated that those isolates were able to inhibit the growth of G. boninense as well as alter its mycelia morphology. The ability of the crude metabolites of these isolates in inhibiting the growth of G. boninense is reported here.

2. Materials and Methods

2.1 Preparation of culture filtrates of rhizosphere bacteria

The rhizobacteria isolates were isolated from rhizosphere of healthy oil palm trees and showed antifungal activity against *G. boninense* on dual culture assay (reported elsewhere). The bacteria were grown and maintained on HPDA (half strength Potato Dextrose Agar, Oxoid) media. To obtain secondary metabolites from the rhizosphere bacteria isolates, ISP2 media (malt extract 10 g, yeast extract 4 g, glucose 4 g, 1 l RO water, pH 7.2 ± 0.2) [10] was used to grow all of isolates. The rhizobacteria isolates were inoculated into the ISP2 media covering all of media surface. It was then incubated at room temperature for 14 days. ISP2 media fully covered with the rhizobacteria isolate was cut into cubes of 5 x 5 mm and submerged with absolute methanol in a 50 ml centrifuge tube. It was shaken overnight to extract the secondary metabolites secreted into the ISP2 agar media. The agar and bacterial cells were removed through Whatman filter paper no 1. The filtrate was filter sterilized using 0.22 μm membrane filters (Millipore). It was then subjected for further test for detecting its ability to inhibit the growth of *G. boninense.*

2.2 Agar well diffusion assay of rhizbacterial crude metabolites

The antifungal activity of the metabolites was conducted on PDA media using agar well diffusion method (Balouiri *et al.*, 2016). A 6 mm diameter well was made using a sterile cork borer by which 30 μl of the filtrate prepared earlier was inoculated into the well. A 6 mm plug of *G. boninense* was inoculated at 3 cm away across the well. The antifungal activity of the metabolites was detected by the growth inhibition of *G. boninense*. Percentage inhibition of radial growth (PIRG) was calculated based on the reading of *G. boninense* radial colony at the direction of filtrate containing well (R2) against the *G. boninense* radial colony in the control plate (R1) [11].

Where, $$PIRG = \frac{R1-R2}{R1} \times 100\% \qquad (1)$$

Morphology of the *G. boninense* following the agar well diffusion assay was observed under microscope. Mycelia of *G. boninense* facing the filtrate-containing well were collected at several spots to observe the effect of the secondary metabolites on the *G. boninense.* The fungal colony diameter was measured and used to calculate the Area Under Growth Colony Curve (AUCGC) using modified formula of Area Under Disease Progress Curve (AUDPC) [12].

$$AUCGC = \sum_{i=1}^{n-1} \frac{y_i + y_{i+1}}{2} \times (t_{i+1} - t_i) \qquad (2)$$

Where y_i is the fungal colony growth diameter at the *i*th observation, t_i is the time of the *i*th observation taken and n is the total number of observations taken. All data was analyzed using ANOVA and Duncan Multiple Range Test on SPSS version 17.0.

3. Result and Discussion

Agar well diffusion assay demonstrated the ability of the crude metabolites secreted by the rhizobacterial isolates in inhibiting the growth of *G. boninense* (Table 1; Fig 1). The diameter colony of *G. boninense* was measured in three days interval. At most of the observation times, the colony of *G. boninense* in plates containing crude metabolite-wells was significantly shorter than the diameter colony of *G. boninense* in control plates, except at the 9th day after inoculation. At the 9th day after inoculation, there were no significant differences of the means of diameter colony of *G. boninense* among treatments. Although in general, the colony of *G. boninense* at control plates was wider than in the crude metabolites treatments. On the control plate, *G. boninense* was able to fully cover media surface at 12 days after inoculation whilst at treated plates, *G. boninese* colony growth was inhibited.

Table 1. Average diameter (cm) colony of *G. boninense* following treatment with crude metabolite filtrates of rhizobacteria isolates.

Treatment	Means diameter colony of *G. boninense* (cm)			
	DAI-3*	DAI-6	DAI-9	DAI-12
BARK4	1.14 a**	1.51 a	2.79 a	2.93 a
BARK5	1.12 a	2.29 ab	3.84 a	4.49 a
BARK6	1.10 a	1.99 ab	3.63 a	4.04 a
BARK7	1.22 a	1.52 a	2.77 a	3.22 a
BARK8	1.08 a	1.48 a	3.02 a	3.28 a
BARK9	1.11 a	2.00 ab	3.09 a	3.32 a
BARK15	1.17 a	1.70 a	3.09 a	3.49 a
BARK 16	1.10 a	2.05 ab	3.13 a	3.37 a
Control	1.67 b	3.15 b	4.76 a	9.00 b

* DAI = day after inoculation
** Means following with different lowercase letters in the same column indicate significant difference at *P*<0.05 level by Duncan Multiple Range Test (*P*< 0.05)

Table 2. Area under colony growth curve (AUCGC) and percentage of growth inhibition of *G. boninese* following crude metabolites treatments.

Treatment	AUCGC*	Growth inhibition (%)
BARK-4	6.34 b	51.63
BARK-5	8.94 b	32.73
BARK-6	8.19 b	38.46
BARK-7	6.50 b	51.09
BARK-8	6.69 b	50.23
BARK-9	7.30 b	43.57
BARK-15	7.12 b	46.29
BARK 16	7.43 b	43.81
Control	13.25 a	-

*Means following with different lowercase letters in the same column indicate significant difference at *P*<0.05 level by Duncan Multiple Range Test (*P*< 0.05)

Mycelia of *G. boninense* after treated with the crude secondary metabolites experienced morphological alteration (Fig 1A-I). Visually, the colony growth of *G. boninense* was suppressed. The mycelia treated

The progress growth of *G. boninense* was observed through AUCGC (Table 2). It can be seen that the colony of *G. boninense* on plates without crude metabolite treatment grew normally as shown by highest AUCGC value compared to its growth on plates with crude metabolite treatments. The data also demonstrated the lower AUCGC value, the better effect of crude metabolites against *G. boninense*. *G. boninense* growth was significantly ($P<0.05$) inhibited by the presence of crude metabolites produced by the rhizobacteria isolates. The growth inhibition of *G. boninense* ranged from 43.57% to 51.63%. with crude metabolites showed centered growth in the middle of the plates. The mycelia avoided the crude metabolites-containing well on each it side. This showed by its upward and denser growth which was concentrated on the center of the plate. Thick white threads were also found on several treatments. Furthermore, mycelia thinning were observed at the edge of the colony where it got closer to the well containing crude metabolites.

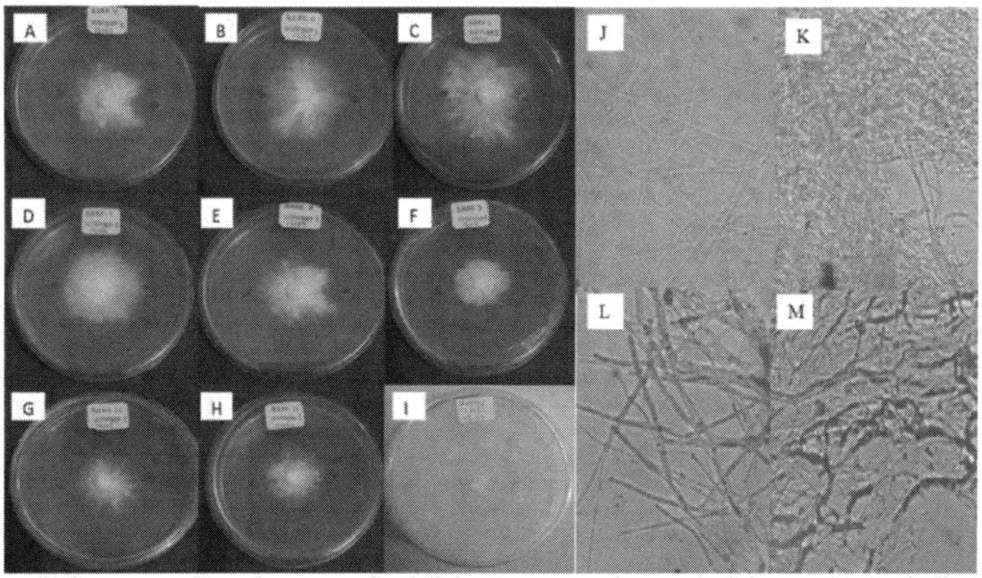

Figure 1. Representative colony growth of *G. boninense* on plates containing crude metabolite-well produced by BARK4 (A), BARK5 (B), BARK6 (C), BARK7 (D), BARK8 (E), BARK9 (F), BARK15 (G), BARK16 (H) and control (I). Macroscopic of *G. boninense* mycelia on control plate (J) and representative of crude metabolites effect on *G. boninense* (K, L and M), magnification 400x.

Under the influence of crude metabolites, morphological alterations were detected on the *G. boninense* mycelia (Fig 1K-M). Mycelia of *G. boninense* were observed lysed (Fig 1K) and in several spots observed to be more compact compared to control (Fig1L, M). An important feature of mycelia is the ability to grow forward at its apex to explore the surrounding environment. Hyphae plays important role in recognizing the plant surface thus trigger infection [13]. Therefore, when its growth was inhibited, it will affect the virulence of the fungal pathogen. Antimicrobial compounds produced by the rhizobacteria secreted during stationary phase [14]. These can act against the fungal pathogens through various mechanisms including ceased the fungal growth, inhibited conidia germination, lysed of fungal mycelia and/or trigger the fungal death [15]. The rhizobacteria isolates demonstrated biocontrol potential to be developed as biocontrol agents on *G. boninense* of oil palm tree. However, the antimicrobial compounds need to be optimized to achieve maximum production. Indigenous

Conclusion

Control of pathogen by indigenous biocontrol agents is enviromentally advantageous. The rhizobacteria isolates demonstrated promising ability to control basal stem rot pathogen, *G. boninense*.

The isolates were able to produce antifungal compunds which were capable to inihibit the growth of *G. boninense* and triggered mycelia mophological alteration.

Acknowledgement
The research was finacially supported by internal fundamental research grant (contract no 855/UN6.3.1/PL/2017) provided by Universitas Padjadjaran.

References

[1] Chong KP, Dayou J, Arnnyitte A 2017 Detection and Control of *Ganoderma boninense* in Oil Palm Crop. Springer Briefs in Agriculture.

[2] Hushiarian R, Yusof NA, Dutse SW 2013 Detection and control of *Ganoderma boninense*: strategies and perspectives *SpringerPlus* **2** 555.

[3] Susanto A, Sudharto PS, Purba RY 2005 Enhancing biological control of basal stem rot disease (*Ganoderma boninese*) in oil palm plantation. *Mycopathologia* **159** 153-7

[4] Bivi MSHR, Paiko AS, Khairulmazmi A, Akhtar MS, Idris AS 2016 Control of basal stem rot disease in oil palm by supplementation of calcium, copper, and salicylic acid *The Plant Pathology Journal* **32** 396-406.

[5] Qin S, Xing K, Jiang J-H, Xu L-H, Li W-J 2011 Biodiversity, bioactive natural products and biotechnological potential of plant-associated endophytic actinobacteria *Applied Microbiology and Biotechnology* **89** 457-73.

[6] Tanaka Y, Omura S 1993 Agroactive compounds of microbial origin *Annual Review of Microbiology* **47** 57-87.

[7] Ahemad M, Kibret M 2014 Mechanisms and applications of plant growth promoting rhizobacteria: Current perspective *Journal of King Saud University - Science* **26** 1-20.

[8] Bajpai A, Singh B, Joshi S, Johri BN 2018 Production and characterization of an antifungal compound from *Pseudomonas protegens* strain W45 *Proceedings of the National Academy of Sciences, India Section B: Biological Sciences* **88,** 1081-9.

[9] George E, Kumar SN, Jacob J, Bommasani B, Lankalapalli RS, Morang P, Kumar BSD 2015 Characterization of the bioactive metabolites from a plant growth-promoting rhizobacteria and their exploitation as antimicrobial and plant growth-promoting aAgents *Applied Biochemistry and Biotechnology* **176** 529-46.

[10] Kiranmayi MU, Sudhakar P, Sreenivasulu K, Vijayalakshmi M 2011 Optimization of culturing conditions for improved production of bioactive metabolites by *Pseudonocardia* sp. VUK-10 *Mycobiology* **39** 174-81.

[11] Skidmore AM, Dickinson CH 1976 Colony interactions and hyphal interference between *Septoria nodorum* and phylloplane fungi *Transactions of the British Mycological Society* **66** 57-64.

[12] Simko I, Piepho H-P 2012 The area under the disease progress stairs: Calculation, advantage, and application *Phytopathology* **102** 381-9.

[13] Mendgen K, Hahn M, Deising HB 1996 Morphogenesis and mechanisms of penetration by plant pathogenic fungi *Annual Review of Phytopathology* **34** 364-86.

[14] Trivedi P, Pandey A, Palni LMS 2008 In vitro evaluation of antagonistic properties of *Pseudomonas corrugata* *Microbiological Research* **163** 329-36.

[15] Gould WD, 1990. Biological control of plant root diseases by bacteria. In: Nakas JP, Hagedorn C, eds. *Biotechnology of Plant-Microbe Interactions.* New York: McGraw-Hill Inc., 287-317.

Effects of Dosage and Application Frequency of Microbial Consortium Mixed with Animal Manure on Bacterial Wilt and Late Blight Diseases of Potato

N Istifadah[1], N Fatiyah[1], B N Fitriatin[1] and L Djaya[1]

[1]Agriculture Faculty, Universitas Pdjadjaran, Jl. Raya Bandung- Sumedang. Km 21 Jatinangor. 45363 Indonesia

E-mail : nistifad@yahoo.com

Abstract. Major constraints in potato production includes bacterial wilt disease (*Ralstonia solanacearum*) and late blight disease (*Phytophthora infestans*). To support environmentally-friendly agricultural system, a microbial consortium containing *Trichoderma harzianum, Bacillus subtilis* and non pathogenic *Pseudomonas* sp. has been studied for biocontrol plant diseases in horticultural crops. For application, the microbial consortium was mixed with organic matters (10%, v/v). The objective of this research was to obtain dosage and application frequency of microbial consortia mixed with chicken manure for controlling bacterial wilt disease and their effects on airborne disease, late blight disease, in potato. The experimental design used was Randomized complete design with 10 treatments and three replications. The treatments consisted of application of the mixture in planting hole at the dosage of 25, 50 or 100 g solely or in combination with drenching the water suspension of the mixture weekly or every two weeks. The results showed that the dosage and application frequency of the microbial consortia mixed with chicken manure influenced their abilities in suppressing the bacterial wilt disease in potato. The application of the mixture at planting hole with the dosage of 50 g per plant followed by drenching the water suspension of the mixture every week suppressed the bacterial wilt disease by 75.0 - 81.9%.

1. Introduction

Potato is one of important horticultural crop that are commonly cultivated in many highland areas in West Java, Indonesia. The main constraints in potato production include plant diseases. Bacterial wilt disease caused by *Ralstonia solanacearum* is a devastating soil-borne disease in many plants including potato. The infected plants become wilted and finally died. The latent infection of seed potatoes can spread the disease to other field and areas [1]. In addition to soil-borne disease, airborne disease such as late blight disease caused by *Phytophthora infestans* is also destructive disease in potato worldwide that may lead to significant yield loss. The disease can infect almost all parts of the potato such as the leaves, petiole, stem and potato tubers [2].

The most common control measure for plant diseases is the use of pesticides. However, the continuous use of synthetic pesticides may lead to various negative impacts to the human health and the environment. Concerning this, environmentally-friendly control measures, such as the use of organic matters and biological control need to be developed.

Microbes commonly used for biological control of plant diseases include genera *Trichoderma* [3], *Bacillus* [4,5] and non- pathogenic *Pseudomonas* [6]. Combination of several compatible microbes can improve the efficacy of the control [7,8]. Microbial consortium containing *Trichoderma harzianum, Bacillus subtilis and Pseudomonas* sp. has been developed for biocontrol of plant diseases in horticultural crops. The microbes were compatible and the consortium of the microbes provided better disease control rather than individual microbe [8] For application, the consortium was mixed with organic matters such as compost or animal manure. The mixture has been found to be effective in controlling damping off disease caused by *Rhizoctonia solani* [8] and root knot nematode (*Meloidogyne* spp.) [9] in chili. *Trichoderma harzianum* and *B. subtilis* mixed with chicken manure was also reported to suppress the disease incidence of bacterial wilt disease and reduced the late blight disease intensity in potato [10].

The efficacy of biocontrol agents is influenced by several factors such as the dosage and frequency of application. To control soil-borne disease, biocontrol agents can be applied at the nursery, or at the planting holes. Repeated application of biocontrol agents after planting can improve their effectiveness in controlling the disease [11,12]. [10] also found that repeated application of *T. harzianum* and *B. subtilis* mixed with chicken manure could reduce the dosage of NPK fertilizer up to 50% of the standard dosage. The objective of this study was to obtain dosage and frequency of applications of microbial consortium mixed with chicken manure that lead to the best control of bacterial wilt and to examine if that application can reduce late blight disease in potato.

2. Materials and methods

The experiment was conducted in the field located at Citespong, Lembang, West Bandung, West Java, Indonesia. The experiment was arranged in Randomized Complete Block Design with 10 treatments and three replications. The treatments examined were the application of microbial consortia mixed with chicken manure in planting hole with the dosage of 25 g, 50 g and 75 g and drenching its water suspension every one or two weeks solely or in combination. Each treatment was repeated three times and in each replication there were 20 potato plants. In all treatments chicken manure was applied at the rate of 15 tons per hectare. Concerning that the use of microbial consortium could reduce the dosage of NPK fertilizer up to 50% of the standard dosage [10] in this study, all treatments using the microbial consortium NPK fertilizer was applied at the rate 50% of the standard dosage. In these treatments, part of the chicken manure was mixed with the microbial consortia (in molasses 2% medium) with proportion of 10% (v/v) and incubated for one week. The mixture was applied at planting holes at the time of planting with the dosage depended on the treatment. For drenching the plants, the mixture was added to the water (1:4. v/v) and incubated in the lid-closed container for one week. The plants were drenching with the suspension 50 ml per plant every one or two weeks.

The variables observed were the percentage of plant infected by bacterial wilt disease (disease incidence) and disease intensity of late blight disease. Disease index used for calculating disease intensity of late blight disease were 0: no symptom; 1: percentage of plant infected $0 < x \leq 12\%$; 2 :percentage of plant infected $12\% < x \leq 25\%$;3: percentage of plant infected $25\% < x \leq 50\%$; 4: percentage of plant infected $50\% < x \leq 75\%$; 5: percentage of plant infected $75\% < x \leq 100\%$. The disease incidence or severity was observed weekly. Data disease intensity and incidence were used for calculating *Area Under Disease Progress Curve* (AUDPC). Another variable observed was plant height which was measured every week. Data was statistically analyzed by analysis of variance using SPSS program 20 version. Data were checked for their normal distribution and transformed when it was necessary. The differences between treatments were analyzed using Tuckey's HSD test ($p<0.05$).

3. Results and Discussion

3.1 The effects of dosage and application frequency of microbial consortia mixed with chicken manure on bacterial wilt disease

The results showed that application of microbial consortium mixed with chicken manure significantly suppressed the bacterial wilt in potato by 31.0 - 82.2 %. The dosages and frequency of application influenced the disease reduction. The microbial consortium mixed with animal manure was effective when it was applied at the appropriate dosage or frequency of application. Based on the disease incidence at 7 weeks after planting, the application of the microbial consortium mixed with chicken manure at planting hole the dosage of 25 g or 50 g and or drenching the plants with the mixture suspension every week were effective to reduce bacterial wilt disease in potato by 71.4-75%. However, based on the progress or development of disease incidence during 5 weeks observation, the treatment that led to a relatively highdisease control (81.9% suppression) was application of the microbe-manure mixture at the rate of 50 g at planting holes and drenching its water suspension weekly (Table 1).

The application of the mixture only once at the time of planting also reduced the bacterial wilt disease by 57.1-61.2%, but the rate of application was 100 g per planting hole. It seems that if the dosage of the mixture application in planting hole was low (i.e. 25 g per planting hole) the repeated application by drenching its water suspension could enhance its efficacy. The weekly application of the water suspension of the mixture itself also led to effective bacterial wilt disease control (75.0 - 78.4% disease reduction). [14] also found that drenching the tomato plant by the suspension of *Trichoderma spp.* and *Pseudomonas fluorescens* weekly resulted in 97% plants recovered from bacterial wilt (*R. solanacearum*) infection.

Table 1. The effects of dosage and application frequency of the microbe-chicken manure mixture on bacterial wilt disease (8 Weeks After Planting)

Treatments	Disease incidence (%)	Disease suppression (%)	AUDPC value	Disease suppression (%)
A Control	93.3 d	0.0	1353.3 d	0.0
B Drenching the plant with water suspension of the mixture of microbial consortia and manure every week	23.3 a	75.0	291.7 ab	78.4
C Application of the mixture in planting hole 25g and drenching its suspension every week	26.7 a	71.4	443.3 ab	67.2
D Application of the mixture in planting hole 50g and drenching its suspension every week	23.3 a	75.0	245.0 a	81.9
E Application of the mixture in planting hole 75g and drenching its suspension every week	46.7 abc	50.0	711.7 bc	47.4
F Drenching the plant with water suspension of the mixture of microbial consortia and manure every two weeks	30.0 ab	67.8	443.3 ab	67.2
G Application of the mixture in planting hole 25g and drenching its suspension every two weeks	63.3 c	32.1	921.7 c	31.9
H Application of the mixture in planting hole 50g and drenching its suspension every two weeks	33.3 ab	64.3	443.3 ab	67.2
I Application of the mixture in planting hole 75g and drenching its suspension every two weeks	60.0 bc	35.7	898.3 c	33.6
J Application of the mixture in planting hole 100 g	40.0 abc	57.1	525.0 abc	61.2

Note : Data in one column followed by different letters were significantly different ($p < 0.05$) based on Tuckey HSD test

3.2 The effects of dosage and application frequency of microbial consortia mixed with chicken manure on late blight disease

In addititon to their effects on soilborne disease, application of the microbial consortia mixed with chicken manure also reduced the airborne disease, potato late blight, by 44.4%- 56.8% (Table 2). This results was in line with previous study which also found that application of *B. subtilis* and *T. harzianum* mixed with chicken manure reduced late blight disease by 44.0- 45.4% [10] In this current study, the dosage and application frequency did not affected the abilities of the microbes-manure mixture in reducing the late blight disease. The AUDPC values, that reflected the disease development during the observation, in all treatments using the mixture were not significantly different (Table 2).

Table 2. The effects of microbial consortia mixed with chicken manure on late blight disease intensity

Treatments	AUDPC value	Disease suppression (%)
A Control	595.9 b	0.0
B Drenching the plant with the suspension of the microbial consortia mixed with manure every week	299.8 a	49.7
C Application of the mixture in planting hole 25g and drenching its suspension every week	261.0 a	56.2
D Application of the mixture in planting hole 50g and drenching its suspension every week	276.4 a	53.6
E Application of the mixture in planting hole 75g and drenching its suspension every week	331.3 a	44.4
F Drenching the plant with suspension of the microbial consortia and manure mixture every two weeks	280.7 a	52.9
G Application of the mixture in planting hole 25g and drenching its suspension every two weeks	296.1 a	50.3
H Application of the mixture in planting hole 50g and drenching its suspension every two weeks	285.6 a	52.1
I Application of the mixture in planting hole 75g and drenching its suspension every two weeks	260.5 a	56.3
J Application of the mixture in planting hole 100 g	257.3 a	56.8

Note : Data in one column followed by different letters were significantly different ($p < 0.05$) based on Tuckey HSD test

Even though the microbial consortium mixed with chicken manure was applied in the soil, but it was also able to reduce late blight disease in the leaves. This result indicated the abilities of the mixture to enhance the potato resistance to the disease. Biocontrol agents such as *T. harzianum* and *B. subtilis* have been reported to induce resistance to early blight (*Alternaria solani*) and late blight (*P. infestans*) diseases in tomato [15] as well as to late blight disease in potato [16].

3.3 The effects of dosage and application frequency of the microbes-manure mixture on the potato growth

The potato growth was not only influenced by the availability of nutrient but also affected by the diseases infected the plants. In this study, the plant height was measured on the plants that were not wilted. The growth of the plants was still evenat two weeks after planting (WAP), as the disease infection was still low. In this stage the effect of the mixture of microbe-manure on potato growth could be observed more clearly. At 2 WAP, the application of the microbes mixed with chicken manure enhanced the potato growth (Table 3).

Table 3. Effects of dosage and application frequency of the microbial consortia mixed with chicken manure on the potato growth

Treatments	Average of Plant Height (cm)		
	2 WAP	3 WAP	4 WAP
A Control	43.3 a	53.8 a	63.4 a
B Drenching the plant with of the suspension of the microbial consortia and manure mixture every week	47.4 ab	65.2 a	73.9 a
C Application of the mixture in planting hole 25g and drenching its suspension every week	52.8 bc	63.8 a	71.6 a
D Application of the mixture in planting hole 50g and drenching its suspension every week	56.6 c	67.6 a	73.6 a
E Application of the mixture in planting hole 75g and drenching its suspension every week	53.7 bc	61.4 a	71.7 a
F Drenching the plant with suspension of the microbial consortia and manure mixture every two weeks	53.9 bc	54.8 a	59.5 a
G Application of the mixture in planting hole 25g and drenching its suspension every two weeks	52.0 abc	67.8 a	78.1 a
H Application of the mixture in planting hole 50g and drenching its suspension every two weeks	55.0 bc	58.7 a	66.7 a
I Application of the mixture in planting hole 75g and drenching its suspension every two weeks	53.1 bc	68.1 a	74.7 a
J Application of the mixture in planting hole 100 g	50.7 abc	52.6 a	61.6 a

Note : Data in one column followed by different letters were significantly different ($P < 0.05$) based on Tuckey HSD test

At two week after planting, application of the microbial consortium mixed with chicken manure 50 g at planting hole, followed by drenching its water suspension weekly resulted in better potato growth. [10] also found that combination of application of the microbial-manure mixture at planting hole and drenching the plant with its suspension led to better potato growth.

At 3 and 4 WAP, the potato heights in all treatments were not significantly different (Table 3). At that time, the number of plants infected by late blight disease started to increase; therefore, this might affected the plant growth. The heights of the plants were varied, even in the same treatments, and hence there were no significant difference among the treatments. The observation on potato growth was only up to 4 WAP as after that ≥ 50 % of the plants in the control has already infected by bacterial wilt disease. At that time the plant height was varied considerably so that it was not be observed anymore.

Overall results showed that the microbial consortium containing *T. harzianum*, *B. subtilis*, *Pseudomonas* sp. that was mixed with chicken manure suppressed bacterial wilt disease by 31.0 - 82.2 %. The effectiveness of the mixture can be achieved if the rate or frequency of application is appropriate. Repeated application of the mixture tended to result in better disease control rather than one application, only at the time of planting. [12] found that weekly application of bacterial antagonist (*Pseudomonas putida* 06909-rif/nal) through water irrigation led to significant increase on its population in the soil. The repetitive application could also maintain the high population of biocontrol agent in the soil over time. Application of the mixture 50 g in planting hole followed by drenching its suspension tended to result in better disease control. [17] also found that combination of application of spent mushroom substrate in planting hole and drenching its water extract every two weeks resulted in relatively better suppression of basal rot disease and development of the onion bulb rather than solely method.

Application of microbial consortium mixed with chicken manure in the soil was not only able to control soil borne disease, but it also reduced the destructive airborne disease, the potato late blight disease. This result was in line with another study that also found that *T. harzianum* and *B. subtilis* mixed with chicken manure reduced the late blight disease in potato [10].

To determine the appropriate rate and frequency of application, it needs to consider NPK dosage applied. [10] found the trend that if the application of microbial-manure mixture only once at the time of planting (100 g per planting hole), the NPK fertilizer was needed to be applied at standard dosage. However, if the mixture was applied several times (at planting hole and as regular soil drenching) the NPK dosage could be reduced up to 50% of the standard dosage. Based on this result, the NPK dosage in this study was only 50% of the standard dosage. It means that the mixture of microbial consortium and chicken manure can reduce the synthetic fertilizer and hence it can support Low External Input Sustainable Agriculture (LEISA) system.

4. Conclusion

Based on the results of this study it can be concluded that the mixture of microbial consortium (*T. harzianum*, *B. subtilis*, *Pseudomonas* sp.) and chicken manure was able to suppress bacterial wilt disease in potato. The rate and frequency of application influenced the efficacy of the mixture in reducing the bacterial wilt disease, but it did not affect its ability to reduce late blight disease intensity. Application of the mixture 50 g at planting hole followed by drenching its water suspension weekly resulted in the highest reduction (by 75.0%-81.0 %) of bacterial wilt disease incidence. The application of the mixture also reduced late blight disease by 44.4%-56.8%.

Acknowledgement

This study was part of a series of research funded by RAPID (Collaborative Research between University and Industry) DRPM, Ministry of Research and Higher Education, Directorate General of Higher Education, Republic of Indonesia. We also thank to Ms. Heni Krestini for facilitating the experiment.

References

[1] Meng F 2013 *Ralstonia Solanacearum* Species Complex and Bacterial Wilt Disease. *J. Bacteriol Parasitol* **4:** e119. doi:10.4172/2155-9597.1000e119

[2] Arora R.K., Sharma S., Singh B.P. 2014. Late blight disease of potato and its. *Management Potato Journal*, **41 (1)** 16-40

[3] Ozbay, N. Newman, S.E. 2004. Biological control with *Trichoderma* with Empahsis on *T. harzianum*, Pakistan Journal of Biological Science, 7 (4):478-484

[4] Cawoy, H., Bettiol, W., Fickers, P., Ongena, M. 2011. *Bacillus*-Based Biological Control of Plant Diseases. In: Pesticides in the Modern World - Pesticides Use and Management, Stoytcheva, M (Ed.) InTech : 273-302

[5] Shafi, J. Tian, H., Ji, M. 2017. *Bacillus* species as versatile weapons for plant pathogens: a review, Biotechnology & Biotechnological Equipment, 31 (3): 446-459.

[6] Weller, D. M. 2007. *Pseudomonas* biocontrol agents of soilborne pathogens: Looking back over 30 years. Phytopathology 97:250-256.

[7] Maketon, M., Apisitsantiku, J., Greenhouse, J., Siriraweeku, C. 2008. Evaluation of *Bacillus subtilis* Ap-01 and *Trichoderma harzianum* Ap-001 in controlling tobacco diseases. Brazilian Journal of Microbiology, 39:296-300

[8] Istifadah, N., A. Melawati., P. Suryatmana., B. N. Fitriatin. 2014. The Effectiveness of Consortia of Antagonistic and Biofertelizer Microbes in supressing damping off disease (*Rhizoctonia solani*) in chili. Agricultural Science Journal, 1(4):337-345 (In Indonesian language).

[9] Istifadah, N., Nurhasanah, L. Fitriatin, B.N dan Sunarto, T. 2017. Abilities of biopesticide formulation and organic matters to supress root knot nematode (*Meloidogyne* spp.) in chili.

Proceeding of National Seminar, Indonesian Agrotechnology Association, UNS, Surakarta Indonesia : 546-550 (Indonesian language).

[10] Istifadah, N. Fatiyah, N., Sunarto, T., Djaya, L, B. N. Fitriatin, Krestini, H. 2018. The effects of delivery methods of biocontrol agents and their combination with synthetic fertilizers on potato growth and diseases. *Academic Journal of Science,* 08(01):63–70

[11] Bahme J.B., Schroth, M.N. Van Gundy, S.D. , Weinhold, A.R. and Tolentino, D.M. 1988. Effect of inocula delivery systems on Rhizobacterial colonization of underground organs of potato. *Phytopathology* 78, 534–542.

[12] Steddom, K., Becker, O., Menge, J.A. 2002. Repetitive applications of the biocontrol agent *Pseudomonas putida* 06909-rif/nal and effects on opulations of *Phytophthora parasitica* in citrus orchards. Phytopathology 92:850-856.

[13] Campbell, L.C. Madden, V.L. 1990. Introduction Plant Disease Epidemiology. John Willey and Son. USA.

[14] Yendyo S, G.C. R and Pandey BR. 2018. Evaluation of *Trichoderma* spp., *Pseudomonas fluorescens* and *Bacillus subtilis* for biological control of *Ralstonia* wilt of tomato *F1000Research* , 6:2028 (doi: 10.12688/f1000research.12448.2)

[15] Chowdappa, P., Kumar, S.P.M., Lakshmi, M.J., Upreti, K.K. 2013. Growth stimulation and induction of systemic resistance in tomato against early and late blight by *Bacillus subtilis* OTPB1 or *Trichoderma harzianum* OTPB3. Biological Control 65: 109–117

[16] El-Naggar MA, Abouleid HZ, El-Deeb HM, Abd-El-Kareem F, and Elshahawy IE. 2016. Biological Control of Potato Late Blight by Means of Induction Systemic Resistance and Antagonism. Research Journal of Pharmaceutical Biological and Chemical Sciences, 7 (1): 1338-1348

[17] Yusidah, I. Istifadah, N. 2018. The abilities of spent mushroom substrate to suppress basal rot disease (Fusarium oxysporum f.sp cepae) in shallot. International Journal of Biosciences, 13 (1):440-448

IOP Conf. Series: Earth and Environmental Science **334** (2019) 012039 doi:10.1088/1755-1315/334/1/012039

Performance of Rice Production and Pest in Riau Province

Rustam[1]

[1]Riau Assessment Institute of Agricultural Technology, Jl. KH Nasution No. 341 Pekanbaru, Indonesia

E-mail: rustamriau@gmail.com

Abstract. Production of rice plants is affected by pest attacks. This study aimed to (i) analyze the trend of rice production and pest in Riau Province over the past 10 years, (ii) to predict the trend of pests and rice production of Riau Province for the next 5 years, and (iii) to formulate policy alternatives in order to increase rice production and pest attack control in Riau Province. Secondary data covering production data and paddy pests were obtained from related institutions. For validation and completion of some secondary data, the primary data was collected from direct observation and various competent sources. The data were analyzed in tabulative, regression and simple descriptive to answer the research objectives. The results showed that there has been a decline in rice production in Riau Province for the last 5 years was quite high (23.0%). The decline in production is mainly due to a decrease in harvested area. The decline in production is also caused by pest attacks. The type and extent of rice pest attack show a fluctuating trend. There are 8 types of pests causing rice crop failure (puso), namely bird, rice bug, golden apple snail, stem borer, brown spot disease, blast, and bacteria leaf blight. The total area of rice pests attack by category of puso 161.4 ha. In the next 5 years, Riau rice production is predicted to increase every year while pest attack will decrease every year. Rice production can be increased through increasing harvested area, reducing land conversion, water availability. While pest attack can be suppressed through the use of resistant varieties, early detection, simultaneous control, and intensify counseling.

1. Introduction

The need for food, especially rice every year is always increasing. Increased rice demand follows population growth rate. The total rice requirement reaches 636.680 ton/ year to supplay the needs of Riau Province's population of 6.3 million people. Population growth in Riau Province is quite high with the growth rate estimated at 3.6% while the average rice production over the past 10 years tends to decrease with the production amount per year is only about 285.022 tons (459.13 tons of dry mill rice) so that there is a dificit of rice by 351.658 ton (55%) each year [1].

On the other hand rice production in Riau Province in the last 10 years is decreasing. The average decline in rice production is about 3.6 %/year (13.816 ton) [1]. Ironically, nationally the production of paddy commodity every year keeps increasing, which is about 3.4% [2]. Various factors are accused of causing a decrease in production, including reduced planting area, low productivity, climatic factors, and attack of pest. Pest attack can decrease production significantly. Types of rice pests that often cause heavy damage to puso (crop failure) include: rats, brown plant hopper, stem borer, blast disease, tungro disease, bacterial leaf blight disease.

Nevertheless, efforts to increase rice production in Riau Province continue to be done. Since 2009-2014 through a national program to increase Rice Production has been conducted throught the activities of "Sekolah Lapang Pengelolaan Tanaman Terpadu" (SLPTT) in various food production centers in Riau. Then the Riau Provincial Government itself since 2009-2013 has also conducted "Operasi Pangan Riau Makmur" (OPRM , with 3 main activities, namely increasing the index of cropping, rehabilitation of abandoned paddy fields and expansion of planting area through new rice field printing activities. Furthermore, since 2014 efforts to increase rice production by the Central Government have been continued through "Program Upaya Khusus" of rice, corn, and soybeans (Upsus Pajale). In Upsus Pajale activities, all strategies and efforts are made to increase planting area and pajale productivity in food production center areas including in Riau Province for national food sovereignty to be achieved. Operationalization of Upsus Pajale program is supported by the provision of funds, mobilization, repair of irrigation networks, fertilizer assistance, availability of superior seeds, tractor assistance and agricultural machinery tools and marketing certainty [3].

The increase in production needs to be continuously improved so that the needs of the increasing number of population can be fulfilled. But the challenge of increasing production in the future is also increasingly associated with climate change and the threat of pest attack [4]. Associated with plant pests, it is important to know which plant pests are always dominating along with their level of attack so that the mitigation measures can be anticipated early on.

Based on the above explanation, this study aimed to analyze production performance and rice pest attack in Riau Province in the last 5-10 years.

2. Materials and Methods

This study basically uses secondary data obtained from several related institutions, such as Badan Pusat Statistik (BPS) Indonesia, BPS of Riau Province, Unit Pelaksana Teknis (UPT) Perlindungan Dinas Pertanian of Riau Province, and other relevant institutions. From BPS Indonesia obtained data of production, harvested area, and productivity of paddy, corn and soybean at national and provincial level. From BPS of Riau Province obtained data: (i) production, harvested area, and productivity of rice, corn, and soybean plants at district level; (ii) land area data based on district level type and land use; (iii) population data; and (iv) other relevant data. Meanwhile, from UPT of Protection of Food Crops Department of Riau Province obtained cumulative data of the pest attack covering the type and extent of district-level paddy rice pest attack in the last 5 years.

Each type of pest groups grouped into four levels of attack category, as mild, moderate, severe, and puso. The attack category is based on the level (percentage) of pests or diseases that attack the plant. Here are the details of the attack category and the level of attack of each pest group [5].

Table 1. Category and Level of Attack of rice pests

Category of attack	Attack rate (%)	
	Phytophagous Insect	Disease
Light	≤ 25	≤ 11
Medium	> 25 - ≤50	> 11 - ≤25
Weight	<50 - ≥85	<50 - ≥85
Puso	> 85	> 85

For validation and supplementing some secondary data, primary data were collected from competent sources either through interviews directly or through discussion activities or group meetings on various occasions. At the time of extracting primary data also collected other related field information, such as information about pest control techniques conducted in several rice production centers in Riau Province. The data obtained were analyzed by simple tabulative and descriptive.

3. Results and Discussion

3.1 Performance of Production, Harvests, and Productivity

Production of rice in Riau Province in the period of 2008 -2017 tends to decrease, with average annual production decrease reach 13.816 tons . The decline in rice production was mainly due to the reduced rice harvest areas throughout the year 2008 - 2017. The area of rice harvest in the year 2008 covering 147.796 ha then in 201 7 was reduced to 92.683 ha, or a reduction of 55.113 ha (37%) (Figure 1).

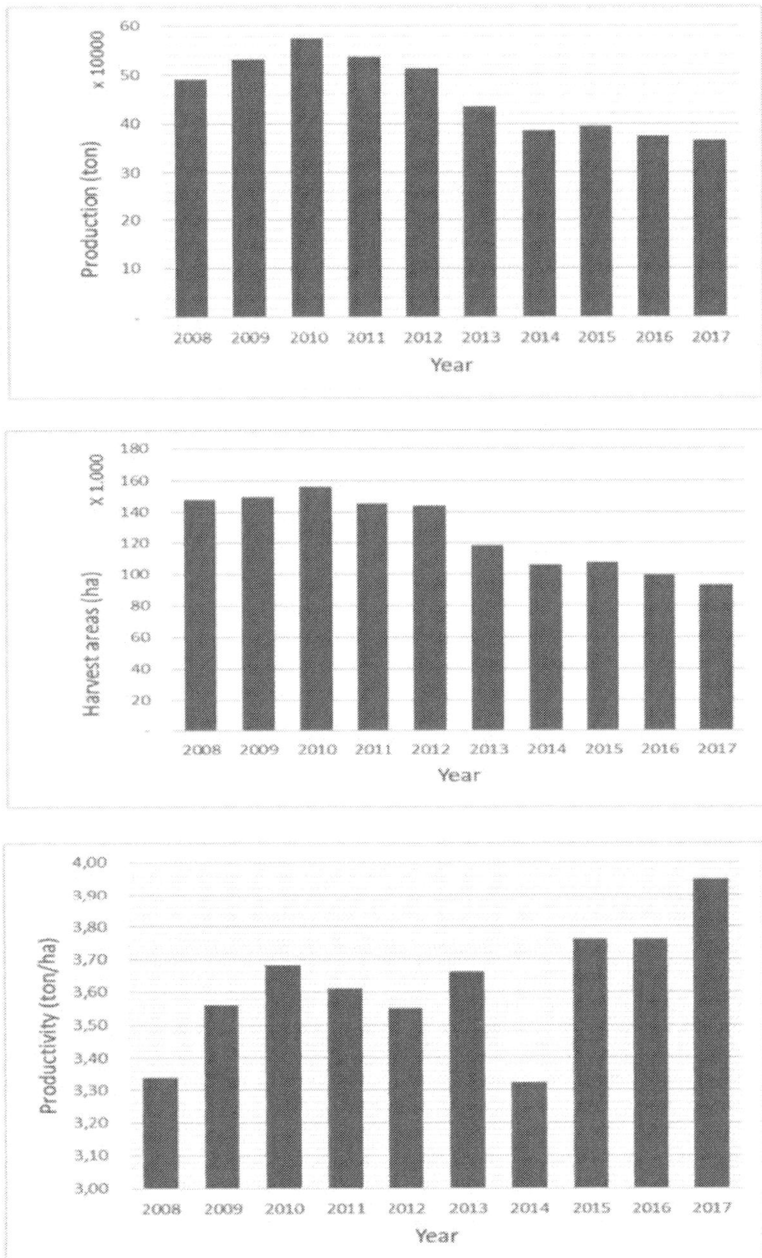

Figure 1. Production (a), harvested area (b), and productivity (c) of rice in Riau Province in 2008 - 2017.

If analyse of rice production per regency in Riau Province in the last 10 years (data not shown), the drastic decrease of production occurred in Rokan Hilir Regency. Rice production in Rokan Hilir regency in 2011 amounted to 158.344 tons and in 2015 fell to 50.056 tons, or decreased 108.288 tons (68.4%) [1]. If the production decline does not occur in Rokan Hilir Regency, in other words, the production in 2015 is the same as the production in 2011, the Riau Province's rice production as a whole will be relatively stable.

Rokan Hilir regency is one of the main production centers of rice in Riau Province so that the decrease of production that occurred in the district has a significant effect on total rice production in Riau Province. From some sources [6,7] it is known that the cause of the decline in rice production in Rokan Hilir Regency is mainly caused by the rampant conversion of food land into non-food land, especially oil palm (. As an illustration, the total increase of oil palm area in Rokan Hilir Regency for the last 5 years is about 165.992 ha (7.4%) [1]

In addition, rice productivity in Riau Province is relatively low. The average productivity per year in Riau Province during 2011-2015 is 3.58 tons/ha [1]. This productivity is well below the national average productivity over the last 5 years of 5.14 tons/ha. Rice productivity in Riau Province is also relatively low when compared with rice productivity in neighboring provinces, including: Jambi Province of 4.32 tons/ha and South Sumatra Province of 4.52 tons/ha [2] In terms of land conditions in the two neighboring provinces are relatively similar to the conditions of land in Riau Province.

The low productivity of paddy in Riau Province is caused by rice production area in Riau consist of 13% of paddy fields and 77% wet rice field [1]. As it is known that the productivity of paddy fields is generally much lower than the productivity of lowland rice so low productivity value will decrease the overall productivity of rice. In addition, wetland land in the form of tidal paddy fields is the dominant rice field type in two main production centers in Riau Province, namely Indragiri Hilir and Rokan Hilir Regency, with productivity of lowland rice is usually also lower than the productivity of dryland paddy.

Another thing that can cause low productivity of rice in Riau Province is the attack of rice pests. During the 2011 - 2015 timeframe has been reported there are 22 types of rice pests that attack rice crops. Of the 22 types of pests, 8 types of pests cause crop failure (puso) with a total area of attack 161.4 ha and 9 types of pests cause severe damage with total area of attack 124.05 ha (Table 2). The details of the type and extent of the respective pest attack are described in the performance of pest section of this paper.

The decrease of production and harvest area in Riau Province in the period of 2011 - 2015 indicate that the programs of increasing the production of rice either coming from the central government, such as "P2BN program" and "self-sufficiency program pajale" or program coming from local government, such as program of "OPRM" and other local program can not be considered successful. In the case of national, "P2BN program" for the year 2011-2014 were considered successful due to an increase in rice production of 65.76 million tons in 2011 to 70.85 million tons in 2014, up about 2%/year. Similarly, the "pajale program" launched by the central government in 2014 has increased rice production from 70.85 million tons in 2014 to 75.40 million tons in 2015, up by 6.4% per year. Although for the pajale program there was also an increase in rice production in Riau, from 385.475 tons in 2014 to 393.917 tons in 2015 [2,1] but within the last 5 years increase in rice production in Riau Province can be considered not successful.

3.2 Rice Pests Performance

From the extensive cumulative data added to the rice pest attack in Riau Province in 2011-2015 issued by TIU of Protection of Food Crops Department of Riau Province, it was reported that there are 22 types of pests that attack rice plants. Of the 22 types of paddy pests, 8 types of pests can cause harvest failure (puso) and 9 types of pests can cause severe damage. The eight types of pest that can cause puso are bird, mouse, sardine, mulberry snail, stem borer, brown spot disease, blast, and leaf blight. Then 9 types of pests that can cause severe damage such as orong-orong pests, earthworms, pigs, fake white pests, brown planthopper, cutworm, midrib blight disease, and iron poisoning (Table 2).

Among the pests reported to attack rice crops, birds are the most destructive pest species in Riau Province. The average area of bird pest attack per year reaches 455.2 ha. During the period of 2011 - 2015, every year bird pests always cause heavy damage to rice crops, with an average damage of 27.8 ha/year. Then bird pest also has caused puso rice plants as much as 3 times, namely in 2011 area of 35.85 ha, the year of 7 hectares of land, and the year 2014 area of 24 ha. In addition, there is a tendency that the extent of bird pest attacks always increases annually (Table 2).

Many species of birds cause rice yield loss. In Riau province there are 12 species of birds belonging to 4 orders, 8 families, and 9 genera that attack rice plants [8]. Spesies of the bird are *Bubulcus ibis*, *Halcyon smyrnensis*, *Hirundo rustica*, *Lonchura leucogastroides*, *L. maja*, *L. punctulata*, *L. striata*, *Padda oryzivora*, *Passer montanus*, *Pycnonotus goiavier*, *Prinia familiaris* and *Streptopelia chinensis*. According to [9] the preferences of a particular type and population of birds that come right to attack rice plants closely related to the amylose content in the rice variety.

Bird pests begin to interfere when the rice is seeded, rice is still young, rice yellowing until ready to harvest. The birds usually nest near the house, the low trees and the bushes around the rice fields. Although the rate of bird consumption per tail is relatively small, such as Bondol Peking and Bondol Jawa which are only 2-2,8 gram/day [10] but because the population is thousands when eating rice that can cause a considerable loss of yield.

Table 2. Extent of rice pests attack in Riau Province in 2011 - 2015.

Pest name	Category of attack	Area of attack (ha)					
		2011	2012	2013	2014	2015	Average
Bird	Light	246.7	250	287.1	381.2	419.7	316.9
	Medium	90.51	29.15	105.8	116.1	143.9	97.1
	Weight	60.5	17.2	2.9	56.5	2	27.8
	Puso	35.85	7	0	24	0	13.4
	Amount	433.6	303.35	395.7	577.8	565.6	455.2
Rats	Light	676.9	670.89	643.5	428.4	487.5	581.4
	Medium	141.1	203.21	217.2	112.5	209.9	176.8
	Weight	7.25	41	46	9.8	21.4	25.1
	Puso	3.8	33.6	0	0	1	7.7
	Amount	829	948.48	906.6	550.8	719.8	790.9
Rice bug	Light	1383	731,78	770.2	382.5	609.9	775.5
	Medium	57.6	137.81	167.9	94.77	139.9	119.6
	Weight	9.3	11.1	12.5	2.1	18	10.6
	Puso	25	5	2	0	0	6.4
	Amount	1475	885.69	952,6	479.4	767.8	912.1
	Light	140.8	122.92	68.8	81.65	98.45	102.5
	Medium	11.65	46.01	28.7	17.3	40.25	28.8

Pest name	Category of attack	Area of attack (ha)					
		2011	2012	2013	2014	2015	Average
Brown spots	Weight	3.2	10.8	0.4	0.5	35.9	10.2
	Puso	10	0	0	0	0	2.0
	Amount	165.6	179,73	97.9	99.45	174.6	143.5
Snails	Light	209.3	128.6	115.1	201	183	167.4
	Medium	80.4	45.5	60.16	119.9	57.7	72.7
	Weight	15.7	4	12.8	2.5	5	8.0
	Puso	7.45	0	0.3	1.8	0	1.9
	Amount	312.9	178.1	188.4	325.2	245.7	250.1
Bacterial leaf blight	Light	400.1	162.2	90.5	110.2	78,53	168.3
	Medium	22	58.25	38.1	20.32	35.9	34.9
	Weight	16	24	0.1	4	0	8.8
	Puso	0	0	0	2.5	0	0.5
	Amount	438.1	244.45	128.7	137	114.4	212.5
Blast	Light	247.7	268.85	173.7	231.6	331.2	250.6
	Medium	47.7	34.2	63.25	77.3	113.5	67.2
	Weight	0.1	0.6	0.1	1.65	22.7	5.0
	Puso	0	0	0	1.5	0	0.3
	Amount	295.5	303.65	237	312	467.3	323.1
Stem borer	Light	506.5	455.7	583.5	427.7	635.9	521.9
	Medium	50.78	87.16	224.7	62.11	210.2	127.0
	Weight	31.7	10.1	55.4	1.8	58.5	31.5
	Puso	0	0.2	0.4	0	0	0.1
	Amount	589	553.18	864	491.6	904.6	680.5
Mole cricket	Light	85.41	89.05	81.84	88,59	80.3	85.0
	Medium	24.24	13.69	22.96	54.89	47.1	32.6
	Weight	0.5	0.4	0.1	1	34.6	7.3
	Puso	0	0	0	0	0	0.0
	Amount	110.2	103.14	104.9	144.5	162	124.9
Black bugs	Light	657.8	303,7	63.3	192	137.2	270.8
	Medium	24.3	41.9	14.4	11.66	15.6	21.6
	Weight	22.7	7.8	0.7	0.3	0	6.3
	Puso	0	0	0	0	0	0.0
	Amount	704.8	353.4	78.4	204	152.8	298.7
Pig	Light	370.3	4	48.78	32.01	16	94.2
	Medium	16.1	0	38.8	0	2.5	11.5
	Weight	22.2	0	0	0.8	1	4.8

	Puso	0	0	0	0	0	0.0
	Amount	408.6	4	87.58	32.81	19.5	110.5
Caseworm	Light	618.1	737.46	565.2	301.4	376.2	519.7
	Medium	43.97	48.9	149.3	48	40.8	66.2
	Weight	0	5	9.4	0.4	3	3.6
	Puso	0	0	0	0	0	0.0
	Amount	662	791.36	723.9	349.8	420	589.4
Brown planthopper	Light	27	97.21	208.1	115	68	103.1
	Medium	10.2	0	32.4	4.9	9.2	11.3
	Weight	0	0	4	2.55	0	1.3
	Puso	0	0	0	0	0	0.0
	Amount	37.2	97.21	244.5	122.4	77.2	115.7
Armyworm	Light	22.1	10.4	22.85	9.5	7.4	14.5
	Medium	17.4	11.8	6	0.1	3	7.7
	Weight	0	2	0	0	0	0.4
	Puso	0	0	0	0	0	0.0
	Amount	39.5	24.2	28.85	9.6	10.4	22.5
	Light	76.1	102.3	25.3	71.3	65	68.0
Sheath Blight	Medium	5.9	6.8	32.6	0	0	9.1
	Weight	0	0.1	0	0	0	0.0
	Puso	0	0	0	0	0	0.0
	Amount	82	109.2	57.9	71.3	65	77.1
Iron Toxicity	Light	7.55	195.5	45.9	60.55	7	63.3
	Medium	0.5	18	2.8	3	1	5.1
	Weight	0	0	5.5	0	0	1.1
	Puso	0	0	0	0	0	0.0
	Amount	8.05	213.5	54.2	63.55	8	69.5

Source: Processed from data of UPT Perlindungan, Dinas Pertanian of Riau Province.

Farmers usually avoid bird pest by way of waiting for the rice fields to expel the birds directly who want to eat rice . Farmers also often use a lat help scarecrows and tin cans are placed to spread the planting area and connected by a rope and then pulled to cause noise. Another way is to close the rice y ang yellowed and ready for harvest with fishing nets. This last way although quite effective but expensive to buy a net enough to cover a vast expanse of rice fields.

For rat pests, the average extent of the attacks during 2011-2015 is 790 ha/year. Every year, rodent pests also cause severe damage to rice crops, with an average annual weight of 25,1 ha per year. In addition, mice was recorded 3 times causing puso rice plants, namely in 2011 area of 3.8 ha, the year of 33.6 hectares in 2012, and 2015 of 1 ha, with an average of 7.7 ha per year.

In the past 10 years (1997-2006), rats have also caused the greatest damage to rice crops [28]. Among the 7 species of mice, the major pest species are *Rattus argentiventer*Rob and Kloss . The mice are capable of causing 50 - 100% damage. Some of the recommended rat control technologies include: (i) Environmental sanitation and habitat manipulation; (ii) technical culture, (iii) physical control, use of light, pumping water or mud into rat nests, ultrasonic sound, mass "gropyokan", trapping, linear trap system, (iv) utilization natural enemies; (v) chemical control (rodenticide, fumigation, *reppelent,* antifertility) [11, 12].

For the pests of the pest, the average area of attack is 912.1 ha / year. Every year within the period of 2011 - 2015, the pest of the pest is also always causing heavy damage to rice crops, with the average area of heavy attack per year reaches 10.6 ha. The pest of Walang is also recorded 3 times causing puso rice plants, namely in 2011 as wide as 25 hectares, in 2012 as wide as 5 ha, and in 2013 area of 2 ha.

According [13] that 5 witches (*Leptocorisa oratorius* F.) on every 9 clumps of plants can reduce yield by 15% and then 10 tails in 9 clumps of plants canreduce yield by 25% . Thus, the pest that causes puso and severe damage to rice plants in Riau Province is estimated to exceed 10 population/9 clumps of plants.

Among the pests belonging to disease (pathogens), brown spots are the most destructive type of rice crop diseases in Riau Province in 2011-2015. The average area attacked by brown spotting disease reached 143.5 ha/year. Every year brown patches also cause severe damage to rice crops, with an annual average of attack area is 10.2 ha. The brown spot disease was noted to have caused serious rice plant damage in 2011 with the area of damage 10 ha.

In Riau Province, some things that can be suspected as the cause of severe damage by brown patch disease in rice plants are many bad drained fields, high acidity so that nutrients are less available, and planting of vulnerable varieties. This is in accordance with the opinion of [14], that severe damage by brown patches is common in rice cultivation on land with poor drainage systems or nutrient-deficient soils, especially potassium elements. In addition, the extent of damage by brown patch disease is also influenced by the type of varieties grown and the level of attack [15].

The golden snail pest (snail) also belongs to the pest that harms rice crops in Riau Province, with an average attack area reaching 250.1 ha/year. Each year within the period of 2011-2015, the mangrove snail always causes severe damage to rice crops, with an average annual attack area reaching 8.0 ha. Mutation of mulberry recorded four times causing puso rice plants, namely in 2011, 2012, 2013, and 2014, with an average attack reached 1.9 ha.

According to [16], snails (*Pomacea canaliculata* Lamarck) can cause severe damage especially when the plants are young with the condition of the land rather stagnant. For that young plants are attacked should be immediately inserted and reduce puddles. Other recommended controls include: egg trap fitting, trap feeding such as taro leaves, duck grazing in affected areas, and chemical control [17].

Bacterial leaf blight (BLB) is the second most destructive type of pest after brown spot disease in rice plants in Riau Province. The average extent of BLB disease attacks reaches 212.5 ha / year. In the period 2011 - 2015, HDB disease caused severe damage as much as 4 times, with the average area of heavy attack per year reached 8.8 ha. BLB disease recorded 1 time causing puso in rice plants, ie in 2014 of 2.5 ha.

BLB disease caused by *Xanthomonas oryzae pv. oryzae* can cause severe damage to dried up because this pathogen can infect rice leaves at all phases of plant growth, ranging from nursery until the harvest. Therefore, control is recommended in an integrated manner with the various ways to suppress the development of the disease, including the use of varietas resistant [18, 19], healthy seeds, chemical, and biological [18].

Blast disease is the most destructive type of pest number 3 after brown spots and HDB disease in rice plants in Riau Province. For blast disease, the average attack area reaches 323,1 ha / year. In the period of 2011 - 2015, blast disease always causes severe damage to rice crops, with an average annual attack area reaching 5.0 ha. Blast disease was recorded only 1 time to cause puso rice plants, namely in 2014 of 1.5 ha.

Blast disease caused by *Pyricularia grisea* fungus Cavara [27] often attacks upland rice in addition to wetland rice. As is known, that 13% of rice production in Riau Province is produced from upland rice so this is also suspected as the cause of high blast disease in Riau Province.

In addition, blast disease attacks are also influenced by the type of varieties grown. For that purpose, resistant varieties are the main components and the most effective, economical, and easy way to control blast disease [20]. A rice variety can be resistant to a type of blast even though its resistance can eventually be broken [21,22] For this reason it is recommended to use resistant varieties alternately and proper fertilization in order to anticipate rapid blast type changes [14].

For stemborer pests, the average extent of the attack reaches 680,5 ha/year. Each year within the period of 2011 - 2015, stem borer pests also always cause severe damage to rice crops, with an average annual attack area of 31.5 ha. The stem borer was recorded 2 times causing puso of rice plants, ie in 2012 and 2013, with average attack area reaching 0.3 ha.

In the field, the presence of stem borer (*Scirpophaga* spp.) is characterized by the presence of moths, the deaths of rice buds (sundeps), death of panicle, and larvae of stem borer. These pests can damage crops at all growing phases, both at the time of the nursery, the seedling phase, and the flowering phase. Until now there are no varieties that resist the stem borer. Stem borer prefers organically managed rice plants [23]. According to Baehaki *et al.* [24] The strategy for stem borer control is a powerful *triangle* implementation *strategy,* namely implementing properly stem borer control (applying the economic threshold is based on monitoring of population moths using light traps, four days after the first moth flight), building community controls in the community, as well as support from central and local government policies and tehir commitments.

Of the 8 types of pests that can cause severe damage to rice plants, pest orong-orong is the type of pest that always causes severe damage each year during the period 2011-2015. The average area of orong-orong pest attacks with severe damage categories reached 7.3 ha / year. Other pests that cause severe damage are 6.3 ha of land cover, 4.8 ha of pigs, 3.6 ha of white pest, 1.3 ha of brown planthopper, 0.4 ha of grayak caterpillar, an area of 0.1 ha, and Fe toxicity of 5.5 ha.

Mole cricket (*Gryllotalpa orientalis* Burmeister) often attack rice plants in tidal land. Orongs attack the root or base of growing rice plants and rice plants in irrigated rice fields, lebak, tides if there are no puddles. Mole cricket pests are usually controlled by farmers through soil processing and carbofuran insecticide use.

Similar to mole cricket, black bugs (*Scotinophara* spp.) often attack rice plants in lowland areas, such as in Riau Province [26]. Black bugs damage rice plant by sucking up the rice plant liquid so that the color of the plant turned reddish or yellow. The population is often abundant so control using pesticides often fails.

References

[1] Badan Pusat Statistik Provinsi Riau 2016 Provinsi Riau dalam Angka. http://www.riau.bps.go.id. [21 Agustus 2016].

[2] Badan Pusat Statistik Indonesia 2016 Pertanian dan pertambangan. Tanaman pangan. http://www.bps.go.id/linkTableDinamis/ view/id. [20 Agustus 2016].

[3] Kementerian Pertanian 2015 Upaya khusus (Upsus) Swasembada Pangan 2015-2017. http://biogen.litbang.pertanian.go.id/index.php/2015/02/upaya-khusus-upsus-swasembada-pangan-2015-2017/. [20 Agustus 2016].

[4] Susanti E, Ramadhani F, Runtunuwu E, Amien I 2012 Dampak perubahan iklim terhadap serangan organisme pengganggu tanaman (OPT) serta strategi antisipasi dan adaptasi. Balai Penelitian Agroklimat dan Hidrologi. Bogor.

[5] [Dirjentan] Direktorat Jenderal Tanaman Pangan 2015 Keputusan Direktorat Jenderal Tanaman Pangan Nomor:55/HK.310/8/2015 tentang Petunjuk Teknis Pemantauan dan Pengamatan serta Pelaporan Organisme Pengganggu Tanaman dan Dampak Perubahan Iklim. Jakarta: Direktorat Jenderal Tanaman Pangan. Pp. 88.

[6] [DPTP] Dinas Pertanian Tanaman Pangan Provinsi Riau 2014 Laporan Gerakan Operasi Riau Makmur. Dinas Pertanian Tanaman Pangan Provinsi Riau. 176 hal.

[7] [DTPHP] Dinas Tanaman Pangan Hortikultura dan Perkebunan. 2017. Rumusan Hasil Rapat Evaluasi dan Rencana tahun 2017. Nomor surat 521.21/Dis-TPH Bun/325, tanggal 30 Januari 2017. 5 hal.
[8] Jurati, Ade FY, dan Dahlia 2015 Jenis-jenis burung (aves) di persawahan Desa Pasir Baru Kabupaten Rokan Hulu Riau. *E-Journal Mahasiswa Prodi Biologi, FKIP Universitas Pasir Pangaraian* **1 (1)** 1-4.
[9] Sumari GD 2011 Identifikasi Burung Pemakan Biji pada Berbagai Kultivar Padi (*Oryza sativa* L.) di Desa Gentan Kecamatan Bendosari Kabupaten Sukoharjo. Skripsi Jurusan Biologi, Fakultas Matematika dan Ilmu Pengetahuan Alam, Universitas Sebelas Maret Surakarta. Surakarta.
[10] Priyambodo S, Ziyadah K 2012 Kemampuan makan dan preferensi pakan pada bondol peking (*Lonchura punctulata* L.) dan bondol jawa (*Lonchura leucogastroides* Horsfield & Moore). Prosiding Nasional Hasil Penelitian Padi 2011. Inovasi Teknologi Padi mengantisipasi Cekaman Lingkungan Biotik dan Abiotik. Editor: I Putu Wardana, Sudir, N Usyati, Made J Mejaya. Balai Besar Penelitian Padi.
[11] Priyambodo S. 1995. *Pengendalian Hama Tikus Terpadu. Penebar Swadaya.* Jakarta.
[12] Sudarmaji, Herawati NA. 2009. Ekologi tikus sawah dan teknologi pengendaliannya. http://www.litbang.pertanian.go.id/special/padi/bbpadi_2009_itp_11.pdf.[4Sept 2015].
[13] Suharto H. Damardjati DS 1988 Pengaruh waktu serangan walang sangit terhadap hasil dan mutu hasil padi IR 36. *Reflektor* **1(2)** 25-28.
[14] Syam M, Suparyono, Hermanto, Wuryandari DS 2007 Masalah Lapang Hama, Penyakit, Hara pada Tanaman Padi. Cetakan ketiga. Pusat Penelitian dan Pengembangan Tanaman Pangan. Bogor
[15] Djunaidi A 2009 Ketahanan padi (Way Apo Buru, Sinta Nur, Ciherang, Singkil, dan IR64) terhadap serangan penyakit bercak coklat (*Drechslera oryzae*) dan produksinya. *Agrovigor* **2 (1)** 8-15.
[16] Suharto H. Kurniawati N 2009 Keong mas dari hewan piaraan menjadi hama utama padi sawah. http://www.litbang.pertanian.go.id/special/padi/bbpadi_2009_itp_14.pdf. [4 September 2016].
[17] Hamidy S. Khalid J. Hamdani MA 2007 Rakitan teknologi pengendalian keong mas. http://nad.litbang.pertanian.go.id/ind/images/dokumen/Rekomtek/07-RAKITAN%20TEKNOLOGI%20PENGENDALIAN%20HAMA.pdf. [4 September 2016].
[18] Kadir TS, Suryadi Y, Sudir, Machmud M 2014 Penyakit bakteri padi dan cara pengendaliannya.
[19] Herlina L, Silitonga TS. Seleksi lapang ketahanan beberapa varietas padi terhadap infeksi hawar daun bakteri strain IV dan VIII. *Buletin Plasma Nutfah* **17 (2)** 80-87.
[20] Khaeruni A, Taufik M, Wijayanto T, Johan EA 2014 Perkembangan penyakit hawar daun bakteri pada tiga varietas padi sawah yang diinokulasi pada beberapa fase pertumbuhan. *Jurnal Fitopatologi Indonesia* **10 (4)** 119-125.
[21] Corea-Victoria FJ, Zeigler RS 1993 Pathogenic variability in *Pyricularia grisea* at rice blast "hot spot' breeding site in Eastern Colombia. *Plant Disease* 77 (10) 1029–1035.
[22] Sudir, Nasution A, Santoso, Nuryanto B 2014 Penyakit blas *Pyricularia grisea* pada tanaman padi dan strategi pengendaliannya. *Iptek Tanaman Pangan* **9 (2)** 85 – 96.
[23] Hadi M, Soesilohadi RCH, Wagiman FX, Suhardjono YR 2015 Populasi penggerek batang padi pada ekosistem sawah organik & anorganik. *Bioma* **17 (2)** 106–117.
[24] Baehaki SE 2013 Hama penggerek batang padi dan teknologi pengendalian. *Iptek Tanaman Pangan* **8 (1)** 1 – 14.
[25] Paendong E. Palealu J. Rimbing J 2011 Penyebaran hama kepinding tanah dan musuh alaminya pada pertanaman padi sawah di sulawesi utara. *Eugenis* **17 (3)** 178 – 186.
[26] Sudarmaji 2007 Pengendalian hama tikus terpadu untuk mendukung P2BN (Peningkatan Produksi Beras Nasional). Direktorat Perlindungan Tanaman. Direktorat Jenderal Tanaman Pangan. Jakarta.
[27] Rossman AY. Howard RJ, Valent B. 1990. *Pyricularia grisea,* the correct name for the rice blast disease fungus. *Mycologia* **82(4)** 509-512.

IOP Conf. Series: Earth and Environmental Science **334** (2019) 012040 doi:10.1088/1755-1315/334/1/012040

The presence of honey bees in intensive pesticides use area: Case study in Pangalengan and Majalengka

N Carsono[1], S Sari[1], Y Saragih[1], I S Amalia[1], A Purdianty[1], Y Hidayat[1] and M Johannis[2]

[1]Faculty of Agriculture, Universitas Padjadjaran, Jatinangor, Sumedang 45363, Indonesia.
[2]Head of Research & Development PT Syngenta Indonesia, Kotabaru, Karawang 41374, Indonesia.

E-mail : n.carsono@unpad.ac.id

Abstract. Pesticide action can leave a residue in plant, including in the nectar and pollen of the flowers. Besides residues, sprays of pesticides could also make unintended possible effect to honey bee and other pollinators. Pollinators have a great impact on plant productivity, so it was important to keep the pollinators in healthy and in high diversity for sustainable of environment, food, and economy. The research was aimed to obtain valuable information regarding on pesticide use in both locations. The survey was divided into two steps. First, we did a preliminary survey to determine representative locations of study, and then followed by the second step, it was in-depth interview of pesticide use directly with 61- and 51 respondents in Pangalengan and Majalengka, respectively. Ninety six pesticide trademarks were applied by farmers in Pangalengan, meanwhile, 83 trademarks were in Majalengka with the variable bioactive compound. Neonicotinoids as bioactive of some pesticides were found only in one farmer from Pangalengan who used Alika (pesticide brand name). Besides, there were 13 active ingredients of fungicides, 15 insecticides, one herbicide, and one molluscicide found in Pangalengan. In Majalengka, there were 9 active ingredients of fungicides, 19 insecticides, and one molluscide. They usually applied herbicides, insecticides, and fungicides at least twice a week during crop cultivation. Although some farmers may be aware of pesticide hazards, adequate protection is hardly taken to minimize the risks. Sixty percent of farmers mentioned the presence of honeybees on their land. Farmers' knowledge in the use of pesticides is appropriate and safe for both the health of pollinators and farmers when applying pesticides in the field.

1. Introduction

Pesticides including herbicides, insecticides, and fungicides are commonly used to prevent or manage pests such as weeds, insects, and plant pathogens, while reducing the amount of labor, fuel, and machinery used for pest control. Vegetable crops were the largest users of pesticides than other crops, accounting for 30% of total sales. This trend has been confirmed in surveys of chili crops in West Java in 2002, where the adequacy of crop protection is a major factor affecting production and profitability of chili crops [1]. However, pesticide action can leave a residue in plant, including in the nectar and pollen of the flowers. Besides residues, sprays of pesticides also could possibly make unintended effect to honey bees and other pollinators [2]. Therefore, it is necessary to maintain the pollinators, especially in areas that have large farming fields of pollinator dependent crops such as mango, rambutan, durian, melon, water melon, pumpkin, and zucchini. In addition to pollinators, the dangers of pesticides also occur to farmers, as in the case of Magelang, as many as 14.3% of farmers as revealed by the study are declared by spraying methods that do not comply with the proper rules and not use personal protection

[3]. The existence of pollinators especially honey bee in farmers' field and application of pesticides by farmers is our concern for sustainable production of crops.

The field survey were conducted in West Java, which has a large agricultural area producing food crops, horticulture, estate crops, industrial and also forestry crops. These two locations having large areas of farming as well as forestry systems are Pangalengan and Majalengka. Both areas supply many agriculture- and forestry-based products, and also cross-fertilizing crops such as cabbage, carrot, cucumber, guava, mango, and many other tropical crops which are depended on pollinators. Among pollinators, honey bee is well known since it has many benefits for human being.

Information regarding on pesticides use which supports pollinator health and pesticide use status in both agriculture and forestry farming systems of Pangalengan and Majalengka is needed in order to get information regarding type, concentration, brand name, and other issues related to pesticide use in farmers' field. there were two necessary activities to get relevant data and information such as: preliminary study for determining relevant site of observation and in depth interview of farmers, face to face interview, related to the application of pesticides (type, dose of application, concentration, etc). These are fundamental for assessing pollinator health and pesticide use status in order to design next experiment and other possible sustainable management action plan.

2. Materials and Method

This study was divided into two steps. Firstly we did a preliminary survey to determine representative location of study. Secondly in depth interview of pesticide use directly at farmers' field.

2.1 Preliminary survey

This activity was done in order to determine representative locations of study. We visited some specific areas both in Pangalengan and Majalangka. We also visited government office such as Office for Forestry and Estate Crops, and Office for Food and Horticulture Crops for obtaining secondary data related to explore pesticides use by farmers and the presence of apiary.

2.2 Interview of Pesticides Use

The activity was performed to assess pesticides use by farmers. The sample location was determined by using purposive sampling in which representative location was deliberately selected. The materials used were as follows: Global Positioning System (GPS), questionnaire, cameras, stationery and other tools. Data were collected then descriptive analysis was performed.

In depth farmers interview (face to face) was directly conducted in famers' field using questionnaire. The data recorded were as follows: farmer personal data, education background, crops they cultivated and pesticides applied, type of pesticide applied, dose, application interval, knowledge of farmers related to pesticide, farmers' perception for the presence of honey bee in the field.

3. Result and Discussion

3.1 Preliminary survey

Preliminary survey showed that the representative locations have been determined according to locations with large areas of farming in Pangalengan and Majalengka. In Pangalengan, some promising locations were Cilaki, Malabar, Ciwidey, and border area of Pangalengan-Garut (Cisewu). In Majalengka some representative locations were Argapura (Sukasari Kidul and Teja Mulya), Raja Galuh and Lemah Sugih.

3.2 Survey of Pesticides Use

Data regarding on pesticides use was obtained by interviewing the farmers directly in the field. Data collected were education background, crop they cultivated, pesticides application, type of pesticide, dose, application interval, knowledge of farmers related to pesticide, and farmers' perception regarding on the presence of honey bee in the field.

3.2.1 Farmer background

Age characteristics are one of the characteristics that have a close relationship with the ability to do daily work. Based on the results of the study it was found that 61 respondents in Pangalengan and 51 respondents in Majalengka, the youngest age was 17 years and farmers using pesticides were dominated by the elderly. The age of farmer ranged from 46 to 55 years old (Table 1). These data should be taken into consideration since the age of farmers is getting old. It is suggested that regeneration of farmers should also be programmed by government though formal and informal education. Public-private partnership to attract more young generation to get involved in agriculture should be conducted. Strategic Plan of Ministry of Agriculture 2015-2019 mentioned that improvement of human resources in agriculture should be implemented soon.

Table 1. Age of respondents in Pangalengan and Majalengka

Age range (years)	Number of Respondents in	
	Pangalengan	Majalengka
17-25 (late adolescence)	5%	6%
26-35 (early adulthood)	13%	12%
36-45 (late adulthood)	23%	12%
46-55 (early age)	33%	33%
56-65 (late age)	15%	25%
> 66 (old)	11%	12%

Phases of Adulthood according to Department of Health of Indonesia (2009) [4]

Farmers with education level graduated from high school in Pangalengan were 39% and elementary schools 31%, while farmers in Majalengka were 65% graduated from elementary school (Figure 1b). It seems likely human resources in agriculture for both agriculture centres need to be improved. It is expected that education of farmers will be soon improved since future agriculture will be more challenging and Indonesia has entered Asean Free Trade Area (AFTA) in which high qualified farmers will be required in order to solve many problems and challenges. Low level of education, as found in Majalengka, might be led to low productivity and less competitive in managing agriculture as reported in Ghana. In this country farmer education has a positive effect on farm productivity [5]. This condition seems to be relevant to Indonesia condition.

In this survey, most farmers (65% in Majalengka and 56% in Pangalengan agree; 27% in Majalengka and 43% in Pangalengan strongly agree, Figure 1a) believed that applying pesticides will guarantee their harvest, even though they also realize that pesticides harmful to other insects (Figure 1a) and consumers (Figure 1c). Most likely the farmers get a lot of information about the safety of pesticide use from pesticide packages, newspapers, television and other media, such as counseling services, and others. It is thought that extension services provided by the government are rarely conducted as revealed by some farmers.

Tabel 2. Education background of respondents in Pangalengan dan Majalengka

Level of Education	Pangalengan	Majalengka
No Formal School	0%	4%
Elementary School	31%	65%
Junior High School	20%	14%
Senior High School	39%	14%
Diploma 1	2%	0%
Diploma 2	0%	0%
Diploma 3	3%	0%
Bachelor's degree	5%	4%
Magister's Degree	0%	0%
Doctoral's Degree	0%	0%

Most respondents agreed or strongly agreed that pesticide use in food production reduces food safety. The respondents were well aware that pesticides are harmful to the environment and human health. The majority of the sample (over 80%) agreed or strongly agreed that pesticide use adversely affects human health. With regard to respondents' perceptions of their personal health risks, most of the farmers thought using pesticides would adversely affect their own health.

Perceptions of farmers can be influenced by their own knowledge either getting information from counseling or others. Fig. 2 reports on farmers' main sources of information regarding pest control and pesticide application. It can be seen that the most important information source regarding pest control and how to apply pesticides is communicated with other farmers, followed by pesticide retailers.

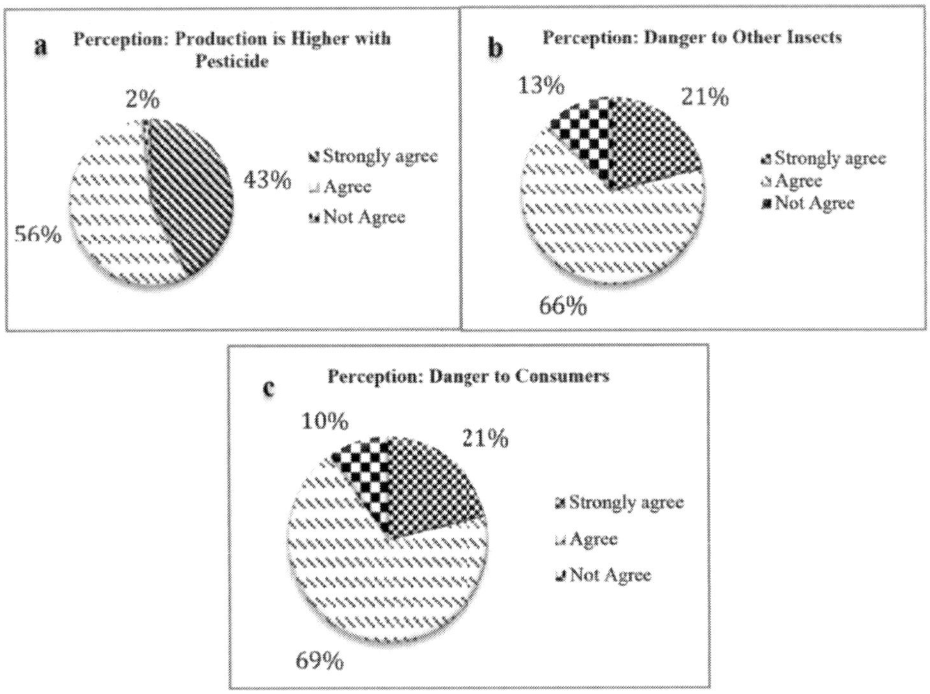

Figure 1. Farmers' perceptions **a**) on pesticide use for securing crop production, **b**) on the impact of pesticides to other insects, and **c**) on Pesticide Risk on Consumers Health

Only 5% and 20% were in Pangalengan and Majalengka, respectively, reported that they had acquired information on pest control and pesticide application from government extension services (Figure 2a). As many as 49% and 41% were Pangalengan and Majalengka, respectively, reported that they believed that the use of pesticides was not dangerous (Figure 2b). Ineffective extension services are considered as a key factor leading to the overuse of pesticides in developing countries. It is likely very few farmers learned about safety use of pesticides via media, e.g., the internet, television, books or newspapers.

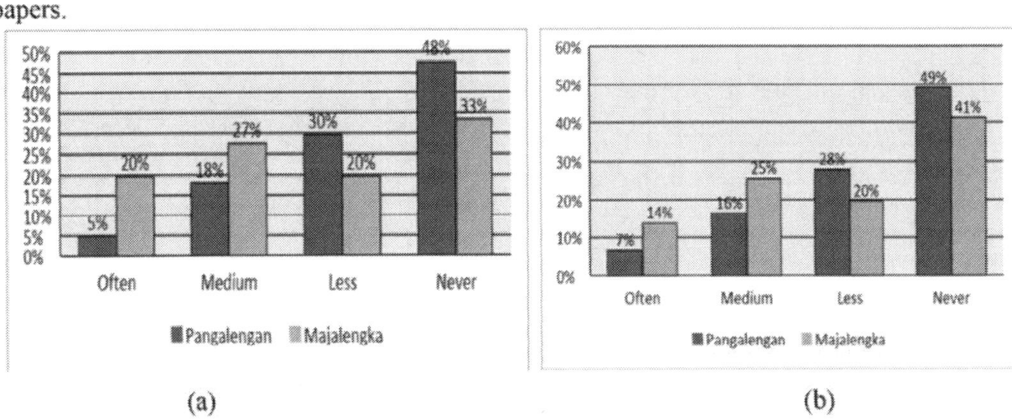

Figure 2. Main source of information on pest control and pesticide use: (a) Agriculture extension on pesticide use; (b) Agriculture extension on danger of pesticide

3.3 Pesticides Use

Horticultural plants that often sprayed by pesticides are potatoes, tomatoes, red chili, shallots, and shallots. Of the 3207 pesticide brands registered in Ministry of Agriculture Indonesia year 2016 [6], there were 96 pesticide trade brands were applied by farmers in Pangalengan, while 83 were in Majalengka with bioactive variable components. There were 14 active ingredients of fungicides, 15 of insecticides, one of herbicide, and one of molluscicide found in Pangalengan. In Majalengka, there were 9 active ingredients of fungicides, 19 of insecticides, and one of molluscicide. Neonicotinoids as bioactive from some pesticides were only found in one ingredient from Pangalengan with the active ingredient imidacloprid. Seed treatment is not carried out by most farmers. It seems that farmers have many choices to apply pesticides for controlling pests and diseases. Training on the safe use of pesticides application in both locations should be more frequently conducted by public or private institutions. Farmer knowledge in safety procedure is very important in order to keep safe for human and environment; however, most farmers (56%) in Faridabad, India, did not implement safety procedures in their farming. This condition is likely to be equal for Indonesia as developing country [7].

Dose application of pesticide in Pangalengan was according to guideline as written in pesticides' package however interval application was very short, within 1 until 2 days. Differ with Pangalengan, farmers in Majalengka, they generally sprayed pesticides two times a week. Farmers in Majalengka found many pollinators than in Pangalengan according to their observation.

UTZ Organization (2015) [8] classifies the types of pesticide active ingredients into two types, namely prohibited pesticides and monitoring pesticides. Pesticides in monitoring scheme are active ingredients that are not prohibited but have the potential to pose serious and/or cumulative risks to human health and/or the environment. The use of active ingredients mentioned in the list is only permitted on certified harvest if: 1) all steps of Integrated Pest Management (IPM) have been followed, 2) alternative pesticides that are not too dangerous are not available, and 3) specific recommendations for reducing or reduce the risk associated with the dangerous nature of the product has been carried out. While forbidden pesticides are pesticides, all uses are prohibited by final legislation, to protect human health or the environment.

3.4 Insecticides

In Figure 3, five pesticides were banned i.e., beta siflutrin, carbofuran, carbosulfan, permethrin, and cypermethrin, meanwhile five of them are monitoring pesticides, namely abamectin, deltamethrin, fipronil, imidacloprid, and thiamethoxam, while others are not classified between these two.

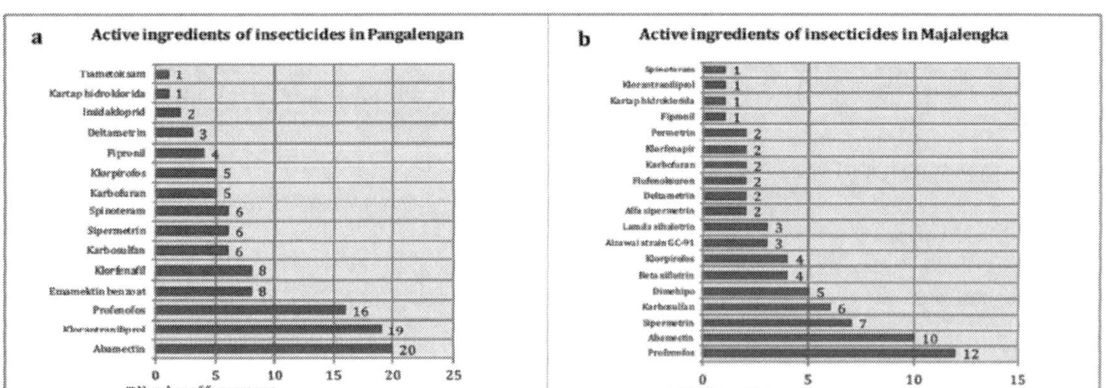

Figure 3 Active ingredients of insecticides in a) Pangalengan and b) Majalengka

Abamectin and profenofos were the most widely used active ingredient of pesticide in Pangalengan and Profenofos in Majalengka. Abamectin is an active ingredient of pesticides used to treat pest caused by *Spodoptera*, *Liriomyza*, *Thrip*, etc. Usually, they found in many chilli, onion, potato, etc. Profenofos is an active ingredient of pesticides used to treat pest caused by *Spodoptera*, *Plutella*, *Thrip*, etc. It is a contact and stomach poison. Usually, they applied this pesticide for controlling pest on chilli, onion, potato, etc. The active ingredient abamectin is very risky if inhaled by human [8].

3.5 Fungicides

There were 16 active ingredients of fungicides in Pangalengan and Majalengka (Figure 4). Chlorotalonyl was the most widely used fungicide's active ingredient in both locations. Chlorotalonyl is an active ingredient of pesticides used to control diseases caused by phytoptora, colletroticum, altenaria, etc. Usually, they found in many chili and tomato. The active ingredient Chlorotalonyl is very risky if inhaled by humans [8]. Of the two types of classification of the list of pesticides [8], in the active ingredient of fungicides commonly used there are four lists of pesticides in monitoring, while the list of prohibited pesticides does not exist.

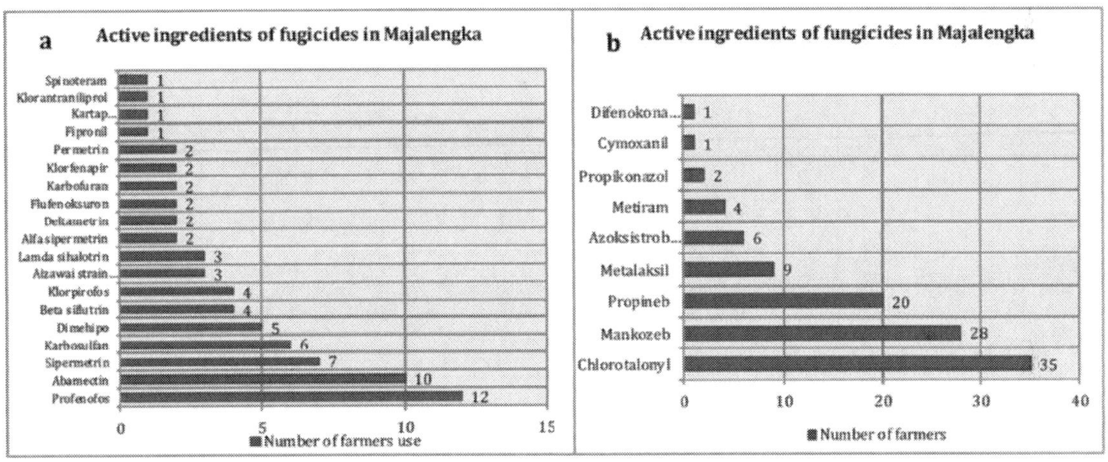

Figure 4. Active ingredients of insecticides in a) Pangalengan and b) Majalengka

3.6 Herbicide and Mollucicides

In Pangalengan, there were one herbicide and one molluscicide, whereas, in Majalengka, there was only one molluscicide. Propaquizafop is systemic herbicides (Table 5). It used to control weeds, ex: Echinocloa sp., Cynodon sp. Metaldehyde has molluscicided that control Achatina sp. in cabbage. Fentin acetate used to control Pomacea canaliculata in rice.

3.7 Farmer's perception for the presence of honey bees in the field

The results of the interview showed that 60% of farmers said that the presence of honey bees on their field was abundance (Figure 5). General information about honey bee from interview showed that *Apis cerana* was more plenty than *A. mellifera* and *A. dorsata*. It is likely that land covered by large forestry area had much honey bee population. During the rainy season, honeybees were much easier found in the field, while during the dry season they could be found also in many agricultural crops. When it rains, honey bees will stop foraging and will only stay in the hive, but in the dry season they will gather water to keep the colony cool [9, 10]. These qualitative findings should be more explored to verify the real fact regarding on honey bee populations and health in the field.

Table 3. Active ingredients of herbicide and molluscicides in Pangalengan and Majalengka

Active ingredients of herbicide in Pangalengan	**Number of farmers**
Propaquizafop	1
Active ingredients of moluscide in Pangalengan	**Number of farmers**
Metaldehyde	1
Active ingredients of moluscide in Majalengka	**Number of farmers**
Fentin acetate	1

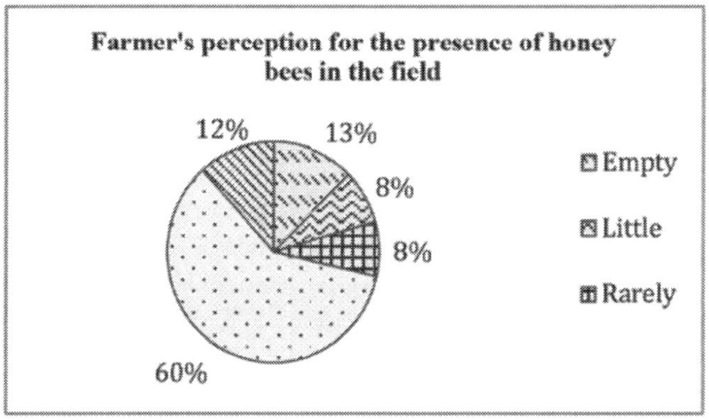

Figure 5. Farmer's perception for the presence of honey bees in the field

4. Conclusion

The respondent farmers used a wide variety of pesticides, such as insecticides, herbicides, fungicides as well as mixed formulations of fungicides and insecticides. The majority of the respondents reported that they usually apply herbicides, insecticides and fungicides at least twice a week during crop season. Although some farmers may be aware of pesticide hazards, but adequate protection is hardly taken to minimize risks. Pollinators perceived by farmers to be presence in their field in Pangalengan and Majalengka. Further confirmation on the abundance and diversity of pollinators using the most pesticide applied that found from this survey should be performed.

Acknowledgment The authors would like to express our sincere gratitude to PT Syngenta Indonesia and Universitas Padjadjaran for supporting this research.

Reference

[1] Mustafa U, Ali M, Kuswanti H (2006) Indonesia. In: M. Ali (ed.). Chili (Capsicum spp.) food chain analysis: setting research priorities in Asia. Technical Bulletin No. 38, AVRDC Publication 06-

678. Shanhua, Taiwan: AVRDC - The World Vegetable Center. pp. 146-195. http://www.avrdc.org/pdf/TB38/TB38-Indonesia.pdf

[2] Lawrence T, Sheppard WS (2013) Neonicotinoid Pesticides and Honey Bees. Washington State University Extension

[3] Samosir K, Seataini O, Nurjazuli N (2017) Hubungan Pajanan Pestisida dengan Gangguan Keseimbangan Tubuh Petani Hortikultura di Kecamatan Ngablak Kabupaten Magelang. **16(2):** 63-69

[4] Department of Health of Indonesia (Departemen Kesehatan RI) (2009) Kategori Usia. Dalam http://kategori-umurmenurut-Depkes.html.

[5] Afari E (2001). The effect of farmer education on productivity and income in Ghana: Implication for food security. Magister Thesis. Univeristy of Ghana. http://ugspace.ug.edu.gh.

[6] Ministry of Agriculture (Kementrian Pertanian Indonesia). 2016. Pestisida Pertanian dan Kehutanan 2016

[7] Tyagi H, Gautam T, Prashar P (2015) Survey of pesticide use patterns and farmers' perceptions: A case study from cauliflower and tomato cultivating areas of district Faridabad, Haryana, India. *International Journal of MediPharm Research* **1:** 03139.

[8] UTZ Organization (2015) Daftar Pestisida Terlarang dan Pestisida dalam Pantauan. Online : https://www.utz.org

[9] Le Conte Y, Navajas M (2008) Climate change: impact on honey bee populations and diseases. *Rev. sci. tech. Off. int. Epiz.*, **27 (2)**: 499-510.

[10] Puškadija Z, Lužaić R, Ozimec S, Bošković I, Opačak A, Jelkić D (2009) Influence of weather conditions on honey bee visits (*Apis mellifera* carnica) during sunflower (*Helianthus annuus* L.) blooming period. Faculty of Agriculture in Osijek. ISSN 1330-7142.Burges, H.D. 1998. Formulation of mycoinsecticides. In: Formulation of Microbial Biopesticides, Beneficial Microorganisms, Nematodes and Seed Treatments (H.D. Burges, Ed.), pp. 131-185. Kluwer Academic Publishers, Dordrecht, the Netherlands.

IOP Conf. Series: Earth and Environmental Science **334** (2019) 012041 doi:10.1088/1755-1315/334/1/012041

Rheological Properties of Honey and its Application on Honey Flow Simulation through Vertical Tube

N Bambang[1], M Ikhsan, Tensiska, N Sukri and Mahani

[1] Faculty of Agric. Industrial Technology, Universitas Padjadjaran, Indonesia

E-mail: bnhnur@gmail.com

Abstract. Honey is believed as functional food. Honey is very viscous fluid and make it difficult to flow. Rheological properties of honey is an important parameter to control and process design in food industry where honey used as food ingredients. Rheology is the study of flow and deformation properties of a material when pressure given to it intentionally or naturally. The majority of fluid foods had flow behavior best modelled with power law model which divide fluid into three models: Newtonian and non-Newtonian (Dilatant and Pseudo plastic). The purpose of this study was to examine the rheological characteristics of honey and the effect of temperature on viscosity as well as knowing how accurate the rheological model used (Power Law) to analyze the fluidity properties of honey. The method used is an experimental method with a descriptive analysis. The results showed that the honey used in this study has showed pseudo plastic flow properties with a flow behavior index value of 0.95 ($\tau = 6318.46\dot{\gamma}^{0.947}$). Effect of temperature on the viscosity of honey showed that the viscosity of honey decreases with increasing temperature and best modeled with Arrhenius equation ($\mu = 1.05\times10^{-10}\,e^{\frac{76.01\,1}{8.314\,T}}$). The results of honey flow simulation through the vertical glass tube transport system showed a very high degree of accuracy when compared with the actual results (99.57 %).

1. Introduction

Honey is a natural sweet substance produced by honey bee (*Aphis* sp.) from floral nectar or other part of plant [1]. According to [2], honey can be classified based on type of honey source, floral and extra floral honey. Extra floral honey refer to source of nectar from other than flower such as leaves and trunk. Honeydew honey is honey whose raw material obtained from Hemipterous insect such as *Aphidina, Coccina* and *Cicadina* [3]. Mono floral honey mean its nectar source dominated by only one source of origin while multi floral honey refer to various nectar sources.

Honey has been known as a functional food since a long time ago honey has been used in traditional medicine by many societies in the world. Honey has antioxidant properties and it is believed that the darker the color of honey, the more antioxidant capacity own by the honey [4]. Honey has been also used in treatment to some disease like diabetic ulcers, gastrointestinal disorder, cardiovascular diseases and cancer. Honey is also used as food or food ingredient to give sweet taste. Honey mostly contain sugar fructose and glucose as the main component and the other sugar is also found in very small amount [4]. Protein, vitamin, essential oil and mineral are also found in honey [4].

Rheology is a study the flow and deformation of material under given pressure. Viscosity is the main rheological properties of honey. Honey is usually used in liquid form with high viscosity. Viscosity is simply correlated to the easiness to flow, the higher the viscosity the more difficult the fluid to flow. Honey has viscosity several times (thousand times) of water viscosity and make it difficult to flow. Honey viscosity is mainly affected by water content, temperature and honey composition [5,6]. Composition of honey is affected by the source of nectar and bee types.

Power law model is the simplest and most widely used equation to describe the properties of fluid. The model can be expressed as:

$$\tau = K\left(\dot{\gamma}\right)^{n} = K\left(\dot{\gamma}\right)^{n-1}\dot{\gamma} \quad (1)$$

$$\mu_{app} = K\left(\dot{\gamma}\right)^{n-1} \quad (2)$$

In Power law model, rheological properties is time independent and no yield stress. The yield stress is the minimum stress needed to make the fluid flow. The Newtonian fluid (ideal viscous fluid) has n value equal to 1 and K is equal to viscosity which mean viscosity is not affected by its shear rate. Meanwhile for non-Newtonian fluid ($n\neq 1$), the viscosity is a function of shear rate and if the viscosity value decrease as the shear rate increase is called as pseudoplastic (shear-thinning) fluid ($n<1$) and in contract to Dilatant (shear-thickening) fluid ($n>1$) where viscosity increase as its shear rate increase [7,8]. Thus the two parameters (n and K) is important parameter to describe the flow properties of honey. The natural flow of honey is caused by the force related to gravitational force and shear stress affected by rheological properties of honey with opposite direction to gravitational force. Honey flow through vertical tube is affected by the two forces, gravitational force and shear force due to internal friction within the honey. The thickness of honey attached to the wall surface during flow through vertical tube is affected by the time and honey height from the bottom where honey flowing out the tube [8]. The differential equation which relate the thickness of honey when flowing through vertical tube can be expressed as [8].

$$\frac{\partial\left(\overline{v}\delta\right)}{\partial z} = -\frac{\partial\delta}{\partial t} \quad (3)$$

The analytical solution of this differential equation is needed to determine the thickness of honey attached to the tube wall and the solution is

$$\delta = \left\{\left(\frac{K}{\rho g}\right)^{\frac{1}{n}}\frac{z}{t}\right\}^{\frac{n}{n+1}} \quad (4)$$

From the equation above, the thickness of honey attached to the wall is affected by rheological properties of honey. The total volume of honey attached to the wall is determined by integrating the volume over the height range and the total volume of honey left after flowing at certain time will be [8]:

$$q = \int_{0}^{z}\pi\left(r^2 - \left(r-\delta\right)^2\right)\partial z \quad (5)$$

The current study is to determine the rheological properties of honey and its dependency to temperature and to predict the total volume (total mass) of honey left after some flowing times and compare the prediction with the actual value.

Nomenclature List

τ	Shear Stress	Pa
$\dot{\gamma}$	Shear Rate	1/s
μ	Viscosity	mPa s
T	Temperature	°C or K
n	Flow Behavior Index	-
K	Consistency Index	Pa s^n
R	Gas Constant (8.314)	J/mole K
n	Number of rotation	-
δ	Gap distance between spindle and cup wall	m
μ_o	Constant in Arrhenius Model	
Ea	Activation Energy	kJ/mole
v	Point velocity, its value depend on location	m/s
$\bar{v}$	Average velocity	m/s
ρ	Density	Kg/m^3
g	Gravitational constant	9.8 m/s^2
α	Angle formed between the surface where fluid flow and Y-axes. For vertical flow, $\alpha = 0$	
x	The height of fluid film from the surface, $x \leq \delta$	m
z	Height of fluid	m
W	Width of fluid	m
r	Radius	m
m	Mass	Kg or g
L	Length of fluid flow	m
q	Total mass of honey left in the tube	m^3

2. Material and Methods

2.1 Material

Honey was obtained from Indonesian forest company (*PERHUTANI*) with water content of 19.79 %. Honey was kept in a closed container and stored at ambient temperature for further analysis. Density of honey was determined by weighing the honey that previously put in 100 ml graduated cylinder.

2.2 Rheological Properties Determination

Rheological properties of honey was determined using rotational viscometer (Raypa RP-1 L Trade, Spain). Spindle L4 with radius of 0.0015 m and cup radius of 0.0235 m were used. The gap distance between the spindle and the cup was 0.022m. Four different rotational rate of spindle were used namely, 20; 30; 50 and 60 RPM. The experiments were replicated twice. Two data were collected, those are rotational spindle rate as independent variable and apparent viscosity as dependent variable. Rheological properties of honey was determined based on Power Law model. The two rheological parameters, n and K, were determined by linearization of equation 2.

$$\ln \mu_{app} = \ln K + (n-1)\dot{\gamma} \quad (6)$$

Shear rate ($\dot{\gamma}$) was calculated using equation provided by [13]

$$\dot{\gamma}_w = \frac{2\pi RN}{\delta} \quad (7)$$

The linear regression analysis between $\ln\mu_{app}$ and shear rate $\dot{\gamma}$ was used to determine the value of n and K.

2.3 Temperature Effect on Honey Viscosity

Honey was kept at different temperatures (25, 30, 35 and 40 °C) in water bath (Clifton, UK) before being analyzed with rotational viscometer (Raypa type RP-1 L trade, Spain). Spindle L3 with radius 0.006 m and the gap distance of 0.0172 m and 30 RPM was set to measure viscosity of honey. Arrhenius model was used to relate the dependency of viscosity to Temperature.

$$\mu = \mu_o e^{\frac{E_a}{RT}} \quad (8)$$

After linearization, equation 8 is becoming

$$\ln\mu = \ln\mu_o + \frac{E_a}{R}\frac{1}{T} \quad (9)$$

Two parameter μ_o and Ea were calculated by linear regression between ln μ and (1/T).

2.4 Flow Simulation through Vertical Tube

Honey was poured into vertical tube with diameter of 0.03 m. The height of honey in tube was fixed 0.23 m. The tube was open to let honey flowing out the tube at several flowing times 5, 10, 15 and 20 seconds. At each flowing time, the honey left in the tube was weighed as actual mass of honey. Flow simulation was used to compare the real mass of honey left and the predicted one. The model developed to predict fluid flow was explained detail in [8]. The thickness of fluid flow over the surface (δ) was expressed as :

$$\delta = \left(\frac{\overline{v}(2n+1)}{(\rho g\cos\alpha)^{\frac{1}{n}} n}\right)^{\frac{n}{n+1}} \quad (10)$$

The schematic of fluid flow over the surface was presented in Figure 1.

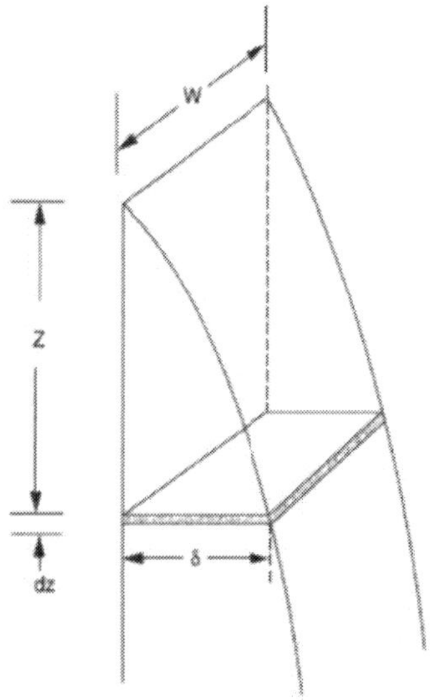

Figure 1. Schematic Flow of Fluid (Source : [13])

Mass balance of fluid entering the system was set up between the mass of fluid entering the system and leaving the system plus accumulation of fluid mass that might occurred.

$$\bar{v}\delta W\rho\big|_z = \bar{v}\delta W\rho\big|_{z+dz} + \frac{\partial \delta\rho W}{\partial z}dz$$

$$\bar{v}\delta W\rho = \bar{v}\delta W\rho + \frac{\partial(\rho W\bar{v}\delta)}{\partial z}dz + \frac{\partial \delta\rho W}{\partial z}dz \qquad (11)$$

The density (ρ) and width (W) are independent to height, z. thus

$$-\frac{\partial(\rho W\bar{v}\delta)}{\partial z}dz = \frac{\partial \delta\rho W}{\partial z}dz$$

$$-\frac{\partial(\bar{v}\delta)}{\partial z} = \frac{\partial \delta}{\partial z} \qquad (12)$$

From the equation above, the thickness of fluid flow over the surface is the function of height (z) and time (t). The solution of the differential equation above can be seen in appendix A.2.

3. Result and Discussion

3.1 Rheological Properties of Honey

Power law was used to study the rheological properties of honey as reported by [6,5]. In power law model, yield stress was set to zero and viscosity is a function of only shear rate and time independent characteristic [7]. The two parameters in power law model (n and K) would describe the rheological properties of honey. The two parameters were determined by linear regression after linear modification of equation which relate apparent viscosity and shear rate (equation 6). The set up condition for the rheological properties of honey was presented in Table 1. All the units were in SI units.

Table 1. Supported Condition for Rheological Measurement

Specification	Value
Spindle	L4
Spindle radius (R_D)	0,0015 m
Radius of honey container	0,0235 m
Gap distance between spindle and sample container (δ)	0,022 m
π	3,14
Rotational rate of spindle (RPS*)	1 ; 0.83; 0.5; 0.33
Water content	19.79 %

*RPS= Rotation per Second

The water content of honey need to be stated due to its main influence on the viscosity of honey apart from other factors such as honey composition [9]. Viscosity of honey is affected by sugar content and sugar content [10]. The calculation and result for determining rheological properties of honey was presented in Table 2. From Table 2, it can be seen that the higher the shear rate (derived from spindle rotational rate), the lower the viscosity of honey which indicate the tested honey had a pseudo plastic behavior. In literature, honey was reported having a pseudo plastic behavior [6,1] and also Newtonian properties [6]. As can be seen in Table 2, the n value of honey was very close to 1 (n = 0.947) which indicating pseudo plastic behavior. The parameter K in Newtonian model fluid is the viscosity itself while in pseudo plastic model fluid, the viscosity was the function of K and shear rate and called as apparent viscosity. The K value obtained was close to the viscosity range of honey, 6540-7010 mPa s.

Table 2. Data and Result in Determining Rheological Properties of Honey

RPM	RPS	$\dot{\gamma}$	μ (mPas)	ln $\dot{\gamma}$	n-1	ln K	n	K	R^2
60	1	0.42	6540	-0.84	-0.05	8.75	0.947	6312.2	0.83
60	1	0.42	6620	-0.84					
50	0.83	0.35	6610	-1.03					
50	0.83	0.35	6750	-1.03					
30	0.5	0.21	6890	-1.54					
30	0.5	0.21	6980	-1.54					
20	0.33	0.14	6910	-1.94					
20	0.33	0.14	7010	-1.94					

3.2 Effect of Temperature on Viscosity

Viscosity of fluid is dependent on temperature and generally increasing temperature of fluid will decrease its viscosity [11,12]. Figure. 2 showed viscosity of honey at different temperature. It can be seen that the viscosity of honey decreased as temperature increased. [4] reported that at temperature above 40 °C, honey viscosity is not affected by temperature any more.

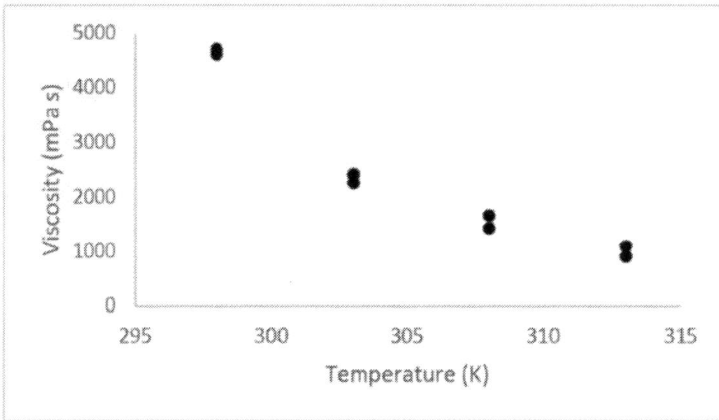

Figure 2. Viscosity of Honey at Different Temperature

Arrhenius equation is the simplest and common model to relate the effect of temperature on physical or chemical properties of food system [12]. The temperature dependency of viscosity of honey were also reported following Arrhenius model [13]. The parameter determination of Arrhenius model was presented in Table 3. The activation energy of honey obtained was 77.69 kJ/mole and this value was higher than the activation energy of some liquid food as reported by [12] such as 5.0 kJ/mole for apple sauce and 59.5 kJ/mole for depectinized apple juice. The activation energy was affected by the temperature range where the measurement done [12]. The model to relate the dependency of viscosity was also depend on the range of temperature and [6] reported that the William-Landel-Ferry (WLF) model is better model to relate the dependency of temperature on honey viscosity in the temperature range of Tg (glass temperature) + 100 °C compared to Arrhenius model and Vogel-Tammann-Fulcher (VTF) model.

Table 3. Data and Result in Determining the Effect of Temperature on Honey Viscosity

T (°C)	T (K)	1/T	μ (mPa s)	ln μ	ln μ_o	μ_o	Ea/R	R (J/mole.K)	Ea (kJ/mole)	R^2
25	298	0.00336	4630	8.44	-22.98	1.05E-10	9344.65	8.314	77.69	0.903
25	298	0.00336	4710	8.46						
30	303	0.00330	2290	7.74						
30	303	0.00330	2430	7.80						
35	308	0.00325	1450	7.28						
35	308	0.00325	1690	7.43						
40	313	0.00319	920	6.82						
40	313	0.00319	1110	7.01						

3.3 Flow Simulation of Honey through Vertical Tube

Rheological properties of honey was used to simulate the flow of honey through vertical tube and compared the prediction with the actual result. The analytic solution of fluid flow through vertical tube was explained in [8]. The vertical tube flow of honey is a resultant between gravitational force and resistant flow related to rheological properties of the fluid. For simplification, friction force due to contact of fluid to surface is negligible. The schematic flow of honey through vertical tube was presented in Figure 3.

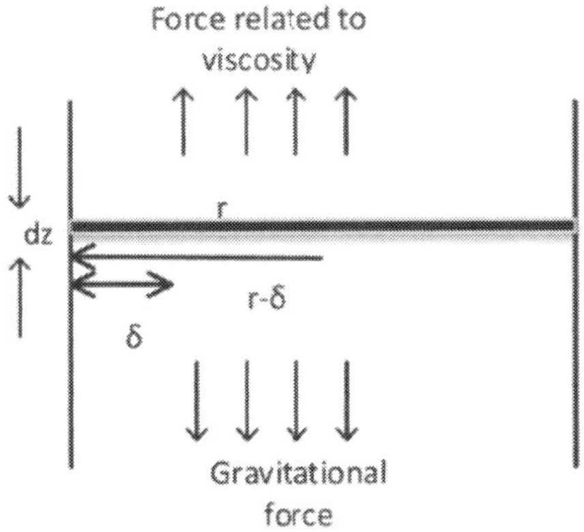

Figure 3. The Schematic Flow of honey through vertical tube

From Figure. 3, when honey flows through vertical tube, the thickness of honey attached to the tube wall varies with the height. If the height of honey in vertical tube was z, the volume of honey, dq over an increment of height, dz was expressed as:

$$\partial q = \pi\left(r^2 - (r-\delta)^2\right)\partial z \quad (13)$$

Thus to determine the total volume of honey attached to the tube wall (q) during flowing through vertical tube was

$$\int \partial q = \int_0^z \pi\left(r^2 - (r-\delta)^2\right)\partial z$$

$$q = \int_0^z \pi\left(2r\delta - \delta^2\right)\partial z \quad (14)$$

$$q = \int_0^z \pi 2r\partial z - \int_0^z \pi\delta^2 \partial z$$

Since

$$\delta = \left\{\left(\frac{K}{\rho g}\right)^{\frac{1}{n}} \frac{z}{t}\right\}^{\frac{n}{n+1}}$$

Thus

$$q = 2\pi r\left(\left(\frac{K}{\rho g}\right)^{\frac{1}{n}}\right)^{\frac{n}{n+1}} \left(\frac{1}{t}\right)^{\frac{n}{n+1}} \int_0^z z^{\frac{n}{n+1}}\partial z - \pi\int_0^z \left\{\left(\left(\frac{K}{\rho g}\right)^{\frac{1}{n}} \frac{z}{t}\right)^{\frac{n}{n+1}}\right\}^2 \partial z$$

$$q = 2\pi r\left.\left(\left(\left(\frac{K}{\rho g}\right)^{\frac{1}{n}}\right)^{\frac{n}{n+1}}\left(\frac{1}{t}\right)^{\frac{n}{n+1}}\frac{n+1}{2n+1}z^{\frac{2n+1}{n+1}}\right|_0^z - \pi\left(\frac{K}{\rho g}\right)^{\frac{2}{n+1}}\left(\frac{1}{t}\right)^{\frac{2n}{n+1}}\frac{n+1}{3n+1}z^{\frac{3n+1}{n+1}}\right|_0^z \quad (15)$$

$$q = 2\pi r\left(\left(\frac{K}{\rho g}\right)^{\frac{1}{n}}\right)^{\frac{n}{n+1}}\left(\frac{1}{t}\right)^{\frac{n}{n+1}}\frac{n+1}{2n+1}z^{\frac{2n+1}{n+1}} - \pi\left(\frac{K}{\rho g}\right)^{\frac{2}{n+1}}\left(\frac{1}{t}\right)^{\frac{2n}{n+1}}\frac{n+1}{3n+1}z^{\frac{3n+1}{n+1}}$$

Thus the total mass of honey left after allowing honey flow through vertical tube at certain time would be:

$$m_{honey} = q \times \rho \quad (16)$$

The data needed to calculate the total mass of honey left in the tube could be seen in Table 4.

Table 4. Data Needed in Simulation of Honey Flow through Vertical Tube.

Specification	Value
Mass of Honey	300 g
Radius of Tube	0.0155 m
Height of honey in tube	0.23 m
Honey Density	1600 kg/m^3
Gravitational constant	9.8 m/s^2
Flow behavior index (n)	0.946
Consistency Index (k)	6.3172 mPa s

The flow of honey through vertical tube was tested at some series of flowing times namely, 5; 10; 15 and 20 seconds. The calculated mass of honey left after flowing at certain time was shown for flowing time of 5 second as follow:

$$q = 2\times\pi\times 0.0155\left(\left(\frac{6.3172}{1600\times 9.8}\right)^{\frac{1}{0.947}}\right)^{\frac{0.947}{0.947+1}}\left(\frac{1}{5}\right)^{\frac{0.947}{0.947+1}}\frac{0.947+1}{2\times 0.947+1}0.23^{\frac{2\times 0.947+1}{0.947+1}}$$

$$-\pi\left(\frac{6.3172}{1600*9.8}\right)^{\frac{2}{0.947+1}}\left(\frac{1}{t}\right)^{\frac{2\times 0.947}{0.947+1}}\frac{0.947+1}{3\times 0.947+1}0.23^{\frac{3\times 0.947+1}{0.947+1}} \quad (17)$$

$$q = 6.081\times 10^{-5} - 5.964\times 10^{-6}\,m^3 = 5.485\times 10^{-5}\,m^3$$

$$total\ mass\ of\ honey\ left\ in\ the\ tube = 5.485\times 10^{-5}\,m^3\times 1600\frac{kg}{m^3} = 0.08775\ kg = 87.75g$$

The comparison of calculated mass of honey left and real mass of honey for all time of flow tested was presented in Table 5.

Table 5. Mass of Honey Left (Actual vs Predicted)

Time (s)	Mass of Honey Left (g)	
	Actual	Predicted
5	90	87.75
5	88	87.75
10	66	64.59
10	64	64.59
15	52	53.74
15	54	53.74
20	46	47.1
20	47	47.1

Figure 4 showed the closeness (accuracy) of calculated value and actual value of honey mass attached to the tube at different flowing times. The linier model with the intercept set to zero, thus the slope will describe the accuracy of the predicted values to actual values. The slope 1 mean all (100 %) the actual value equal to the predicted value. As can be seen from Figure. 3, the obtained slope was 0.9957 which mean the model developed to predict the mass of honey left comply with the actual value correctly at accuracy level of 99.57 %. The rheological properties of honey is an important input in developing the model and it showed to us that the rheological properties could be used as an important component in developing control and process design in food industry where honey used as one of food ingredients.

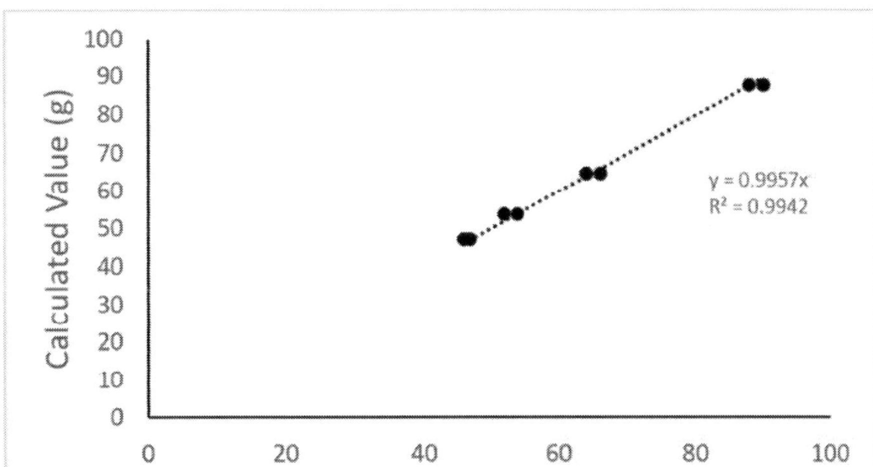

Figure 4. Accuracy of Mass of Honey Left (Actual vs Predicted)

4. Conclusion

Honey showed pseudoplastic fluid behavior with n and K value 0.946 and 6318.46 respectively. The effect of temperature on honey viscosity was best modelled with Arrhenius model ($\mu = 1.05 \times 10^{-10} e^{\frac{76.011}{8.314T}}$). Simulation flow of honey through vertical tube showed good accuracy between the predicted value and actual value (99.57 %). The rheological properties of honey and equation developed were proven to be accurate in predicting the honey flow through vertical tube.

References

[1] SNI 3545 2013 Madu *Badan Standardisasi Nasional* Jakarta
[2] Gojmerac W L 1983 Bees, Beekeeping, Honey and Pollination *The AVI Publishing Co. Inc.* Westport Connecticut.
[3] Maurizio A 1979 How Bees Make Honey dalam Honey: A Comprehensive Survey Heinenann London.
[4] Chmielewska R H 2004 *Honey* dalam Tomasik P (Ed.) Chemical and Functional properties of Food Saccharides *CRC Press* Boca Ration USA.
[5] Ahmed J, Prabhu S T, Raghavan G S V, Ngadi N 2007 Physico-Chemical, Rheological, Calorimetric and Dielectric Behavior of Selected Indian Honey *Department of Food Science and Agricultural Food Chemistry* Bioresource Engineering McGill University Canada.
[6] Sopade P A, Haley P, Bhandari B, D'Archy B, Doebler C, Coffin N 2002 Application of the Williams-Landel-Ferry Model to the Viscosity-Temperature Relationship of Australian Honeys *Journal of Food Engineering* **56**: 67-75.
[7] Steffe J F 1996 Rheological Methods in Food Process Engineering *Freeman Press* USA.
[8] Toledo R T 2007 Fundamental of Food Process Engineering 3rd Edition *Springer* New York.
[9] Stansell D 1997 Sugar and Chocolate Confectionery in Food Industries Manual 24th edition Ranken M D, Kill R C, Baker C G J (eds) *Blackie Academic and Profesional* UK.
[10] Gomez-Diaz D, Navaza J M, Quint'ans-Riveiro L C 2005 Rheological Behaviour of Galician Honeys *European Food Research Technology* **222**: 439-442.
[11] Goodrum J W, Geller D P, Adams T T 2002 Rheological characterization of yellow grease and poultry fat *J. Am. Oil Chem. Soc.* **79**: 961-964.
[12] Rao M A 2005 Rheological Properties of Fluid Foods in Engineering Properties of Foods 3rd edition Rao M A, Rizvi S S H, Datta A K (eds) *Taylor and Francis* Boca Raton.
[13] Lazaridou A, Biliaderis C G, Bacandritsos N, Sabatini A G 2004 Composition, Thermal and Rheological Behaviour of Selected Greek Honeys *Journal of Food Engineering* **64**: 9-21.

Antimicrobial Activity of Set Yoghurt Probiotic from Milk as a Functional Food

C Hartati[1], R L Balia[1], L Suryaningsih[1], E Wulandari[1] and W S Putranto[1]

[1]Department of Animal Husbandry Technology, Faculty of Animal Husbandry, Universitas Padjadjaran, Bandung, Indonesia

E-mail : wendrysp @ yahoo.co.id

Abstract. Milk is generally a product that is quickly damaged by microbes due to the complete and balanced nutritional content of milk so that microbes are very easy to grow. Preservation technology is needed to increase durability of milk and one of them by fermentation using Lactic Acid Bacteria (LAB). Setyoghurt products with mung bean extract supplementation and BAL starter used with *Lactobacillus casei* ALGHCH 212 combination can effectively inhibit pathogen bacteria growth (*E coli, Listeria, Pseudomonas, S aureus*) in 24 hours incubation compared with control of antibiotic amphicillin with a concentration of 31.25 mg/ml.

1. Introduction

Milk is a potential livestock commodity developed into food products of high economic value. Innovation in producing new dairy products is needed as an effort to increase food diversification and support national food security. Livestock commodities are rapidly damaged primarily by microbes, as well as with milk. Fermentation by utilizing lactic acid bacteria is an attempt to preserve and produce a processed product known as yoghurt.

Yogurt with raw material of cow's milk is quite popular in the community. Based on different methods of manufacture, yogurt type is divided into two namely, Stirred Yogurt is Yogurt which fermentation process is done on a large tank / container and after incubation then the product is packed in small packaging, thus allowing the coagulum is broken or broken before cooling and packaging finished and Set Yogurt is yogurt which at the time of incubation / fermentation of milk is in the small packaging and coagulum characteristics are not changed. Yogurt that spread on the market today is yogurt type Stirred Yogurt, while the Yogurt Set is still rarely developed and people are less familiar with Yogurt Set when compared to Stirred Yogurt, Set Yogurt has better consistency. As an effort to diversify food product, a combination of milk and vegetable extract in yoghurt is made.

Mung beans (*Vigna radiata L*) have balanced nutrients and are rich in antioxidants [1]. Mung bean extract (Mung Bean Milk) fermented with *Lactobacillus plantarum* B1-6 potentially developed as a functional food [2]. Use of 30% green beans juice in whey fermented tofu with starter Lactic Acid Bacteria provides best lowering lactose levels, increasing acidity and raising total solids [3].

This research is expected to produce a new product of probiotic fermentation beverage which is good for health. It is expected to find a combination of milk and vegetable extracts and the right starter formulation to produce yoghurt products that meet the standards as a probiotic drink.

2. Materials and Method

2.1. Growth curve of Lactic Acid Bacteria

The observation of BAL growth curve is *Streptococcus thermophillus* (ST), *Lactobacillus bulgaricus* (LB), *Lactobacillus acidophillus* (LA), *Lactobacillus casei* ALGHCH.2.12 (LC ALG 2.12) on MRS

Broth media and incubation temperature 370 C, non shaking and bacth system. Sampling dilakuan every 5 hours and measured absorbance at 600 nm wavelength.

2.2. Making Mung bean Extract
The cleansed mungbeans are soaked in hot water for 8 hours, then boiling and adding water (1: 6). Furthermore, screening and extract of green beans was obtained.

2.3. Specific Growth Rate
Observation on logarithmic phase of BAL isolate on MRSB media with Mung bean extract supplementation with ratio (1: 1).

2.4. Setyoghurt of Mung bean Extract and Starter Combination
The treatment includes the following combinations:

Groups	Combinations of starter		
	P1	P2	P3
Ys	YsP1	YsP2	YsP3
KH	KHP1	KHP2	KHP3

Ys = Milk (50 ml)
KH = Milk : Mung bean extract (1:1) (25 ml : 25 ml)
P1 = 5% (2.5 ml) (ST : LB : LA) (1:1:1)
P2 = 5% (2.5 ml) (ST:LB:LA) (1:1:1) = 1.875 ml + LC ALG 2.12 0.625 ml)
P3 = 5% (2.5 ml) (ST:LB:LA) (1:1:1) = 1.875 ml + BF 0.625 ml)

2.5. Bacterial Resistance Testing with Agar Diffusion
Testing of antimicrobial activity using well diffusion method against some pathogenic bacteria and BAL. The culture of the first test bacterium was cultured on Nutrient Broth (NB) 10 ml overnight at 37 °C, then calculated the number of bacterial cells up to 107 per ml. Further inoculated as much as 0.1% into 20 ml of Nutrient Agar medium, was carried out to homogeneous so that the test bacteria was chalked flat, then poured into sterile petri dish, and left to solid then made well with diameter 6 mm (5 wells per cup). The sample set of 50 µl yogurt was fed into the well then incubated at 37 °C for 24 hours. Observations were made by measuring the zone of inhibition based on the diameter of the clear area formed around the well.

3. Result and Discussion
Observation of the growth curve through the absorbance value of BAL isolate culture on MRSB at 600 nm wavelength. In MRSB media all BAL isolates showed logarithmic phase at 5 th to 20 th incubation period at 37 °C. Thus, logarithmic phase information is used as reference in conducting analysis of fermentation efficiency by considering the value of Specific Growth Rate (µ) of each isolate of BAL. The increase in value (µ) also shows an increase in fermentation efficiency as a result of extrinsic factor manipulation such as the use of prebiotics. The growth curve begins with the initial phase (lag) which is the period of microbial adjustment. At that phase enzyme synthesis occurs by cells used for metabolism. After the initial phase is complete, new cellular reproduction begins. Cellular concentration increases, initially slowly and then increasingly until one day the rate of cellular growth or reproduction reaches the maximum and logarithmic or exponential growth occurs. Furthermore, after certain substrate or compound is required for bacterial growth in the culture

medium approaching and there is accumulation of inhibiting products, there is a decrease of bacterial growth rate. The decline phase is characterized by a decrease in the number of living cells (viable) in the media due to mortality (mortality) [4].

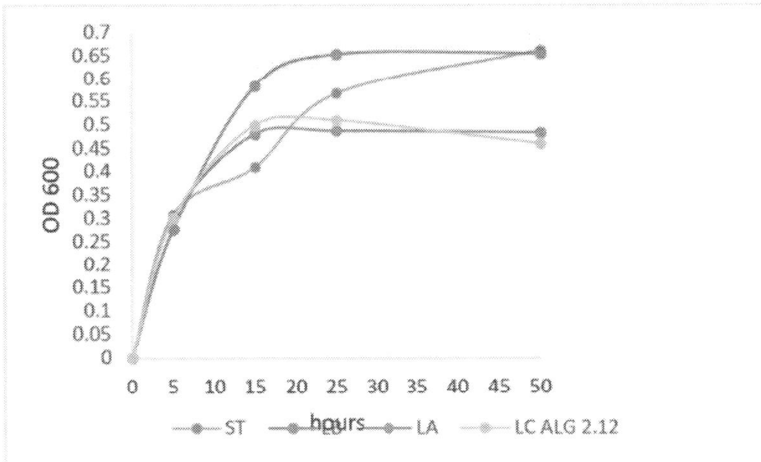

Figure 1. Growth rate curve of *Streptococcus thermophillus* (ST), *Lactobacillus bulgaricus* (LB), *Lactobacillusacidophillus* (LA), *Lactobacillus casei* ALG.2.12 (LC ALG 2.12), in MRS Broth at 37 °C, *non shaking* dan *bacth system*.

3.1. *Mung bean Extract Supplementation*

At 37 °C incubation the entire isolate undergoes a logarithmic phase at 5 to 20 hours, so we can analyze the specific growth rate at that time span with the value (μ) = hour-1 as follows:

Table 1. Specific growth rate (μ) of lactic acid bacteria

Isolat Bakteri Asam Laktat	Specific growth rate (μ) = hours^{-1}	
	MRS	MRS + Mung bean extract
Streptococcus thermophillus (ST)	0.1503	-
Lactobacillus bulgaricus (LB)	0.2415	0.2621
Lactobacillus acidophillus (LA),	0.1230	0.1547
Lactobacillus casei ALG.2.12 (LC ALG 2.12)	0.1382	0.1641

The use of green bean extract was able to increase *Lactobacillus bulgaricus* (LB) Specific Growth Rate, *Lactobacillus acidophillus* (LA), *Lactobacillus casei* ALG.2.12 (LC ALG 2.12).

3.2. *Set Yoghurt Mung bean Extract*

One of the milk processing technology is by fermentation using lactic acid bacteria and the product we know is yoghurt. The fermentation process with starter of lactic acid bacteria causes the milk to coagulate until it is obtained by its acidity, distinctive odor and flavor, with visions of viscous liquid until semi-solid. Dry matter content of yogurt at least 12% [5] and pH yogurt 4.4-4.5 [6]. The method of making yoghurt in the absence of stirring and the addition of flavor and flavor at the end of the yogurt making process is known as the Set yoghurt. The yoghurt coagulum characteristics were unchanged, since the Set yoghurt were not stirred after incubation such as Stirred Yoghurt, but incubation after packaging in a sealed packet that could be consumed immediately [6].

Figure 2. Set yoghurt extract green beans with various starter formulations BAL

3.3. Inhibitions of Set Yoghurt Green Bean Extract Againts Pathogenic Bacteria

The set of green bean yoghurt extract that has been produced by bioassay to know its resistance to pathogen bacteria.

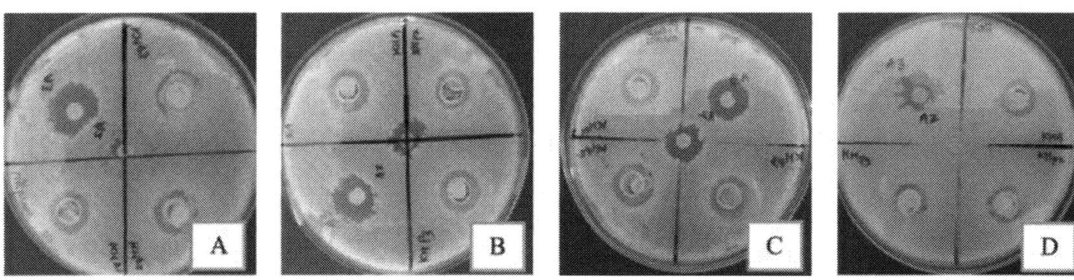

Figure 3. Inhibition set yoghurt extract green beans on growt phatogenic bacteria with agar diffusion method at 37 ^{0}C incubation during 24 hours

Lactic acid bacteria in the fermentation of food in addition to providing a distinctive flavor, this bacteria also extend durability due to its ability to produce metabolite products that can inhibit the growth of bacterial decay and bacterial pathogens. The antimicrobial properties possessed by lactic acid bacteria are due to their suitable conditions with the available nutrients, which can lead to competition with other bacteria especially pathogenic bacteria [7]. Lactic acid bacteria (BAL) produce lactic acid as the end of sugar metabolism (carbohydrate). Lactic acid produced in such a way will decrease the pH value of its growth environment and cause a sour taste. BAL ability to produce proteolytic enzymes around cell walls, cytoplasmic membranes, or inside cells and it is known that *Lactobacillus acidophilus* has the ability to produce extracellular proteases [8]. The antimicrobial properties possessed by lactic acid bacteria are due to their suitable conditions with the available nutrients, which can lead to competition with other bacteria especially pathogenic bacteria [7].

Table 2. Diameter of the resistor zone set of yoghurt extract of green beans compare with antibiotic control.

Sample	Inhibitions (mm) at incubation 24 hours			
	E coli	*Listeria*	*Pseudomonas*	*S aureus*
YsP1	14.5	15.4	11.2	15.3
YsP2	14.7	15.2	12.4	13
YsP3	15.8	17.9	11.7	14.8
KHP1	12.9	13.1	13.9	15.1
KHP2	12.8	15.1	14.7	12.9
KHP3	12	16.3	14.6	14.7
Control Antibiotic 31.25 mg/ml	all antibiotic controls are largely unable to inhibit the growth of tested bacteria			

4. Conclusions

The use of mung bean extract was able to increase *Lactobacillus bulgaricus* (LB) *Lactobacillus acidophillus* (LA), *Lactobacillus casei* ALG.2.12 (LC ALG 2.12) spesific growth rate. Set yoghurt with mung bean supplementation and using Lactic Acid Bacteria starter with *Lactobacillus casei* ALG212 combination (LC ALG 2.12) can effectively inhibit pathogen bacteria growth (*E coli, Listeria, Pseudomonas, S aureus*) in 24 hours incubation compared with control antibiotic amphicillin with a concentration of 31.25 mg / ml.

Acknowledgements

We would like to thank Rector University of Padjadjaran, who has funded this research through *Academic Leadership Grant* (ALG) 2017.

References

[1] Shi Z, Yao Y, Zhu Y, Ren G 2016 Nutritional composition and antioxidant activity of twenty mung bean cultivars in China *The Crop Journal* **4**: 398-406.

[2] Wu H, Rui X, Li W, Chen X, Jiang M, Dong M 2015 Mung bean (*Vigna radiata*) as probiotic food through fermentation with *Lactobacillus plantarum* B1-6.LWT-

[3] Nawangsari D N, Legowo A M, Mulyani S 2012 Kadar laktosa, keasaman dan total bahan padat whey fermentasi dengan penambahan jus kacang hijau *Jurnal Aplikasi Teknologi Pangan* **1(1)**.

[4] Mangunwidjaja D, Suryani A 1994 Teknologi Bioproses *Penebar Swadaya* Jakarta.

[5] Standar Nasional Indonesia (SNI) 1992 *SNI 01-2981-1992* Yogurt *Pusat Standarisasi Industri* Jakarta.

[6] Bylund G 1995 Tetra Pack Processing Systems *Dairy Processing Handbook* AB Lund. Sweden pp. 383-393.

[7] Codex *STAN* 2003 Codex Standar For Fermented Milk (Codex STAN 243-2003) *Codex Alimentarius*.
[8] Tamime A Y, Deeth H C 1980 Yoghurt. Technology and Biochemistry *J. Food Protection* **43(12)**: 939-977.

Chemical and Microbiological Characteristics of Fermented Cheese Whey Beverages With Soymilk Powder Addition

H Chairunnisa[1], R L Balia[1] and Eka Wulandari[1]

[1] Laboratory of Animal Products Technology Faculty of Animal Husbandry, Universitas Padjadjaran, Indonesia

E-mail: hartati.chairunnisa@unpad.ac.id

Abstract. This research was aimed to determine the chemical and microbiological characteristics of fermented cheese whey beverages which was made with soymilk powder addition. The research was done experimentally using Completely Randomized Design (CRD), three treatments and replicated six times. Parameters taken were total plate count, pH, lactic acid content, lactose content and viscosity. The impact of the treatments significance was determined by ANOVA and the difference between treatments was determined by Tukey test. The best soymilk powder addition that produced desirable fermented cheese whey beverages properties was 12.5 % with properties as follows : the total plate count of 175 x 10^9 colonies/g and pH of 3; Lactic acid content of 0.675 %; Lactose content of 7.2623 %; and Viscosity of 24 823.33 cP.

1. Introduction

Cheese whey is byproduct obtained from the precipitation and elimination of milk casein during the production of cheese [1]. Whey represents 85 – 95 % of the milk volume and preserve 55 % of the milk nutrients. The composition of bovine whey are fats 0.1 %, casein protein <0.1 %, whey protein 0.7 %, ash 0.5 %, lactose 4.9 % and total solid 6.3 % [2,3]. The green-yellowish color observed of whey is caused by the presence of riboflavin (vitamin B2) [4].Every 1 kg of cheese made, it will produce 9 kg of whey [5]. The production of cheese whey reaches 800 tons per year and considered as byproduct because of its low concentration of milk components. As an example, one of Gouda and Cheddar cheese producer which located in Sumedang, West Java, produces 3,000 kg of cheese whey per month and it's never been utilized. Meanwhile, the utilization of whey could reduce disposal high organic matter into stream such as biochemical oxygen demand (BOD) of 30-50 g/L, chemical oxygen demand (COD) of 60-80 g/L and minimizing wastes up to 62.1% [6,7].

Cheese whey can be utilized into food products through fermentation to produce yoghurt. Low concentration of total solid from cheese whey produces yoghurt with undesirable viscosity. As an alternative, addition of high amount of total solid substances is needed. Soy milk powder, a source of galactooligosaccharides (GOS) namely raffinose (3.04 %) and stachyose (8.87 %); also most importantly, it contains 92 % protein content of 22% (it is not clear what the author mean?) [8,9]. Galactooligosaccharides in soy milk in the form of stachyose and raffinose consist of prebiotic components that can be used as the source of carbon and energy for soygurt starter bacteria [9].Thus, the purpose of the research was to determine the characteristics of fermented cheese whey beverages with soy milk powder addition, the properties examined were lactose concentration, total lactic acid bacteria, pH, lactic acid content and viscosity.

2. Materials and Methods

2.1 Cheese Whey Beverages Fermentation

Liquid whey cheese (6.13 % with 2.27 % of protein and 3.27 % of lactose) was preheated at 60 °C, Melilea Soy Powder (DM ??? 92 % with 22 % of protein, 12 % of lipid and 60 % of carbohydrate) was added in various concentration: A1 (7.5 %), A2 (10 %), A3 (12.5 %). 5 % (w / v) of skimmed milk powder was added while stirring and heated it up to 90- 95 °C for 5 minutes. The skim milk was then Cooled down to 37 °C and bulk culture of Freeze-dried Yogurmet yoghurt starter (*L. bulgaricus, S. thermophilus, L. acidophilus* with ratio 1: 1: 1 as much as 5 % (v / v). Was inoculated and then Incubated in 37 °C for 16 hours. Fermented whey cheese beverages were stored in refrigerator at 3-5 °C for 24 hours (Modification of [10]).

2.2. Total Lactic Acid Bacteria
Lactic acid bacteria were isolated and enumerated by pour plate method on Plate Count Agar (PCA). Serial dilutions of 10^{-1} to 10^{-6} were prepared by diluting 1 mL of sample into 9 mL of sterilized distilled water. One milliliter from each of the dilutions were inoculated into petri dish which were previously added with PCA. The plates were then incubated at 37 °C for 24 hours. After incubation, viable colonies were counted and recorded, which expressed as colony forming unit per gram (cfu·g-1).

2.3. Lactose Concentrations
Samples was diluted up to 1:200 dilution, then 2.5ml of sample was poured into capped test tube 0.2 mL of $ZnSO_4$ 5 % and $Ba(OH)_2$ 4.5 % was added. Samples were then centrifuged at 1000 rpm until the pellet appears. 1 mL of supernatant was poured into a capped test tube with addition of Telles Reagent, and boiled in water bath for 6 minutes. Aquades was added to the mixture until the volume became 12.5 mL continued by homogenizing the mixture [11].
Samples result was read into spectrometer using the formula below:

$$Lactose\ concentration = \left(\frac{x\left(\frac{gram}{100ml}\right)}{B(gram)}\right) \quad (1)$$

Information: χ : Result of conversion absorbant sample
B : weight of sample
It is not clear

2.4. pH Values
pH values of samples were measured by pH-meter (MP-120, Mettler-Toledo, Switzerland). Samples were poured into respective tubes, measured for pH, and recorded.

2.5. Viscosity
Viscosity measurement for all samples were done using Brookfield DV-E Viscometer : Spindle No 3 at 20 rpm [12].

The experiment was done by using Randomized Complete Design (CRD) with 3 different amount of soybean powder addition, those are 7.5 %, 10 %, and 12.5 % respectively. The significancy of treatment was tested by Analysis of Varian and the difference between treatment was tested by Tukey Test.

3. Results and Discussion

The amount of powdered soybean addition affected the growth of lactic acid bacteria. As the amount of powdered soybean increases, total lactic acid bacteria were also increased (Table 1).

Table 1. Total Lactic Acid Bacteria in Various Amount of Powdered Soybean

Treatment	Number of Lactic Acid Bacteria (colonies·g-1)
7.5% Soy powder	28.33 x 10^8c
10.0% Soy powder	20.86 x 10^9b
12.5% Soy powder	175 x 10^9a

*Description: Different letters in the direction of the columns show significantly different

The increase might be caused by the higher amount of nutrients contained such as lactose, protein, and amino acid that is enough to be used for starter bacteria growth and activities. Lactose as the main carbohydrate found in milk is well utilized by starter bacteria as a source of carbon and energy. Skim milk added acted as the source of protein β-casein for the growth of starter bacteria [13].

The addition of powdered soymilk resulting significantly increased the number of lactic acid bacteria in fermented cheese whey beverages. This was in line with the opinion of [9], the prebiotic elements of GOS contained in soybeans in the form of stachyose and raffinose are a handy energy for the activity and growth of lactic acid bacteria in the fermentation process.

Stachyose and raffinose can be selectively exploited by Lactobacillus acidophilus as the source of carbon into glucose, galactose, and fructose with the help of α-galactooxidase enzyme, so it can be utilized by *S. thermophilus* and *L. bulgaricus* bacteria to increase the population of lactic acid bacteria in the product [14].

The addition of powdered soymilk in the making of fermented cheese whey beverages can increase the growth of starter bacteria and the breakdown of lactose in raw material using lactase enzyme produced by lactic acid bacteria to form glucose and galactose.

Prior to incubation, the lactose content of the product was 10.53 % while during incubation, it decreased due to the breakdown of lactose by lactic acid bacteria. So, the average lactose content available ranged from 7.0069 % - 7.2623 %. It happened by glycolysis, where lactose is converted into glucose and galactose, then both monosaccharide were converted into pyruvate and forms lactic acid and acetaldehyde. One of the factors that can affects the amount of lactose, such as the activity of starter bacteria in an optimal growth conditions [15]. It is suspected that lactic acid bacteria not only breakdown lactose, but also breakdown other sugar components such as galactooligosaccharides that existed in the product of fermented cheese whey beverages.

The pH value achieved in a fermentation product was influenced by several factors such as time, starter amount, the optimum condition of bacterial growth and number of bacteria [16]. From Table 2., it is shown that the pH of the fermented whey cheese beverages increases with the higher concentration of powdered soymilk added. pH value of fermented cheese whey beverages increased with the increasing of soymilk powder concentration dissolved in the product [17]. It was known that soymilk powder has basic properties, so the higher amount of powdered soybean added can produce a higher pH of fermented beverages [18].

The fermentation process which involves lactic acid bacteria could accumulate organic acids such as lactic acid, formic acid, pyruvic acid and acetic acid followed by the decreasingof pH into 3.65 - 4.40 [8]. The acidity of fermented milk is caused by the rate of in lactic acid production with the decreased pH [19].

Table 2. Various Soybean Powder Concentration Towards Product Properties

Treatment	pH	Lactic Acid (%)	Lactose levels (%)	Viskositas (cP)
7.5% Soy powder	4.27a	0.558c	7.0069a	8685.33c
10.0% Soy powder	4.28b	0.634b	7.1146b	15294.16b
12.5% Soy powder	3.00c	0.675a	7.2623a	24823.33a

*Description: Different letters in the direction of the columns show significantly different

Production of lactic acid in yoghurt is associated with lactose hydrolysis by the activity of lactic acid bacteria or the ability of yogurt starter to be used. Lactose were hydrolyzed by starter into lactic acid and other organic acids such as acetic acid and pyruvic acid.

Table 2. showed that the viscosity of the fermented cheese whey beverages increased along with the increasing amount soymilk powder added. Soybean contains a globulin protein that has an isoelectric pH that is almost equal to the isoelectric pH of milk protein which is 4.5 [20]. Protein from cheese whey raw materials, skim milk, and soymilk powder that contained in the fermented cheese whey raw materials, will curdled due to the increased of lactic acid levels and the formation of acidic atmosphere with pH range between 4.27 to 4.30; Which in turn the concentration of hydrogen ions in the raw material approaches the isoelectric point of the protein and when the isoelectric point is reached the protein will be completely curdled, wherein the contained protein has zero electrical charge. In addition, the presence of exopolysaccharides (EPS) from bacterial fermentation that breakdown lactose into glucose and galactose increasingly EPS formation activity results in higher viscosity of the resulting fermented beverage [21]. This is in accordance with the opinion of [20] that the higher protein content and presence of EPS will increase the viscosity of fermented whey cheese beverages in all treatments.

4. Conclusion

The addition of 10 % and 12.5 % powdered soymilk in the production of the fermented cheese whey beverages resulted in the desired product with criteria as follow an average number of lactic acid bacteria of 208.58 x 10^8 colonies·g-1 and 175 x 10^9 colonies·g^{-1}· pH 4.28 and 4.30, and viscosity 15294.16 cP and 24823.33 cP, respectively.

References

[1] Siso M I G 1996 The Biotechnological Utilization of Whey cheese: A Review *Bioresource Technology* **57 (1)**: 1-11.

[2] Zadow J G 1994 Utilization of Milk Components Whey in Robinson R K (eds) Modern Dairy Technology *Chapman and Hall* London **1994**: 313-317;

[3] Smithers G W, Ballard F J, Copeland A D, De Silva K J, Dionysius D A, Francis G L 1996 New Opportunities from The Isolation and Utilization of Whey Proteins *Journal of Dairy Science* **79(8)**: 1454-1459.

[4] De Wit J N 2001 Lecturer's handbook on whey and whey products, 1st edition *European Whey Products Association* Brussel **15**.

[5] Kosikowski F V 1979 Whey Utilization and Whey Products *Journal of Dairy Science* **62 (7)**: 1149 – 1160.

[6] Utama G L, Benito T B A, Sunardi, Balia R L 2016 The Isolation and Identification of Stress Tolerance Ethanol-Fermenting Yeasts from Mozzarella Cheese Whey *International Journal on Advanced Science Engineering Information Technology* **6(2)**: 252-257.

[7] Utama G L, Benito T B A, Sunardi, Balia R L 2017 Reducing Cheese-Making By-Product Disposal Through Ethanol Fermentation and The Utilization of Distillery Waste for Fertilizer *International Journal of GEOMATE* **13(3)**: 103-107;

[8] Kim S, Kim W, Hwang I K 2003 Optimization of The Extraction and Purification of Oligosaccharides from Defatted Soybean Meal *International Journal of Food Science and Technology* **38(3)**: 337-342.

[9] Espinosa-Martos, Ruperez P 2006 Soybean oligosaccharides. Potential as new ingredients in functional food *Nutr. Hosp.* **21(1)**: 92-96.

[10] Bylund G 1995 Dairy Processing Handbook Tetrapack Processing System *AB Lund* Swedia pp. 243-257.

[11] Teles F F F, Cheryl K Y, Stull J W 1978 A Method for Rapid Determination of Lactose *Journal of Dairy Science* **61(4)**: 506-508.

[12] Denin Djurdjević J, M aćej O D, Jovanović S 2001 The Influence of Dry Matter and Heat Treatment on The Viscosity of Set-style Yogurt Produced from Reconstituted Skim Milk Powder *Journal of Agriculture Science* **46(2)**: 123-135.

[13] Tarboush A H M 1995 Comparison of Associative Growth and Proteolytic Activity Yogurt Starters in Whole Milk from Camels and Cows *Journal of Dairy Science* **79(3)**: 366-371.

[14] Usmiati S, Utami T 2008 Pengaruh Bakteri Probiotik Terhadap Mutu Sari Kacang Tanah Fermentasi *Jurnal Pasca Panen* **5(2)**: 27-36 ;

[15] Helferich W, Westhoff D 1980 All About Yogurt *Prentice-Hall Inc.* Englewood Cliff New Jersey.

[16] Tamime A Y, Deeth H C 1980 Yoghurt: Technology and Biochemistry *Journal of Food Protection* **43**: 939-977.

[17] Sawitri M E 2011 Kajian Penggunaan Ekstrak Susu Kedelai terhadap Kualitas Kefir Susu Kambing *Jurnal Ternak Tropika* **12(1)**: 15-21.

[18] Kumala N T, Ratna S, Ari S 2004 Pengaruh Konsentrasi Susu Skim dan Madu Terhadap Kualitas Hasil Yogurt Kedelai (*Glycine max* (L.) Merr.) dengan Inokulum Lactobacillus casei *Bio SMART : Journal of Biological Science* **6(1)**: 15-18.

[19] Jay J M 2000 Modern Food Microbiology 6th edition *Aspen Publisher, Inc.* Gaithensburg Maryland.

[20] Purbowatiningrum R S, Nies S M, Agustina LN, Aminin, Wuryanti 2006 Profil Kandungan Protein dan Tekstur Tahu Akibat Penambahan Fitat Pada Proses Pembuatan Tahu *JSKA* **8(1)**.

[21] Zubaidah E, Yusnita L, Ella S 2008 Produksi Eksopolisakarida Oleh Lactobacillus plantarum B2 Pada Produk Probiotik Berbasis Buah Murbei *Jurnal Teknologi dan Industri Pangan* **9(1)**: 59-68.

IOP Conf. Series: Earth and Environmental Science **334** (2019) 012044 doi:10.1088/1755-1315/334/1/012044

Isolation and Biochemical Characterization of Extracellular Proteolityc Enzym From Lactic Acid Bacteria Isolated From Various Indonesian Traditional Fermented Product (Bakasam)

K Suradi[1], H Chairunnisa[1], E Wulandari[1] and W S Putranto[1]

[1] Laboratory of Animal Products Technology Faculty of Animal Husbandry, Universitas Padjadjaran, Sumedang, Indonesia

E-mail: kusmajadi@unpad.ac.id

Abstract. Local isolates of lactic acid bacteria isolated from Bakasam had potential to producing extracellular proteases. The research was carried out with the purpose of knowing the biochemical character of protease produced from isolates of lactic acid bacteria. The results showed extracellular protease isolates BAL 2 (*Lactobacillus acidophillus*) has a specific activity i.e. of 0.89 U/mg. Biochemical characterization of proteases indicates optimum pH 5.5, and is inhibited by chelating agent EDTA. The result of the purification with Sephadex G-100 column showed a band on SDS-PAGE with the size of molecular weight approximately 90 kD.

1. Introduction

Food processing industry in Indonesia is heavily depends on importing enzyme protease, so it inhibits the development of the livestock product food industry. Development of enzyme technology to replace the independence is necessary to support national food industry. The enzyme protease produced by the microbes can be used in lieu of enzymes from plant sources. One of the microbes that can be utilized is lactic acid bacteria. Indonesia has a germplasm microorganisms which are priceless. Lactic acid bacteria (LAB) as "friendly bacteria" has a great potential in the development of biotechnology-based food industry.

LAB has the ability to produce proteolytic enzymes around the cell wall, membrane, the cytoplasm within the cell, and *Lactobacillus acidophilus* has the capability of generating extracellular protease [1]. *Lactobacilus acidopillus*, during the fermentation process, utilize peptides or free amino acids for its growth, this bacteria are also capable of hydrolysing casein by using protease which is excreted in the vicinity of the surface of the cell wall [2].

Some research has been conducted to find out the biochemical characters of intracellular protease of some lactic acid bacteria such as *Lactobacillus bulgaricus* and *Streptococcus thermophilus*, while biochemical characters of intracellular protease as well as extracellular of isolates originating from Indonesia have not been much researched. Protease is an enzyme capable of hydrolysing peptide bonds in proteins. Based on Nomenclature Committee of The International Union of Biochemistry, protease are classified in hydrolase enzymes (EC.3.4). Some commercial proteases producers are genus *Bacillus* (*B. cereus, B. pumilus, B. subtilis, B. licheniformis, B. stearothermophilus*, and *B. polymixa*. Another group of bacteria are *Aeromonas, Lactobacillus, Pseudomonas, Serratia, Streptomyces*, and *Staphylococcus*. The aim of the study was to Isolate and biochemical characterization of extracellular proteolytic enzyme from lactic acid bacteria isolated from various indonesian traditional fermented product (Bakasam).

2. Material and methods

2.1. Screening the proteolytic activity of LAB Isolates
Isolates grown in MRS media that had been added 2% skimmed milk. Subsequently incubated 37 °C for 24 hours. Isolates that had a high activity of extracellular proteolytic will show a clear zone surrounding colonies. Isolates that show width clear zone will be taken as isolates to do testing of their proteolytic activity quantitatively and their protein levels.

- Purification and Biochemical
- Characterization of Extracellular
- Proteolytic Enzymes

2.2. Protein Concentration Assay (Bradford, 1976)
Protein concentration was measured using a standard protein Bovine Serum Albumin (BSA) with the concentration of 0.1 to 1 mg/ml. As many as 100 µl of the sample plus 5 ml of Bradford reactant, homogenized and incubated for 5 minutes at 37 °C and then its absorbance was measured with a spectrophotometer at 595 nm.

2.3. Proteolytic activity assay (Walter,1984)
The activity of the protease was measured based on the method [3], single unit activity is expressed in number of enzymes that are able to generate one micro mol tyrosine per minute.

2.4. Ammonium Sulfate Precipitation
To precipitate protein, gradually added ammonium sulfate to supernatant results of centrifugation while stirring with a magnetic stirrer with a concentration (30%, 40%, 45%, 60%) and left on for one night at 4oC.

2.5. Gel Filtration Chromatography
Equilibrated gel filtration chromatography column of Sephadex G-100 with the stream buffer Tris HCL 10 M as much as 200 ml and rest overnight at a temperature of 4 ·C. 2 ml Protease solution added with dropper vertically into column right above the surface of the gel. Enzymes were allowed to flow slowly until the entire enzyme was under the surface of the gel. The column is filled with buffer Tris HCl pH 8 and 10 mM as elution. As many as 20 oil fraction by as much as 70 drops on each fraction using the fraction collector.

2.6. The optimum pH of extracellular protease enzymes
Determined by changing the universal buffers to the desired pH on 50 °C. pH value tested between 6-10. Enzyme activity is measured by [3].

2.7. Determination of Protease Molecular Weight
Molecular weight of highest activity protease and high activity antimicrobial proteins determined with SDS-PAGE (Sodium Dodecyl Sulphate-Poly Acrilamide Gel Elektrophoresis)

2.8. Identification with 16S rDNA
The identification was done molecularly using specific primer and amplification of DNA fragments using PCR (Polymerase Chain Reaction). Electrophoresis has done on the results of amplification fragments using agarose and sequencing was done on isolated fragments. Analysis has done on the result of the sequencing with BLASTsoftware..

IOP Conf. Series: Earth and Environmental Science **334** (2019) 012044 doi:10.1088/1755-1315/334/1/012044

3. Results and Discussion

3.1. Screening the proteolytic activity of LAB Isolates

Isolation and purification of LAB isolates from various animal food source which was a protein-rich substrates, so it was expected that obtained LAB isolates have a high potential in generating extracellular protease. Isolation from fermented food products namely Bakasam (made from beef and chicken meat). Retrieved two LAB isolates potential with proteolytic activity, BAL 1 (source isolation Bakasam beef) and BAL 2 (source isolation Bakasam chicken meat).

Skim Milk Agar (SMA) *Test* LAB *Isolate*

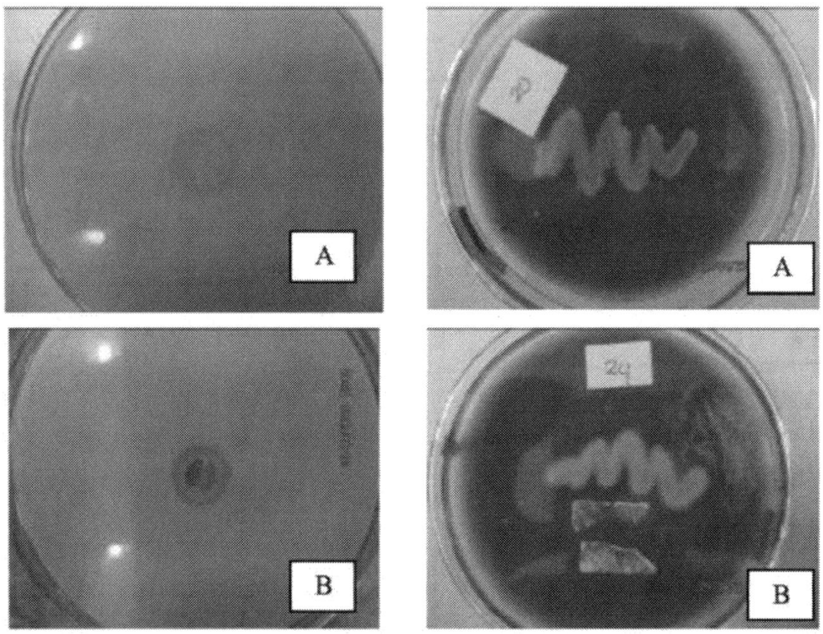

Figure 1. (A) LAB Isolate 1; (B) LAB Isolate 2.

3.2. The Growth of The BAL 1 Isolates and Its Proteolytic Activity

Growth and protease production, observations was carried on the cell suspension turbidity data (wavelength 600 nm) measured in every certain time lapse, beside that, activity of proteases and protein levels on cell suspension was measured.

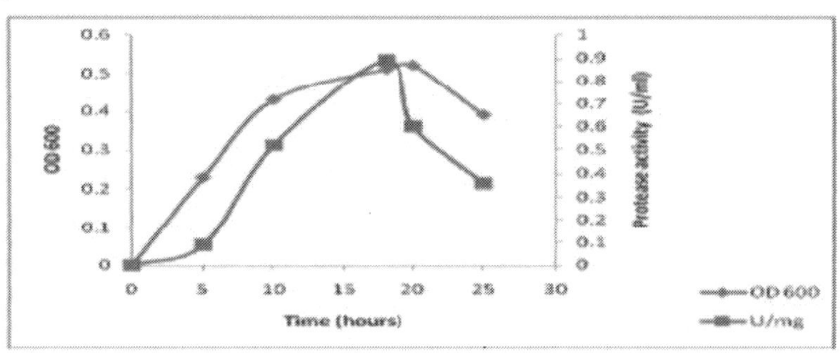

Figure 2. (A) Isolate Growth BAL 2 And The Specific Activity Of Proteases Produced.

Based on the data in Figure 2. It was seen that the growth curve beginning with the initial phase (lag) which was a period of microbes adjustment. On that phase enzyme synthesis was occurred by cell used for metabolic pathway. After the initial phase was completed, cellular reproduction started to happen. Cellular concentration was increasing, slowly it was increasing until reaching maximum and logarithmic growth occurred or exponential. After substrate and certain compounds which was required for bacterial growth in media culture run out and product inhibitor occurred, then the rate of the growth of bacteria declined. Decline phase was characterized by a reduced number of live cells (viable) in the media due to the occurrence of death (mortality). BAL 2 isolates execrated extracellular protease with the highest activity after 18 hour (0.83 U/mg) of the bacteria growth period (logarithmic growth phase). The data production time of data can be used to do enzymes production in a larger scale.

3.3. Ammonium sulfate precipitation and Dialysis
Before purification by gel filtration , protease was concentrated using ammonium sulfate. Precipitation used salt based on the solubility of proteins which polar interact with water molecules, ionic interactions of proteins with salt, and repulsive force from same charged protein. The solubility of proteins (in particular temperature and pH) increasing in the concentration of salt (salting in). The increasing in the solubility of the protein will increase the power of ion solution. The addition of certain salts would cause protein solubility decrease (salting out). Water molecules that bind with the salt ions was increasing that eventually led to the withdrawal of the sheath that surrounded the water surface, thus causing the protein, aggregated, and then settled. Ammonium sulfate is a salt that is most often used to precipitate proteins because it has high solvency in water, relatively inexpensive.

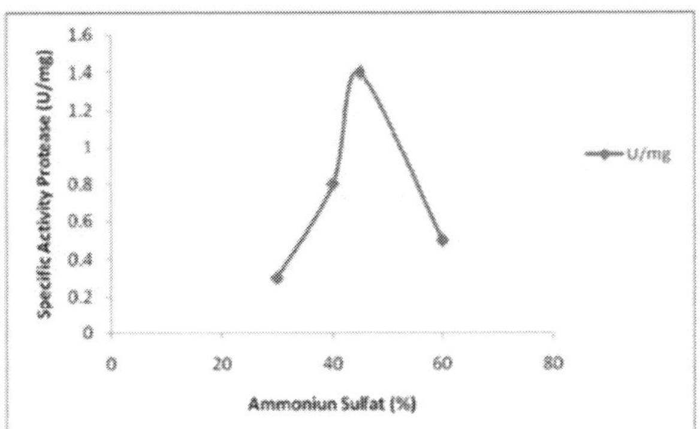

Figure 3. The Precipitation of Protease Isolates BAL 2 with Ammonium Sulfate.

Protease BAL 2 Isolates which was result from 45% ammonium sulfate precipitation; showed the highest specific activity of proteases, namely 1.42 U/mg, so that was used for purification. The impact of mineral salts and excess of ammonium sulfate in the enzyme was removed by using the dialysis method.

3.4. Gel Filtration Chromatography
The selection of the column chromatography method based on the nature of separated enzyme proteins. Information about the characterization of the enzymes was needed, such as the approximate molecular weight, degree of hydrophobicity, and the sulfidril bond.

Results of Sephadex G-100 column produced a widen protein peak. The fraction that had the highest protease activity was the 7th fraction with a specific activity of 4.10 U/mg. Purification of protease BAL 2 Isolates with Sephadex G-100 column gives the level of purity to 4.60 times rough extract with the rate of 32.8%. The rate of under 50% estimated because the levels of a protein that had a protease activity is lower than the other cell excreted protein. Protein loss during purification could occur because autolysis (Scopes, 1987), and this happens because of the effect of dilution of the enzyme.

Table 1. Purification of protease Isolates BAL 2.

Purification Step	Volume (ml)	Total protein (mg)	Total Activity (Unit)	Specific Activity (Unit/mg)	Purity	Yield (%)
Crude	100	29.67	26.4	0.89	1	100
$(NH4)2SO4$ 45%	15	13.17	18.7	1.42	1.59	70.8
Gel Filtration	2	2.11	8.67	4.1	4.60	32.8

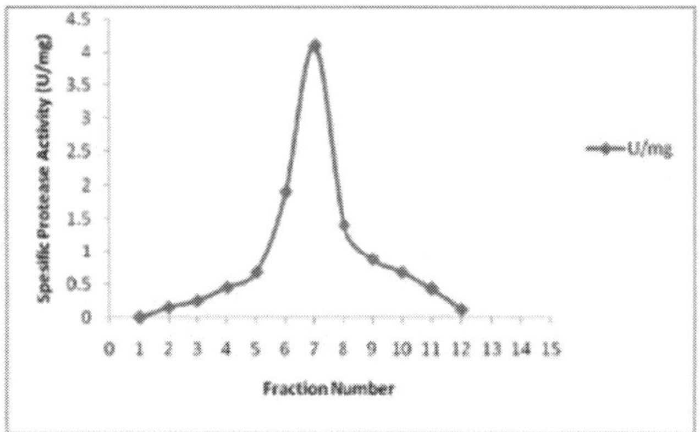

Figure 4. The Precipitation of Protease Isolates BAL 2 with Ammonium Sulfate.

3.5. SDS-PAGE

The level of purity of the enzyme could be known by using the technique of polyacrylamide gel electrophoresis, SDS-PAGE. Gel prepared by acrylamide and N,N'-methylene-bis-acrylamide which polymerizes through a free radical mechanism with the help of catalysts for N,N,N',N'-tetrametilen diamine (TEMED) and initiator amoniumpersulfat (APS). Principle analysis of SDS-PAGE is protein separation that is based on the size of the molecule.

On the well is the result of purification with Sephadex G-100 column that shows a Ribbon with molecular weight 90 kD.

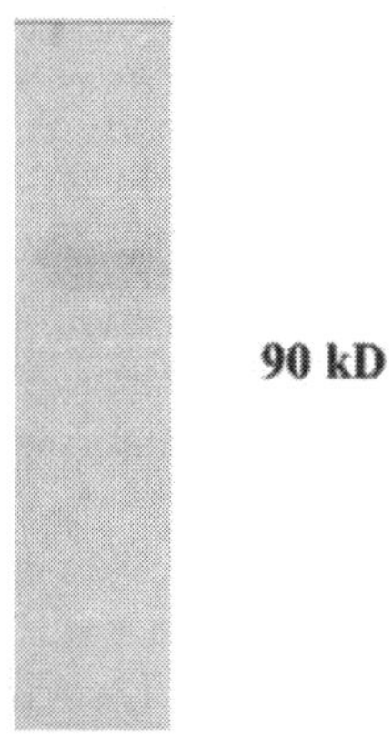

Figure 5. *SDS - PAGE protease Isolates BAL 2*

3.6. The optimum pH of extracellular protease enzymes

pH which changes on a small deviation scale could lead to decreasing enzyme activity due to ionization changes of its functional group. Ionic bonds played an important role in maintaining the active side of the conformation of the enzyme to bind and change the substrate into product. Changes onn large deviation scale, changes in pH will result in enzyme undergoes denaturation due to disruption of various non-covalent interactions that maintain the stability of the three-dimensional structure of the enzyme [4].

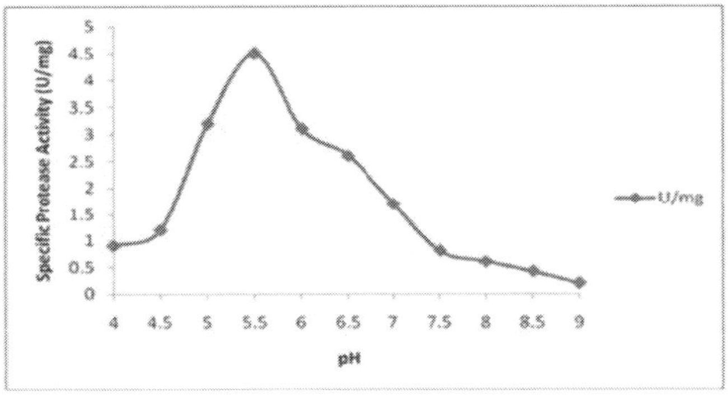

Figure 6. Effect Of pH On The Activity Of Protease Isolates BAL 2 (Temperature of 37°C).

3.7. BAL 2 Isolates Identification

BAL 2 Isolates: *Lactobacillus acidophillus* (identity 99%). Identifications results showed that BAL 2 isolates *Lactobacillus acidophillus*.

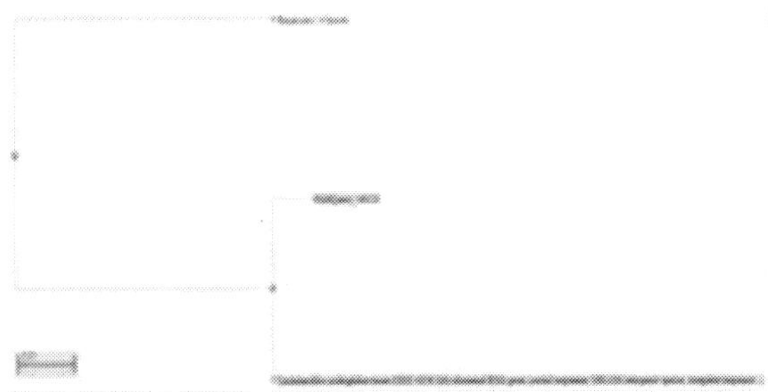

Figure 7. *Lactobacillus acidophillus* (identity 99%).

4. Conclusion

Extracellular proteases BAL 2 isolates (*Lactobacillus acidhophillus*) had a specific activity i.e. of 0.89 U/mg. Biochemical characterization of proteases indicated of optimum pH 5.5, and was inhibited by chelating agent EDTA. The result of the purification with Sephadex G-100 column shows a Ribbon on SDS-PAGE with the size of molecular weight approximately 90 kD. Based on the results of the characterization could provide information for further research in the field of food biotechnology and further studied were needed to review more about the character of the gene that encodes the protease.

Acknowledgement

We would like to thank Ministry for Research Technology and Higher Education, Directorate General for Strengthening Research and Development, which has funded this research through *Penelitian Unggulan Perguruan* (PUPT) 2015.

IOP Conf. Series: Earth and Environmental Science **334** (2019) 012044 doi:10.1088/1755-1315/334/1/012044

References

[1] Zhao R, Sun J, Mo H, Zhu Y 2007 Analysis of Functional Properties of Lactobacillus acidophillus *World J. Microbiol. Biotechnol.* **23**: 195–200

[2] Thomas T D, Pritchard G G 1987 Proteolytic enzymes from dairy starter cultures *Fed. Eur. Microbiol. Soc. Microbiol.* Rev. **46**: 245

[3] Walter H E 1984 Method with haemoglobin, casein, and azocoll as substrate In. Bergmeyer HU (ed) Methods of enzymatic analysis *Verlag Chemie* Deerfield Beach Florida Basel.

[4] Hames B D, Hooper N M 2000 Biochemistry *The instant Notes* Ed ke-2. Hongkong Spinger Verlag. pp. 83-84

Non-Destructive Method for Predicting Sapodilla Fruit Quality Using Near Infrared Spectroscopy

Kusumiyati[1], S Mubarok[1], W Sutari[1], Farida[1], J S Hamdani[1], Y Hadiwijaya[2] and I E Putri[2]

[1] Lecturer of Agronomy Studies Program, Agriculture Faculty, Padjadjaran University, Jalan Raya Bandung Sumedang KM.21, Hegarmanah, Jatinangor, Kabupaten Sumedang 45363, Indonesia
[2] Alumni of Agronomy Studies Program, Agriculture Faculty, Padjadjaran University, Jalan Raya Bandung Sumedang KM.21, Hegarmanah, Jatinangor, Kabupaten Sumedang 45363, Indonesia

E-mail: kusumiyati@unpad.ac.id

Abstract. The non-destructive method was used to predict the moisture content, firmness, total dissolved solids, and color values (including L*, a*, b*, h^{θ} and C*) of sapodilla fruit using near infrared spectroscopy with wavelength of 312 - 1050 nm. Irradiated and measured sapodilla samples were analyzed using partial least squares (PLS) algorithm of multivariate calibration technique to build the calibration models. Validation analysis was performed to build validation models and verify the calibration models. The accuracy of the validation models were defined by the ratio performance deviation (RPD) and root mean squares error of prediction (RMSEP) values on each parameters. The RPD and RMSEP values of moisture content, firmness, total dissolved solids, L^*, a^*, b^*, h^{θ} and C^* were 1.35 (0.01), 3.82 (2.08), 2.04 (1.70), 3.32 (1.30), 3.20 (0.82), 3.41 (1.38), 3.64 (2.03) and 3.11 (1.32), respectively. The results showed that non-destructive method using near infrared spectroscopy was able predict the quality of sapodilla fruit with high accuracy and low error.

1. Introduction

Sapodilla is fruit originating from Central America. It grows wild in the forests of Central America and Mexico. During colonialism, Spanish introduced sapodilla to the Philippines. Nowadays sapodilla spreads to other countries including Indonesia where sapodilla growing commercially [1]. Quality of fruits could be describe by the value of color, firmness, and sugar content [2]. Generally, most researchers assess the quality of fruit by doing destructive analysis. This method damages the fruit, hence the fruit can not be sold afterwards. Destructive analysis also requires some chemical materials and consuming a lot times and costs [3].

Near infrared spectroscopy is new non-destructive method to predict fruit quality. It is a spectroscopic method based on atomic vibrations of the molecules. The infrared spectrum is generated by passing infrared radiation on the sample and determining the absorbed radiation. The advantages of using near infrared spectroscopy that this method is non-destructive, and also many sample forms can be investigated such as fluids, solvents, pastes, powders, fibers, and gases can be identified by appropriate sample selection techniques [4]. In addition, near infrared spectroscopy technique is environmentally friendly.

The near infrared specrtoscopy technique was first used in agriculture to measure the moisture content of grain [5]. Nowadays, this technique has been done to determine the total soluble solids, firmness, acidity and dry matter of mangoes [6, 7]. In another study near infrared spectroscopy technique was used to predict soluble solids content, moisture content, firmness and color of tomato fruits on-tree and after-harvesting [8, 9]. Other reserach reported that this technique was perfomed to predict the quality of bitter gourd [10]. Storage is one of the important things in post harvest handling of fruits, because the quality of the fruit may change during storage. The factors that affecting fruit quality changes are storage duration and storage temperature. This research aimed to predict the quality of sapodilla fruit, which stored for 0, 5 and 10 days at room temperature.

2. Materials and Method

2.1. Raw Mateials

Sapodilla fruit for this study were obtained from Situraja, Sumedang, Indonesia. Each sapodilla sample was numbered before the analyis, then stored for 0, 5 and 10 days at room temperature (25$^{\varnothing}$C). A total of 300 fruit samples were splitted into 2 groups, first group used as calibration set and second group used as validation set.

2.2. Spectra Data Acquisition

Sapodilla fruit samples were irradiated using near infrared spectrometer (NirVana AG410) with wavelength of 312-1050 nm. Each sample was irradiated 3 points on the front and 3 points on the back vertically starting from the top to the bottom of the fruit. After the spectra data were obtained, multivariate calibration of chemometric was performed to extract the information from near infrared spectrum data [11]. Then the spectra of data along with reference data were used to build the calibration models using partial least squares (PLS) algorithm.

2.3. Reference Data Measurement

The water content of sapodilla is calculated by calculating fresh wet weight compared with dry fruit weight. The sapodilla fruits were sliced then dried in the oven with temperature of 80$^{\varnothing}$C. Dried samples were calculated every day until the weight were constant. The water content value was expressed by percentage (%) of water content. Firmness of sapodilla was calculated using AD-4932A-50N tension gauge. The firmness value of sapodilla fruit was expressed with Newton (N). TDS (total dissolved solids) were measured using PR-101α refractometer (Atago, Model 41325, Japan). Sapodilla fruits were extracted for the analysis. The obtained data was expressed as %Brix.

The reference data of color was measured using chromameter (Konica Minota CR-400). The obtained data were the values of L*, a* and b*. L* shows the color brightness of the sapodilla fruit, the value of 0 means black while 100 means white. a* is a gradation of green to red, negative a* indicates green while positive a* denotes red. b* is a gradation of blue to yellow, negative value of b* defines blue while positive b* describes yellow. Hue (h$^{\varnothing}$) is color appearance parameter or color dimension, it describes whether the color is red, blue, yellow, green, brown, purple, orange, and other colors. Chroma (C*) is the color intensity, it explains the strength and weakness of the color. The formula of h$^{\varnothing}$ and C* are described in the following equations :

$$h^{\varnothing} : tg^{-1}\left(\frac{b^*}{a^*}\right) \quad (1)$$

$$C^* : \sqrt{a^{*2} + b^{*2}} \quad (2)$$

2.4. Chemometrics Analysis

Chemometrics is a science of extracting data using mathematics and statistics by analyzing chemical data. Chemometrics in spectroscopic analysis consists of three aspects, pre-processing of spectral data, model building and transfer model. One of the most important chemometric analysis is to construct a calibration model through a pattern recognition technique to identify the major similarities and patterns of data. Furthermore, this model is used to predict unknown samples [12]. Nowadays chemometrics is used to analyze multivariate data that involving multiple variables and each variable can be mutually correlated.

2.5. *Partial Least Squares*

Partial least squares is one of the most common used multivariate calibration technique in chemometrics due to the quality of calibration model and the ease of application. PLS is able to analyze a large number of data, having high level of collinearity, a lot number of X variables and some Y response variables [13]. This is dimensional data reduction method to find the most relevant factors in predicting and interpreting data. PLS reduces predictor data that is not relevant to the data diversity. In addition, this method can also be applied for classification purposes. PLS analysis was done using second derivative absorbance (d2a) spectra data of each quality parameter to build the calibration models.

2.6. *Model Accuracy*

Root mean squares error (RMSE) is a method to evaluate the forecasting technique that is used to assess the accuracy level of forecasting models. RMSE measures the difference of values predicted by a model and the actual measured values. Root mean squares error of calibration (RMSEC) is RMSE for calibration model, on the other hand the root mean squares error of prediction (RMSEP) is the RMSE for validation model. RMSE informs the average inexactness that can be expected for predictions of impending samples. The equations of RMSEC and RMSEP are defined in the following explanation [14]:

$$\text{RMSEC}=\sqrt{\frac{1}{n_c}\sum_{i=1}^{n^c}(\hat{y}i-yi)^2} \tag{3}$$

$$\text{RMSEP}=\sqrt{\frac{1}{n_p}\sum_{i=1}^{n^p}(\hat{y}i-yi)^2} \tag{4}$$

where:
$\hat{y}i$:near infrared prediction,
yi :actual data,
n^c :amount of data in the calibration set,
n^p :amount of data in the validation set.

RPD (Ratio performance deviation) is determined by dividing the standard deviation of reference data in validation set with root mean squares error of prediction. RPD value ranged from 1.5 to 2 describes the model is able to distinguish low from high values of the reference data; value ranged from 2 to 2.5 defines the decent quantitive model prediction; RPD value above 2.5 indicates the excellent model prediction [15, 4]. RPD is calculated using the following formula:

$$\text{RPD} = \frac{\text{SD}}{\text{RMSEP}} \tag{5}$$

3. Result and Discussion

The spectra data were preprocessed by second derivative pretreatment to examine the peaks of sapodilla fruit samples. The second derivative (d2a) spectra were used to reduce the ratio of noise . The d2a spectra showed the peaks at around 360, 640 and 900 nm (Figure. 1).

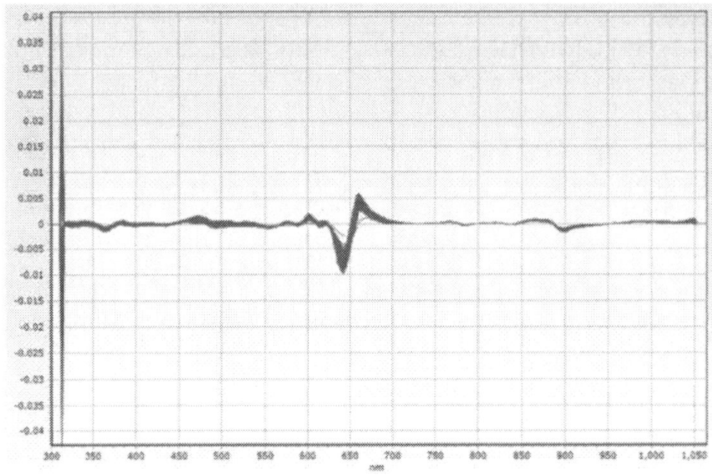

Figure 1. Typical spectra data (d2a) of sapodilla in wavelenght of 312 – 1050 nm.

Reference data were obtained from the destructive analysis. 6 reference data and 6 spectra data were collected from each samples. The amount of fruit used in this study was 300 fruit samples. Data selection was then performed for calibration and validation purposes. Validation set used different data from the calibration set. The data used in calibration set were 600 reference data and 600 spectra data, as well as the validation set. Table 1 shows the reference data used in the calibration and validation set.

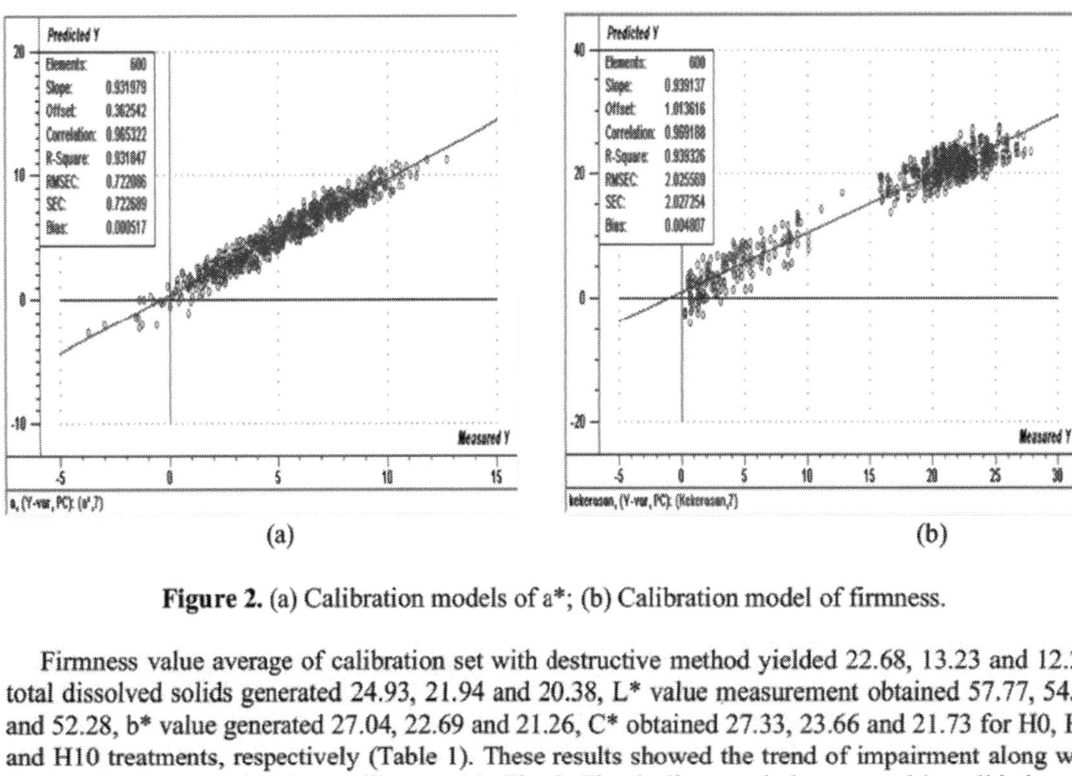

(a) (b)

Figure 2. (a) Calibration models of a*; (b) Calibration model of firmness.

Firmness value average of calibration set with destructive method yielded 22.68, 13.23 and 12.21, total dissolved solids generated 24.93, 21.94 and 20.38, L* value measurement obtained 57.77, 54.63 and 52.28, b* value generated 27.04, 22.69 and 21.26, C* obtained 27.33, 23.66 and 21.73 for H0, H5, and H10 treatments, respectively (Table 1). These results showed the trend of impairment along with the length of storage duration as illustrated in Fig. 2. The decline trend also occured in validation set. This revealed that the length of storage duration had effect in decreasing the quality of sapodilla on firmness, TDS, L*, b* and C* parameters. This is affirmed by [16] that reported there were firmness and total dissolved solids loss of mature dutch eggplant fruit in storage treatment of 5 and 10 days at room temperature.

3.1. Calibration models

The results of PLS analysis for calibration set showed the R^2 and RMSEC values were 0.57 (0.01), 0.93 (2.02), 0.82 (1.47), 0.91 (1.32), 0.93 (0.72), 0.92 (1.35), 0.93 (2.00), and 0.93 (2.00) for the moisture content, firmness, total dissolved solids, L*, a*, b*, h^{θ} and C* values, respectively (Table 2). The highest value of R^2 were shown in a* and firmness (Figure 2.) . High R^2 value indicates the calibration models are sufficiently good and feasible to be analyzed in validation set. In accordance with [17] statement, which states that the value of R calibration as 0.71 (R^2=0.5) or higher indicates good value for calibration screening and also to be proceeded for validation analysis.

Table 1. Reference data of quality parameters of the calibration and validation set.

Quality	Calibration (n=600)			Validation (n=600)		
Parameter	Range	Mean	SD	Range	Mean	SD
MC						
H0	0.66-0.75	0.70	0.01	0.66-0.75	0.70	0.01
H5	0.63-0.78	0.71	0.02	0.63-0.78	0.71	0.02
H10	0.65-0.75	0.71	0.02	0.65-0.75	0.71	0.01
Firmness						
H0	18.03-27.81	22.68	1.93	15.59-27.81	22.42	1.99
H5	0.62-25.57	13.23	8.30	0.62-25.57	14.06	8.14
H10	0.23-25.60	12.21	8.42	0.23-26.13	12.45	8.56
TDS						
H0	20.4-32.20	24.93	2.21	20.5-30.60	24.92	2.19
H5	15.9-29.40	21.94	3.02	12.3-28.80	21.93	3.00
H10	11.7-29.20	20.38	3.77	11.6-28.00	20.36	3.86
L*						
H0	47.63-65.62	57.77	2.49	50.22-64.18	57.59	2.30
H5	45.30-63.00	54.63	3.63	42.45-63.47	54.75	4.07
H10	33.65-61.33	52.28	5.27	37.30-60.48	52.74	4.57
a*						
H0	(-1.55)-11.81	3.94	2.84	(-1.34)-11.06	4.27	2.39
H5	(-3.01)-12.72	6.55	2.50	(-1.17)-11.34	6.30	2.62
H10	(-3.72)-10.55	5.10	2.37	(-2.25)-10.34	4.92	2.48
b*						
H0	19.98-33.69	27.04	2.51	21.46-32.37	26.91	2.15
H5	11.44-31.31	22.69	4.48	12.01-31.22	22.92	4.34
H10	10.29-30.38	21.26	5.03	10.22-32.70	21.56	5.30
h^{θ}						
H0	66.70-89.79	80.43	4.93	62.92-89.36	80.69	4.71
H5	60.01-88.00	73.64	6.75	60.63-89.47	73.40	6.69
H10	47.95-88.89	73.90	8.72	56.30-88.82	74.89	8.13
C*						
H0	21.94-33.37	27.33	2.13	21.31-32.26	27.57	2.12
H5	13.71-31.43	23.66	4.04	14.56-31.27	23.69	3.68
H10	11.65-30.74	21.73	4.79	12.83-30.53	22.40	4.32

MC – Moisture content; TDS – Total dissolved solids; SD – Standard deviation

The calibration models that have been built need to be reassessed to verify the calibration accuracy, therefore validation analysis was required. Validation method used in this research was internal validation. The data were divided into 2 groups, calibration and validation set. The calibration models that have been built using calibration set was then applied to the validation set to build the validation models.

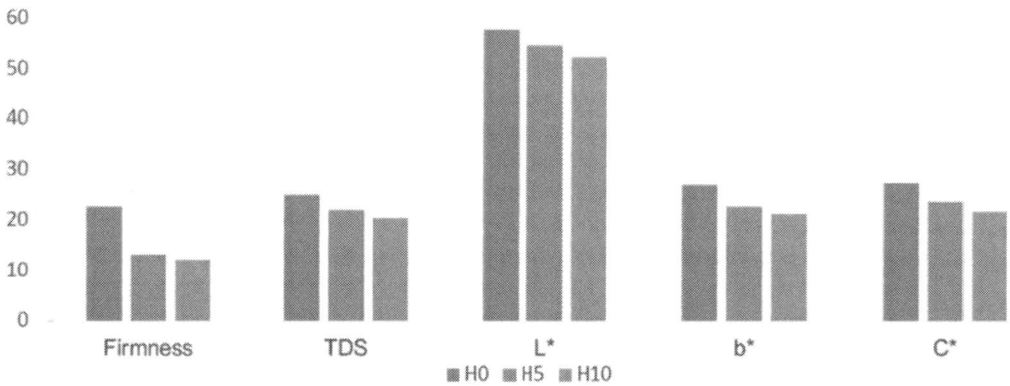

Figure 3. Reference data of firmness, TDS, L*, b* and C* values in calibration set

3.2. Validation models

Validation models that have high correlation coefficient value (closed to 1) indicate the high validity. The values of R^2 and RMSEP of validation set were 0.46 (0.01), 0.93 (2.08), 0.76 (1.70), 0.90 (1.30), 0.90 (0.82), 0.91 (1.38), 0.92 (2.03), and 0.89 (1.32) for moisture content, firmness, total dissolved solids, L*, a*, b*, $h^{\emptyset}$ and C*, respectively (Table 2). The obtained results showed the validation models didn't differ much from the calibration model. This accords with [15], which explain a good model should have a high R value in calibration and validation set, as well as having low RMSEC and RMSEP values. In addition, RMSEC and RMSEP are expected to have almost the same value. RPD value of moisture content (1.35) had the lowest accuracy compared to the other parameters, followed by TDS (2.04), C* (3.11), a* (3.20), L* (3.32), b* (3.41), $h^{\emptyset}$ (3.64) and firmness (3.82). The use of RPD to evaluate fruit color was also investigated in tomato with feasible values of L* (3.66), a* (4.76), b* (1.77), $h^{\emptyset}$ (2.50) and C* (2.06) [9]. RPD values alone can not be used as benchmark for calibration potential, [3] suggest to consider other statistical analysis results such as R^2 value of validation set, RMSEP, and the number of factors used in the calibration set.

Table 2. Calibration and validation model by partial least square (PLS)

Elements		Calibration (n=600)		Validation (n=600)		RPD
		R^2	RMSEC	R^2	RMSEP	
Moisture Content		0.57	0.01	0.46	0.01	1.35
Firmness		0.93	2.02	0.93	2.08	3.82
Total Dissolved Solids		0.82	1.47	0.76	1.70	2.04
Color	L^*	0.91	1.32	0.90	1.30	3.32
	a^*	0.93	0.72	0.90	0.82	3.20
	b^*	0.92	1.35	0.91	1.38	3.41
	$h^{\emptyset}$	0.93	2.00	0.92	2.03	3.64
	C^*	0.89	1.40	0.89	1.32	3.11

R^2 – Coefficient of determination; RPD – Ratio performance deviation; RMSEC – Root mean standard error of calibration; RMSEP – Root mean square error of prediction

4. Conclusion

The results showed that non-destructive method using near infrared spectroscopy had high accuracy values to predict the firmnes and color of the sapodilla fruit (including L*, a*, b*, h^0, and C*). In addition, good accuracy values were shown on moisture content and total dissolved solids parameters. Firmness, TDS, L*, b*, and C* parameters presented the trend of quality impairment along with the longer storage duration. Non-destructive method using near infrared spectroscopy technique was able to predict the quality of sapodilla which stored for 0, 5 and 10 days storage at room temperature.

Acknowledgement
This research was financially supported by LPPM of Padjadjaran University.

Conflicts of interest
he authors declare that there is no conflict of interest regarding the publication of this paper.

References

[1] Lakshminarayana S, Rivera M A M 1979 Proximate characteristics and composition of sapodilla fruits grown in Mexico *Proceedings of the Florida State Horticultural Society* **92**: 303-305.
[2] Shewfelt R L 1999 What is quality? *Postharvest Biology and Technology* **15**: 197-200.
[3] Ignat T, Schmilovitch Z, Fefoldi J, Steiner B, Alkalai-Tuvia S 2012 Nondestructive measurement of ascorbic acid content in bell peppers by VIS–NIR and SWIR spectroscopy *Postharvest Biology and Technology* **74**: 91–99.
[4] Stuart B 2004 Infrared spectroscopy : Fundamental and Applications *Saunders College Pub.* Philadelphia.
[5] Norris K H 1964 Design and development of a new moisture meter *Journal of Agricultural Engineering* **45**: 370-372.
[6] Schmilovitch Z, Mizrach A, Hoffman A, Egozi H, Fuchs Y 2000 Determination of mango physiological indices by near-infrared spectrometry *Postharvest Biology and Technology* **19(3)**: 245–252.
[7] Nagle M, Mahayothee B, Rungpichayapichet P, Janjai S, Müller J 2010 Effect of irrigation on near-infrared (NIR) based prediction of mango maturity *Scientia Horticulturae* **125(4)**: 771–774.
[8] Kusumiyati, Akinaga T, Yonemori S, Kawasaki S, Tanabe T 2007 Evaluation of tomato quality on tree and after harvesting using portable NIR spectroscopy *Journal Society of Agricultural Structures Japan* **38(2)**: 117-126.
[9] Kusumiyati, Akinaga T, Tanaka M, Kawasaki S 2008a On tree and after harvesting evaluation of firmness, color, and lycopene content of tomato fruit using portable NIR spectroscopy *Journal of Food Agriculture and Environment* **6(2)**: 132-137.
[10] Kusumiyati, Akinaga T, Yonemori S, Shikanai T, Okamoto H, Tanabe T, Kawasaki S 2008b Internal quality sensor of fruit jagged surface: preliminary study of bitter gourd (*Momordica charantia* Linn) *Acta Horticulturae* **768**: 391-398.
[11] Stchur P, Cleveland D, Zhou J, Michel R 2002 A review of recent applications of near infrared spectroscopy, and the characteristics of a novel PbS CCD array-based near infrared spectrometer *Applied Spectroscopy Reviews* **37**: 383–428.
[12] Africia S M 2007 Kombinasi Metode Spektrofotometri dan Kalibrasi Multivariat untuk Penentuan Simultan Natrium Benzoat dan Kalsium Sorbat Bogor Indonesia *Thesis* Bogor Agricultural University.
[13] Kuno A, Matsuo M 2000 Nondestructive specification of solid mixtures by multivariate calibration of x-ray absorption near edge structure using artificial neural networks and partial least square *Analytical Sciences* **16**: 597-602.
[14] Yahaya O K M, MatJafri M Z, Aziz A A, Omar A F 2015 Visible spectroscopy calibration transfer model in determining pH of Sala mangoes *Journal of Instrumentation* **10**.
[15] Nicolaï B M, Beullens K, Bobelyn E, Peirs A, Saeys W, Theron K I, Lammertyna J 2007 Nondestructive measurement of fruit and vegetable quality by means of NIR spectroscopy: a review *Postharvest Biology and Technology* **46**: 99–118.
[16] Julianti E 2011 Pengaruh tingkat kematangan dan suhu penyimpanan terhadap mutu buah terong belanda (*Cyphomandra betacea*) *Jurnal Hortikultura Indonesia* **2(1)**: 14-20.
[17] Williams P C 1987 Variables affecting near-infrared reflectance spectroscopic analysis In: Williams P, Norris K (ed.) Near-infrared technology in the agricultural and food industries

American Association of Cereal Chemists Inc. Saint Paul pp. 143–167

[18] Viscarra R R V, McGlyn R N, McBratney A B 2006 Determining the composition of mineralorganic mixes using UV–vis–NIR diffuse reflectance spectroscopy *Geoderma* **146**: 403-411.

[19] Williams P 2007 Application to agricultural and marine products In: Ozaki Y, McClure W F, Christy A A (Ed.) Near-Infrared spectroscopy in food science and technology *John Wiley and Sons, Inc. Pub.* New Jersey. pp. 165–218.

[20] Gómez A H, He Y, Pereira G 2006 Non-destructive measurement of acidity, soluble solids and firmness of Satsuma mandarin using Vis/NIR-spectroscopy techniques *Journal of Food Engineering* **77(2)**: 313–319.

IOP Conf. Series: Earth and Environmental Science **334** (2019) 012046 doi:10.1088/1755-1315/334/1/012046

The Effect of Ozone to The Characteristic of Peeled Onion during Storage at Low Temperature

M Djali[1], I S Setiasih[1], E Sukarminah[1], T Rialita[1], Y Cahyana[1], R Andoyo[1], E Mardawati[1], I Hanidah[1] and M M Ihsan[1]

[1] Department of Food Technology Industy, Faculty of Agricultural Industry, Universitas Padjadjaran, Sumedang, Indonesia

E-mail: muhamadmaftuhihsan @ gmail.com

Abstract. The physical, chemical, microbiological and organoleptic characteristics of peeled onion have decreased during storage. The alternative technology to maintain the characteristics of peeled onion was ozonation technology. The aim of this research was to determine the characteristic of peeled shallot ?? (shallot or onion, should be consistent) which treated with or without ozone and its characteristics during storage at 5±2°C. The ozonation used bubbling method with ozonizer TIP-01. The research method used was regression and correlation analysis. This experiment consisted of 4 treatments, without ozonation which was not packaged and packed with clingwrap plastic and 0,8 ppm ozonation result which was not packaged and packed with clingwrap plastic (should improve the English and rewrite again). The analysis was carried out for 15 days with intervals of time every 3 days. The results showed that the peeled red onion with ozone concentration of 0,8 ppm and packed in plastic clingwrap changed its physical characteristics, chemical, microbiological and organoleptic slower than the peeled onion without ozonation and packed with plastic clingwrap. The peeled Onion with ozonation of 0,8 ppm and packed with clingwrap plastic lost its weight by 4,51%. The color was bright brownish red (L*: 52,52, a*: 12,65, b*: -3,2), VRS levels decreased by 10,29 µgr/g and the total number of microorganisms decreased by 1,17 log CFU/g. The overall color, aroma, and appearance were not favored by panelists with consecutive values of 2,80; 2,87; 3,0 respectively. The peeled Onion without ozonation that was not packed with clingwrap plastic experienced weight loss by 12.30%. The color was bright red brown (L*: 60, a*: 8,34, b*: -065). VRS levels decreased by 7,78 µg/g and the total number of microorganisms increased by 9,38 log CFU / g. The overall Color, aroma, appearance was not favored by panelists with consecutive values of 1,73; 1,87; 1,73 consecutively.

1. Introduction

One of the cultivated horticultural commodities in Indonesian is onion *(Allium ascalonicum L.)*. The Onion is one of the vegetable commodities that have a significance value for the community, both seen from the high economic value and the nutritional content. The onion production in Indonesia continues to increase, the production of shallots or onion in 2016 reaches up to 1225 million tons. In 2017 it reaches 1,256 million tons and it always increases every year, so it is predicted that the year 2018 will reach 1,287 million tons and in 2019, it will reach 1,318 million tons. This condition is occured because the onion is often used by the community for raw ingredients to manufacture spices.

Along with the increasing of activities leads to various demands in the processing and suppling of foodstuffs including practical spices. One effort to fulfill the consumer's demand, especially for the consumption of onion as a spice is by providing in the peeled form. Peeled onion is one of the alternative options that can be done to speed up and facilitate the processing. However, the process of stripping on shallots causes the onion is easy to damage during storage, so that it give an effect to the physical characteristics and other characteristics that will cause a decrease in quality.

Cooling at refrigerator temperatures is the simplest method to preserve and prolong the shelf life of onions. [1] conducted a study on low temperature storage that can be used for vegetable products is 7 ^{0}C. [2] conducted research on storage of onion at 5 ^{0}C resulting in the best characteristics of onion. However, the cooling method can only inactivate microorganisms so that a cooling combination with a handling effort that can reduce total microorganisms to maintain peeled onion quality in order to provide safe and quality products to the consumers.

One of the efforts of handling fresh food that is currently being used, namely ozonation. Ozone (O_3) is one type of sanitizer that is widely used and safer than chlorine because ozone leaves no residue so it does not pollute the environment. Based on the description, it is necessary to study the effectiveness of ozonation of shallot onion which is stored using low temperature (5±2 ^{0}C) to reduce the rate of onion quality reduction.

2. Materials and research methods

The materials used are roots of Menteng cultivar onion that has 70-80 days old harvested from the farmers of Ciparay , Bandung Regency. The onion was cut for the neck, roots, outer skin, and cleaned of dirt that stick without washing with water. The peeled onion is divided into 2 parts each 250 grams. 1 part of peeled onion was ozonated in aquades with the temperature 10 ^{0}C (1: 8 w/v) with bubbling ozone gas method with flow rate 2.5 L/min for 6 minutes to get the concentration 0.8 ppm. The concentration is maintained for 5 minutes. 1 part peeled onion was soaked in aquades at the temperature 10 ^{0}C for 5 minutes without ozone treatment. This ozonation technology performed using TIP-01 Ozonizer l (specification of equipment such as the country origin and other details). The sample was packed in styrofoam with size 15x10 cm^2 covered by clingwrap plastic, and then stored for 15 days at the temperature of 5±2 ^{0}C.

The method used was descriptive experimental method continued by regression analysis and correlation. This result was presented in graph, where the independent variable (x) was the length of storage peeled onion of given treatments, while dependent variable (y) was the main results observation of form shrinkage weights , colors (L*, a* and b*) , VRS levels, total microorganisms and organoleptic (color, aroma, and overall appearance) . The treatment was done by ozonating the peeled onion until reached the desired concentration, the concentrations used was as the following :

- A= Peeled Onion Without Ozonation and packed with Clingwrap plastic (BMTOK)
- B= Peeled Onion without ozonation Not Packed with Clingwrap plastic (BMTOT)
- C= Onion peeled with the concentration of ozonation 0.8 ppm packed Clingwrap plastic (BMOK)
- D= Onion peeled with the concentration of ozonation 0.8 ppm not packed Clingwrap plastic (BMOT)

The best treatment was determinized by the lowest slope (coefficient regression), where it shows that on this concentration the decline of quality decreases slowly, so it could maintain the characteristics physical, chemical, microbiology and organoleptic peeled onion .

2.1. Shrink weight testing

The sample weight was measured used analytical balance (merk, specification and country origin) by noted the weight of the packaging then weigh the onion peeled on the storage in the day of 0, 3 , 6 , 9, 12 , and 15 . The each weight of the sample obtained with this below formula :

$$W_{onion\ peeled} = W_{overall} - W_{styrofoam} \quad (1.1)$$

The shrink weight was calculated at the end of the observation (day 15). Shrink weight could be calculated as follow :

$$Shrink\ weight = \frac{W_{initial} - W_{final}}{W_{initial}}\ x\ 100\% \qquad (1.2)$$

Information:
W1 = Initial weight of sample (it is not matched with the above equation)
W2 = The final weight of the sample

2.2. *Analysis color L*a*b* testing*

The color of peeled onion analyzed by measuring CIE L*,a* and b* with CR-400 chromameter (detail of specification). The values of L*, a* and b* referee to white/dark, reddish/greenish, and bluish/yellowish from each sample respectively. The maximum values fol L* is 100 which indicated white/light color and the minimum value of L* is 0 for black/ dark. While for values a* and b* were not has a value limit, positive a* indicated red and negative b* show green, meanwhile positive b* showed yellow and negative b* showed blue color.

2.3. *VRS content*

The 20 grams Onion was weighed and then ground by using the grinder in cold sample condition then added with 30 ml of aquades. Subsequently, the onion mixture and aquades were put into VRS flask and Erlenmeyer flask previously filled with 25 ml of $KMNO_4$ 0.02 N solution. Then the distillation was run until the volume of fluid in the Erlenmeyer flask reaching 150 ml, and then 25 ml $KMnO_4$ and 5 ml of $H_2\ SO_4$ 6N and 3 ml of KI 20% were added into the flask, and Titration was done with a titrant of $Na_2\ S_2\ O_3$ 0.02 N until the color turns to light yellow. Next to the Erlenmeyer flask was added 3 drops of amylum indicator and the titration was continued until the color becomes white. The blank experiments were done with distilled water instead of samples. VRS content was calculated:

$$\text{VRS content } (\mu\ \text{grek/g}) = \frac{(B-C) \times N\ Na_2S_2O_3 \times 1000}{W\ sample} \qquad (2.1)$$

Description :
B = V $Na_2\ S_2\ O_3$ 0.02 N Solution for titration blank
C = V $Na_2\ S_2\ O_3$ 0.02 N Solution for titration sample
N = Normality solution of $Na_2\ S_2\ O_3$ 0,02 N

2.4. *Microbiological content*

Samples for microbiological testing were made by dilution method up to 10^{-7}, then taking 1 mL of the 3 highest dilutions and poured in a subsequent petri dish for total microorganism testing in poured PCA (Plate Count Agar) medium. The total colony is calculated by calculating the BAM (Bacteriological Analytical Manual) method. The formula for calculating the total of micro colonies by BAM method :

$$N = \frac{\Sigma c}{[(1 \times n1)+(0,1\ x\ n2)]\ x\ (d)} \qquad (2.2)$$

Information:
N = a number of product colonies, expressed in colonies per mL or colony per g
ΣC = a number of colonies on all calculated bowls
n1 = a number of saucers on the first dilution calculated
n2 = a number of saucers on the second calculated calculation
d = the first calculated compensation

2.5. *Organoleptic Testing*

This sensory testing of peeled onion red with and without ozonation was carried out by hedonic test using 15 panelists. The quality parameters tested include color, aroma, and overall appearance. Samples were tested by assessing the onion quality based on the panelist preferences on onion characteristics.

IOP Conf. Series: Earth and Environmental Science **334** (2019) 012046 doi:10.1088/1755-1315/334/1/012046

3. Results and discussion

3.1. Physical Characteristics

3.1.1. Shrink Weight

The result analysis in trial this show that their relationship between storage periods to shrinkage weight of peeled onion (please should be peeled onion). The Images of the relationship curve was shown in Figure 1. The percentage of weight loss onion peeled results which packed in plastic clingwrap was small decline in VRS compared with the treatment of onion peeled using ozonation and without ozonation results that were not packed with plastic clingwrap. The weight of onion peeled result of ozonation 0.8 ppm which is packed with clingwrap plastic is thought to be due to the ozone molecule oxidizing the respiratory enzymes (phosphorylase, amylase, transglucosylase, and maltase) to the material. Ozone gets in contact with the cell wall and oxidizes the cell wall which further causes lysis. This results in respiratory enzymes (phosphorylase, amylase, transglucosylase, and maltase) contained in cells, especially on the membranes and nuclei of the cells in the cytoplasm, will be partially inactive, and the respiratory process involving ozone will be inhibited. The inhibited respiratory process inhibits the process of water loss due to evaporation and loss of carbon during respiration resulting in shrinkage of small weight loss compared to the treatment without ozonation [3]. [4] states that this type of packaging also affects the weight loss because it affects the amount of environmental air entering into the material or the amount of air released from the material. The type of packaging relates to the permeability so that appropriate packaging is required to regulate the air coming in and out so that the weight loss can be suppressed.

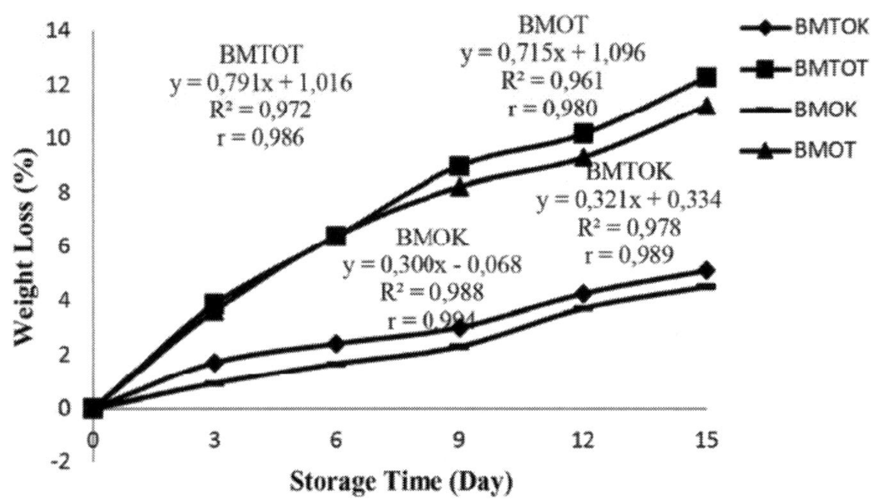

Figure 1. Curve Relationship between Storage Duration and Percentage of Weight Red Onion Shoots Results Ozonation and Without Ozonation

Color. Known that the red color on onion originated from pigment anthocyanin. Result analysis show that there were relationship between storage periods with the value of L* (brightness), a* (red), and b* (yellow) on onion peeled. The change value L* (brightness) of onion peeled for storage could see on Picture 2. The onion peeled using ozonation and without ozonation results which are packed in plastic clingwrap has a slight increase in color (L*) compared with the treatment of the onion peeled result using ozonation and without ozonation ozonation results that were not packed with plastic clingwrap. The increase in brightness on the peeled onion treatment is suspected that the anthocyanin pigment in onion decreases as it is degraded during the storage. Ozone is a powerful oxidizing agent that could remove the organic dyes (discoloration), so the color of the ozonated material will be maintained compared with the material without ozonation [5].

Giving ozone gas to potato tubers can slow the course of respiration. The absence of ozone bubbling in the onion without ozonation causes respiration to run faster than the ozone treated onion bubbling onion. This results causes a rapid changes in brightness due to the anthocyanin degradation also in the treatment without ozonation, so the onion color is getting pale and resulting an increase in the brightness intensity value (L*).

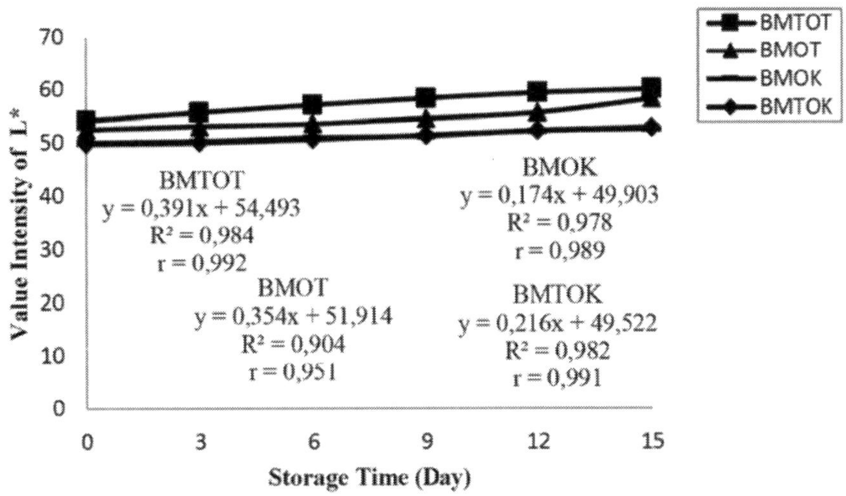

Figure 2. Curve Relationship between Storage Duration and L * (Brightness) Red Onion Peel Ozonation and Without Ozonation

The decline of brightness was undesirable for the stored peeled onion because signifies a decline quality from rottenness materials, so if seen from the coefficients regression value of each treatment, the onion result using ozonation and without ozonation results which are packed in plastic clingwrap show a good brightness compared with ozonation result and without ozonation which is not packed with clingwrap plastic.

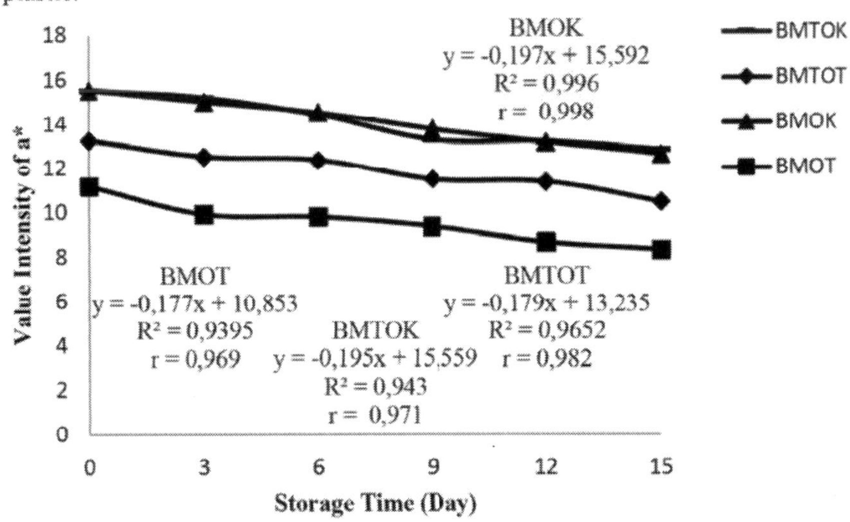

Figure 3. Curve Relationship between Storage Duration with Value a* Onion Peeled Ozonation Results and Without Ozonation

Different with value a* (red), which indicates that onion peeled without ozonation and result ozonation experience decline a* value (Figure 3). When viewed from the value of the regression coefficient, the ozonation result of the onion and onion results without ozonation that is not packed with clingwrap plastic has a small (a*) color reduction compared to the treatment of shallot peeled ozonation results and without ozonation packed with clingwrap plastic. Color a* (red) onion peeled ozonation results 0.8 ppm allegedly due to ozone oxidizing mitochondria in the cell. This results in respiratory enzymes (phosphorylase, amylase, transglucosylase, and maltase) contained in the cell, especially the mitochondria in the cell organelle, will be partially inactive, and the respiratory process involving ozone will be inhibited. The inhibited respiratory process inhibits the process of mitochondrial damage resulting in a lack of energy supply for cell metabolism, resulting in cell death which affects a decrease in a* (red) color [6].

The same result showed by the value of b* (yellow), the decrease of color b* (yellow) during storage is caused by the degradation of anthocyanin pigment and also the occurrence of decay so that the red color of the outer shell is partially pale. Anthocyanin degradation can occur during the process of extraction, food processing, and storage. The factor that cause anthocyanin degradation are light, oxygen, and enzymes. The anthocyanin pigment is very sensitive to these factors so that in this observation the peeled onion is tarnished and finally yellowing. In this study, the condition of onion storage peeled generally can still be exposed to light even if stored in the refrigerator. In addition, packaging treatment with styrofoam and clingwrap still allows the exchange of oxygen into the material because it has permeability capability. During storage, the reactions to the oxygen, light, and enzymes make the peeled onions change color.

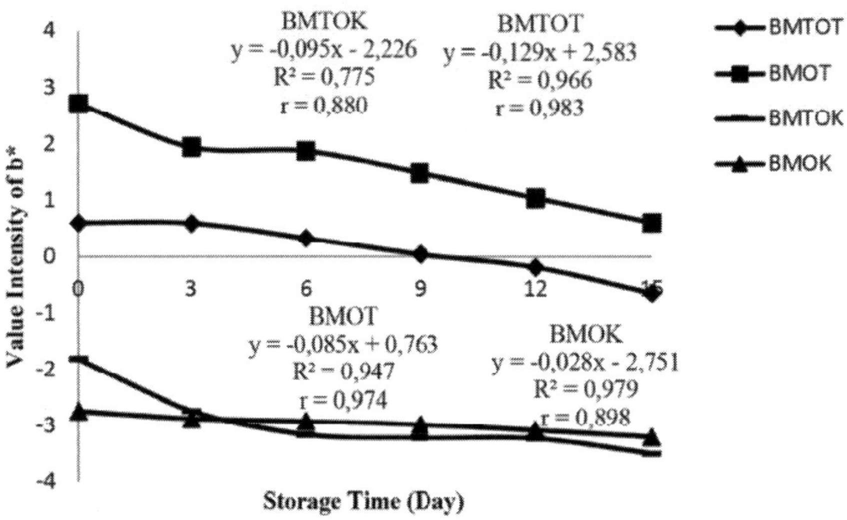

Figure 4. Curve Relationship between Storage Duration with Value b * Red Onion Peel Ozonation Results and Without Ozonation

3.2. Chemical Characteristics

Content Analysis VRS. The results analysis in this trial showed that there are relationship between storage periods to the VRS onion peeled. The onions peeled without ozonation packed with clingwrap plastic have a small decrease in VRS compared to the treatment of shallot peeled ozonation and without ozonation that is not packed with clingwrap plastic. Figure curve relationship can be seen in Figure 5. Ozon can work as deodorant. By deodorizing ozone can eliminate the odor caused by various aromatic yellow cents and microorganisms, so the smell on the treated material ozonation will be less than the ozonation-treated material [3]. In addition the use of clingwrap has a high water vapor permeability of 800 cm^3/ cm^2/mm/s/cmHg. This high permeability of water vapor allows the respiration of vegetables and fruits in the package to remain smooth, so that the quality is maintained [7].

The change of VRS content of onion during storage is affected by enzyme activity. The enzyme that plays a role in this flavor change is the peptidase enzyme. The flavor change of red onion resulted from the activity of peptidase enzyme and flavor precursor of the onion *L-glutamil peptidase* [8]. This is supported also by [9], the higher activity of VRS-forming enzyme then the VRS level on the tuber will increase, otherwise if enzyme activity decreases then the VRS level will also decrease. Onion flavor may be affected by several factors such as the state during growth including plant life, soil sulfur content and soil water content, conditions during storage and further processing after harvest.

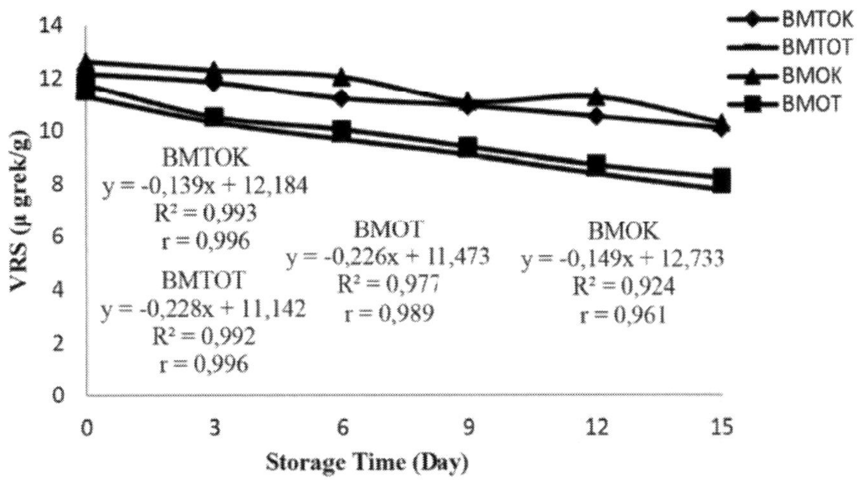

Figure 5. Curve Relationship between Storage Long Time and VRS Value Red Onion Peel Ozonation and Without Ozonation

3.3. *Characteristics of Microbiology*

Analysis Microbiology. For storage, the onions peeled is very susceptible to damage, one of them was by growth microorganisms. Ozonation is one of the handling postharvest method that have already used for reduce amount microorganisms on onion peeled. Total number of microorganisms onion peeled for storage could seen on Figure 6.

The peeled Onion with ozonation and packed with clingwrap plastic and not packed with clingwrap plastic showed lower total number of microorganisms at the beginning of storage up to the 6th day of storage and subsequently increasing until the 15th day of storage. When compared with peeled shallot or onion without ozonation, on the day of storage of the 6th, the onion resulted from ozonation which is packed with clingwrap plastic and not packed with clingwrap plastic can reduce the total number of microorganisms as much as 1.17 log CFU/g and 2.08 log CFU/g. Onion peeled without ozonation increase in total microorganisms during storage. This is due to the absence of water delivery with *bubbling* method that can reduce the total microorganisms. The ozonation result of the onion decline the microorganism first, then curve has an increase in total of microorganism. The decrease in the total number of microorganisms on the 0 th day is due to the fact that ozone effectively kills microorganisms through oxidation of the cell wall membrane [10]. Ozone spreads rapidly and penetrates through bacterial cell walls causing bacteria to be inactivated and die. The effectiveness of ozone on microorganisms depends not only on the amount of ozone given to a material, but also depends on the effectiveness of the given ozonation method, the type of material, and the type of microorganism. The increase in the total number of microorganisms on day 9 because of the ozone effect for sterilization has been lost because ozone leaves no residue in the samples given the immersion treatment in the water of

IOP Conf. Series: Earth and Environmental Science **334** (2019) 012046 doi:10.1088/1755-1315/334/1/012046

the ozone [11]. Although ozone is capable of reducing total organisms, the total microorganisms will rise again during storage. This is because the longer the storage time the damage from the material will increase so that trigger the growth of microorganisms and other damages. The use of low temperature and the type of packaging can inhibit the growth of microorganisms during storage. Packaging clingwrap has transparent, high tensile strength, not easily torn, resistant to acids and bases except oxidizing acids, resistant to oil, alcohol and petroleum solvents so as to inhibit the growth of microorganisms.

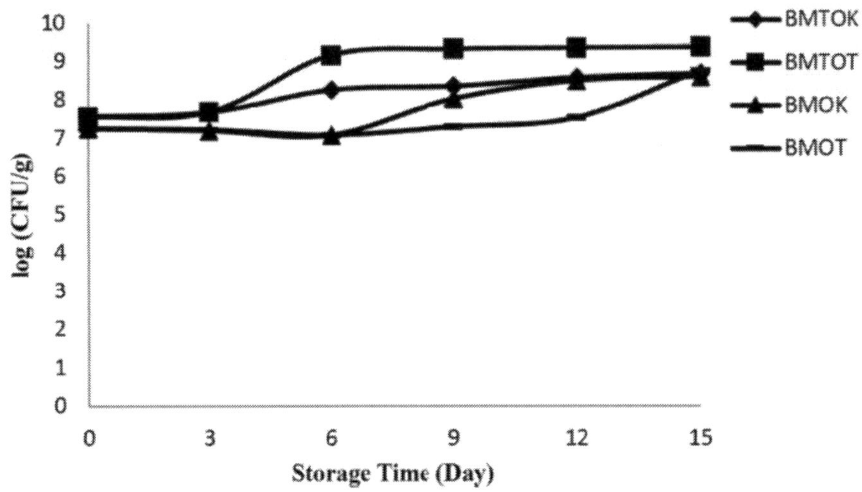

Figure 6. Curve Total Amount of Red Onion Microorganisms Peel Ozonation and Without Ozonation

3.4. Organoleptic Characteristics

Color. Based on observations presented in Figure 7, the peeled onion ozonation results which are packed in plastic clingwrap has a small decline compared to the onion peeled using ozonation and without ozonation results were not packed with plastic clingwrap during 15 day storage. The decrease preference for color values correspond to color values a * (red) as shown in Figure 3. The decline in the value of the color, a * (red) during storage at each treatment possibilities for onion is still experiencing metabolism during storage, in which the respiration activity is still happen. Respiration generates heat which causes the increase of heat, so the process of decline such as water loss, withered, and growth of microorganisms will increase [12].

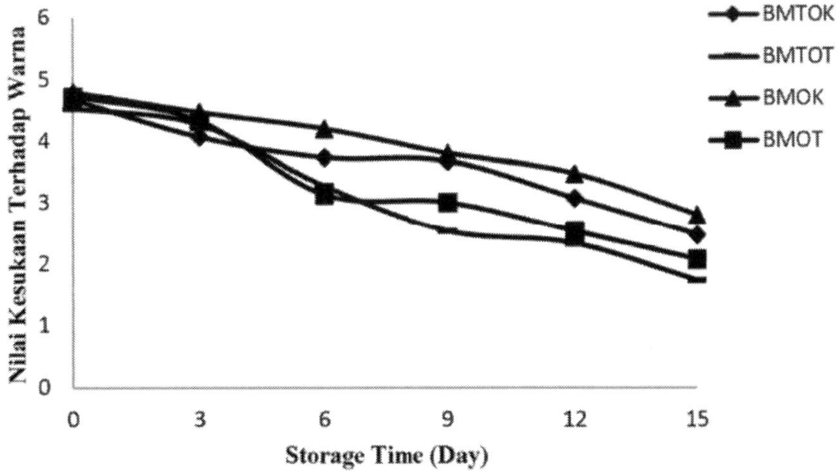

Figure 7. Curve Level Curve on Red Onion Color Result of Ozonation and Without Ozonation

Aroma. Based on observations presented in Figure 8, onion peeled ozonation results are packed in plastic clingwrap small decline compared to treatment color red onions peeled and without ozonation results were not packed with plastic clingwrap during 15 day storage . The decrease in the preferred value to the scent corresponds to the VRS value as shown in Figure 5. The decrease in VRS levels can be caused by increasing damage as the length of storage such as the sprout bulbs. [8] argues that the sulfur precursor of flavor can be used for metabolic activity and shoot growth so that the number decreases and causes a decrease in flavor production. The typical aroma of onion will undergo changes during storage depending on the amount of allicin compounds. The aroma will increase as allinase enzyme activity increases and decreases with decreasing allinase enzyme activity [2].

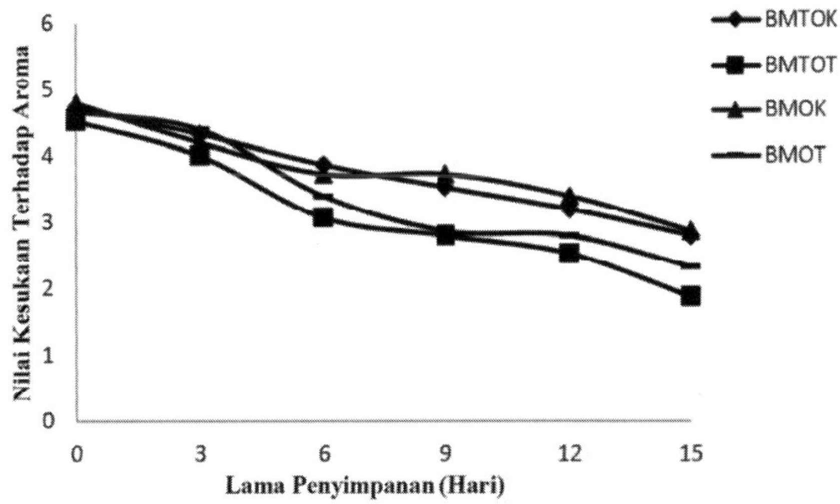

Figure 8. Curve level curve on onion scent result of ozonation and without ozonation

Overall Appearance. Based on observations presented in Figure 9, the peeled onion ozonation results are packed in plastic clingwrap has a small decline compared to treatment color red onions peeled and without ozonation results were not packed with plastic during storage clingwrap in 15 day. The decrease in the preferred value to the overall appearance corresponds to the a * (red) and VRS color values as shown in FIGS. 3 and 5. Clingwrap has transparent, high tensile strength, is not easily torn, resistant to acids and bases except oxidizing acids, resistant to oil, alcohol and petroleum solvents, this packaging has a thickness of 0.01 inches. Styrofoam has the properties to prevent leakage and maintain the form of food when held, maintaining the freshness and integrity of packaged and lightweight materials [13]. Ozone treatment can inhibit the occurrence of damage if used in the right concentration, but if used with excessive concentration can damage the structure of the network. Ozone with the right concentration is able to inhibit the occurrence of damage to vegetables, both in terms of color, texture, aroma, but ozone with very high concentration can cause damage to tissue on vegetable surface due to oxidation reaction. The use of clingwrap has a function to maintain the freshness of vegetables and vegetables. Packaging of fresh vegetables will be better if combined with minimal processing.

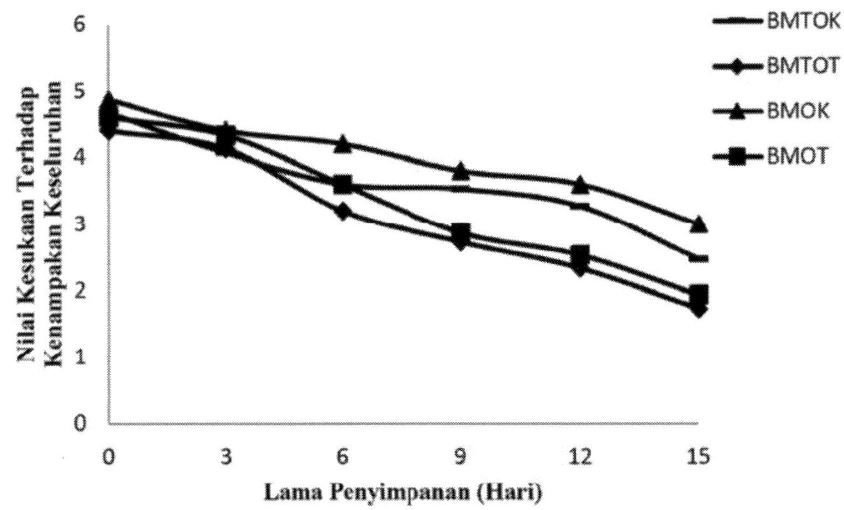

Figure 9. Curve of readability level on overall appearance of peeled onion by ozonation and without ozonation

4. Conclusion

During the storage for 15 days at 5±2 ^{0}C, the peeled onion of 0.8 ppm ozonation which is packed with clingwrap plastic showed less changes in physical, chemical, microbiological, and organoleptic characteristics than that of the peeled onion without ozonation packed with plastic clingwrap. peeled Onion with ozonation and packed with clingwrap plastic had shrink weight of 4.51%. The color is bright brown (L*: 52.52, a*: 12.65, b*: -3.2), the VRS level decreases by 10.29 µgrek/g and total number of microorganism noodles decreased by 1.17 log CFU/g. The overall color, aroma, and appearance are not favored by panelists with consecutive values of 2.80; 2.87; 3.0. The peeled Onion without ozonation that is not packed with clingwrap plastic experience weight loss of 12.30%. The color is bright red brown (L*: 60. a*: 8.34, b*: -065). VRS levels decreased by 7.78 µgr / g and the total number of microorganisms increased by 9.38 log CFU / g. Color, aroma, overall appearance is not favored by panelists with consecutive values of 1.73; 1.87; 1.73.

Acknowledgments
The research carried out for this study was funded by ALG Research Team and Rector Unpad

References

[1] Koswara S 2009 Teknologi Pengolahan Sayuran dan Buah-Buahan *e-Book Pangan*.

[2] Mutia A K, Purwanto Y A, Pujantoro L 2014 Perubahan Kualitas Bawang Merah (*Allium ascalonicum L.*) Selama Penympanan pada Tingkat Kadar Air dan Suhu yang Berbeda *Jurnal Pascapanen* **11(2)**: 108-115.
[3] Asgar A, Sugiarto A T 2009 Pengaruh Pencucian dengan Larutan Ozon Terlarut Kualitas Sayuran Selama Penyimpanan *Balai Penelitian Tanaman Sayuran* Bandung.
[4] Pradhana A Y, Hasbullah R, Purwanto Y A 2013 Pengaruh Penambahan Kalium Permanganat Terhadap Mutu Pisang (CV. Mas Kirana) pada Kemasan Atmosfer Termodifikasi Aktif *Jurnal Pascapanen* **10(2)**: 83-94.
[5] Miller F, Silva C, Brandao T 2013 Ozone Based Treatment for Fruit and Vegetables Preservation *Food Eng. Rev 5* pp. 77-106.
[6] Muchtadi D 1992 Petunjuk Laboratorium Fisiologi Pasca Panen Sayuran dan Buah-buahan *Departemen Pendidikan dan Kebudayaan Direktorat Jenderal Pendidikan Tinggi Pusat Antara Universitas Pangan dan Gizi* Institut Pertanian Bogor Bogor.
[7] Buckle K A, Edward R A, Fleet G H, Woodon N 1985 Ilmu Pangan. Terjemahan Hari Purnomo dan Adiono *UI Press* Jakarta.
[8] Mutia A K 2015 Penyimpanan Bawang Merah (*Allium ascalonicum L.*) pada Suhu Rendah dan Tingkat Kadar Air Awal yang Berbeda *Disertasi* Sekolah Pasca Sarjana Institut Pertanian Bogor.
[9] Endrasari R, Prayudi B 2011 Teknologi Penanganan Pascapanen Bawnag Merah dan Aspek Teknologi Produk Olahannya untuk Peningkatan Nilai Tambah di Kabupaten Brebes *Risalah Hasil Pengkajian* Jawa Tengah (ID) Balai Pengkajian Teknologi Pertanian Jawa Tengah.
[10] Cataldo F 2003 On The Action of Ozone on Proteins *Polymer Degradation and Stability* **82**: 105–114.
[11] Graham D M 1997 Use Of Ozone For Food Processing *Food Technology* **51**: 72-75.
[12] Kitinoja L, Kader A A 2003 Praktik-Praktik Penanganan Pascapanen Skala kecil: Manual untuk Produk Holtikultura (Edisi ke 4) Penerjemah: Utama I M S *Udayana-press* Bali.
[13] Sulchan M, Nur E 2007 Keamanan Pangan Kemasan Plastik dan Styrofoam Available at http://indonesia.digitaljournals.org/. Created on 27 August 2017.

IOP Conf. Series: Earth and Environmental Science **334** (2019) 012047 doi:10.1088/1755-1315/334/1/012047

A Study of Bread Yeast and Sucrose to the Successful Process of Cocoa Beans Fermentation

S T Sahannur

Study Program of Agrotechnology, Faculty of Agriculture, Universitas Muslim Indonesia, Jl. Urip Sumoharjo Km 5, Makassar, Sulawesi Selatan, 90231 Indonesia

E-mail: siti_sabahan@yahoo.com

Abstract. This research aims to know the effect of the concentration of sucrose and bread yeast to the successful process of cocoa beans fermentation. The research was conducted in the form of Completely Random Design of Factorial Pattern, two Factors. The first factor is sucrose consisted of 0% (control), 1%, 2%, and 3% sucrose. The second factor was the bread yeast consists of two (2) concentrations: 0,5% and 1 %, and each repeated twice to obtain 16 experimental units. The observed parameters were fermentation index, slaty beans and the pH of cocoa beans. The result of the research showed that there was an interaction between sucrose and bread yeast in the successful process of cocoa beans fermentation in small scale. The treatment without sucrose and adding 1% of bread yeast had a better effect to the fermentations index value of 1,72; salty beans were 1,56 and pH beans was 5,30.

1. Introduction

Cocoa is an industrial crop that has potential as a producer of foreign exchange in non-oil and gas sector. The economic value of cocoa beans is significant in its contribution to the life of society and the source of foreign exchange. Therefore, the development of cocoa is continuing both in cultivation and post-harvest aspects. Indonesia is the third largest producer and exporter cacao in the world, after Ivory Coast and Ghana, with a total area of 1.65 million hectares of plantations and the total production of 440,000 tons in 2010/2011; 87 percent is produced in local farms [1]. The fundamental disadvantage in the production of local cocoa is low quality and failure to achieve the properties required in international standards of trades [2] .

The farmers mostly produce the production of cocoa seeds in Indonesia. Around 965.000 farmers are involved in cocoa farming. In 2005, there was 887.735 ha (89,45%) cocoa plantation in Indonesia as smallholdings of cocoa farmers. Meanwhile private extensive plantation covering 54, 737 ha (5,51%) and state extensive plantation only 49,976 ha (5,04%). Thus, the cocoa farmers contribute about 90% of national production. However, from the existing cocoa plantation in Indonesia, the value of its productivity is still low: 897 kg/ha/year, whereas the potential productivity usually reaches more than 2,000 kg/ha/year [3].

Even though it has a good marketing prospect, the commodities of Indonesian cocoa are still facing a problem on the image of poor quality that could affect to the competitiveness of product in the international market. It is because the cocoa seed still dominates the production of Indonesian cocoa has not been fermented. It still appeared in dirt seeds and contaminated with insect and fungi. Processing cocoa, in essence, is an attempt to process cocoa into dried cocoa seeds that fulfill the standard of quality and can elicit the distinctive characteristic of cocoa, especially the taste. Stage of processing that considered as the most influence to the quality of dried cocoa seed is fermentation. Fermentation of cocoa seeds aims to destroy the pulp and cultivate the condition for biochemical reactions in the seeds, which contribute to the formation of the taste precursors and the color of chocolate. The crushed pulp will be smooth to escape from the seed, forming pulp fluid dripping out from the seeds [4].

Lopez and Dimic stated the fermentation of cocoa beans is the process of transforming the sucrose into alcohol and then into organic acids performed by a consortium of fermenting microbes [3]. Those organic acids will induce the reaction of enzymatic in the seeds; it results in biochemical change that will form a compound that gives a smell, taste, and color to the cocoa beans [5].

The non-fermented dry cocoa beans have the different characters to the wet cocoa beans when the fermentation is used. The dried cocoa beans have lost some content of water and substrate which is the requirement in the fermentation of cocoa. The water will combine the enzyme and substrate in the seed, so the hydrolysis and oxidation process of compound taste, color, and the smell of cocoa will happen there. The content of water required in cocoa fermentation is more than 35%. The substrate is the material that is recognized by microbes during the fermentation process. Substrates in fermenting cocoa seeds are sucrose and citric acid that contained in the pulp [6]. During the process of fermentation, the pulp will be transformed by microbes into organic acid. Acids will diffuse into the beans and induce enzymatic reactions to form the compounds of taste, aroma, and color [7]. To optimize the process of fermentation, it is required conditions that help the process of fermentation such as dissolved substrates and microbes.

Therefore, it is necessary to research about the use of bread yeast and sucrose that is more applicable so that optimization efforts can be implemented mainly by the farmers. Successful fermentation of cocoa beans using bread yeast , and sucrose is determined by its concentration. The Research Objective is to study the influence of bread yeast with sucrose in optimizing the process of fermentation of cocoa beans.

2. Research Method

The experiment was conducted by applying Complete Randomized Design (CRD) with a factorial pattern in two (2) factors. The first factor was sucrose (P) with concentration: 0% (p0), 1% (p1), 2% (p2), 3% (p3). The second factor of bread yeast (R) is 0.5% (r1), and 1% (r2). The total of eight treatment combinations was repeated twice, so there were 16 experimental boxes.

2.1. Materials and Tools

The ingredients used in the study are: Lindak type cocoa, bread yeast (*fermipan*), sucrose, fermentation box size 40 cm x 30 cm x 30 cm, oven, thermometer, and gunny sack. Chemicals for analysis include: ethanol 70%, N-hexane, methanol 90%, and NaOH 0.1 N. The tools used are: pH-meter, blender, analytical scale, measuring flask, UV-Vis Type spectrophotometer, measuring cylinder, electric oven, pipette, Petridis plate, magnetic stirrer, Whatman 42 filter paper, test tube, Erlenmeyer, and glassware.

2.2. Observation

2.2.1. The Fermentation Index

The dried cocoa beans are meased and sieved with size> 35 mesh, then weighed as much as 0.5 grams. The sample was put into a100 ml breaker glass, add a fermentation index solution (FI) (97 parts of methanol ±three parts 30% (HCl) of 50 ml. The FI solution obtained from 3 ml of HCl was inserted into a 100 ml measuring flask, then added methanol to the limit and shaken until homogeneous. 100 ml breaker glass containing cocoa samples and FI solution covered by aluminum foil for refrigeration for one night after storage. The solution is filtered with filter paper to be separated by raffinate, and the filter is prepared in cuvette then measured. The absorbance value was measured by using a spectrophotometer at a wavelength of 460 nm and 530 nm to calculate its fermentation index.

$$FI = \frac{abs\ \lambda\ 460\ nm}{abs\ \lambda\ 530\ nm}$$

2.2.2. Salty beans with the cut test method
The test is done by observing the changes color visually and subjectively. A total of 100 cocoa beans were split longitudinally right in the middle into two (2). From 100 parts of the beans were observed one by one the color of cocoa beans based on its classification [8]. In this study, the classification classified into three (3) classes where the *slat* color was inserted into unfermented beans classes, the dominant purple to brown color is classified into the underfermented bean's class, and the dominant brown enters the fermented bean's class. The percentage of the three (3) classifications were *slaty*, purple, and brown calculated the percentage with the following formula*(9,10)*:

$$\%\ Unfermented\ beans\ = \sum \frac{slices\ of\ slaty\ beans}{total\ part\ of\ cocoa\ beans} \times 100\ \%$$

$$\%\ Underfermented\ beans = \sum \frac{slices\ of\ purplr\ beans}{total\ of\ cocoa\ beans} \times 100\%$$

$$\%\ Fermented\ beans = \sum \frac{slces\ of\ chocolate\ beans}{total\ of\ cocoa\ beans} \times 100\%$$

2.2.3. pH of dried beans
Implementation procedure, sample test is taken as many as 12 *beans* up to 20 *beans*, outer skin separated, then milled using a blender. The test sample is weighed 10 g into the cup glass, add 90 ml of hot distilled water (70° C to 80° C), stir slowly until a suspension is formed which must be free of lumps. The filtrate is filtered and cooled till the room temperature is (27-29)° C and the filtrate pH is determined as soon as possible at that temperature. The results are expressed according to the readings indicated by the pH-meter for the filtrate.

3. Result and Discussion

3.1. Result
Fermentation Index. Based on the result of research showed that giving sucrose had a significant effect, while the addition of bread yeast and the interaction of sucrose and yeast did not significantly affect the fermentation index. The average fermentation index at various concentrations of sucrose is presented in Table 1.

Table 1. The average of fermentation index

Sucrose Concentration	Bread yeast Concentration		Average	HSD 0.05
	r1 (0.5%)	r2 (1%)		
p0 (control)	1.71	1.73	1.72^{a}	0.23
p1 (1%)	1.67	1.33	1.50^{a}	
p2 (2%)	1.15	1.14	1.14^{b}	
p3 (3%)	1.10	1.10	1.10^{b}	
Average	1.41	1.33		

Note: the number in the line followed by different letters (a,b,c) means significantly different with the Honestly significance Difference (HSD) 0.05

Based on the Honestly Significance Difference (HSD), results on Table 1 shows the index fermentation value in the treatment of the average addition of sucrose >1. The treatment without the addition of sucrose, fermentation index value was 1.72 but not significantly different from the sucrose concentration of 1% (1.50), and it was significantly different in 2% and 3% (1.14) of the sucrose concentration with 1.10 of fermentation index.

3.1.1. Slaty Beans

Based on the research result showed that sucrose adding was not significant. While the bread yeast is adding and the interaction of sucrose and bread yeast had a very significant effect on the number of slaty beans. The average of slaty beans in various concentrations of sucrose and yeast is presented in Table 2.

Table 2. The average of the number of *slaty* beans

Sucrose Concentration	Bread Yeast Concentration		Average	HSD 0,05
	rl (0.5%)	r2 (1%)		
p0 (control)	2.70^{a}_{x}	1.56^{b}_{x}	2.13	
pl (1%)	1.40^{b}_{x}	1.70^{a}_{x}	1.55	0.75
p2 (2%)	1.70^{b}_{x}	1.20^{b}_{x}	1.45	
p3 (3%)	1.60^{b}_{x}	1.70^{a}_{x}	1.65	
Average	1.85	1.54		

Note: The numbers on the line followed by different letters (a,b) means significantly Honestly significance Difference (HSD) 0.05)

Based on the result of Honestly Significance Difference (HSD) results in Table 2 shows that interaction of 2% sucrose treatment with 1% bread yeast addition resulted in the lowest *slaty* beans of 1.20 significantly different from the addition of 0.5% bread yeast in all sucrose concentrations. The average number of most *slaty* beans obtained in cocoa beans that are not given sucrose with only 0.5% of bread yeast was 2.70.

3.1.2. pH Cocoa beans

The results showed that the addition of sucrose had no significant effect, while the addition of bread yeast had a significant effect and the yeast and sucrose interaction had a very significant effect on the pH of cocoa beans. The average pH of cocoa beans at various concentrations of bread yeast and sucrose is presented in Table 3.

Based on the result of HSD 0,05 in Table 3 shows that 1% of sucrose addition and 0.5% of yeast bread added result lower pH (4.50) of cocoa beans significantly different with 1% of bread yeast concentration in (5.45). Higher average pH values were obtained in non-sucrose cocoa beans in either 0.5% or 1% of bread yeast with values of 5.20 and 5.30 respectively.

Table 3. The pH average of Cacao beans

Sucrose Concentration	Bread Yeast Concentration		Average	HSD 0,05
	r1 (0.5%)	r2 (1%)		
p0 (control)	5.20^{a}_{x}	5.30^{a}_{x}	5.25	
p1 (1%)	4.50^{a}_{x}	5.45^{a}_{x}	4.98	0.49
p2 (2%)	5.15^{b}_{y}	4.90^{b}_{x}	5.03	
p3 (3%)	4.95^{a}_{x}	5.05^{a}_{x}	5.00	
Average	4.95	5.18	5.06	

Note: The numbers on the line followed by different letters (a, b) means very different in Honestly Significance Difference (HSD) (0.05).

3.2. Discussion

The purpose of fermentation is to release the pulp from the seeds to facilitate the drying process, and the seed shell is easily removed from seed pieces. Also, fermentation aims to kill seeds and provide opportunities for the process leading to the formation of colors, flavors, and aroma, and lower levels of non-fat ingredients so that the relative fat levels will increase [11].

The cocoa seeds fermentation process generally takes place naturally by microorganisms present in the fermentation atmosphere which lasts for six days with reversals on day two and at every 24 hours [12, 13, 14, 7]. At the beginning of 24 hours, bread yeast fermentation dominates the fermentation that will break the sugar component inside the pulp into alcohol [5, 12].

The results showed that the addition of sucrose and bread yeast at the beginning of fermentation of cocoa seeds significantly affect the fermentation index, slaty seed, and pH of cocoa beans. The addition of sucrose and bread yeast at the beginning of fermentation yields cocoa seeds with a fermentation index greater than 1 (Table 1). This shows that cocoa seeds have complete fermentation due to the addition of yeast at the beginning of fermentation. The cocoa fermentation process generally takes place naturally by microorganisms found in the fermentation atmosphere [12, 13, 14, 7]. At the time the fruit is broken, the pulp is immediately contaminated by microorganisms in the surrounding air so that the pulp fermentation process will happen soon. High sugar content with high acidity (pH 3.5) because the content of citric acid is an ideal condition for the growth of microorganisms.

Dry cocoa beans with FI values (fermentation index) equal to one and above one (FI $\geq$1) are considered perfectly fermented [15, 16]. Fermentation of cocoa beans with fermentation index values below one (FI<1) indicates imperfect fermentation. The fermentation index is a brownish size of the cocoa beans to ensure the level of fermentation of the beans [17]. According to Kongor et al., the potential of cocoa beans flavor can be recognized from the quality of fermentation through a color index called fermentation index [18]. Qualitatively the perfection of the fermentation process can be seen from the color change of cocoa bean pieces. The measurement of fermentation index by spectrophotometry is the ratio of brown pigment absorbance (λ: 460 nm) and purple (λ: 530 nm). The measurements were based on flavonoid levels that gave brown color with anthocyanin giving purple color [15, 19].

Table 2 shows that there is an interaction between sucrose and bread yeast with the number of slaty seeds. By adding 2% of sucrose and 1% of bread, yeast is the best treatment for the number of slaty seeds because the lowest value is 1.20. This happens because the addition of 2% sucrose can provide energy for yeast growth to perform activities optimally during the fermentation process. According to the Indonesian National Standard (SNI 2323-2008), the number of slaty beans is a maximum of 3% (quality I-B) [8]. Slaty beans meant unfermented beans showed half or more slices of grayish-gray slices, such as; slate or grayish-blue and textured stable and reliable like cheese. Slaty beans will give astringent flavor and excessive bitter and low taste of cocoa beans [20]. Polyphenol compounds are the determinants of the color of the beans. The presence of polyphenols is not only responsible for the formation of bitter taste and astringent flavor but also causes the characteristic of brown color from fermented cocoa beans [20].

The quality of fermented cocoa beans is determined mainly by the acidity (pH). The cocoa processing industry requires seed pH between 5.2-5.8 to produce quality cocoa butter [21]. The results of the research on the pH of dry cocoa beans showed that there was an interaction between giving bread yeast and sucrose to the pH of the cocoa beans. The highest average pH value of 5.45 is found in the addition of 1% sucrose and 1% yeast. The excellent pH value of cacao beans is close to neutral (pH> 6) so that Brown-specific compounds can form intensively and acidic beans have a pH <5.0 [22]. The fermentation process made a pH change of 3.5 on fresh beans before fermentation to about 4.8 on the 3rd day of fermentation and eventually to about 5.5 in dried beans. The addition of yeast increases the number of microbes in the seed pulp, thus increasing the ability of yeast to break down sucrose into alcohol. The higher the concentration of yeast added the more carbohydrates are converted into glucose, alcohol, lactic acid, and other compounds. pH is closely related to total acid since in general the increase in total acid is followed by a decrease in pH.

4. Conclusion

Based on the result of the research, it can be concluded that there is an interaction between sucrose and bread yeast with the success of the cocoa beans fermentation process. The treatment without sucrose and the addition of 1% bread yeast had a better effect on the fermentation index value of 1.72, the number of slaty beans was 1.56, and the pH of the beans was 5.30.

Suggestion

To optimize the process of fermentation of small-scale cocoa beans it is recommended to add bread yeast at the beginning of fermentation with a concentration of 1%, but still, it needs to do more research by increasing the yeast concentration above 1% to obtain optimal concentration.

Acknowledgments

The authors would like to thank the Ministry of Research and Technology and Higher Education who have funded the research in the Competitive Grant scheme so that this research can be done.

References

[1] Standard and Trade Development Facility (STDF) 2014 Panduan Teknis Pelatihan Bagi Fasilitator Utama."Cocoasafe" *Pembangunan Kapasitas dan Kerjasama Pengetahuan Tentang Sanitari dan Fitosanitari Kakao di Asia Tenggara* (STDF/PG/381) Indonesia Edition.

[2] Sabahannur St, Mursalim, Asrul L, Bilang M 2015 Use of High Performance Liquid Chromatography (HPLC) for the Analysis of Amino Acid of Sulawesi and Local Clone Cocoa Bean Fermentation *Journal of Food Research* **4(4)**: 120-126

[3] Wahyudi T 2009 Panduan Lengkap Kakao Jakarta Timur *Penebar Swadaya* pp. 364

[4] Bangkit D W, Ganda P, Wayan I A 2013 Pengaruh Penambahan Ragi Tape dan Waktu Fermentasi terhadap Karakteristik Cairan Pulpa Hasil Samping Fermentasi Biji Kakao *Jurusan Teknologi Industri Pertanian* Fakultas Teknologi Pertanian Universitas Udayana.

[5] Lopez A S, Dimick P S 1995 Cocoa Fermentation. In: Reed, G., Nagodawithana TW. (Eds), Enzymes, Biomass, Food, and Feed *Weinheim: VCH* pp. 561–577

[6] Wood G A R, Lass R A 2001 Cocoa. 4th ed. London Longman.

[7] Lehrian D W and Patterson G R 1983 Cocoa Fermentation, In G. Reed (ed) Biotechnology

Comprehensive Treatise **5**: 529-575

[8] Efendi S, Hardjosuwito B 1988 Penetapan Derajat Fermentasi Biji Kakao Berdasarkan Indeks Fermentasi dan Uji Organoleptik Bogor *Menara Perkebunan.*

[9] Senanayake M, Errol R J, Buckle K A 1996 Effect Of Different Mixing Intervals on the Fermentation of Cocoa Beans *J. Sci. Food Agric.* **74**: 42-48

[10] Kustyawati M E, Setyani S 2008 Pengaruh Penambahan Inokulum Campuran Terhadap Perubahan Kimia dan Mikrobiologi Selama Fermentasi Cokelat *Jurnal Teknologi Industri dan Hasil Pertanian* **13(2)**: 73-84

[11] Misnawi, Jinap S, Nazamid S, Jamilah B 2002 Activation of Remaining Key Enzymes in Dried Under-Fermented Cocoa Beans And Its Effect On Aroma Precursor Formation *Food Chemistry* **78**: 407–417

[12] Yusianto, Winarno H, Wahyudi T 1997 Mutu dan Pola Citarasa Beberapa Klon Kakao Lindak *Pelita Perkebunan* **13(3)**: 171-187

[13] Schwan RF 1998 Cocoa Fermentations Conducted With A Defined Microbial Cocktail Inoculum *J. Microbiol* **14**: 1477-1483

[14] Hii C L, Rahman R A, Jinap S, Man Y B C 2006 Quality of Cocoa Beans Dried Using A Direct Solar Dryer T Different Loadings *J. Sci. Food Agric* **86**: 1237-1243

[15] Hashim P, Selamat J, Muhammad S K S, Ali A 1998 Effect of Mass and Turning Time on Free Amino Acids, Peptide-N, Sugar, And Pyrazine Concentration During Cocoa Fermentation *J. Sci. Food Agric.* **78**: 543-550

[16] Gourieva K B, Tserevitinov O B 1979 Method of Evaluating the Degree of Fermentation of Cocoa Bean USSR Patent No: 646254

[17] Alamsyah T S 1991 Peranan Fermentasi dalam Pengolahan Biji Kakao Kering *Suatu Tinjauan Berita Perkebunan* **1(2)**: 97-103

[18] Takrama J F, Aculey P C, Aneani F 2006 Fermentation of Cocoa With Placenta: A scientific study *Proceedings of 15th International Cocoa Research Conference* (Costa Rica) **2**: 1373-1379

[19] Kongor J E, Takrama J F, Budu A S, Brown H M, Afoakwa E O 2013 Effects of Fermentation and Drying on the Fermentation Index and Cut Test of Pulp Pre-Conditioned Ghanaian Cocoa (*Theobroma cacao*) Beans *Journal of Food Science and Engineering* **3**: 625-634

[20] Ilangantileke S G, Wahyudi T, Bailon M A G 1991 Assessment Methodology To Predict Quality Of Cocoa Beans For Export *J. Food Qual.* **14**: 481-496.

[21] Misnawi, Jinap S, Nazamid S, Jamilah B 2002 Activation of Remaining Key Enzymes in Dried Under-Fermented Cocoa Beans and Its Effect On Aroma Precursor Formation *Food Chemistry* **78**: 407–417

[22] Sulistyowati 1998 Keasaman Biji Kakao dan Masalahnya *Pelita Perkebunan* **3** 151- 158

Fresh Cheese Probiotic with Local Isolate *Lactobacillus casei* 2.12 as starter in fermentation

W S Putranto[1], M T Suhartono[2], H D Kusumaningrum[2], P EGiriwono[2], A Z Mustopa[3], K Suradi[1] and H Chairunnisa[1]

[1]Faculty on Animal Husbandry, Universitas Padjadjaran
[2]Department of Food Science and Technology, Bogor Agricultural University, Indonesia
[3]Research Centre for Biotechnology, Indonesian Institute of Science (LIPI), Cibinong Bogor

E-mail : wendrysp@yahoo.co.id

Abstract. The local isolates of Lactic Acid Bacteria (LAB) have not been optimally developed in the cheese making process. The aim of this research was to know the clotting activity and the percentage and characteristics of curd produced during milk fermentation using local isolate of BAL *L casei* 2.12. The results showed that during 12 hours incubation the value of milk clotting activity (MCA) was 8.33 SU / mL, the curd percentage was 12.5% (w / v) with curd firmness (++++) very good and Total Plate Count 1.14. 10^9 cfu / g. These results indicate the use of *L casei* 2.12 as a starter LAB probiotic capable of causing coagulation of casein and produce fresh cheese probiotic.

1. Introduction

Cheese is a dairy product that is very popular with Indonesian people, even though it is not an authentic Indonesian food. Cheese was found in the fertile region of the Euphrates and Iraq Tigris rivers 8,000 years ago at Revolution Agriculture. Many kinds of food products are added to cheese to add flavor, so the demand for national cheese is quite fantastic reaching 8 million tons per year. National cheese production has not fulfilled this demand with various obstacles faced, especially the availability of milk-clotting enzymes namely rennin or rennet which is a proteolytic enzyme.

Protease is an enzyme that is able to hydrolyze peptide bonds in proteins. Based on the Nomenclature Committee of The International Union of Biochemistry, proteases are classified as hydrolase enzymes (EC.3.4). Protease is a very important enzyme in the food and non-food industries. Utilization of proteases in the food industry is to reduce turbidity in the beer industry, reduce gluten in the bread industry, and to agglomerate milk in the cheese industry. Protease enzymes can be obtained from plant, animal and microbial tissue. The limitations of the ability of animals and plants to fulfill protease requests have led to the development of microbial proteases.

The production of rennin from calves is decreasing in the world, resulting in high prices of these enzymes, so efforts are made to find alternative enzymes (substitute). Protease enzymes other than rennin which has rennin-like activity, namely milk clotting activity (Milk Clotting Activity), are known as Rennin Like Protease (RLP). RLP sources can be obtained from plant and microbial sources. Plant sources have characteristics with high proteolytic activity.

There are proteases produced by microbes that have characteristics similar to rennin from the stomach of the calf and produce curd products from milk as produced by animal rennin so that the protease is known as Rennin Like Protease (RLP). *Rhizomucor miehei, Rhizomucor pusillus, Endothia parasitica, Aspergillus oryzae,* and *Irpex lacteus* have been commercially produced to produce RLP [1]. The Milk Clotting Enzymes (MCE) was produced from *Bacillus* [2,3,4]. *Bacillus stearothermophillus* showed the milk-clotting activity of 24.23 SU/ml [5].

Microbes that are native Indonesian germplasm are still not optimally developed into biological agents in industries based on biotechnology, so this study is expected to contribute to the development of local microbial utilization, especially Lactic Acid Bacteria (LAB) for the production of Rennin Like Protease enzymes (RLP) needed by the national cheese industry and reducing the import of these enzymes. Local Lactic Acid Bacteria (LAB) isolated from traditional Indonesian fermented food products have not been developed and mass produced to produce rennin.

2. Materials and Methods

2.1. Screening Skim Milk Agar

L casei 2.12 was screened using MRS (deMan Rogosa and Sharpe) agar medium containing 3% skim. *L casei* 2.12 colonies capable of coagulating casein.

2.2. Milk Clotting Activity (MCA)

As much as 2.5 mL of the substrate (10% skim milk in 10 mM $CaCl_2$) was incubated for 5 min at 37 °C followed by adding of 0.5 mL enzyme extract. Measurement of time length was started from the addition of enzyme extract to the formation of the first particles. MCA was calculated as: SU = 2400×5×D/T×0.5 (1) as described by Kawai and Mukai (1970), in which T is milk-clotting time (s), and D is dilution of the enzyme. One Soxhlet unit (SU) of milk-clotting activity was defined as the amount of enzyme required for clot formation of 1 mL of substrate in 40 min at 35 °C.

3. Results and Discussion

The use of local LAB isolates in the process of curd formation in cheese making is still not widely used. Isolate *L casei* 2.12 which was isolated from the milk of Ettawa goat had the ability to coagulate milk (casein) with the MCA activity of the crude enzyme produced was 18 SU / mL. This activity is not much different from the MCA activity of *Enterococcus faecium* 1.15 which is 20 SU / mL [6]. MCA testing is quantitative testing of quantitative enzyme activity. The main factor that determines the value of MCA is the velocity of milking (skim). The faster coagulation occurs, the higher the MCA value. A total of 5 ml (10% skim solution in 0.01 M $CaCl_2.2H_2O$) was added an enzyme to be tested as much as 0.5 ml. Next, incubate at 37 °C in the water bath and record the formation time of curd (t).

Table.1. Milk clotting activity crude enzyme *L casei* 2.12

No	Isolate	Sources	*Clotting zone* (cm)	Milk Clotting Activity (SU/mL)
1.	2.12	Ettawa Goat Milk	1.5	18

Coagulation of casein can occur due to acids or enzymes, namely rennin. Enzymatic coagulation of milk is caused by the breakdown of the peptide bond between phenylalanine and methionine (105th sequence with 106) from the kapa casein polypeptide, resulting in k-casein and macropeptide [7,8]. Macropeptides produced from these cuts are soluble in water, while the casein will settle. Instability in casein miselae results from the breakup of Phe and Met bonds so that other casein fractions settle. The mechanism of casein clotting due to enzymatic activity (rennin) is different from casein clotting due to acid. Clumping because rennin in the initial phase (primary) occurs the interaction between rennin and k-casein (a), then the peptide bond breaks between phenylalanine and methionine (105th sequence with 106) of the casein kapa polypeptide, producing para-k-casein and macropeptide (b), and then macropeptide aggregation

3.1. Making Fresh Cheese with Fermentation

Isolate *L casei* 2.12 is a class of LAB which is a safe bacteria, so that in the process of forming curd can be used in the form of a starter with a fermentation process.

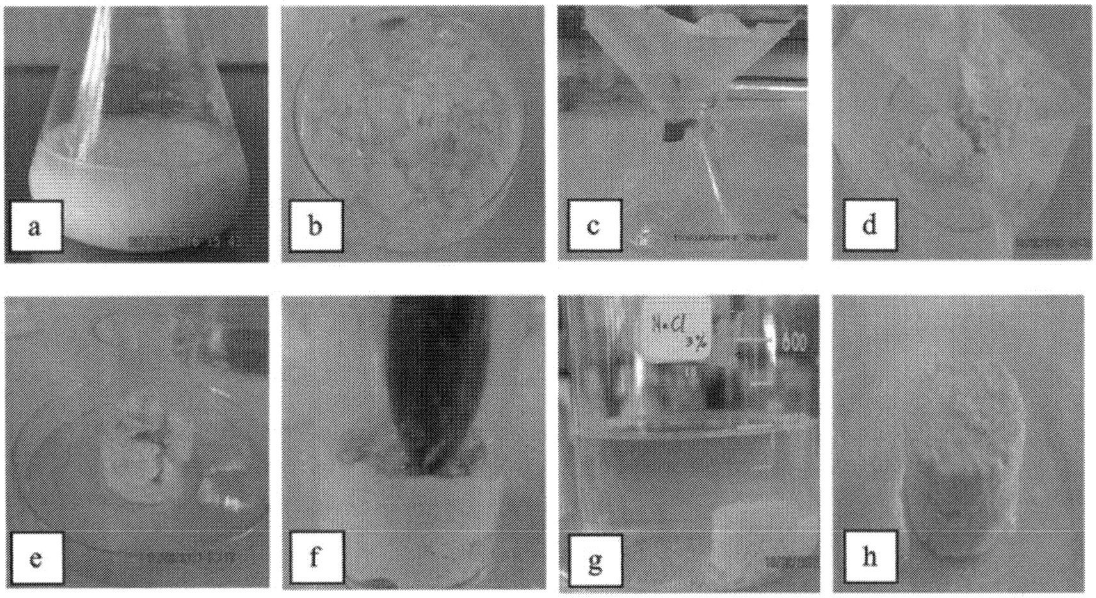

Figure 1. Processing of fresh cheese : fermentation with starter L casei 2.12 (1%) at 37^{0}C, 12 h (a), characteristic of curd (b), pressing (c, d, e, f), salting (3%) (g), fresh cheese (h).

The results showed that during 12 hours incubation the value of milk clotting activity (MCA) was 8.33 SU / mL, the curd percentage was 12.5% (w / v) with curd firmness (++++) very good and Total Plate Count 1.14. 10^9 cfu / g.

4. Conclusion
The use of *L casei* 2.12 as a starter in the process of making cheese is very potential with curd production reaching 12.5% (w/v).

References
[1] Jacob M, Jaros D, Rohm H 2011 Recent advances in milk clotting enzymes *Int. J. Dairy Technol.* **64**: 14–33.
[2] Wu F C, Chang C W, Shih L L 2013 Optimization of the production and characterization of milk-clotting enzymes by *Bacillus subtillis* natto *Springer Plus* **2**: 33-42.
[3] An Z, He X, Gao W, Zhao W, Zhang W 2014 Characteristic of miniature cheddar-type cheese made by microbial rennet from *Bacillus amyloliquefaciens*: A comparison with commercial calf rennet *J. Food Sci.***79**: 214-221.
[4] Lemes A C, Pavon Y, Lazzaroni S, Rozycki S, Brandelli A, Kalil S J 2016 A new milk-clotting enzyme produced by *Bacillus sp*. P45 applied in cream cheese development *LWT-Food Science and Technology* **(66)**: 217-224.
[5] Ahmed S A, Wehaidy H R, Ibrahim O A, Hasan S A, El-Hofi M A 2016 Novel milk-clotting enzyme from Bacillus stearothermophillus as a coagulant in UF-white soft cheese *Biocatalysis and Agricultural Biotechnology* http://dx.doi.org/10.1016/j.bcab.2016.06.011.
[6] Wendry S P, Kusmajadi S, Hartati C, Apon Z M, Puspo E G, Harsi D K, Maggy T S 2017 Enterococcus faecium 1.15. isolated from bakasam showed milk clotting activity *Annales Bogoriensis* **21**: 9-14. http://dx.doi.org/10.14203/ab.v21i1.293
[7] Suhartono M T 1989 Enzim dan Bioteknologi *Departemen Pendidikan dan Kebudayaan* Direktorat Jendral Pendidikan Tinggi Antar Universitas Bioteknologi IPB Bogor.
[8] Fox P F, Guinee T P, Cogan T M, McSweeney P L H 2000 Fundamentals of cheese science *Aspen Publishers Inc.* Gaithersburg MD.

Quality Protein, Viscosity, Gel Strength and Structural Morphology of Sheepskin Gelatin Catalyzed HCl With Different Concentrations

M Hasdar[1], Y D Rahmawati[2], Y Erwanto[3] and Rusman[3]

[1]Departement of Food Science and Technology, Faculty of Agriculture, Muhadi Setiabudi University
[2]Departemen of Science of Nutrition, Faculty of Healt Science, Muhadi Setiabudi University
[3]Departement of Animal Product Technology, Faculty of Animal Science, Gadjah Mada University

E-mail : hasdarmuhammad@gmail.com

Abstract. Sheep skin has a high content of collagen and can be an alternative to gelatin making. Utilization of gelatin increasingly widespread so that it becomes life style. gelatin is often used for food or non-food. This study aims to determine the effect of HCl concentration on the quality of sheepskin gelatin. Sheep skin obtained from slaughterhouses in Brebes district. Parameters in this study were yield, protein, viscosity, gel strength and gelatin morphological structure. Different HCl concentrations (0.5%, 1% and 1.5%) are used to catalyze collagen from sheepskin. Hydrolysis of sheep skin collagen was performed for 4 hours with temperature 55-66°C. The results showed no significant effect on quality of protein and yield but significant effect on viscosity and gel strength. The observations by scanning electron microscopy (SEM) show a very fine morphological structure of gelatin produced with 1% and 1.5% HCl catheters. Gelatin catalyzed with HCl 1% and HCl 1.5% was appropriate standard GMIA.

1. Introduction

Gelatin is a type of collagen protein obtained by extracted from the skin. One of the alternative ingredients of gelatin processing is sheepskin. Sheepskin has a high content of collagen and can be an alternative to make gelatin. Gelatin includes a unique protein because it is capable of forming a thermo-reversible gel and is easily soluble in water [1]. Utilization of gelatin increasingly widespread to become a lifestyle. Gelatin is often used for food or non-food. Sheepskin gelatin should have the appropriate characteristics if it will be applied in the food or non food industry and should be in large quantities and the price is also cheap so it can be produced continuously. The use of chemical solutions to help the process of skin collagen extraction. Usually use acid or alkaline chemical solutions.The quality of gelatin catalyzed with an acidic chemical solution is preferable when compared to a gelatin catalyzed with alkali chemicals. The use of a solution of hydrochloric acid as a catalyst of sheep skin collagen protein can convert triple helices of collagen into a single chain in a short time and produce more collagen [2]. Immersion in an acid solution to collagen can produce gelatin polymers with glycine as the main constituent. Setting the concentration of hydrochloric acid solution in catalyzing collagen can affect the quality of gelatin produced.This study aims to determine the effect of HCl concentration on the quality of sheepskin gelatin.

2. Materials and Methods

The raw materials are prepared through a series of preliminary beam house operations from raw sheep skin, weighing, washing, soaking, discarding, discarding wool, disposing of fat to a clean sheep's skin (no hairs) and conditioned in a neutral atmosphere. Sheepskin without hair is cut into small pieces 2x2 cm. These pieces of sheepskin are used as raw materials of gelatin. Small cut sheepskin was immersed in 0.5%, 1% and 1.5% (v/v) HCl solutions for 4 hours. After the immersion process is complete, the skin is washed with running water repeated three times until the pH is neutral. The washed skin is then extracted temperature 55-60°C for 4 hours. Comparison of sheepskin: aqueous solution = 1: 3 for each treatment. Then do the filtration process to get collagen solution. The gelatin solution obtained for each ± 300 ml was poured into a 30.5 cm x 30.5 cm container, then dried in a 60° C oven for 48 hours. The gelatin obtained is then mashed using a blender. The parameters used in this study were yield, protein content [3], gel strength and viscosity [4], and Scanning electron microscopy [5]. This research uses Completely Randomized Design (RAL) unidirectional pattern. The treatments were concentration of HCl solution 0.5%, 1% and 1.5% (v / v) and replication 5 times. If there is a marked difference between treatments, then a real difference test according to Duncan's Multiple Range Test [6].

3. Results and Discussions

3.1. Yield

Based on the result of data analysis in Table 1. showed the difference of HCl concentration in sheep sheep catalysis there was no difference (P> 0,05) to sheep skin gelatin value. The average yield value generated in this study ranged from 23.10-23.33%. this shows the difference in HCl concentration having the same effectiveness in producing gelatin. In principle, the acid solvent causes the structure of collagen protein in sheepskin expands and opens so as to facilitate the process of collagen hydrolysis into gelatin [7].Not the yield difference may be caused by long time soaking in an acid solution and the same 4-hour long extraction is also the same 4 hours. This process occurs due to the number of H^+ ions hydrolyze more collagen, while the longer the extraction time can lead to the disintegration of collagen which is more into gelatin. Extraction serves as the advanced stages of the hydrogen bonds between molecules damage tropocollagen that at the time of preparation not previously been decomposed by the acid. H_2O molecules can denaturant hydrogen bonds in tropocolagen [2].

Table 1. Average of yield, protein content, gel strength, viscosity of sheepskin gelatin with different concentration HCl treatment

Parameters	Concentration of HCl			Commercial Gelatin	GMIA Standard
	0,5%	1%	1,5%		
Yield (%)	23.21 ± 0.21[a]	23.10 ± 0.11[a]	23.33 ± 0.15[a]	-	-
Protein content (%)	90.93 ± 0.02[a]	90.59 ± 0.03[a]	90.98 ± 0.01[b]	90.02 + 0.01	-
Gel strength (Bloom)	371.908 ± 0.637[a]	411.269 ± 0.586[b]	64.821 ± 0.750[c]	191.416 ± 0.362	50 – 300
Viscosity (cP)	13.55 ± 0.05[a]	4.44 ± 0.06[b]	2.51 ± 0.01[c]	29.15 ± 0.06	1.5 - 7.5

Different letters within same column denote significant differences (P < 0.05). Mean ± SD from duplicate determinations

3.2. Protein Content

Different concentrations of HCl have the same effectiveness in producing protein content. This is seen in the results of statistical analysis showing that protein content in this study did not have differences (P> 0.05) between treatments, protein content of this study is directly proportional to yield. The average value of the resulting content proteins is 90.59% - 90.98% even still higher when compared with protein content of commercial gelatin. This shows the process of HCl catalysis of sheepskin collagen is very good, so as to make the sheep skin to expand which ultimately facilitate the process of collagen hydrolysis. In principle, the quality of the gelatin protein is strongly influenced by the hydrolysis process, the extraction process and the use of chemical solutions as the main treatment [2].

3.3. Gel *Strength*

The gel strength of sheepskin gelatin produced in this study showed a significant difference (P>0.05) between treatments. Sheepskin gel treated with HCl 1.5% solution had a median strength of 64,821 Bloom and still met the GMIA standard, whereas sheepskin gel gelled with 0.5% HCl solution resulted in average gel strength 371,908 Bloom and sheepskin gelatin which catalyzed with a 1% HCl solution having an average gel strength of 411,269 Bloom. Gelatin gel strength of 0.5% HCl solution catalyst and 1% HCl solution did not meet GMIA standard. This phenomenon occurs due to HCl solution that is able to convert triple helical sheep skin collagen into single chain but display the amino acid chain different between treatments. The main factor that distinguishes the quality of gel strength is the length of its amino acid chain, and if its collagen condition has been perfectly hydrolyzed, resulting in a long polypeptide chain of hydroxyproline that increases gel strength [8].

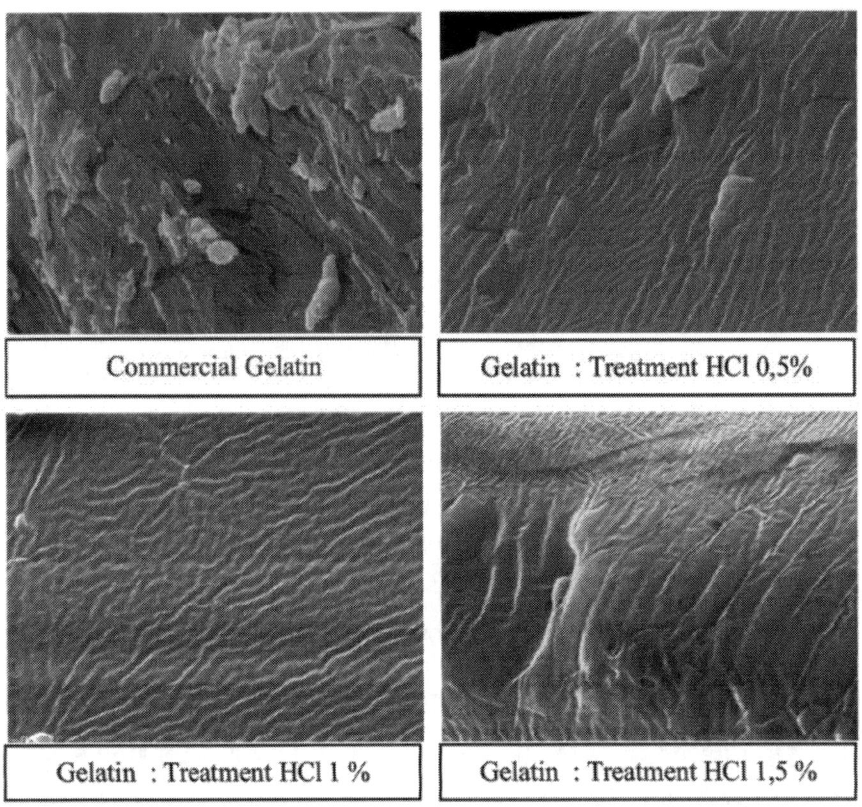

Figure 1. Structural Morphology of sheepskin gelatin with Scanning Electron Microscopy (SEM) 5000x magnification

3.4. *Viscosity*

The results of viscosity analysis presented in Table 1. showed a significant difference (P <0.05) between treatments, which showed a decrease in the mean value of viscosity as the concentration of HCl solution as catalyst increased. Visible viscosity values of GMIA are sheepskin gelatin catalyzed with 1% HCl and 1.5% HCl. The mean value of sheep skin gelatin viscosity catalyzed by HCl solution in this study is still better when compared with the value of commercial gelatin viscosity. This event is caused by the suspected decrease in the viscosity value of gelatin produced due to the decrease of pH in the addition of HCl concentration causing the amino acid chain to be shorter so that the viscosity of the gelatin becomes lower [9]. The high concentration of HCl in the acid process will have an effect on the sheep skin collagen change. Collagen structure is the main compiler is hydroxyproline swelling and become spread so causing the resulting viscosity value changes. Changing collagen chain structure causes a decrease in the molecular weight of gelatin [8].

3.5. Structural Morphology

Based on the results of Scanning electron microscopy (SEM) can be seen differences in morphological structure of sheepskin gelatin. The gelatin morphological structure in the catalyst with 1% HCl solution is noticeably smoother and denser without the presence of collagen protein clumps. In contrast to the gelatin SEM results categorized with 0.5% HCl there is still a lump of protein and there are still cracks. Results of gelatinized SEMs categorized with 1.5% HCl were still better than 0.5% catalytic gelate of HCl. The result of commercial SEM gelatin still shows the number of collagen protein clumps and uneven surfaces. The SEM results provide a reinforcing explanation that the sheepskin gelatin that catalyzes with 1% HCl solution is better.

Clumps of protein will make gelatin morphological structure becomes uneven, while cracks caused by not compact the collagen into gelatin protein binding. This morphological structure will describe the choice of using gelatin to be applied to the next stage.

4. Conclusion

Gelatin catalyzed with HCl 1% and HCl 1.5% was appropriate standard GMIA. Gelatin catalyzed with HCl 1% showed a finer morphological structure based on SEM results.

Acknowledgements
Acknowledgments to DRPM DIKTI through research grants of Higher Colleges Cooperation in 2018 so that this research works well.

References

[1] Arnesen J A, Gildberg A 2002 Preparation and characterization of gelatin from the skin of harp seal (*Phoca groendlandica*) *Bioresource Technology* **82**: 191–194
[2] Hasdar M, Rahmawati Y D 2017 The variance of strong acid and long time Soaking solutions on quality of pH and protein gelatin of sheep skin *Journal Agrisaintifika* **1(2)**: 88 – 96
[3] Badan Standarisasi Nasional 1992 Cara uji makanan dan minuman Jakarta
[4] Gómez-Guillén M C, Turnay J, Fernández D M D, Ulmo N, Lizarbe M A, Montero P 2002 Structural and physical properties of gelatin extracted from different marine species: A comparative study *Food Hydrocoll* **16**: 25-34
[5] Goldstein J I, Djeil E N, Philps E, Diel C J, Andrean D R, Chonseta E L, Cerozan F, Lifshin 1992 Scanning electron microscopy and X-ray microanalysis : A text for biologist, materials Scientist, and cytologists, 2nd ed. *Plemun Press* New York pp. 820
[6] Steel R G D, Torrie J H 1991 Principles and procedures of statistics *McGraw-Hill Book Co.Inc*, New York
[7] Gudmundsson M, Hafsteinsson H 1997 Gelatin from cod skins as affected by chemicaltreatments *Journal of Food Science* **62(1)**: 37–39
[8] Aewsiri T, Benjakul S, Vinessanguan W, Tanaka M 2008 Chemical compositions and functional properties of gelatin from pre-cooked tuna fin *International Journal of Food Science and Technology* **43**: 685-693
[9] Sukkwai S, Kijroongrojana K, Benjakul S 2011 Extraction of gelatin from bigeye snapper (Priacantuhus tayens) skin for gelatin hydropolysate production *International Food Research Journal* **18(3)**: 1129-1134

IOP Conf. Series: Earth and Environmental Science **334** (2019) 012050 doi:10.1088/1755-1315/334/1/012050

Antibacterial Activity of Ethanol Extract of *Nigella Sativa* L. Seed Against *Streptococcus mutans*

T Rostinawati[1], S Karipaya[1], and Y Iskandar[1]

[1]Faculty of Pharmacy, Department of Pharmacy Biology, Padjadjaran University, Jl. Raya Bandung Sumedang km 21, Jatinangor, 45363, West Java, Indonesia

E-mail: t.rostinawati@unpad.ac.id

Abstract. Dental caries or cavities are most commonly found in the oral cavity. Dental caries is caused by plaque-forming bacteria such as *Streptococcus mutans*. Getting dental care with a professional dental expert every two weeks or three-monthly intervals can prevent caries but it is expensive and difficult for most individuals. So an alternative is sought, especially in herbal plants. Black cumin (*Nigella sativa* L.) from the *Ranunculanceae* family has been used as a medicinal plant for various diseases for more than 2000 years. This study aims to determine the antibacterial effect of *Nigella sativa* L. seed ethanol extract on *Streptococcus mutans* and to find out at what concentrations *Nigella sativa* L. seed ethanol extract provides antibacterial activity. This research was carried out by extracting *Nigella sativa* L. seed using ethanol solvent. Activity tests are carried out using paper disc diffusion method. The results of inhibition zones obtained from concentrations of 400, 600, and 800 mg/mL of black cumin extract were 13.3, 16.4 and 20.3 mm, respectively. Furthermore, the minimum inhibitory concentration test (MIC) for *Streptococcus mutans* bacteria was obtained at a concentration of 380 mg/mL.

1. Introduction

Dental caries is a disease of the hard tissues of the teeth most commonly found in the oral cavity. Caries experience varies greatly between countries, depending on age, lifestyle, diet and socio-economic conditions. Results of Basic Health Research in 2013 in the field of dental and oral health showed that the active caries prevalence of the Indonesian population was 43.4% [1].

Dental caries or tooth decay is a progressive and irreversible microbial disease characterized by fermentation of bacterial carbohydrates that produce acid production causing destruction of hard tissue in the teeth. Among various oral microorganisms, *Streptococcus mutans* (*S. mutans*) has been identified as a plaque-forming bacteria that is capable of producing dental caries in humans. Several attempts have been made to eliminate *S. mutans* from oral flora. There is no convincing evidence that careful mechanical debridement by patients has reduced the incidence of tooth decay, even though the procedure can reduce the amount of dental plaque and inflammation level of the gums (inflammation of the gingiva or gum). However, if mechanical debridement is performed by a dental professional once every two weeks or quarterly intervals, and includes fluoride treatment, both decay and periodontal disease basically stop happening. The level of dental debridement that is carried out professionally is very labor intensive so that the cost is expensive and unable to afford by most individuals[2].

Black cumin (*Nigella sativa* L.) is from the *Ranunculanceae* family has been used as a medicinal plant for more than 2000 years. This plant has light purple or white flowers and the fruit is capsule shaped, with small white and trigonal seeds. Once cooked, the capsule opens and the seeds turn black [3]. *Nigella sativa* L. seed contains essential oils, saponins, polyphenols, fatty oils, fatty acids and bitter nigelon, nigelin and thymokinon substances. Black cumin seeds are widely used as a medicine for various diseases. These seeds have been shown to strengthen the immune system, prevent blockages in blood vessels, reduce cholesterol, improve heart performance, improve memory and concentration, increase hormone bioactivity, neutralize toxins, overcome sleep and stress, warm the stomach, improve the digestive tract, anti-inflammatory, launch breast milk, rejuvenate skin cells and have anti-tumor activity [4]. Various studies have shown that *Nigella sativa* L. seed have activity against several bacteria such as *E. coli, P. aeruginosa, S. flexneri, S. typhimurium, S. enteritidis and S. aureus* [4]. In previous study performed *Nigella sativa* L. seed extracted with ethanol showed antibacterial activity against Methicillin resistant Staphylococcus aureus (MRSA) [4]. Therefore, this study was conducted to complement previous research by conducting antibacterial activity of ethanol extract of black cumin seeds against *S. mutans* causing dental caries. Therefore, the aim of this study was to determine total phenolic and total flavonoid contents in water, ethanol and chloroform extracts of *Baeckea frutescens* and also establish their potential effects on health promotion as antioxidant and antimicrobial agents.

2. Materials and Methods

2.1. Bacteria

Streptococcus mutans bacteria (ATCC 35668) was obtained from the Faculty of Dentistry, Padjadjaran University.

2.2. Preparation of Nigella sativa L seed extracts

Nigella sativa L. seed was extracted using 96% ethanol by maceration method. The fine powder of black cumin seeds was dissolved in 96% ethanol solvent for 24 hours and the solvents were replaced every 24 hours for three times then a thick extract was obtained. Organoleptic and phytochemical screening of black cumin seed extract was carried out as a qualitative measurement in determining the secondary metabolites of *Nigella sativa* L. seed extract included with water content testing.

2.3. Antibacterial activity assay of Nigella sativa L. against S. mutans

Antibacteial activity of *Nigella sativa* L. against *S. mutans* was performed with agar diffusion method. Bacterial suspension of *S. mutans* with optical density which was equal to 0.5 Mc Farland standard was incubated in Muller Hinton Broth (MHB) at 37 °C for 24 h. The sterile cotton swab was dipped in bacterial suspension in several times press heavily on the inside wall of the tube above the fluid level. Bacteria was inoculated in the dried surface of a Mueller-Hinton agar (MHA) plate by streaking the swab over the entire sterile agar surface. Then, paper disc was put on the MHA dan loaded with 20 µL of 200 mg/mL, 400 mg/mL, 600 mg/mL and 800 mg/mL. Plate was inverted and incubated at 37 °C for 24 h. Antibacterial activity was determined by the inhibition zone formed around the disc paper containing extract of *Nigella sativa* L. seed. The inhibitory zone was determined in millimeters (mm) using the calipers [5].

2.4. Minimum Inhibition Concentration assay of Nigella sativa L. against S. mutans

Minimum inhibition concentration was performed with agar dilution plates. The extract was incorporated into the agar medium (MHA), with each plate containing a different concentration of the extract i.e. 360, 370, 380, 390 and 400 mg/mL. The inoculum can be applied rapidly and simultaneously to the agar surface. A total of 100 uL of 107 CFU/mL bacterial suspensions was placed on the surface of the agar (10 cm diameter Petri dish) and spread evenly using a spreader. Allow the inoculated plates to stand at room temperature until the moisture in the inoculum spots has been absorbed into the agar, until the spots are dry, but no more than 30 mins. Then, the plates were inverted and incubated at 35±2 °C for 18 h. MIC was noted as the lowest concentration that inhibit the growth of bacteria [6].

2.5. Equality test of Nigella sativa L. extract activity against amikacin in growth inhibition of S. mutans

This equality test used MHA media. A number of 100 uL of 10^7 CFU/mL bacterial suspensions was evenly distributed on the surface of agar media. Inoculated plate was left at room temperature for no more than 30 min. Then, the paper discs were placed on the inoculated plate surface. The paper discs were 20 µL loaded with serial concentration of amikacin i.e. 100 µg/mL, 200 µg/mL, 300 µg/mL, 400 µg/mL and 500 µg/mL and *Nigella sativa* L. extract i.e. 400 mg/mL, 600 mg/mL, 800 mg/mL, 1000 mg/mL and 1200 mg/mL. The Petri dish was incubated 37 °C for 24 h. The diameters of the inhibiton zone were measured with the caliper.

3. Results and Discussion

3.1. Nigella sativa L seed extract

Organoleptic test result was performed in Table 1. Whereas the organoleptic of *Nigella sativa* L. seed oil extract showed yellowish brown, aromatic oder and agreeable taste [7]. Water content of *Nigella sativa* L. ethanol extract was 5% (v/v). The water content in the extract must not be more than 10% [8].

Tabel 1. Organoleptic of *Nigella sativa L.* seed extract

Organoleptic properties	
Color	Blackish red
Taste	Bitter
Odor	Typical fragrance
Form	Viscous liquid

The ethanol extract of *Nigella sativa* L. seed contained the component some secondary metabolites i.e. alkaloids, tannin, terpenoids, saponin and steroids. Secondary metabolites such as alkaloids, tannin, terpenoids, saponin and steroid were also found in black seed of Nigella sativa L. [9].

3.2. Antibacterial activity assay of Nigella sativa L. against S. mutans

The *Nigella sativa* L. seed extract had performed bacterial activity with diameter of inhibition zones i.e 13.3+0.54, 16.4+0.75, and 20.3+ 0.35 for extract concentrations 400, 600 and 800 mg/mL respectively 'figure 1'. From this result to determine the MIC value, the concentration range was started between 200-400 mg/mL.

Figure 1. Antibacterial activity of *Nigella sativa* L. seed ethanol extract against S. mutans

3.3. Minimum Inhibition Concentration assay of Nigella sativa L. against S. mutans

MIC value of *Nigella sativa* L. seed extract was 380 mg/mL. At this concentration *Nigella sativa* L. seed extract inhibited the bacterial growth. The extracts that are from herb can be categorized as inhibitors of bacterial growth are extracts that have less than 100 mg/mL of MIC, moderate 100-500 mg/mL, weak 500 -1000 mg/mL [10]. So, the extract of *Nigella sativa* L. seed had moderate bacterial activity to *S. mutans*. Whereas the diethyl ether extract of *Nigella sativa* L. seeds (25-400 μg extract/disc) had bacterial activity to *Staphylococcus aureus, Pseudomonas aeruginosa, Escherichia coli* and *Candida albicans* [11].

Tabel 2. Minimum inhibitory concentration of *Nigella sativa* L. seed extract against *S. mutans*

Extract concentration (mg/mL)	Growth of *S.mutans*
360	+
370	+
380	-
390	-
400	-

+ = bacterial growth, - = no bacterial growth

Tabel 3. Equality test of *Nigella sativa* L. extract activity against amikacin in growth inhibition of *S. mutans*

Extract concentration (mg/ml	Log (C)	Diameter of inhibition zone (mm)	Amikacin concentration (μg/ml)	Log (C)	Diameter of inhibition zone (mm)
400	2.6	13.8 ±0.61	100	2	13.8 ±0.61
600	2.78	16.8 ±0.32	200	2.3	16.8 ±0.32
800	2.9	21.9 ±0.55	300	2.48	21.9 ±0.55
1000	3	24.8 ±0.55	400	2.6	24.8 ±0.55
1200	3.08	26.6 ±0.66	500	2.7	26.6 ±0.66
y = 26.2x-54.1 r2 = 0.9951			y = 13.4x-14.7 r2 = 0.9936		

3.4. Equality test of Nigella sativa L. extract activity against amikacin in growth inhibition of S. mutans

Equality assay of *Nigella sativa* L. seed extract activity compared to Amikacin performed that 1 part of extract activity was equal to 3 x 10-4 part of Amikacin. This revealed that extract activity was still lower than Amikacin activity.

4. Conclusion

The ethanol extract of *Nigella sativa* L. seed had bacterial activity against *S. mutans* with MIC value i.e. 380 mg/mL and equality of activity of 1: 3x10-4 against Amikacin.

References

[1] Riset Kesehatan Dasar (Riskesdas) 2013 Pedoman Pewawancara Petugas Pengumpul Data Jakarta *Badan Litbangkes*, Depkes RI.

[2] Tanzer J 2001 The Microbiology of Primary Dental Caries in Humans *J Dent Educ,* **65(10):** 1028-1037

[3] Donmez A A, Mutlu B 2004 A new species of Nigella (*Ranunculaceae*) from Turkey *Bot J Linean Soc* **146:** 251-55

[4] Iran J 2014 Black cumin (*Nigella sativa*) and its constituent (*thymoquinone*) a review on antimicrobial effects *Basic Med Sci* **17(12):** 929–38

[5] Clinical Laboratory Standards Institute (CLSI) 2006 Performance standards for antimicrobial disk susceptibility tests; Approve Standard - Ninth Edition Document M2-A9 *Wayne PA* USA: *National Committee for Clinical Laboratory Standards* pp. 7-11

[6] Clinical Laboratory Standards Institute (CLSI) 2012 Methods for Dilution Antimicrobia Susceptibility Tests for Bacteria That Grow Aerobically Approved Standard—Ninth Edition Document M07-A9 Wayne PA USA: *National Committee for Clinical Laboratory Standards* pp. 13-161

[7] Youssef M K E, Eshak N S, Hana R S 2013 Physicochemical characteristics, nutrient content and fatty acid composition of Nigella sativa oil and sesame oil *Food Pub Health* **3:** 309–314

[8] Departemen Kesehatan Republik Indonesia (DepKes RI) 1977 Materia Medika Indonesia *Jakarta Departemen Kesehatan Republik Indonesia* Jakarta **1**: 138

[9] Desai S D, Shaik H S, Kusal K D, Haseena S 2015 Phytochemical Analysis of Nigella sativa and it`s Antidiabetic Effec *J Pharm Sci Res* **7(8):** 527-532

[10] Dall'Agnol R, Ferraz A, Bernardi A P, Albring D, Nor C, Sarmento L, Lamb L, Hass M, von Paser G, Schapoval E E 2003 Antimicrobial activity of some Hypericum species *Phytomedicine* **10(6-7)**

[11] Hanafi M S, Hatem M E 1991 Studies on the antimicrobial activity of *Nigella sativa* seed (black cumin) *J Ethnopharmacol* **34(2-3):** 275-8

IOP Conf. Series: Earth and Environmental Science **334** (2019) 012051 doi:10.1088/1755-1315/334/1/012051

1-Methylcyclopropene as an Effective Ethylene Inhibitor to Extend *Musa acuminata* Colla 'Muli' Postharvest Quality

S Mubarok[1], N Suwali[1], E Suminar[1] and NN Kamaluddin[2]

[1]Department of Agronomy, Faculty of Agriculture, Universitas Padjadjaran Jl. Raya Jatinangor Km 21 Bandung, Indonesia
[2] Department of Soil Science, Faculty of Agriculture, Universitas Padjadjaran Jl. Raya Jatinangor Km 21 Bandung, Indonesia

E-mail: syariful.mubarok@unpad.ac.id.

Abstract. *Musa acuminata* Colla 'Muli' is one of the most popular banana variety in Indonesia. As a commodity that is consumed fresh, is important to regulate the ethylene production of climacteric fruit in order to extend the shelf life and reduce deterioration. 1-MCP (1-Methylcycloprophene) is a potential chemical that can effectively regulate ethylene production in Muli bananas. In this study, 1-MCP in different concentrations (0, 0.25, 0.50 and 1.00 μL L^{-1}) were applied at two maturation stages to observe the effects. Lower 1-MCP concentration was effective in delaying fruit ripening on mature green stage shown by fruit firmness maintenance and improved shelf life up to 4.5 days longer than control. The application of 1-MCP was effective in delaying fruit ripening if applied prior to climacteric stage.

1. Introduction

Bananas are a type of horticulture crop famous as a consumable fruit in Indonesia. One of the most famous banana fruit in Indonesia is *Musa acuminate* Colla 'Muli'. It is a native banana in West Java and the most popular local banana in this area. It is similar with the cavendish banana however it characterized by a small fruit size, a yellow to orange fruit peel, and more sweetness than cavendish banana. Even though this fruit is small, it has a good taste and aroma and also contains high levels of macro and micronutrient such as vitamins and minerals that are important for human health and dietary. Muli banana has an economically important for farmer in Indonesia, however it has many problems such as in maintaining postharvest live quality. Compared to cavendish banana, Muli banana has not been investigated preciously, therefore the study on Muli banana is very important to resolve its problems.

For market purposes, many banana fruits have a problem such as the short of fruit shelf life, where the fruit shelf life relatively fast and finally it can't be sold freshly in the market. This effect directly reduces the fruit price and consumer acceptance. The most important thing in reducing banana fruit quality is ethylene, where it can accelerate fruit ripening therefore the fruit shelf life will be short. Ethylene is gaseous plant hormones that accelerates fruit ripening and senescence. Besides ethylene has an effect in accelerating fruit ripening, it has effect in regulating many plant growth and development process such as the plant response to the environmental conditions and stress [1]. Moreover, ethylene effects on seed germination, flower senescence, leaves drop, formation of root nodule [2].

Based on post-harvest fruit characteristics, bananas belong to the group known as climacteric fruit, where the respiration is increased followed by higher levels of ethylene production that occur at the onset of the fruit-ripening process [3]. During ripening process, changes in biochemical and physiological process occur that influence fruit quality and flavor, color, aroma, and texture [4]. In many fruit species, ethylene accelerates fruit ripening, which contributes to changes in the nutrient content and other compounds that has a positive effect to accelerate for fruit to be consumed, however, it has negative effects to accelerate quality deterioration by shortening the shelf life of the fruit. In banana fruit, fruit softening is one of the important factors for determining fruit quality and also harvesting time that are regulated by ethylene [5]. Chaves and Mello-Farias [6] stated that process of fruit ripening in climacteric fruit is a complex process.

To maintain banana fruit quality and to improve fruit shelf life, therefore the effect of ethylene must be minimized especially during transportation and storage. Some strategies have been conducted to minimize and prevent the ethylene effect on postharvest life of fruits such as by inhibiting ethylene biosynthesis or ethylene perception. To inhibit ethylene biosynthesis, some chemical compounds such as *Aminoethoxyvinylglycine* (AVG), Diazocyclopentadiene (DACP), Silver Thiosulfate (STS), etc. and also atmosphere modification have been widely used. Inhibiting ethylene biosynthesis is only effective in preventing endogenous ethylene, but it is not effective in preventing exogenous ethylene that affect in fruit ripening. 1-Methylcycloproene (1-MCP) has been found as a potential chemical compound that inhibits the ethylene perception of both endogenous and exogenous ethylene by inactivating ethylene receptor.

1-Methylcyclopropene (1-MCP) was developed in the early 1990's. It is not toxic and environmentally friendly, therefore it is potential used for edible crops such as fruit and vegetable. In many studies showed that 1-MCP have been widely used for improving postharvest life and quality of some horticulture crops such as rose, apple, banana, pears, onion, pelargonium, chrysanthemum and plums [7-10]. The affectivity of 1-MCP to prevent the ethylene effects depends on several factors such as cultivars, concentration, duration of exposure, temperature, stage of development and plant maturity [11, 12]. For instance in flower, the application of 1-MCP on young flowers is more effective than on old flowers in increasing the postharvest life of *Pelargonium* [7, 13]. High-quality of postharvest life of fruit is important characters for banana, therefore in this study we reported the strategy to improving postharvest life *Musa acuminata* 'Muli' fruit by the application of 1-MCP. The objective of this study was to evaluate the effect of different concentration of 1-MCP (0; 0.25; 0.5; 0.75 and 1 $\mu L.L^{-1}$) on two stages of banana fruit maturation (mature green and mature)

2. Materials and Methods

2.1. Fruits preparation and 1-MCP application

The experiment was conducted from February until March, 2017 at Laboratory of Horticulture, Faculty of Agriculture, Universitas Padjadjaran, Indonesia. Muli banana fruit were prepared and collected based on the similarity of the stages of fruit maturation, mature green and mature. These fruit were collected from the farmer, which is cultivated with traditional cultivation technique. Those fruit were treated with five different 1-MCP concentrations (0; 0,25; 0,5; 0,75 and 1$\mu L\ L^{-1}$). The application of 1-MCP was prepared according to the manufactures instruction. 1-MCP powder was placed in sealed chamber that containing banana fruits for 6 hours of incubation. After 1-MCP application fruits were collected and placed on the laboratory bench under the following conditions: temperature, 22 ± 2°C; humidity, 80% for fruit shelf life analysis.

2.2. Analysis of fruit firmness

Fruit firmness is an important characteristic for determining banana fruit quality. To analyze the effect of 1-MCP in the fruit firmness, it was tested using an Ultrasonic Hardness Tester model FHR-5 (Nippon Optical Works Co., Ltd., Tokyo, Japan). The measurement was done according to the manufactures instruction.

2.3. Analysis of total soluble solid (TSS)

Analysis of the Total Soluble Solids (TSS) was measured to estimate the sugar level in the banana fruits. It was measured using a digital refractometer DRBO-45nD type (B&C 32145, Germany). The sample was grinded and the fruit liquid was dropped into the sensor of refractometer. The estimation of TSS value was represented by the Brix value. The measurement was conducted two times at the beginning of storage or 0 days post storage (DPS) and 12 DPS.

2.4. The fruit weight lost

This analysis was evaluated over 12 days of post-harvest storage. At the beginning of storage (0 DPS), designated W_0, the fruits were weighed. This process was repeated at 12 DPS (W_n). The weight lost (W) by the banana fruit during post-harvest storage was calculated using the following equation:

$W = W_0 - W_n$ .... (1)

2.5. Percentage of fruit flesh

The fruit flesh and peel were weighted separately at the beginning of treatment (0 DPS) and 12 DPS. Based on these data the percentage of fruit flesh was calculating according the following equation:

$$\text{Percentage of flesh fruit (PF)} = \frac{\text{Flesh weigh}}{(\text{Fruit skin+flesh weight})} \times 100\% \text{ ....(2)}$$

2.6. Fruit shelf life analysis

To analyze fruit shelf life, fruits were stored on the laboratory bench under temperature at 22 ± 2°C and humidity at 80%. The fruit shelf life was determined by counting the days from the beginning of storage (0 DPS) when the fruit in the grade 0 until the fruit firstly decay as a grade 3. The characteristics of each grade as a following: 0 = Fruit still fresh, the color is green, and the black spot have not been appeared yet. 1 = Fruit still fresh, the color is turn to yellow, and the black spot have not been appeared yet. 2 = The base of fruit was dried, the color is yellow, and the black spot have not been appeared yet. 3 = Fruit starting to decay, the black spot have appeared and fruit still can be eat. 4 = Many black spots on the skin of the fruit have appeared, the fruit becomes soft, severe damage, and the fruit cannot be eat prepared fruit waste

2.7. Statistical data analysis

The statistical analysis was performed to analyze the effect of 1-MCP concentration for each fruit maturation stage. The completely randomized design was used in this experiment, which consisted of three biological replicates, and each replicate consisted of four individual fruit. For statistical data analysis, one factors analysis of variance (ANOVA) was conducted and followed by Duncan test at $p < 0.05$ to compare differences among investigated treatment.

3. Result and Discussion

3.1. 1-MCP was effective used to maintain the fruit firmness of mature green banana fruit

Fruit softening is one of the fruit characteristics that occurred during fruit maturation and ripening. It is represented by a firmness value, which is an important characteristic for fresh banana fruit. Based on the statistical data analysis of 12 days of post-harvest storage the fruit firmness of mature green fruit was harder than that of mature fruit, where the firmness value of the control from mature fruit was 11.50 kgf and that of mature green fruit was 0.30 kgf (Figure 1). These data indicated that the ripening process was correlated with a reduction in fruit firmness or an increase in fruit softening (ethylene regulates this process). To inhibit the effect of ethylene at maintaining fruit firmness, the application of 1-MCP was not effective when it was applied to mature fruit. It was effective on mature green fruit up until 12 days of post-harvest storage (Figure 1).

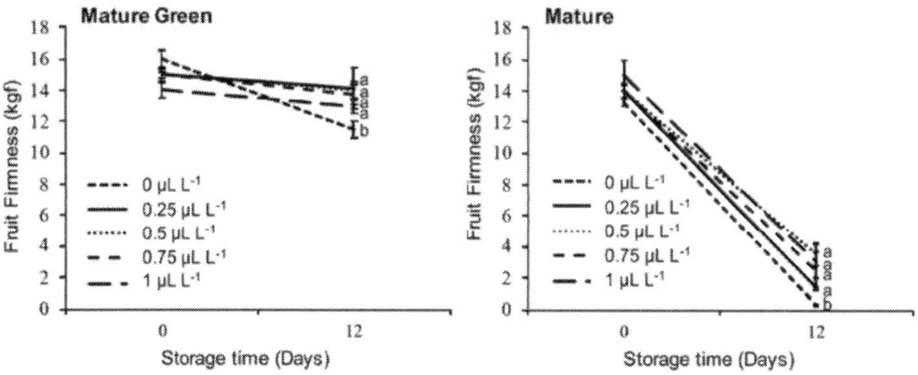

Figure 1. Effect of 1-MCP on fruit firmness of two stages of banana fruit maturity during 12 days of postharvest storage. The mean values in each stage of fruit maturation followed by the same letters are not significantly different according to the Duncan test at $p < 0.05$.

3.2. Total soluble solid of 1-MCP treated-fruit was lower than control

Our data showed that in all treatments at the beginning of storage (0 DPS), the TSS value of mature green fruit was lower than that of mature fruit; the TSS of mature green fruit was 5.40–6.00 Brix and that of mature fruit was 14.80–15.50 Brix (Figure 2). The application of 1-MCP significantly inhibited the accumulation of Brix content during 12 days of storage in both mature green and mature fruit. Our data (Figure 2) showed that during 12 days of post-harvest storage the TSS of 1-MCP-treated mature fruit was higher than that of 1-MCP-treated mature green fruit. The application of 1-MCP in all concentrations significantly reduced TSS over 12 days of post-harvest storage. A concentration of 0.25 µL L^{1} was sufficient to inhibit the increase TSS in both fruit maturation, mature green and mature (Figure 2).

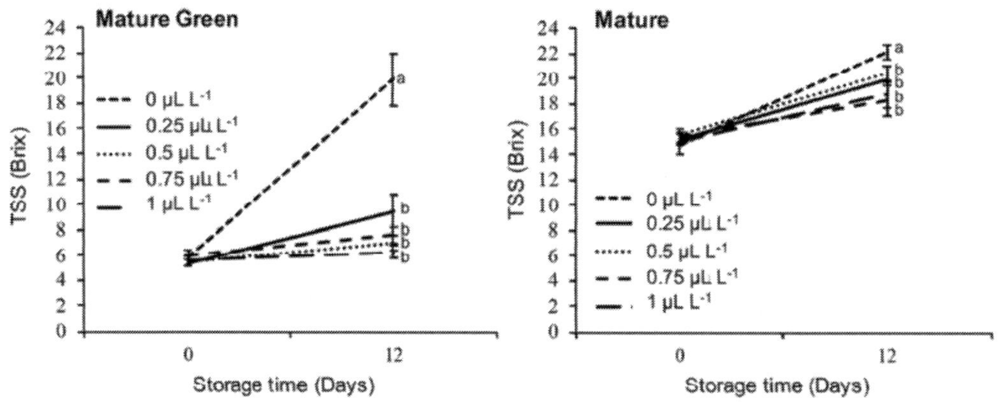

Figure 2. Effect of 1-MCP on TSS of two stages of banana fruit maturity during 12 days of postharvest storage. The mean values in each stage of fruit maturation followed by the same letters are not significantly different according to the Duncan test at $p < 0.05$.

3.3. 1-MCP did not affect the ratio of fruit flesh and peel in mature fruit

In this study we investigated the percentage of fruit flesh as an effect of 1-MCP application during postharvest storage. Based on the statistical data analysis showed that in the low concentration of 1-MCP application was only effective in mature green fruit for inhibiting fruit ripening by delaying the increasing the percentage of fruit flesh, where 0.25 µL L^{-1} has been enough and effective in delaying the increasing the percentage of fruit flesh 8.12% smaller than control at 12 days of postharvest storage that was significantly different compared with control according Duncan test at $p < 0.05$. However, in mature banana fruit, the application of 1-MCP up to 1 $\mu L\ L^{-1}$ was not effective in delaying the increasing the percentage of fruit flesh (Figure 3). Based on this value we assumed that high value of percentage of fruit flesh indicated that the diameter of banana fruit flesh bigger than fruit skin and the ethylene affect the ripening process of banana naturally.

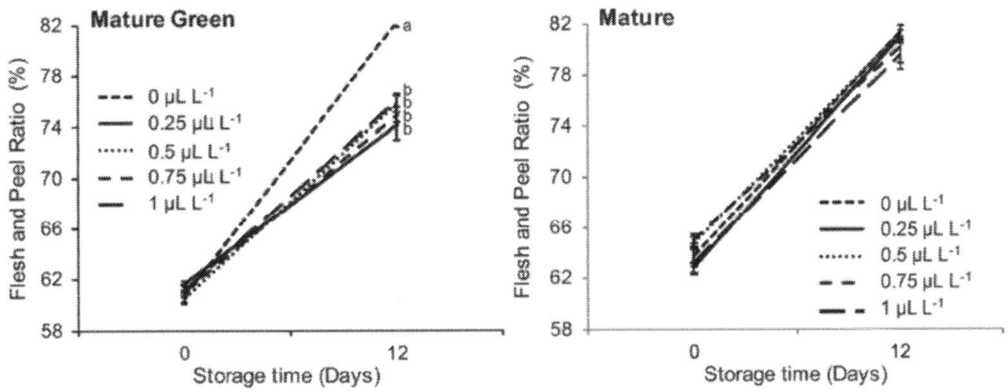

Figure 3. Effect of 1-MCP flesh and peel ratio of two stages of banana fruit maturity during 12 days of postharvest storage. The mean values in each stage of fruit maturation followed by the same letters are not significantly different according to the Duncan test at $p < 0.05$.

3.4. 1-MCP inhibited the weight loss of banana during postharvest storage

Our study demonstrated that the amount of weight lost during 12 days of storage was affected by fruit maturity. Mature fruit lost more weight (14.98 g) than mature green banana fruit (9.58 g). Water lost during post-harvest storage can be reduced by the application of 1-MCP. Our study demonstrated that for both types of banana fruit maturity, mature green and mature, 1-MCP significantly reduced the amount of water lost over 12 days of post-harvest storage. Furthermore, a low concentration $0.25\ \mu L.L^{-1}$ of 1-MCP was sufficient and effective at preventing water loss; the differences between the controls and mature green fruit and mature fruit were statistically significant. This concentration reduced water loss by 1.23 g (mature green) and 2.13 g (mature) compared with the controls during 12 days of storage (Fig. 4).

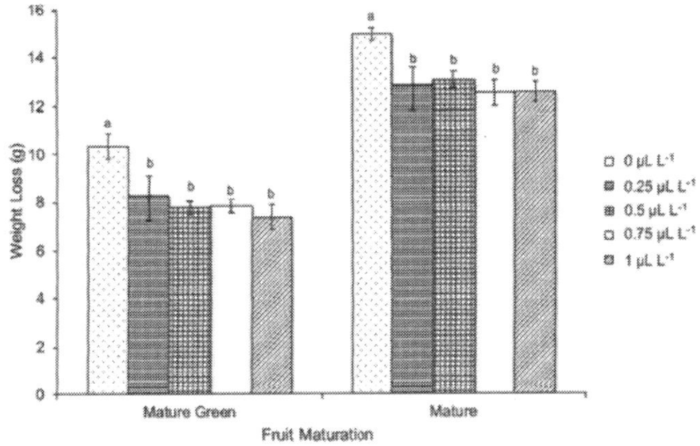

Figure 4. The effect of 1-MCP on the weight lost of two stages of banana fruit maturity during 12 days of postharvest storage. The mean values in each stage of fruit maturation followed by the same letters are not significantly different according to the Duncan test at $p < 0.05$.

IOP Conf. Series: Earth and Environmental Science **334** (2019) 012051 doi:10.1088/1755-1315/334/1/012051

3.5. The application of 1-MCP significantly improved postharvest life of Muli banana in both stages of fruit maturity, mature green and mature.

The application of 1-MCP was effective at increasing the post-harvest life of both stages of the Muli banana mature green and mature (Figure 5). The old stage of fruit maturity (mature stage) results in a shorter shelf life. We demonstrated that the mature-stage controls (0 μL L^{-1} of 1-MCP) decayed faster: they lasted for 2 days post-harvest; mature green fruit treated with 1-MCP lasted for 11 days (Figure 5). For mature green banana fruit, 1-MCP was also effective at improving fruit shelf life. We observed, however, that 1-MCP had a negative impact: it reduced skin color quality because the fruit never became yellow. The reduction in the post-harvest quality of the Muli banana can be inhibited by the application of 1-MCP, but the effect of 1-MCP depends on the stage of fruit maturity: 1-MCP was effective at improving post-harvest fruit shelf life. Based on statistical data analysis, we showed that for both types of fruit maturity, mature green and mature, the application of 1-MCP at a concentration of 0.25 μL L^{-1} was sufficient to delay fruit maturity and fruit decay by improving post-harvest shelf life by the average value of 4.5 days (in mature green fruit) and 3.5 days (in mature fruit) compared with the control fruits (Figure 5).

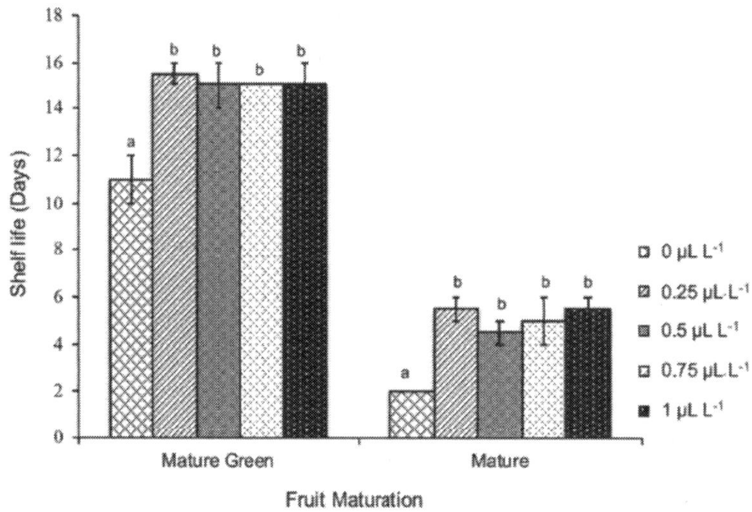

Figure 5. Effect of 1-MCP on the postharvest life of two stages of banana fruit maturity. The mean values in each stage of fruit maturation followed by the same letters are not significantly different according to the Duncan test at $p < 0.05$.

In many climacteric fruits, ethylene accelerates fruit ripening. For market, the present of ethylene must be minimized to improve postharvest life of fruits. To inhibit ethylene effects, 1-MCP most popular to be used, however its effectivity depend on several factors such as the fruit maturation. In this study we investigated the effect of 1-MCP on two stage of fruit maturation of Muli banana. This study showed that, 1-MCP not effective in mature fruit. Infectivity of 1-MCP in mature fruit indicated that the application 1-MCP after climacteric stage cannot totally inhibit the ethylene action due to the ethylene biosynthesis reached the peak and the ethylene has been bound to the ethylene receptor. Our result correspond with Martino, Fabio, & Golding [14], who showed that ethylene production is not affected and inhibited by the application of 1-MCP over 24 hours after the initiation of fruit ripening. In tomato, 1-MCP increases ethylene production when it is applied in breaker stage and also ethylene biosynthesis increases in the stage transition of banana fruit from green to yellow in the color [15].

That the effect of 1-MCP depends on the stage of banana fruit maturity has also been presented in other studies; 1-MCP was more effective on pre-climacteric fruit than on climacteric fruit [11, 16]. From our statistical data analysis, we showed that a treatment of 1-MCP of 0.25 μL.L^{-1} was able to delay fruit softening or maintain fruit firmness in mature green fruit (Figure 1). Therefore, we indicated that 1-MCP totally inhibits fruit ripening by inhibiting ethylene action in the pre-climacteric stage of bananas; the application of 1-MCP after the initiation of fruit ripening did not affect the ethylene-binding process. In other fruit species, 1-MCP has an effect of inhibiting fruit softening (e.g., in apples and pears) [17].

TSS is used to predict sugar level in the fruit and it is commonly associated with fruit sweetness [18]. In many mature fruits such as banana, sugar is the main constituent of dissolved solids and it is one of the important characters for determining postharvest fruit quality of banana [19]. The application of 1-MCP inhibit the change of TSS until 12 days of storage in mature green fruit, but in mature, the change of TSS occurred. Based on this study, therefore we suggest that increased the TSS in mature stage of banana due to the initiation of fruit ripening. Dadzie & Orchard stated that the variation of sugar level appeared in different cultivar and also fruit maturity [19]. We suggest that 1-MCP inhibit sugar accumulation due to the inhibition of ethylene perception in the receptor, therefore the ripening process of banana fruit was delayed or inhibited. Similar symptoms in reducing sugar content due to inhibition of ethylene perception occurs in *Nr, nor* and *rin* tomato mutant, these mutants exhibit the reduction of sugar level in the mature fruit [20]. Contrasting result reported by Mir et al, they stated that 1-MCP did not affect in the change of TSS of tomato fruit [21] and also in the hybrid line of *Sletr1-2* tomato fruits [22].

In many fruits such as banana, the reduction of fruit skin width and increasing diameter of flesh occurred during ripening process. Many physiological processes occur such as the movement of water from the skin to the flesh [14], therefore the width of fruit skin decline and the diameter of flesh increases. The movement of water from skin to flesh would affects the change in the percentage of flesh and peel ratio of bananas and finally increases the size of flesh. In this study the change of percentage of fruit flesh observed at the beginning and after 12 days of postharvest storage. In mature fruit the effect of 1-MCP was not significantly changed the percentage of flesh fruit compared with control, therefore we suggested that it occurred because the morphology of Muli banana which does have fruit skin relatively thin when compared with other types of bananas, where the average of skin thickness around 1 mm. So we supposed that the transfer of water from the banana skin to the flesh of the fruit during the ripening process did not significantly affect the percentage of skin and flesh ratio of Muli banana.

Weight loss is one indicator of a reduced water content in fruit. Reducing the water content in fruits results in wrinkling, the first symptom of loss of quality in banana fruit. In many fruits, weight loss or water loss always occurs during post-harvest storage, and the amount of weight lost varies as a function of fruit maturity. Storage time affected the rate of water loss: longer storage times increased water loss [23]. Our study demonstrated that water loss during post-harvest storage in two fruit maturations of bananas can be reduced via the application of 1-MCP. Similar report found that 1-MCP significantly maintained fruit freshness over 10 days of storage at a storage temperature of 20°C [24].

Ethylene is a major problem for climacteric fruit such as bananas. It accelerates fruit decay by reducing post-harvest fruit shelf life. In many horticultural crops, 1-MCP is effective at extending the quality of post-harvest storage [11]. Low concentrations of 1-MCP or CP (0.7 nL L^{-1}) were effective at preventing bananas from ripening for 12 days at 24°C. Inhibiting ethylene can delay a loss of quality in melons for up to 15 days after harvest [5], and the combination of ethylene and 1-MCP for 24 hours was effective at reducing the growth of sprouts up to 25 weeks after harvest in onions [8]. 1-MCP was more effective than other treatments [16]. This author showed that 1-MCP was more effective at modifying atmosphere storage and increasing fruit shelf life than control. 1-MCP works as an ethylene inhibitor to inhibit the binding process of ethylene to its receptor, and it can be used to prevent the effects of ethylene on many horticultural crops [11].

4. Conclusion

All of the results show that 1-MCP exhibited had a positive effect in delaying Muli banana fruit deterioration by improving fruit shelf life with the average value of 3.5 to 4.5 days longer than control and also 1-MCP was effective in maintaining fruit firmness. We conclude that the application of 1-

MCP in Muli banana has potential as a treatment to maintain postharvest fruit quality by improving postharvest live of fruit

Acknowledgement
We thank all members of our laboratory for helpful discussions throughout the work.

References

[1] Zhao Q, Guo H W 2011 Paradigms and paradox in the ethylene signaling pathway and interaction network. *Mol. Plant.* 1-9.
[2] Guo H, Ecker J R 2004 The ethylene signaling pathway: new insights. *Plant Biol.* **7**: 40-49.
[3] Elitzur T, Vrebalov J, Giovannoni J J, Goldschmidt E E, Friedman H 2010 The regulation of MADS-box gene expression during ripening of banana and their regulatory interaction with ethylene. *J. Ex. Bot.* 1 - 13.
[4] Kitagawa M, Ito H, Shiina T, Nakamura N, Inakuma T, Kasumi T, Ishiguro Y, Yabe K, Ito Y 2015 Characterization of tomato fruit ripening and analysis of gene expression in F1 hybrids of the ripening inhibitor (rin) mutant. *Physiol. Plant.* **123**: 331-338.
[5] Nishiyama K, Guis M, Rose J K C, Kubo Y, Bennett K A, Wangjin L, Kato K, Ushijima K, Nakano R, Inaba A, Bouzayen M, Latce A, Pech J 2007 Bennett, A.B. Ethylene regulation of fruits softening and cell wall disassembly in charenteis melon. *J. Ex. Bot.* **58**: 1281-1290.
[6] Chaves A L S, De Mello-Farias P C 2001 Ethylene and fruit ripening: from illumination gas to the control of gene expression, more than a century of discoveries. *Genet. Mol. Biol.* **29(3):** 508-515.
[7] Cameron A C, Reid M S 2001 1-MCP blocks ethylene induced petal abscission of *Pelargonium peltatum* but the effect is transients. *Postharvest Biol. Technol.* **22**: 169 – 177
[8] Cools K, Chope G A, Hammond J P, Thompson A J, Terry L A 2011 Ethylene and 1-methylcyclopropene differentially regulate gene expression during onion sprout suppression. *Plant Physiol.* **156**: 1639-52
[9] Elfing D C, Drake S T, Reed A N, Visser D B 2007 Preharvest application of sprayable 1-methylcyclopropene in the orchard for management of apple harvest and postharvest condition. *HortScience.* **42**: 1192 – 1199.
[10] Serek M, Woltering E J, Sisler E C, Frello S, Srikandarajah S 2006 Controlling ethylene responses in flower at the receptor level. *Biotechnol. Adv.* **24**: 368 – 381.
[11] Blankenship S M, Dole J M 2002 1-Methylcyclopropene: a review. *Postharvest Biol. Techol.* **28**: 1-25.
[12] Dole J M, Wilkins H P 2005 Floriculture: Principle and Species 2nd edition. Pearson/Prentice Hall. New Jersey. ISBN: 0-13-046250-0. 726 – 739.
[13] Evensen K B 1991 Ethylene responsiveness changes in *Pelargonium* X *domesticum*. *Physiol. Plant.* **82**: 409 – 412.
[14] Martino G D, Fabio M, Golding J B 2007 Preliminary investigation into the uneven ripening of banana (*Musa* sp.) peel. *N. Z. J. Crop Hortic. Sc.* **35**: 193-199.
[15] Pelayo C, de B Vilas-Boas E V, Benichou M, Kader A A 2003 Variability in responses of partially ripe bananas to 1-methylcyclopropene. *Postharvest Biol. Technol.* 28: 75-85.
[16] Oz A T 2011 Combined effects of 1-methylcyclopropene (1-MCP) and modified atmosphere packaging (MAP) on different ripening stages of persimmon fruit during storage. *Afr. J. Biotech.*, **10**: 807-814.
[17] Mattheis J, Fan X, Argenta L 2002 Factors influencing successful use of 1-MCP. Proc WSU-TFREC Postharvest Information Network. Available in http://www. postharvest.tfrec.wsu.edu/PC2002E.pdf.

[18] Defilippi B G, Dandekar A M, Kader A A 2004 Impact of suppression of ethylene action and biosynthesis on flavor metabolites in apple (*Malus domestica* Borkh) fruits. *J. Agric. Food Chem.* **52**: 5694–5701.
[19] Dadzie B K, Orchard J E 1997 Routine post-harvest screening of banana/plantain hybrids: criteria and methods. INIBAP Technical Guidelines 2. International Plant Genetic Resources Institute, Rome, Italy; International Network for the Improvement of Banana and Plantain, Montpellier, France; ACP-EU Technical Centre for Agricultural and Rural Cooperation, Wageningen, The Netherlands
[20] Osorio S, Alba R, Damasceno C M B, Lopez-Casado G, Lohse M, Zanor M I, Tohge T, Usadel B, Rose J K C, Fei Z, Giovannoni J J, Fernie A R 2011 Systems biology of tomato fruit development: combined transcript, protein, and metabolite analysis of tomato transcription factor (*nor, rin*) and ethylene receptor (*Nr*) mutants reveals novel regulatory interactions. *Plant Physiol.* **157**: 405–425.
[21] Mir N, Canoles M, Beaudry R, Baldwin E, Pal Mehla C 2004 Inhibiting tomato ripening with 1-methylcyclopropene. *J. Am. Soc. Hortic. Sci.* **129**: 112–120.
[22] Mubarok S, Okabe Y, Fukuda N, Ariizumi T, Ezura H 2015 Potential use of a weak ethylene receptor mutant Sletr1-2, as breeding material to extend fruit shelf life of tomato. *J. Agric. Food Chem.* **63**: 7995–8007.
[23] Mubarok S, Okabe Y, Fukuda N, Ariizumi T, Ezura H 2016 Favorable effects of the weak ethylene receptor mutation Sletr1-2 on postharvest fruit quality changes in tomatoes. *Postharvest Biol. Technol.* **120**: 1–9.

A Study of Effect Agribusiness Micro Finance for Beef Cattle Farmers in South Sulawesi Indonesia

A Asnawi[1], A A Amrawaty[1] and Nirwana[2]

[1]Department of Socio Economics of Animal Science, Hasanuddin University, Makassar, Indonesia
[2]Department of Accounting, Faculty of Economics and Business, Hasanuddin University, Makassar, Indonesia

Email: aslinaasnawi@yahoo.com

Abstract. The low access to finance institution is one of the obstacles faced by beef cattle farmers in the rural area. The cause were the high interest rates, collateral, lack of information, inadequate credit supply, long procedures and a relatively long distance between farmers and finance institution. The existence of Agribusiness Micro finance Institutions in Indonesia is one of the steps taken to overcome the constraints access financing faced by the community in the rural area. The study examined the effect of Agribusiness Micro finance Institution beef cattle farmers in rural areas in South Sulawesi. The study used descriptive research. The convenient sampling procedure was used to select eighty beef cattle farmers. Data were generated through questionnaire and interview, then used descriptive and descriptive statistics analysis. The findings in this study were agrobusiness microfinance institutions increasing access to capital for beef cattle farmers, empowering farmer group members and increasing their involvement in farmer group activities, raising farmers awareness to allocate their savings for farmers group interest. Thus the micro agribusiness microfinance institution has had a positive impact on the economic life of the community farmer in the rural area.

1. Introduction

The agricultural sector has a very big role in economic development in Indonesia. In addition to contributing to the increase in national income, it also becomes a buffer of community life in rural areas where most of Indonesia's population is in rural areas. They rely on the agricultural sector as a source of income and become their livelihood. One of the sub sectors in the agricultural sector that is very potential to be developed is the livestock business, especially the beef cattle farming. Along with the increasing population of the animal protein needs also tend to increase so that the cattle breeding business plays a very strategic role in supporting national food security. Therefore, the government is now paying enormous attention to the development of cattle farming in rural areas.

One form of government support is to look for solutions to capital constraints faced by most farmers in rural areas other than due to problems of market and technology and farm organizations are still weak. The financial constraints faced are: (1) the existence of formal financial institutions in rural areas is still very limited, (2) procedures and requirements are considered difficult and heavy, (3) relatively high interest rates and described earlier research [1] that (1) low information on sources of financing, (2) lengthy procedures, (3) requirements difficult to be met by mainly collateral breeders, and (4) farmers' relatively remote location from formal financing institutions such as banks. [2] mentioned that some barriers to accessing credit in banking are the high interest rate, collateral requirement required by the bank, procedure and longtime related to the proposal of credit.

Farmers generally use their own capital to finance their livestock business as well as used to finance their daily needs. As explained [3] that almost 90% of self-employed farmers use their own capital to finance their agricultural business. In addition, in general, farmers need credit that the amount is not too high that is only to fix the cage if the intensive maintenance, buying feed and increase the number of cattle maintained.

Based on some of these conditions is the reason the government took a policy through the Ministry of Agriculture since 2008 has issued Rural Agribusiness Development Program. Furthermore, the implementation of that program refers to the basic pattern as stipulated in Regulation of the Minister of Agriculture No.16/OT.140/3/2009 [3] and refined through Regulation of the Minister of Agriculture No. 06 /OT.140/2/2015 on General Guidelines of Agribusiness Development Rural areas of education and training for business development, assistance and granting of business capital assistance facilities to member farmers, both owner farmers, farm workers and farming houses coordinated by the Joint Farmers' Group. In its development, Farmers Group as the recipient of Rural Agribusiness Development Program is the forerunner of the establishment of Agribusiness Micro Finance Institution in rural areas.

According to the Asian Development Bank (ADB) definition, the MFI serves as an institution that provides deposit services, loans, payment services and money transfers aimed at the poor and small entrepreneurs. If referring to [4] on Microfinance Institutions are defined as financial institutions specifically established to provide business development services and community empowerment, either through loans or financing in micro-enterprises to members and communities, savings management, business development consultation that is not solely for profit.

The existence of MFI can be an alternative financial institution that is very close to the farmers both physically because of the proximity of location and psychological and socio-cultural aspects so it is expected that MFI able to overcome the problems of farmers against low access to finance. Therefore, this study aims to determine the impact of Agribusiness Micro Finance Institution to beef cattle farming.

2. Materials and Methods

The study was conducted in Sinjai district of South Sulawesi. This area is chosen as a research location because there are several Agribusiness Micro Finance Institution *(Lembaga Keuangan Mikro Agribisnis/LKMA)* that still exists until now. In addition, that MFI since receiving Rural Agribusiness Development Program funds until now remain concentrated in helping members of farmer groups incorporated in farmers group to get access to capital although small in accordance with the needs of each member. This study was conducted for fourth months.

The number of farmers analyzed in this study was 80 people taken from 40 people per village. This is because the number of members of each unity of farmers group is almost the same, namely 140 people in Patalassang Village and 130 people in Biroro Village.

The research activities include: 1) conducting preliminary survey, 2) identifying farmers group having Agribusiness Micro Finance Institution; 3) separating farmers group channeling Rural Agribusiness Development Program funds for beef cattle farm with other business; 4) collecting data from combined farmers group spread in some districts using questionnaires and conducting in depth interviews to the Animal Husbandry and Animal Health Service, Food Crops, Horticulture and Plantation Office, Head of Livestock Farming Group and Chairman of farmers group. In addition to some members of the farmer's group receiving that loan to obtain data on the perceived impacts of Agribusiness Micro Finance Institution in rural areas. Secondary data were collected from both related agencies to determine farmers group which will be the object of this research. In addition, data from combined farmers group related to the description of members, loan amount, development of distribution of Agribusiness Micro Finance Institution and other data required in this study. Furthermore, the data were analyzed using descriptive statistical analysis. The effect of agribusiness microfinance institutions for beef cattle farmers is measured by the level of awareness of farmers to save, the involvement of group members in all of the group activities, ease of access to finance and in general in overcoming poverty in the rural area.

IOP Conf. Series: Earth and Environmental Science **334** (2019) 012052 doi:10.1088/1755-1315/334/1/012052

3. Results and Discussion

3.1. Description of Beef Cattle Farmers

The number of farmers analyzed in this study is 80 people. The number of selected farmers is based on the completeness of the data collected and can already be considered to represent each farmers group because of almost half of the farmers. The description of the beef cattle farmers is illustrated in Table 1.

Table 1. shows that the number of farmers group members in this study is almost the same, namely 40 and 30 people, respectively. In general, when viewed from the characteristics of the beef cattle farmers, shows that the majority of the age is in the productive age, which is an average of 45 years. The level of education is also quite high because the average has education up to high school that is 10 and 11 years. These conditions strongly support the ability of group members who influence their motivation in groups to joint in all of farmers group activities. The majority of farmers are also very experienced in livestock because on average they run their livestock business over 20 years. The family burden that is not too large will certainly support the willingness and ability of members to save their funds in agribusiness microfinance institutions.

3.2. Description of Farmers Group Combination

Sinjai Regency has 80 units of a combination of farmers group that focus on the distribution of funds for food, horticulture, plantation, cattle and off farm activities. There is 11 unit which channeling funds for livestock activities from 80 farmers group combination in that area. The average farmer's group combination consists of several farmers group. Each farmers group of livestock consists of approximately 20 people. This shows that farmers' involvement in combined farmers group is considerable. Especially in Sipakainga, Patallassang Village, East Sinjai District, the number of members is 140 people (8 groups) and Biroro Village as many as 130 people.

Table 1. Description of Beef Cattle Farmers

Farmers Group	Beef cattle Farmers as a sample (people)	Beef Cattle Farmers Characteristics	Maximum	Minimum	Average
I (Patalassang Village)	42	Age (year)	69	20	45
		Education (year)	12	6	10
		Farm Experiences (year)	40	3	28
		Family Burden (people)	10	2	4
II (Biroro Village)	38	Age (year)	66	25	43
		Education (year)	12	6	11
		Farm Experiences (year)	41	3	30
		Family Burden (year)	9	3	4
Total	80				

3.3. Effect Agribusiness Micro Finance Institution on Beef Cattle Farmers

Microfinance institutions are generally needed to support economic activities in rural areas, especially to facilitate the financing of the beef cattle farmers. This is because most farmers face financing problems. After the existence of Agribusiness Micro Finance Institution then the perceived impact is:

3.3.1. Awareness to save is increasing

This is reflected in the increasing number of principal savings and mandatory savings recorded in the combined farmer's group. Breeders set aside their money to be kept in Agribusiness Micro Finance Institution will then be lent to their needy members. This means that indirectly the involvement of members in depositing their funds in the form of principal savings and mandatory savings has helped and subsidized group members who need it. Deposit mandatory members of IDR 5.000,00 per month, principal, and voluntary savings depends on the ability of each member of the group.

3.3.2. Awareness of members' involvement in groups

This is evident from the increasing number of groups belonging to the combination of farmers group, for example in Pattalassang village in 2018 the number of its members increased to 140 people (7 groups) which previously only 120 people (6 groups) and in Biroro village of 110 people increased to 130 people. The number of members every 20 people. This is because other groups see that after joining a combination of farmers group then there is the opportunity to get a loan even though relatively small funds in accordance with the needs of breeders. This condition has supported one of the targets of Rural Agribusiness Development Program in Agribusiness Micro Finance Institution is the development of farmer group owned and managed by farmers to become an economic institution.

3.3.3. Provide access to capital more easily and quickly

Agribusiness Micro Finance Institution in rural areas provides an opportunity for any member of his group to obtain a loan. With a relatively low interest rate of 1.5% per year with a loan amount of 5-7 million per person, this can be utilized by group members to finance their needs in raising livestock. In addition to loans in the form of money, can also be in the form of cattle in accordance with the price if obtaining a loan in cash. The process of returning the funds is usually done if there are sales made by members of the group. If this is done continuously then the capital obstacles that have been complained about by the breeders can be overcome. Moreover, if it is viewed from the needs of its capital is relatively small so if it must obtain loans from banking institutions is not efficient because it takes a long time, relatively high opportunity costs, and administrative costs take substantial funds.

The presence of Agribusiness Micro Finance Institution in rural areas will accelerate and increase access capital of farmers in accordance with the needs and conditions of business activities undertaken. It confirmed [5] that some of the potentials played by MFIs in the context of rural economic development are dominated by the agricultural sector. First, MFIs are generally located or at least close to rural areas so they can be easily accessed by farmers/economic actors in the village. Second, the farmers/villagers prefer a short process and without much procedure. Third, the characteristics of farming generally require a credit limit that is not too large so that in accordance with the financial capability MFI. Fourth, the proximity of the farm location so that it can disburse credit in a timely manner and the right amount, and Fifth, the existence of socio-cultural linkage and the personal-emotional relationship is expected to reduce the moral hazard in credit repayment.

3.3.4. Alleviation Poverty

Capital obtained from loans to Agribusiness Micro Finance Institution is commonly used by farmers to buy cattle so that the number of cattle ownership can increase. As the number of cattle raises increases, there will be an increase in income for farmers, thereby increasing their business capacity and reducing poverty levels for rural communities, especially beef cattle farmers. If this is sustainable then the objective of establishing Agribusiness Micro Finance Institution originally started from the Rural Agribusiness Development Program is in line with the government's expectation that is to reduce the number of farmers in general. As the expected benefit and impact indicator of that program in accordance with Regulation of the Minister of Agriculture of the Republic of Indonesia No. 01/OT.140/1/2014 are: 1) the development of agribusiness in rural areas, 2) combined farmers group functioning as a farmer economic institution in rural areas owned and managed by farmers, and 3) reduced number of poor farmers in the rural area. Microfinance has thus become very strategic in poverty reduction strategies because of the recognition worldwide that conventional lending institutions do not provide avenues for the advancement of small and micro financial intermediation for micro enterprises and poor non collateralized rural dwellers [6]. [7] revealed that microfinance played a significant role in the economy, as it helped in reducing poverty by providing financial services to the economic active poor, in generating employment opportunities and also provide loans to grow small scale businesses. [8] showed that the average income of participating farmers in the study was higher than those of non participating farmers. The study also established that participating in the UNDP microfinance programme had a positive impact on the income and profit level of the farmers.

The empowerment of MFIs, including informal financing institutions, is the right step in alleviating poverty and developing the people's economy. Informal institutions are more flexible than formal institutions because they do not require complicated administrative procedures, are more accessible and usually based on the principle of trust because they already know each other between the debtor and the creditor such as relatives, neighbors, partners and other kinship relationships, can be obtained as soon as needed [9]. The availability of credit can motivate the community farmers to use and apply modern technologies, and procure inputs for farm use, thus leading them to a greater level of productivity and increasing and increasing their income [10].

3.4. Barriers of Agribusiness Micro Finance Institution

Funds of Rural Agribusiness Development Program is the embryo of Agribusiness Micro Finance Institution formation by channeling funds of IDR 100,000,000.00 each combination of farmers group which is then distributed to groups that need the fund. Some Agribusiness Micro Finance Institution still survive and keep the practice of saving and loan to members of his group but it is not uncommon to find that some of the previously formed Agribusiness Micro Finance Institution are not running. Some of the barriers of Micro Finance Institution in the rurual area were:

3.4.1. The MFIs not supported by full member participation.

It can be illustrated that some members of the group only use Agribusiness Micro Finance Institution when they need funding. However, continuity in depositing mandatory savings and principal savings do not continue so as to reduce the ability of Agribusiness Micro Finance Institution in serving other members who need a loan. Therefore, group consciousness is necessary.

3.4.2. There are some farmers not refund borrowed funds.

The cause is also diverse because they are experiencing calamities such as stolen or dead animals but there are also for reasons that are not clear.

3.4.3. Commitment from the board of farmers group.

Not infrequently there are also unscrupulous administrators who do not distribute the existing funds in Agribusiness Micro Finance Institution in accordance with the allocation. For example, spend these funds to finance activities outside the cattle breeding business. This could be the cause of Agribusiness Micro Finance Institution is not sustainable. Therefore, it is necessary for the management of the group to have a sense of responsibility and good character so as to avoid the occurrence of such things.

4. Conclusion

The existence of an Agribusiness Microfinance Institution for beef cattle ranchers has increased awareness of breeders putting their funds aside for savings, group members increasingly involved in group activities, have increased the accessibility of financing to beef cattle farmers quickly and easily because of their presence in rural areas and funds collected by and for members groups, and indirectly has alleviation of poverty in rural areas.

References

[1] Asnawi A 2013 Determinant of Funding Accessibility and its Impacts to the Performance of Beef-Cow Breeding Enterprises in South Sulawesi Province, Indonesia *European Journal of Business and Management* **5(9)**: 77-84.

[2] Hyz A B 2011 Small and Medium Enterprises (SMEs) in Greece – Barried in Access to Banking Services, An Empirical Investigation *International Journal of Business and Social Science* **2(2).**

[3] Syukur M, Haastuti E L, Supriatna A, Supadi, Sumedi, Wcaksono B W D 2009 Final Report on Agricultural Financing Study Supports Agribusiness and Agro-industry Development in Rural Areas *Center for Agricultural Socio Economic Research and Development* Bogor.

[4] Directorate A 2015 Development of Agribusiness Micro Finance Institution at Combined Farmers Group and Rural Agribusiness Development Program 2015 *Directorate of Agricultural Financing Directorate General of Infrastructure and Agricultural Facility Ministry of Agriculture* Jakarta.

[5] Ashari 2006 Potential of Micro Finance Institution (MFI) in Rural Economic Development and Development Policy *Agricultural Policy Analysis* **4(2)**: 146-164.

[6] Mihyo P 1995 Credit and employment activities: A case study of Ghana *Policy Seminar on confronting the challenges of poverty and employment in Accra* Accra: GIMPA.

[7] Yahaya K A, Osemene, Omar F, Abdulraheem A 2011 The effectiveness of Microfinance Bank in Alleviating Poverty in Kwara State Nigeria *Global Journal of Management and Business Research.* **11 4(1)**.

[8] Kudi T M, Odugbo S B, Banta A L, Hassan M B 2009 Impact of UNDP Microfinance Programme on Poverty Alleviation Among Farmers in Selected Local Government Areas of Kaduna State, Nigeria *International Journal of Sociology and Anthropology* **1(6)**.

[9] Krisnamurti B 2005 Development of Microfinance for Indonesia's Development *Media Information of Rural Banks* The 4th Edition March.

[10] Llanto G 2003 A Microfinance Promise to Provide the Poor Access to Finance Services *Philipine Institute of Development Studies* Makati.

Strengthening the Beef Cattle Farmer-Based Organization in Partnership System of Maiwa Breeding Center

S N Kasim[1*], A Abdullah[1*], Herry S[1], Tanrigiling[1] and Aminawar[1]

[1]Animal Science Faculty, Hasanuddin University, Street Perintis Kemerdekaan Km. 10, Makassar 90245, Indonesia

E-mail: sofyannurdinkasim@gmail.com; abdullah_ina@yahoo.com

Abstract. This research aimed to study improvement of beef cattle farmer-based organization in partnership system of Hasanuddin University Maiwa Breeding Centre. The research population was 75 farmers with a sample of 43 respondents whose determination was carried out in descriptive statistics using the Slovin formula. The sampling technique was carried out by simple random sampling. Observation data retrieval techniques are direct observation on MBC Unhas locations and in-depth interviews using structured questionnaires. The analysis was carried out quantitatively descriptively using descriptive statistics (frequency distribution table). As a result, the improvement strategy on resource (R), organization (O), and norm (N) aspects was perceived as less effective in achieving strengthen beef cattle farmer-based organization in partnership system of Maiwa Breeding Center.

1. Introduction

Beef cattle has been regarded as important commodity in animal farm sub-sector. This can be represented by a growing demand in beef meat. To date, Indonesia has partially supplied meat for domestic need; thus, importation of both meat and live cattle from other countries such as Australia and New Zealand is considered as the most potential solution. To deal with this problem, meat self-sufficiency program was released in 2010 as one of the promising efforts. In this case, the effective approach should consider government and community aspirations from those primarily involved in beef cattle farm if it is to be successful. Unfortunately, current beef cattle industries have been at worrying situation, proven by depletion of beef cattle population up to 4.10% [1].

The livestock industries particularly fattening sector faced numerous constraints in the inadequacy of feeder cattle supply, including its quality, quantity, and continuity. This situation allows to import feeder cattle from other countries. In order to tackle this problem, there is a necessity to formulate further strategies through establishment of science and technology-based breeding center capable of producing high quality cattle for both breeding and fattening sectors.

The program began with collaboration between higher education, i.e. Hasanuddin University and relevant partners, i.e. LIPI and Ministry of Research and Technology, establishing center for beef cattle development, i.e. Maiwa Breeding Centre (MBC). Located in Enrekang Regency with total area of ±250 Ha, it is plotted as Technopark area designed to be one of the centers for research, science and technology. Besides, MBC is also a business unit distributing high quality cattle to farmers under partnership agreement (profit and loss sharing). In fact, we found that MBC activity was less effective since, in term of organizational aspect, there was unclear job description related to the partnership organization. Therefore, a proper solution is inexorable to ensure MBC existence in the future. We proposed the organizational improvement of MBC with aid of social power. [2,3] stated that social power as complementary factor referring to institutional dimension, emerging relationship, and norms that shape quality and quantity means not just a range of institutions or groups supporting the social life, but collectively guards the unity of members. Based on this background, this study is set up to establish effective organizational performance using social media to ensure the sustainability of Unhas Maiwa Breeding Center.

2. Materials and Methods

This research was conducted in Enrekang Regency, South Sulawesi Province, Indonesia. Three-months study period (February-April 2018). This research was carried out with a descriptive quantitative approach which is a description or description of the variables of MBC Unhas institutional strengthening in the partnership system of beef cattle farmers groups. The study population was 75 farmers with a sample of 43 respondents whose determination was carried out in descriptive statistics using the Slovin formula. The sampling technique was carried out by simple random sampling. Observation data collection techniques are direct observation on MBC Unhas locations and in-depth interviews using structured questionnaires. The analysis was carried out quantitatively descriptively using descriptive statistics (frequency distribution table). Due to the qualitative type of data, the qualitative type of data will be transformed into quantitative by using a Likert scale by creating categories and each category given a score with three categories, namely:

1. High category (score 3), interval 87-129
2. Medium category (score 2), interval 44-86
3. Low category (score 1), interval 0-43

Determination of intervals is based on the Likert scale, namely:

- Highest score = 3 x 43 (number of respondents) = 129
- Lowest score = 1 x 43 (number of respondents) = 43

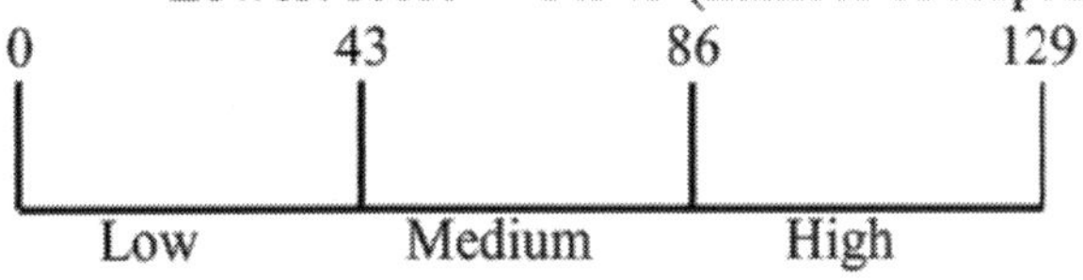

3. Results and Discussion

The organizational improvement of beef cattle farmers in Unhas MBC was assessed using 3 aspects: (1) resource (R) consisting of natural, human and financial resource, (2) organizational aspect (O), and (3) norm aspect (N).

3.1. Resource aspect

3.1.1. Natural resource

In this aspect, the assessment was referred to field ownership status, cattle ownership status, and housing system. The result indicated that farmer's ownership to these natural resources was still limited, as presented in Table 1.

Table 1. Assessment of natural resource ownership of beef cattle farmers

No.	Description	Interval Score	Category
1.	Field ownership	35	Low
2.	Cattle ownership	28	Low
3.	Housing system	40	Low
	Total	103	
	Average	34	Low

Source: Primary data, 2018.

The low ownership was due to limitation in field area owned by farmers, with average area of 0.5 Ha. As a consequence, livestock farmers had limited area to plant crops, leading to inadequacy of forage. In general, the number of cattle owned was less than 5 cows, originating from their own and MBC. Furthermore, MBC's supports for livestock farmers were definitely low, ranging from 1-4 cattle. Hence, there was a big burden for farmers to perform the partnership agreement. In short, this partnership is economically less profitable. Meanwhile, cattle housing was still performed in traditional method, in which cattle shed was constructed close to their house, made of woods and did not meet technical standards. [4] argue that natural resource was considered as engine of growth, transforming resources to man-made capital which was in turn could produce a higher productivity in the following periods.

3.1.2. Human resource

Human resource was evaluated from education level, knowledge, skill, and experience. The assessment result was presented in Table 2.

Table 2. Assessment of human resource of beef cattle farmers.

No.	Description	Interval Score	Category
1.	Education	40	Low
2.	Knowledge	32	Low
3.	Skill	24	Low
4.	Experience	19	Low
	Total	115	
	Average	29	Low

Source: Primary data, 2018.

The result exhibited that overall category for human resource aspect was at low category. This was clear since education level of most farmers was graduation of junior high school, which also contributed to their knowledge for improving animal's productivity. The skill was also categorized as low, since they did not obtain proper supervision from MBC. Meanwhile, their experience was obtained from their predecessors. [5] found that human resource could be a central factor in organization. Related to function, it was established according to vision for human needs and in realization, the mission was managed and controlled by human. Thus, human constitutes a strategic element in institutional activities.

3.1.3. Financial resource

Financial resource of the farmers was evaluated according to own finance and financial support provided by MBC. The result was presented in Table 3.

Table 3. Assessment of financial resource of beef cattle farmers.

No.	Description	Interval Score	Category
1.	Own finance	42	Low
2.	Financial support from partner	39	Low
	Total	81	
	Average	41	Low

Source: primary data, 2018.

The result proved that financial resource of the farmers was categorized as low. This condition seemed to be caused by poor financial management, in which their budgeting was not only allocated for farm, but also for daily expenditure and education. On the other hand, the 1-4 cattle from MBC were still less significant in rising their income. This condition leads to an increasing cost for farmers, which makes them more difficult to run their livestock business. [6] stated that financial resource was fund allocated for organizing the business activities; thus, the business goals could be achieved.

3.2. Organization Aspect (O)

In this aspect, we assessed 3 elements, i.e. organization structure, implementation of Articles of Association (AD/ART), and administrative completeness. The result was presented in Table 4.

Table 4. Assessment of organization aspect of beef cattle farmers.

No.	Description	Interval Score	Category
1.	Organization structure	38	Low
2.	Implementation of ADRT	34	Low
3.	Administrative completeness	20	Low
	Total	92	
	Average	31	Low

Source: primary data, 2018.

Assessment result on organization aspect of farmer group exhibited that this aspect was classified as low. Although the organizational structure was successfully established, it did not run properly. Similarly, the existing AD/ART was not consistently implemented. For instance, disobedience by members and leaders was not punished. Administratively, we found that this farmer organization had poor administrative completeness in some important documents such as saving and loan book, meeting book, and financial report. [7,4] stated that organization defined as distribution of duties and their connectivity among people who hold the position, collectively and consistently directed to achieve the goals of an organization.

3.3. Norm Aspect (N)

Norm aspect was assessed through 3 elements, i.e. return of loan fund, punishment for individual transgression, and member's participation on group activities. The result was exhibited in Table 5.

Table 5. Assessment of norm aspect of beef cattle farmers.

No.	Description	Interval Score	Category
1.	Return of loan fund	36	Low
2.	Punishment	26	Low
3.	Member's participation	19	Low
	Total	81	
	Average	27	Low

Source: primary data, 2018.

The result showed that norm aspect in beef cattle farmer group was also categorized as low. We found that organization regulation was made by leaders, without involving members. Besides, the regulation was then orally communicated to members. Because of this situation, most members considered that the rule was established to meet administrative requirements for accessing the support (cattle), regardless its sustainability.

Additionally, punishment for individual transgression was not entirely implemented; thus, participation of members in organization events was low. We reported that organizational improvement in beef cattle farmer group of Unhas MBC in term of resource, organization and norm aspects was definitely unsatisfied and ineffective to achieve the goals of organization. To deal with this constraint, the organizational aspects could be strengthened effectively and efficiently using social power. [9] stated that the strengthening efforts in an organization involved improvement of individual behavior, organizational behavior and social system in order to reach goals determined effectively and efficiently, including: (a) increasing individual performance in knowledge, skill and behavior; (b)

IOP Conf. Series: Earth and Environmental Science **334** (2019) 012053 doi:10.1088/1755-1315/334/1/012053

elevating organizational performance in organization and management, finance, and structure; (c) improving social independence.

4. Conclusion

Improvement of beef cattle farmer-based organization in partnership system of Unhas Maiwa Breeding Centre according to resource, organization and norm aspects was still weak, which is unable to support group in achieving the goals of organization to create better member welfare. Therefore, institutional development should consider member awareness in trust, collaboration, and consistent regulatory compliance through involving social power.

Acknowledgment

The authors express sincere gratitude to the Rector of Hasanuddin University and the Institute for Public Service and Policy Research (IPSPR) Hasanuddin University for funding this study through the Research Program Project Grant of Hasanuddin University within the Fiscal Year 2018.

References

[1] Central Statistics Agency 2017 Beef Cattle Production, 2009-2016 *Directorate General of Livestock and Animal Health* Ministry of Agriculture.
[2] Coleman, James S 1988 Social Capital in the Creation of Human Capital, in Falk I, Harrison L 1998 Community Learning and Social Capital: Just Having a Little Chat *Journal of Vocational Education and Training* **50(4):** 609–627.
[3] Burt R S 2002 Expert From The Social Structure Of Competition In Structure Holes: The Social of Competition Cambridge MA and London: Harvard University.
[4] Wursanto 2005 Fundamentals of Organizational Science *Andi Publisher* Yogyakarta.
[5] Mudiyatmoko J 2007 Sociology: Understanding and Assessing Society *Grafindo Media Pratama Publisher*, Bandung.
[6] Yazdani, Yoghoubi 2011 The Relationship Between Social Capital and Organizational Justice *Europan Journal of Economics, Finance and Administrative Science* **30** ISSN 1450-2275.
[7] Damsar 2009 Introduction to Economic Sociology Jakarta: *Kencana*.
[8] North D C 1990 Institusions: Institusional Change and Economic Performance *Cambridge University Press* Cambridge.

IOP Conf. Series: Earth and Environmental Science **334** (2019) 012054 doi:10.1088/1755-1315/334/1/012054

The Determinant Factors of The Dynamics of Agribusiness Behavior of the Mango Farmers in Greged Sub District, Cirebon District

E Rasmikayati[1], G W Mukti[1] and B R Saefudin[2]

[1] Fakultas Pertanian, Universitas Padjadjaran
[2] Pusat Riset Pangan Berkelanjutan DRPMI UNPAD

E-mail: elly.agri@yahoo.co.id

Abstract. Cirebon has already been known as one of the mango-producing areas. However, until now, mango agribusiness in the area has not been able to maintain the quality and guarantee the availability of mangoes throughout the year. The problem was caused by the dynamic behavior of the farmers. This indicated that the development of mango agribusiness in the era of globalization should be based on the improvement of agribusiness behavior of mango farmers. The purpose of this paper was to analyze the factors that determine the improvement of agribusiness behavior of mango farmers. The research method employed in this research was survey technique in Greged district of Cirebon regency using simple random sampling technique involving 130 mango farmers. Data were analyzed using path analysis. The results of the study showed that the most powerful factors in improving agribusiness behavior of mango farmers were resource factor (9.2%), institutional factor (8.2%), cultural factor (7%) and technological factor (4.8%). This indicated that the development of sustainable mango agribusiness model in Greged district of Cirebon Regency should be focused on optimizing its resources, increasing its institutional role, preserving the mango cultivation of its farmers and increasing the role of the use of mango farming technology.

1. Introduction

West Java, as one of the mainstay areas of mango development in Indonesia, was able to contribute the third highest mango production after East Java and Central Java (Table 1). Although only ranks third in Indonesia, mango productivity in West Java was higher than the other four provinces. The average productivity of mango in West Java within 5 years was 11.7 tons/ha while East and Central Java were only 9.5 tons/ha and 10.3 tons/ha, respectively. The number of diversification of superior mango varieties in West Java and production of certified mango seeds were higher than in East and Central Java. Five major mango varieties developed in West Java were Arumanis, Gedong, Gedong Gincu, Dermayu, and Golek.

Currently, West Java Provincial Government is cooperating with the Japanese government in developing the market of Gedong Gincu due to the high preference of Japanese people to this type of mango fruit from West Java. Mango-producing areas in West Java spreaded over 27 districts and cities. There were 5 districts in West Java becoming mango production centers that cultivated various types of mango. These areas were the districts of Indramayu, Cirebon, Majalengka, Kuningan, and Sumedang. In addition, the production of certified mango seeds in West Java reached more than 1 million trees. On the other hand, East and Central Java were only 178,061 trees and 29,601 trees .

Greged sub-district is a potential mango producer in Cirebon District (Table 2), although the number of trees and land area only rank third, but the varieties and programs of mango commodity development conducted by the government in Greged Sub District tend to be more diverse. In 2010, the area of Gedong Mango in Cirebon District was 2,929 Ha2. There was about 2% or 50 a^2 had been certified in the Ministry of Agriculture. Certification of mango garden was very important to penetrate export market since importers required mango fruit with clear origin in order to know the quality guarantee. However, not all mango gardens could be registered. In addition, farmers must also understand and apply Stop/GAP if they wanted to have a minimum of one hectare of land .

Table 1. Mango production per ton by province in Indonesia

Province	Year					Share
	2012	2013	2014	2015	2016	(%)
East Java	840.316	799.410	922.727	806.644	655.692	37
Central Java	422.992	404.443	459.669	396.636	334.596	18
West Java	**344.205**	**327.070**	**321.482**	**310.226**	**260.106**	**14**
South Sulawesi	158.006	148.118	161.829	117.205	86.081	6
Nusa Tenggara Barat	137.689	110.637	118.427	120.696	109.069	5
Sum	1.903.208	1.789.678	1.984.134	1.751.407	1.445.544	80
Indonesia	2.376.333	2.192.928	2.431.330	2.178.826	1.814.540	100

Description: Share is ratio between total production of province with total production of Indonesia
Source: Central Bureau of Statistics and Directorate General of Horticulture (processed)

Table 2. Production center of gedong gincu mango in cirebon district

No	Sub-District	Tree number	Large (Ha)
1.	Sedong	130.805	1.308
2.	Dukupuntang	82.500	825
3.	Greged	69.000	690
4.	Lemahabang	49.259	492
5.	Astanajapura	15.700	157
6.	Source	7.070	71

Source: Agriculture, Plantation, Livestock and Forestry Office of Cirebon District

In marketing activities, the purpose of the mango gedong market was determined by grade. The classification of the quality (grading) of the harvest was conducted by the collector based on the size, shape, maturity and damage of the fruit. Mango grade A/B (a combination of grade A and B) was a widely marketed grade to markets outside the region, fruit shops/stalls, supermarkets, and a small portion in exports. While off-grade mango or grade C was only distributed to traditional markets in Cirebon and surrounding regencies such as Majalengka, Sumedang, Indramayu and Bandung. The supply chain of mango chains in Cirebon district was mostly through the fruit dealer. The dealers distributed mangoes to traditional markets (80-83%), modern markets (17%) that were divided into supermarkets by 15% and exports 2%, and a small portion to processing industries (1-3%) [1].

The percentage of Gedong mango production in Cirebon district in 2010 was 60 %, with grade A/B 41.9 percent and grade C of 18.1 percent. Meanwhile, mango Gedong Gincu was 40 percent, with grade A/B was 28 percent and grade C was 12 percent. Factors affecting the farmer's decision to harvest the mango in the form of Gedong Gincu were market price and deal with collector. If the selling price was still low, farmers delayed the harvest time or waited for the price to increase before harvesting.

Although mango was one of the tropical horticulture commodities favored nationally and had the potential to increase the welfare of farmer's family because it had high economic value, the economic potential had not been explored since there were still many technical and non-technical obstacles that caused the production of national mango became unstable and had low quality [2]. The instability of national mango production was due to the low adoption of off season technology in Indonesia which was 17.92% [3]. Production of Gedong Gincu mango in Cirebon district had decreased in 2010 and

2013. The decrease of mango production in 2010 was caused by the unpredictable weather, which is the rainy season that lasted throughout 2010. Extreme weather also occured in 2013. The success of mango farming still depended on climate and weather conditions, high rainfall became the failure factor of mango production.

Production instability and low quality caused the percentage of mango export volume in Indonesia to be below 0.08 percent. In fact, with the exports, farmers could increase their revenue because the selling price of exported mangoes were higher than the mango marketed in the country. For the government, the export of mango was also beneficial as it could increase foreign exchange earnings for the State. Some internal factors that affect the low export volume were the low number of mangoes that meet the export quality criteria, the seasonal shifts result in harvest delays or crop failure, small scale business scale, limited application of post-harvest mango technology, exporter promotion capabilities, knowledge of limited market characteristics, and fruit flies.

The high diversity of characteristics of mango and its farmers reflects the dynamics of mango agribusiness behavior in Greged Sub District, Cirebon District. The problem is how to encourage the dynamics of these behaviors to move in a positive direction and ultimately can improve the competitiveness and prosperity of mango farmers themselves in the current era of globalization. Thus, the purpose of this paper is to investigate the factors encouraging mango farmers to improve the dynamics of agribusiness behavior.

2. Materials and Methods

The research was conducted by Survey-explanatory method using simple random sampling. Based on field observations, the number of farmers group in Greged sub District were 10 groups with the average number of 18-23 peoples per group, so that the estimated population of mango farmers were 180-230 people. Based on expert formulation, then the number of samples in this study was determined as many as 130 people. The amount was sufficient for the minimum sample size for quantitative research.

Table 3. Operationalization of research variables

<table>
<tr><th>No.</th><th>Variable</th><th>Indicator</th><th>No.</th><th>Variable</th><th>Indicator</th></tr>
<tr><td>1.</td><td>Mango Farmer Agribusiness Behavior</td><td>1. Cultivation of mango trees
2. Use of technology
3. Land expansion
4. Sorting / grading
5. Diversification of production</td><td>5.</td><td>Resource Factor</td><td>1. Production Facilities
2. Individual Farmer's Ability
3. Skilled Workers
4. Natural Potential</td></tr>
<tr><td>2.</td><td>Social Factor</td><td>1. Discussion / Sharing
2. Farmer Cooperation
3. Conflict Resolution</td><td rowspan="2">6.</td><td rowspan="2">Institutional factor</td><td rowspan="2">1. Incorporated in Group of Farmers
2. Partnership in Marketing
3. Partnership in Capital
4. Government Support</td></tr>
<tr><td>3.</td><td>Economy factor</td><td>1. Access to Capital
2. Mango Demand
3. Mango Selling Price</td></tr>
<tr><td>4.</td><td>Technology factor</td><td>1. Pest Management Technology
2. Sustainable Fertilization System
3. Use of ZPT</td><td>7.</td><td>Cultural factor</td><td>1. Hereditary business
2. Business inheritance
3. The role of the family</td></tr>
</table>

The sample of mango farmers selected from the sampling frame was based on the listing of the mango farmer population from BPS, which was renewed every 10 years. BPS divided the area by census block, which was a smaller division of the village. Each census block consisted of 80-120 families. BPS defined a mango farmer as a farming family with 4 mango trees or more.

In order to analyze the factors that determine the improvement of agribusiness behavior of mango farmers in Greded sub-district of Cirebon district, multivariate statistical analysis tools (path analysis) was used. The dependent variables were agribusiness behavioral variables of mango farmers, whereas the independent variables, which were the factors that allegedly influence the agribusiness behavior of mango farmers were social factors, economic factors, technological factors, resource factors, institutional factors and cultural factors. Each variable was measured using some relevant indicator by using Likert scale. Operationalization of these variables was presented in Table 3.

Estimation stage is the stage of making econometric model of factors that determine the improvement of agribusiness behavior of mango farmers in Greded sub-district of Cirebon district in the era of globalization as follows:

$$Y = \rho_1 X_1 + \rho_2 X_2 + \rho_3 X_3 + \rho_4 X_4 + \rho_5 X_5 + \rho_6 X_6 + \varepsilon \quad (1)$$

Keterangan:

Y = Agribusiness Behavior of Mango farmers

X_1 = Social factor

X_2 = Economy factor

X_3 = Technology factor

X_4 = Resource factor

X_5 = Institutional factor

X_6 = Cultural factor

ρ_i = Path coefficient of independent variable X_i $(i = 1,2,\ldots,6)$

ε = *Error*

Furthermore, all ordinal scale (likert) scales were transformed into interval scale measurements using the method of successive interval in order to perform algebraic operations that added up all the values of the indicator to be the value of each variable. Then, the validity test was conducted, where each invalid indicator was not included in the next test that was the reliability test. In the reliability test, any unreliable variable was not included in the next test, which was the test of classical assumptions. Thus, after all the classical assumptions were met, the model fit test (goodness of fit test (GoF Test) was performed. Next, once found that the estimation model was good and appropriate, partial test could be conducted for subsequent interpretation of estimation results and discussion.

3. Results and Discussion

3.1 Data Validity Analysis

The validity test was conducted to find out whether the measuring instrument had measured what needs to be measured. Validity test in this research was conducted by correlating each question with total score for each variable. Correlation technique used was product moment correlation technique using the formula as follow:

$$r = \frac{n\sum PQ - \sum P \sum Q}{\sqrt{[n\sum P^2 - (P)^2][n\sum Q^2 - (\sum Q)^2]}} \quad (2)$$

Keterangan:

r = Product moment correlation Cooficient;

n = Sample Size;

X = Indicator score of variable i;

Y = total score of variable i

The statistically obtained correlation value should be compared to the critical r table. If r arithmetic> r table, the data was significant (valid) and feasible to be used in testing the research hypothesis. On the other hand, if r arithmetic <from r table, the data was not significant (invalid) and would not be included in testing the research hypothesis.

Table 4. Data validity analysis result

No	Variable	Indicator	*Product Moment Pearson* (r)	Status
1.	Agribusiness behavior of mango farmer	Cultivation of mango trees	0.646**	Valid
		Use of technology	0.613**	Valid
		Land expansion	0.549**	Valid
		Sorting/ grading	0.519**	Valid
		Diversification of production	0.530**	Valid
2.	Social factor	Discussion/Sharing	0.751**	Valid
		Farmer Cooperation	0.703**	Valid
		Conflict resolution	0.603**	Valid
3.	Economy factor	Access to Capital	0.667**	Valid
		Mango Demand	0.465**	Valid
		Mango selling Price	0.556**	Valid
4.	Technology factor	Pest management Technology	0.670**	Valid
		Sustainable fertilization system	0.754**	Valid
		The use of ZPT	0.755**	Valid
5.	Faktor sourcedaya	Production facilities	0.564**	Valid
		Individual Farmer's ability	0.712**	Valid
		Skilled Workers	0.564**	Valid
		Natural Potential	0.591**	Valid
6.	Institutional factor	Incorporated in Group of farmers	0.738**	Valid
		Partnership in marketing	0.587**	Valid
		Partnership in capital	0.676**	Valid
		Government Support	0.661**	Valid
7.	Cultural factor	Government Support	0.791**	Valid
		Hereditary business	0.690**	Valid
		Business inheritance	0.543**	Valid

Source : Primary Data, processed 2018

According to Table 4, the Product Moment Pearson (r) value of all indicators in each variable was valid. Thus, all of these indicators could be included in the data reliability test.

3.2 Analisis Reliabilitas Data

The reliability test is performed on a valid indicator to test whether the indicators were stable in revealing the data behavior. Testing can be conducted externally or internally. Externally, testing can be conducted with test and retest. Meanwhile, internally, it could be tested by analyzing the consistency of the items.

Based on Table 5, it was found that the Cronbach Alpha value of all indicators on each variable was above 6. Thus, it was declared reliable. Thus, all of these indicators could be included in the classical assumption test.

Table 5. Data Reliability Analysis Result

No	Variable	Indicator	*Product Moment Pearson* (r)	Status
1.	Agribusiness behavior of mango farmer	Mango Tree Cultivation	0.715	Reliable
		Technology use		
		Land expansion		
		Sorting/ grading		
		Diversification of production		
2.	Social factor	Discussion/Sharing	0.765	Reliable
		Farmer Cooperation		
		Conflict resolution		
3.	Economy factor	Access to Capital	0.656	Reliable
		Mango Demand		
		Mango selling Price		
4.	Technology factor	Pest management Technology	0.790	Reliable
		Sustainable fertilization system		
		The use of ZPT		
5.	Resource factor	Production facilities	0.725	Reliable
		Individual Farmer's ability		
		Skilled Workers		
		Natural Potential		
6.	Institutional factor	Incorporated in Group of farmers	0.760	Reliable
		Partnership in marketing		
		Partnership in capital		
		Government Support		
7.	Cultural factor	Government Support	0.760	Reliable
		Hereditary business		
		Business inheritance		

Source : Primary Data, processed 2018

3.3. Classical Assumption Test

3.3.1. Multicollinearity Test

A good path analysis model should be free of multicollinearity or have no high correlation between independent variables. The best way to detect the presence or absence of multicollinearity is to look at the tolerance and Variance Inflation Factor (VIF). If the tolerance value is> 0.5 and the VIF value <5, it can be concluded that there is no multicollinearity among the independent variables in the regression model. The following was the result of SPSS calculation for multicollinearity test.

Table 6. Multikolinieritas test

Variable	*Collinearity Statistics*	
	Tolerance	VIF
X1	0.595	1.680
X2	0.861	1.161
X3	0.664	1.506
X4	0.654	1.530
X5	0.629	1.589
X6	0.945	1.058

Source : Primary Data, processed 2018

Table 6 shows that the tolerance value of each of the independent variables was> 0.5 and tended to be close to 1. Similarly, the VIF value of each free variable did not exist that exceed 5 even tended to close to 1. Thus, it was concluded that the resulting model did not contain multicollinearity.

3.3.2 Heterocesdacity Test

The heteroscedasticity test aimed at testing whether the regression model had a variance inequality of residuals in one observation of another. If the variance of the residuals was fixed then it was called Homocedasticity and if it was different, it was called Heteroscedasticity. A good regression model was a model that did not have Heteroscedasticity. Heteroscedasticity test was using graphical method by looking at the presence or absence of certain pattern on scatterplot of dependent variable. If there was no particular pattern there was no heteroscedasticity and vice versa. Results of analysis using SPSS is presented in Figure 1

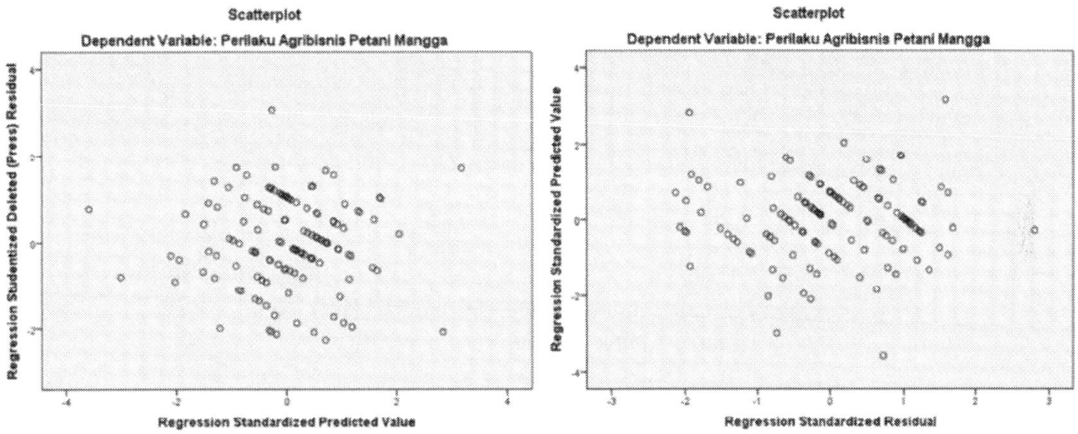

Figure 1. Scatterplot Result
Source : Primary Data, processed 2018

Based on Figure 1, it could be concluded that the path analysis model had no symptoms of heteroscedasticity because there was no pattern and the points were randomly distributed below and above the zero point.

3.3.3 Normality Test

Normality test aimed at testing whether both dependent variable and independent variable had a normal distribution. A good regression model had a normal or near-normal residual distribution. Normality test could be computed using SPSS by looking at histogram and P-P plot. If the histogram of residual value formed a pattern such as a bell shape with a mean value of 0, it indicated that it had a normal distribution. Normality test using P-P Plot was said to be normally distributed if the points were along the line. Normality test results through histogram and P-P Standardized Residual Plot of Regression were presented in figure 2.
In Figure 2, it was indicated that the histogram pattern formed the bell shape. In addition, on the normal P-P Plot chart, the residual points were along the linear (diagonal) lines. Thus, it could be concluded that the path analysis model had met the assumption of normality.

3.3.4 Autocorrelation Test

The autocorrelation test aims to test whether there is a correlation between errors in a given period. One technique for detecting autocorrelation is the Durbin-Watson (D-W) test with the following criteria:

a. If D-W value < -2, there was positive autocorrelation
b. If D-W value > -2, there was negative autocorrelation

IOP Conf. Series: Earth and Environmental Science **334** (2019) 012054 doi:10.1088/1755-1315/334/1/012054

c. If 2D-W value was between 2 and -2, there was no autocorrelation

Based on the analysis, the D-W value is 1,541, so it can be concluded that there is no autocorrelation in the path analysis model.

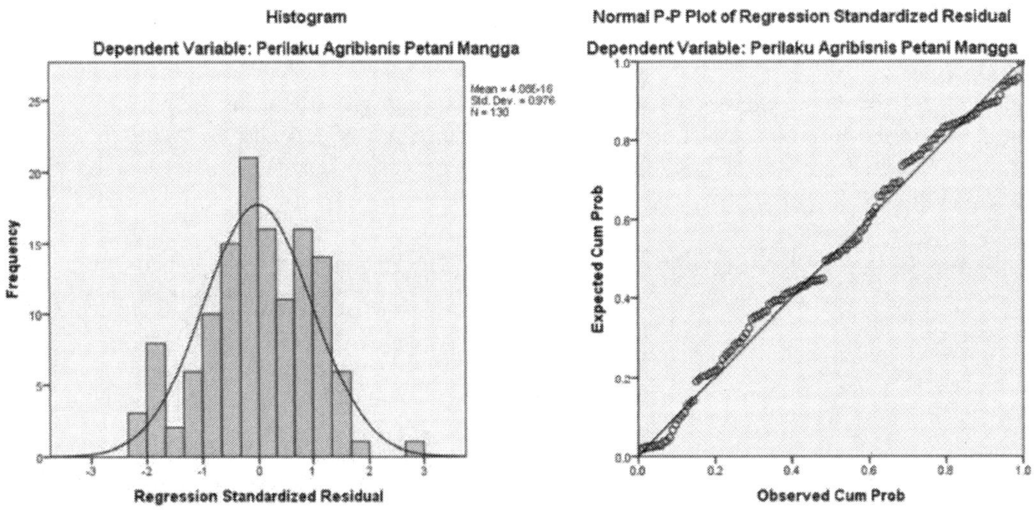

Figure 2. Histogram (Left) and Normal P-P Plot (Right)
Source : Primary Data, processed 2018

3.4. Goodness of Fit (GoF) Test

3.4.1 Coefficient of Determination (R2)

The coefficient of determination (R2) essentially measures the model's ability to explain variations of independent variables. The coefficient of determination is between zero and one. The small value of R2 means that the ability of independent variables to explain the variation of the dependent variable is very limited.

Based on the results of the analysis, R2 was 0.317, which indicated that the variation of the ups and downs of agribusiness behavior variables could be explained as much as 31.7% by the 6 factors / independent variables in this study, while the rest wass determined by other factors not investigated in this study.

3.4.2 F test

F test is used to test the null hypothesis which states that the coefficient of determination of compound in population, R2, is equal to zero. Significance tests include testing the significance of the overall regression equation as well as the specific partial regression coefficients. The whole test can be done using F statistics.

Based on the analysis, the value of F was 9.497 or if converted into an opportunity value was 0,000 $<0.05 = \alpha$. Thus, the null hypothesis is rejected and the alternative hypothesis states that the coefficient of determination in the population, R2, is not equal to zero, is accepted. Thus it can be concluded that the model of path analysis is significant (fit) and simultaneously good.

3.5 Result of Factor Analysis Determining Enhancement of Agribusiness Behavior of Mango Farmer

The results of the path analysis model estimation, as shown in Table 7, indicated that there were 4 factors that have real and positive influence in determining the increase of agribusiness behavior, namely: 1) Technological factors (X3); 2) Resource factor (X4); 3) Institutional factors (X5); and 4) Cultural factors (X6). While social factors (X1) and Economic Factor (X2) have no significant effect.

Table 7. Path Coefficient

Variable	Path Coefficient	t	Sig	Status
Social factor (X1)	-0.019	-0.199	0.842	not significant
Economic factor (X2)	0.110	1.369	0.173	not significant
Technology factor (X3)	0.153	1.672	0.097	Significant*
Resource factor (X4)	0.239	2.594	0.011	Significant**
Institutional factor (X5)	0.253	2.693	0.008	Significant**
Cultural factor (X6)	0.231	3.014	0.003	Significant**

* Significant at 10%
** Significant at 5%
Source : Primary Data, processed 2018

Technological factor (X3) had positive and significant influence with 90% confidence level in determining agribusiness behavior of mango farmer (Y) with coefficient value of 0,153. The farmers in the sub district of Greged had been using a lot of integrated pest management technology to control the disease, especially the disease on the mango. Pesticide spraying was essential starting from flower stage until the harvest time. In addition, farmers also used the technology of fruit fly flies to control fruit fly pest attack. In terms of maintenance of mango trees, farmers applied sustainable fertilization system by combining the use of organic fertilizers and chemical fertilizers to obtain good crops. Then, the use of ZPT had also been applied by many mango farmers, especially to get the harvest outside the season because at that time the price of mango was high.

Resource factor (X4) had positive and significant influence with 95% confidence level in determining agribusiness behavior of mango farmer (Y) with coefficient value of 0,239. The mango farmers in Greged Sub-district mostly had adequate production facilities and infrastructure so that they had the capability to produce a large and quality mango product. In addition, the role of skilled and experienced workers over the years was very important in supporting agribusiness mangoes. Support of natural conditions that were very suitable for mango plants in Greged district also boosted the potential to produce a lot of mango products and quality.

The role of farmer groups in Greged sub-district strongly supported the development of mango farming. Then, in terms of marketing, farmers already had reliable business partners to sell their crops. These business partners also provided access to capital for the sustainability of mango farming. The support and role of the government was also quite good. Educational and extension programs had been widely conducted by local extension agencies for the development of mango farming. Therefore, based on path analysis model, institutional factor (X5) had positive and significant influence with 95% confidence level in determining agribusiness behavior of mango farmer (Y) with coefficient value of 0.253.

Furthermore, cultural factors (X6) also had positive and significant influence with 95% confidence level in determining the increase of agribusiness behavior of mango farmer (Y) with coefficient value of 0.231. Family roles and support were very supportive and played a role in mango farming in Greged district. In addition, the culture of farming mango in Greged district had been implemented for generations long ago. According to farmers, this culture would continue to be passed down to the next generation and would be maintained in the future.

3.6 Direct, Indirect and Total Effect

The direct influence of a factor (independent variable) on the dependent variable was the magnitude of the factor, which directly determined the value changes in the dependent variable. The indirect effect was the magnitude of a factor that indirectly determined the change of values on the dependent variable through its relation to other factors. While the total effect was the sum of the direct and indirect effects of a factor on the dependent variable. The total influence of all factors was the coefficient of determination (R2). The values of direct, indirect and total influence of all 6 factors on agribusiness behavior of farmers are presented in Table 8.

Table 8. Direct, Indirect and Total Effect

	X1	X2	X3	X4	X5	X6
X1	0.000	0.000	-0.001	0.001	-0.003	0.000
X2	0.000	0.012	0.000	0.005	0.007	0.002
X3	-0.001	0.000	0.023	0.017	0.007	0.001
X4	0.001	0.005	0.017	0.057	0.003	0.009
X5	-0.003	0.007	0.007	0.003	0.064	0.004
X6	0.000	0.002	0.001	0.009	0.004	0.053
Direct effect	0.000	0.012	0.023	0.057	0.064	0.053
Indirect Effect	-0.003	0.014	0.025	0.035	0.018	0.017
Total Effect	-0.003	0.027	0.048	0.092	0.082	0.070
R^2			0.317			

Based on Table 8, the direct influence or strength of social factors (X1) and economic factors (X3) that directly determined the changes of farmer's agribusiness behavior (Y) were 0% and 1.2%, respectively. Furthermore, the indirect effect of these two variables through their relationship with the other 5 factors was -0.3% and 1.4% respectively. Thus, their total influences were -0.3% and 2.7%, respectively. But, the numbers were not significant at the population level.

The direct influence or strength of the factors that significantly affect the agribusiness behavior of the mango farmers were technological factors (X3), resource factor (X4), institutional factors (X5) and cultural factors (X6) were 2.3%, 5.7%, 6.4% and 5.3% respectively. While the indirect effects of these four variables that were related to 5 other factors were 2.5%, 3.5%, 1.8% and 1.7% respectively. Thus, the total effect was 4.8%, 9.2%, 8.2% and 7%, respectively. From these values, the coefficient of determination was 31.7%, which indicated that the ability of the six factors studied together in determining the variation of variables agribusiness behavior of mango farmers was 31.7% while the rest was determined by other factors that were not examined in this research.

4. Conclusion

Factors that had significant and positive influence in determining agribusiness behavior of mango farmers in Greged sub district of Cirebon District were: 1) Technological factors; 2) Resource factors; 3) Institutional factors; and 4) cultural factors. However, social factors and economic factors had no significant effect. The most powerful factor in improving the agribusiness behavior of mango farmers was resource factor (9.2%), institutional factor (8.2%), cultural factor (7%) and technological factor (4.8%) respectively. This indicated that developing a model of sustainable mango agribusiness in Greged sub district, Cirebon regency should be focused on optimizing of its resources, increasing its institutional role, preserving the culture of mango cultivation in its farmers and increasing the role of the use of mango farming technology.

Acknowledgement

Thank you to *Kemenristek Dikti* who has financed this research in the scheme of *PUPT (Penelitian Unggulan Perguruan Tinggi)* in fiscal year 2017.

References

[1] Sulistyowati L, Rasmikayati E, Budiman M A, Saidah Z 2014 Pengembangan Kemitraan Usaha dalam Upaya Meningkatkan Komersialisasi dan Pendapatan Petani Mangga. Universitas Padjadjaran

[2] Sulistyowati L, Natawidjaja R, Saidah Z 2013 Faktor-Social factor Ekonomi yang Mempengaruhi Keputusan Petani Mangga Terlibat Dalam Sistem Informal dengan Pedagang Pengumpul *Jurnal Sosiohumaniora.* **15 (3)** 285-293.

[3] Sulistyowati L, Natawidjaja R 2016 Commercialization Determinant Of Mango Farmers In West Java-Indonesia .*IJABER* **11(11)**: 7537-7557.

Study on the Roles and Benefits of Cooperative Producers of Coffee Margamulya in Java Preanger Coffee Agribusiness

E Djuwendah[1], T Karyani[1], A H Sadeli[1] and H Hapsari[1]

[1]Department of Socio-economic Agriculture, Padjadjaran University, Jl Raya Bandung-Sumedang KM 21 Jatinangor, West Java, Indonesia 45363

E-mail : endah.djuwendah@unpad.ac id

Abstract. Cooperative Producers of Coffee Margamulya (CPCM) is one of the cooperatives that help coffee farmers in Pangalengan to access capital and market with other parties. This research aims are : (1) to identify the performance of Coffee Producers Cooperative Margamulya, and (2) to identify the role and impact of cooperative in agribusiness java preager coffee. The research design used is qualitative descriptive with case study method. The research result showed that Margamulya Coffee Producers Cooperative had an important role in the development of Java Preanger coffee farming in terms of procurement of production input, facilitation of production process, product marketing, financial service, business risk minimization, and human resource development. The impact of cooperative's activities on the success of coffee farming are: ensuring the continuous input supply with cheaper price, standardization and improvement of product quality, increasing production volume, establishing effective distribution network, providing price guarantee, reducing the risk of production failure, and increasing the income of farming.

1. Introduction

Coffee is one of the leading commodities of plantation with high economic value for Indonesia. West Java has great coffee development potential because of the supporting natural resources and having a typical coffee known as Java Preanger coffee. Pangalengan Subdistrict of Bandung Regency has several mountainous areas as the center of Arabica coffee development such as Mount Tilu, Mount Patuha, and Mount Malabar with the coffee planting area reaching 1,312.20 hectares [1].

According to the research result of Indonesian Coffee and Cocoa Research Institute (*Balai Penelitian Kopi dan Kakao Indonesia*) of Jember in 2012, Mount Tilu area has the potential of producing high quality coffee of specialty type having the cupping test score 84.08 with excellent category. In 2014, it got cupping test certificate with the score 84.67 [2].

The production of Arabica coffee in Pangalengan is still dominated by smallholders, where small farmers with uneconomic size of farms (<0.5 ha) become the main producers of coffee [3]. On the other hand, farmers face the demands of coffee market dynamics related to the demand for quality products and the products that consider environmental issues. In order to accommodate the agribusiness of Java Preanger Arabica coffee in Margamulya Village, Cooperative coffee producers Margamulya (CPCM) was founded in 2014. CPCM provides services to the cooperative's members, which are Arabica coffee farmers, in the forms of provision of production facilities, development of production process, as well as marketing and facilitation of coffee farmers with financial institutions to meet the capital requirement for coffee farming.

The existence of a professionally managed cooperative in the coffee production center will provide great benefits for the development of Java Preanger coffee farming and coffee agro-industrialization in order to increase the competitiveness of coffee, raise coffee farmers' income, and spur the local economy.

2. Materials and Methods

This research used qualitative method. Qualitative research method is used to understand the phenomena experienced by the research subjects such as behavior, perception, motivation, and action holistically by way of description in the form of words and language in a special context that is natural, and by utilizing various scientific methods [4]. The research technique was case study. Each conclusion is only in certain case, in the sense that it cannot be generalized. The data sources were primary and secondary data. The primary data were obtained from interviews with the cooperative's members and management. The secondary data were obtained through literature study.

3. Results and Discussion

3.1. The Performance of Cooperative coffee producers Margamulya

The people in Pangalengan Subdistrict started to develop coffee in 2001, which was originally planted in Perhutani land. The cultivation of coffee from year to year continued to grow, and in 2003, Pusakasari forest farmer group was formed. Subsequently in 2006, a forest village community institution (*Lembaga Masyarakat Desa Hutan*/hereafter called LMDH) was established. In 2007, LMDH of Margamulya obtained coffee seedlings of 300,000 trees to be distributed to other farmer groups. In 2008, Margamulya farmer group, which was the forerunner to the establishment of the CPCM was founded.

Producer cooperatives are cooperatives where people who produce certain products come together and often find them in the sector of agriculture. Producer cooperatives provide services that support the improvement of the business or the profit of its members. According to Chaerudin Manaf and M. Nuraidi (2016), producer cooperatives assist farmers in providing agricultural production facilities, procurement of tools for cultivation process, providing agricultural crops processing facilities and its marketing, so that farmers can increase the added value and competitiveness of their agricultural products. In turn, they will increase farmers' income.

Cooperative coffee producers Margamulya was officially established through the Notarial Deed No. 9 dated March 18, 2014, and is located on Jl. Raya Bandung-Pangalengan Km 36.5 Margamulya Village. At the beginning of its establishment, the cooperative had 140 members of coffee farmers. In 2016, the members increased to 165 people. The facilities owned by the cooperative were coffee processing plant and crops processing unit (*Unit Pengelolaan Hasil*/UHP), which were obtained from the government in 2011. Other facilities included coffee processing warehouse, pulping machine (pulper), huller, grader, roaster, and espresso machine. With the crops processing facilities, the cooperative was able to collect 300 tons of coffee cherries per harvest season, and to process 70 tons of green beans through wet processing method. The quality of its coffee products was certified by the Indonesian National Standard (*Standar Nasional Indonesia*/SNI), UTZ Certified, and Halal Certificate.

CPCM had a collective business mission, namely to coordinate coffee from the community to be processed, and to help market it, so that it's expected to increase the income of its members. According to the statement from Moch. Aleh, what is meant by the collective system in business activities is as follows:

> "Collective" means working together … also together in processing … one of the examples is that there were those who just stored goods in us … then, we processed them … after processed into green beans, there would be payment to members who stored the coffee … like that …, and they were processed by our colleagues, not outsiders. However, they were also paid. Thus, we are working together; if Pak Aleh is absent, the others keep going on … whoever… this is co-owned. Not controlled only by us …"

Coffee processing activity in Cooperative coffee producers Margamulya applies a wet processing method, consisting of pulping with pulper, and then the coffee beans are fermented for 12 hours to remove mucus in the coffee beans. After that, the resulting coffee beans are called coffee grains, which are then washed by using a washer. After the coffee grains are cleaned from various impurities, then the coffee beans are dried until the water content is ± 30%. After drying, the coffee beans are inserted into the huller machine to remove the horns and the silver skin (epidermis), resulting in green beans. Then,

the green beans are dried again until the water content is close to 12%, and the then coffee green beans are produced. The coffee green beans are then stored in sacks and ready to be sold to consumers.
CPCM's business in coffee distribution and marketing included three activities, namely the purchase of coffee cherries and coffee grains from the members, the consignation, which means selling the products of non-members, and custom working on coffee (*maklunan kopi*). The purchases of coffee cherries and coffee grains were respectively with the price of IDR 7,000 and IDR 23,000 per kg. The purchase system from non-members was the buying-and-selling-off (*jual-lepas*), while the purchase from members with cash-and-back system. This means that the other reciprocities were in the forms of training, assistance, crops production facilities, and so forth. The coffee processing fees from non-members were kept in cash to be used as the cooperative cash reserves and not distributed to the members.

Consignation means the activity of storing coffee of the members and non-members in order that the sales are to be processed and assisted by the cooperative. Consignation fee was IDR 3,000/kg for the members. The processing and selling fees for non-members of the cooperative were 10% of the sales value. From *Maklunan*, a processing of coffee cherries or grain beans (HS) into coffee green beans from non-cooperative members, it received a fee of IDR 3,000/kg of cherries and IDR 1,500/kg of hold Skin (HS).

Coffee selling system conducted by CPCM was "Buying-processing-selling" (beli-olah-jual), meaning that the cooperative immediately processed the coffee cherries bought from the farmers, and then immediately sold them in order that the result of payment would be able to be used to buy the other coffee cherries. This system was applied because of the limited capital owned by the cooperative. This was as expressed by the treasurer of the cooperative, Wahyu Ginanjar.

> "...So, the system was buying-processing-selling.... Come money... then, buying, processing, and selling more... so, if supplied for the stock, they would really appear, but by performing buying-processing-delivering immediately, there will be no stock here. However, in the accumulation of money, the accumulation of production will increase... so is the system... Within 6 months of harvesting it, the sales are done several times. .... The faster the sales the more our production..."

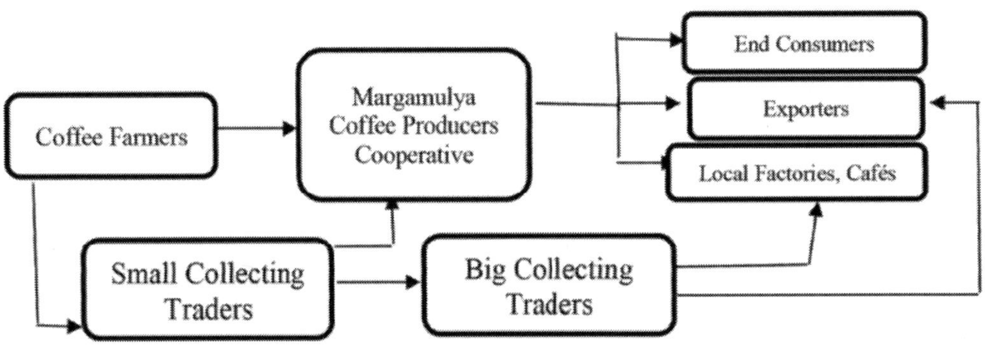

Figure 1. The distribution flow of Java Preanger coffee in Margamulya Village

In the market structure of coffee produced by CPCM, about 70% of the total processing results were sold to the exporter, PT. Taman Delta Indonesia, located in Semarang of Central Java. 30% of the processed coffee beans were sold to cafés. The rejected coffee beans were sold to local coffee processing companies.

Table 1. Green Beans Coffee Sale Price Based on Market Target of 2014

No	Target Market	US$ /kg
1	PT Taman Delta Indonesia	6 dollars
2	Aroma, Glory Morning, other cafés	5 dollars

The specification of Java Preanger coffee beans requested by PT Taman Delta were specialty quality with the UTZ certified, while Aroma, Glory Morning, and other cafes only demanded the requirement of specialty Arabica coffee beans.

3.2. The Role and Impacts of CPCM in the Development of Java Preanger Coffee Farming

To identify the role of cooperative in agroindustry/agribusiness, it can be mapped on the upstream to downstream subsystems (off-farm), so it can provide benefits for members of cooperative in increasing their income. Based on agribusiness conceptualization approach, the role of CPCM is presented in Table 2.

Table 2. The Roles and Impacts of CPCM in Java Preanger Coffee Agribusiness

No	The Roles of Margamulya Coffee Producers Cooperative	The Impacts on the Members
1	Procurement of farming inputs: - Performing quality coffee seedlings measurement - Performing the distribution of free coffee seedlings to coffee farmers - Assisting the distribution of pest repellents to farmers - Providing loans for coffee farming capital	- Availability of quality coffee seedlings - Getting the seedlings for free - Getting free pest repellent - Availability of capital
2.	Production process (coffee farming) - Fostering farmers in coffee farming according to UTZ certified SOP - Facilitation and assistance of the transfer of coffee farming technology, IPM, harvest, and post-harvest	- Ability to carry out cultivation technique according to the Standard Operational Procedure of UTZ Certified - Increased quality and quantity of the member's coffee yields
3.	Coffee processing (agroindustry) - Processing coffee cherries into green beans, roasted beans, and instant coffee (ready to brew) - Conducting custom work on coffee (*maklun*) ordered by non-members - Increasing the quality of the products	- The Increase of added value of coffee so that the price is feasible and they have loyal market shares - Providing job opportunities to members of the cooperative
4	Product marketing - Buying coffee cherries from farmers - Providing price guarantee - Conducting market research - Increasing the product distribution network	- Easy in coffee marketing - Certainty of selling and market prices - Members of cooperatives becoming the agents of distribution and promotion of the products

5	Financial services - Providing recommendations and guarantee of lending money to banks - Providing loans in limited quantities	- Easy access to capital from banks - Getting loan for farming capital from the cooperative
6	Minimization of farming risk - Providing storage facilities (warehouse) - Providing information center of quality, price, and coffee market, in the cooperation with Local Government Office of Cooperative (*Dinas Koperasi*), SMEs, Government Office of Industry and Trade, Indonesian Creative Youth Academy and coffee exporters	- coffee sales able to be done anytime - getting the prices according to quality and price development in the market in order to minimize losses
7.	Facilitation of human resource development - The cooperative as a place of human resource development training - Supervising the transfer of technology and farming skills	- That members of the cooperative can learn to solve the constraints of coffee farming development - Getting guidance and monitoring of coffee technology implementation
8	Research and development of coffee farming business - Building partnerships and business networks with various parties (government, local government offices, associations, entrepreneurs) - Performing market research	- That members of the cooperative can get the latest information related to coffee farming technology and coffee market development

Based on Table 2, it can be said that CPCM has important roles in various activities that support the development of Java Preanger coffee agribusiness. Referring to the traditional roles of cooperatives, that some effective activities that can be done through cooperatives include collective marketing, employment sharing, credit provision, risk management, as well as dissemination of knowledge and technology. The roles of producer cooperatives, namely input procurement, providing production process facilities and financial services, as well as risk minimization. Such roles can be implemented by producer cooperatives in order to provide benefits to the production operations of the members.

At the beginning of coffee development in Pangalengan in 2003, farmers used uncertified seedlings. In 2003, farmers began planting Central Aceh seedlings (*bibit Ateng*) and Kartika 1 seedlings, and in 2013, they began to use *Sigarar Utang* varieties recommended by Coffee and Cocoa Research Center (PUSLITKOKA) for coffee development in Pangalengan.

The CPCM had partnership with the local government to assist in the procurement of quality seedlings by helping the seeding of the coffee seedlings from the government, and distributing them to the farmers in need. CPCM also developed saving and loan activities in the form of coffee saving and loan. Initially, this activity was done when the capital to pay coffee to members was limited, so that the debt was used as saving. When funds received by the members from coffee sales were surplus, they were used as loans for the members. The saving and loan activities were recorded simply in the members' book of savings.

As for the asymmetric information in the supply chain of coffee about the quality and price of coffee, the cooperative played a role in providing clear indicators of the quality. This gives certainty of the coffee selling price to coffee farmers. The cooperative also had low cost in ensuring the quality of coffee demanded as the requirement by the markets because of the activities of assistance and monitoring of coffee farmers.

The factor of easy marketing became the basis of the farmers' consideration to become the members of the cooperative. The cooperative received coffee sales from farmers regardless of the quantity and the time, either during harvest season or not. The price that farmers received when selling to the cooperative was fluctuating, depending on the price the cooperative had received from the buyers. In the harvest season of 2016, the price of coffee cherries ranged from IDR 7,000 to IDR 7,500 per kg, and of coffee grains from IDR 20,000 to IDR 23,000 per kg.

Local coffee marketing trend in the international market is currently responding to a dynamic market. This is demonstrated by the increasing demand for the implementation of product certification related to social and environmental standards. The implementation of social and environmental standards on the coffee economy is important with the consideration that it will bring long-term implications for the environmental quality and sustainability of the coffee economy itself. In this case, Margamulya Coffee Producers Cooperative has an important role in the process of product certification. The certification that has been running is UTZ certified and in 2018, they will try to apply Fairtrade certified. This is done by inspection, monitoring, and guidance of the application of cultivation according to the SOP conducted by the cooperative management with the officers from PT Taman Delta and union certification.

Promotion activities have also been carried out by the Margamulya cooperative, such as following several exhibitions held every year from 2012 until now by the Local Government Office of Agriculture of Bandung Regency and the Local Government Office of Industry and Trade of West Java Province. Currently the cooperative has a website of online store on blibli.com, and have already an official website, javapreasngerkopi.co.id. In addition, the cooperative also prints advertisements in the form of brochures given to buyers or parties who visit Margamulya coffee plant.

According Soverani [6], based on the assessment of the services that had been given by Margamulya Coffee Producer Cooperative, it was obtained the level of satisfaction of the cooperative's members, which was in medium category. The quality of the service viewed from dimensions of tangibility, responsiveness, reliability, assurance, and empathy was in medium category. This indicates the need to increase the services of the cooperative for the members. Several things that can be done by the Margamulya cooperative are the increase of the cooperative capital, so that it can anticipate the risk of late payments from the less disciplined buyer.

4. Conclusion

Cooperative Producers coffee of Margamulya (CPCM) is an institution that accommodates Java Preanger coffee agribusiness activities in Pangalengan Subdistrict, by performing procurement of production facilities, as well as processing and marketing the yields. CPCM business in distribution and marketing of coffee beans is in the form of purchasing coffee cherries and grain coffee beans from the members, selling products of non-members (consignation), and custom work on coffee (*maklunan*). The coffee selling system conducted by CPCM is buying-processing-selling with the consideration of the limited capital of the cooperative.

CPCM played a role in the development of Java Preanger coffee farming, ranging from the procurement of production facilities, production process facilitation, product marketing, financial services, human resource development, and business risk minimization.

References

[1] BPS, 2016

[2]. Xena LG and Trimo L 2014 Potensi pengembangan Preanger coffee *Proceeding Seminar Nasional Pembangunan Inklusif di Sektor Pertanian* Departemen Sosial Ekonomi Faperta UNPAD Bandung 205-212

[3]. Rakotonjanahary J. Audry dan E. Djuwendah, 2018. Analisis Pendapatan Usahatani Kopi Java Preanger pada Kelompok Tani Margamulya Desa Margamulya Kecamatan Pengalengan Bandung. *Agrivet Journal* **6(1)**31-38

[4] Moleong, Lexy 2007 Metodologi Penelitian Kualitatif Bandung : Remaja Rosda Karya

[5] Chaerudin Manaf dan M. Nuraidi, 2016. *Model Pengembangan Koperasi Produsen*. Universitas Pakuan Bogor. *ejournal.unpak.ac.id/download.*

[6] Soverani O and T. Karyani . The Effect of Service Quality to the Satisfaction of the Member of Cooperative Producer of Margamulya Coffee in Subdistrict Pangalengan. https://ejournal.unib.ac.id/index.php/agrisep/article/.../3030/1518. AGRISEP Journal 16(2) 155-164. No. 2 September 2017

IOP Conf. Series: Earth and Environmental Science **334** (2019) 012056 doi:10.1088/1755-1315/334/1/012056

Consumption Patterns of Urban Household: Case in Sarimas Regency, Sukamiskin, Bandung

H Hapsari[1], Y Sukayat[1], E Djuwendah[1], Y Deliana[1] and L Sulistyowati[1]

[1]Faculty of Agriculture, Padjadjaran University Jl. Raya Bandung – Sumedang, Jatinangor KM 21 - 45363

E-mail : hepi.hapsari@unpad.ac.id

Abstract. Red chili was a strategic food commodity. Demand for red chili tends to increase, while production and prices fluctuate so as to affect the inflation rate was quite significant. Information about red chili consumer behavior can be used for product planning and price control. The purpose of this research was to know (1) the characteristics of red chili consumers, (2) the consumption pattern of red chili on household, and (3) the correlation between consumer characteristic and red chili consumption pattern. The research method was descriptive survey. Data collection with structured interviews and observation. Data analysis with cross tabulation and chi-square correlation test. Respondents were 69 housewives. The results showed all respondents used red chili for daily cooking. Average family income was 3-5 million/month. The pattern of red chili consumption was relatively stable on ordinary days and increased on feast days. There was a significant correlation between the number of families and the spicy taste with the consumption of red chili.

1. Introduction

Red chili is a food of commodity strategy's. Red chili has a major impact on the dynamics of the national economy, in which this commodity included in the rank of the national affecting inflation in Indonesia. Based on the data of [1] in February 2014, red chili contributed to Indonesia's economy amounted to 0.10% of total inflation of 6.75%.

Demand for red chili tends to increase, while production and prices fluctuate affecting the inflation rate significantly. Volatility in the price of chili that has occur in Indonesia, could not be separated from the factors of production and demand. Some things that affect the production and the request is the pattern of consumption, production, distribution, and government policies. Based on the consumption pattern Indonesia, most of the Indonesian people (70%) consumed chili fresh while the rest (30%) was consumed for industrial needs. This is because almost all Indonesian cuisine using fresh chilies as a spice. Few who use the processed chili such as chili sauce, instant sauce or chili powder.

Table 1. Red Chili Consumption of Indonesian Households Year 2008-2012 (in tons)

Chili type	2008	2009	2010	2011	2012
Red Chili pepper	15.486	15.226	15.278	14.965	16.529
Cayenne pepper	14.444	12.879	12.984	12.097	14.026

Source: National Social Economic Survey 2008-2012

Big chili has consumed more than small chili as popular as spice in cooking. Chili has a flavor that is not spicy and has an attractive color, so that it can make the display bright and cuisine to be able to increase appetite. The need for big chili has been increasing during the years due to the increasing number of people, variety of dishes, hotel, restaurant, cafe, thriving culinary and lifestyle of parties or celebrations are often held urban communities.

The consumption analysis of red chili on households is expected to represent consumer behavior as a whole red chili. Households have different cultural background (customs), social and economic status. The practice, education, and income are variables that can affect the formation of consumer attitudes and behavior. Information about consumer behavior of red pepper can be used for producing plant planning and price controls. This study aims to: 1) identify the characteristics of consumers of red pepper on the urban households; 2) analyze the consumption pattern of red chili in urban households; 3) analyze the relationship between consumer characteristics and consumption patterns on a large red chili urban households.

2. Methodology

The research method was quantitative descriptive survey, because the study aims to determine the relationship between variables and emphasizes the analysis on numeric data (numbers). Descriptive method was a method in researching the status of a group of people, an object, a system of thought or a class of events in the present. The object of research is the consumption pattern of red chili in urban households. Research was conducted in Sarimas Regency, Sukamiskin Village, Arcamanik Subdistrict, Bandung City. The selection of research area was because of the area has been a residential prototype ethnically diverse region of origin, in which red chili preferences also vary.

This section describes some concepts and the measurement of the variables to be studied :

a) Factors affecting household consumption patterns: revenue (Resource Economics); motivation; attitude; taste; the number of household members; points of purchase
b) Factors that influence the consumption patterns of individual characteristics : age; gender, education; skills; origin ethnic
c) Consumption pattern : total consumption; frequency of consumption

The data consist of primary and secondary data. Primary data were obtained from interviews of respondents using closed and structured questionnaire. Questions in the questionnaire, among other household characteristics, frequency, volume purchasing and consumption patterns of red pepper. Secondary data were obtained from the literature, research reports, the data related agencies, and BPS Bandung. Respondents were interviewed were housewives with consideration that household food purchases are usually done the mother represents the sense of the whole family.

1. Sarimas Regency household population were of 279 families
2. The samples by means of random sampling. The number of samples was determined by Vincent Gaspertz (1991), can be taken 5 percent; 10 percent; or 25 percent of the total population, that is 69 families.

Data consumer characteristics, the red chili purchasing decisions and the factors that influence purchasing decisions, were analyzed using descriptive statistics (crosstab) or tabulation frequency. The frequency of the data was calculated as a percentage as follows :

$$P = \frac{fi}{\Sigma fi} \times 100\% \qquad (1)$$

where :

P: The percentage of respondents who chose a particular category
fi : The number of respondents who chose a particular category
Σfi : Total answers

To test the hypothesis of comparative two independent sample means testing the significance of differences in the value of the two unpaired samples. Nonparametric statistics were used to test the hypothesis of comparative two independent samples when data are nominal is Chi Square.

Chi Square general formula used is as follows : $\chi^2 = \sum_{i=1}^{b} \sum_{j=1}^{k} \frac{(f_{ij} - fh_{ij})^2}{fh_{ij}}$ [2]

where :

f_{ij} : The number of observations for cases that fall within the i-th row j-th column.

fh_{ij} : The number of cases expected under H0 to be categorized in the i-th row j-th column.

Chi Square Table = Chi Square [0,05; (k-1)(b-1)]

Testing Criteria :

Reject Ho if the Chi Square test > Chi Square table, received in another case

Reject Ho if significance < α, Received in another case

Measurement variable attitudes, opinions and perceptions of a person's social phenomenon, using a Likert scale is ordinal scale of 1, 2, 3, 4, 5. The greater the value, the better or positive means a person's perception [3]

3. Results and Discussion

3.1. Characteristics of Respondents

Respondent characteristics include age, education, income, cooking skills, and regional origin. The average age of respondents is 34-42 years, and consumed chili without health violated. Average education respondents graduated from high school and skilled enough to cook. The average family income of 3-5 million per month, more than the standard minimum wage of Bandung 3.2 million rupiah per month. The price increase does not affect the amount of the purchase chili for a sizable family income. Chili price increase occurred because production is reduced due to the rainy season, high demand during Ramadhan, Idhul Fitri, Idhul Adha, Christmas and New Year. Motivation of respondents buy chili, especially to increase appetite and seasoning in Indonesian cuisine.

Number of family members is on average 3-5 people. More and more members of the family, the more consumption of chili. Respondents generally family tastes like spicy, especially for sauce. An assortment of local origin, most of Sundanese who liked the menu chili sauce and vegetable mix.

Table 2. The relationship between motivation purchase and the amount of consumption of red chili

Motivations for Purchase	Total consumption						Total		χ2	P(5%)
	< 1 Kg		1 - 2 Kg		> 2 Kg					
	f	%	f	%	F	%	f	%		
To find out the benefits of red chili	3	4.35	8	11.59	2	2.90	13	18.84	17.743	0.023
Red chili prices affordable	13	18.84	3	4.35	6	8.70	22	31.88		
Bought as offered traders	2	2.90	0	0.00	1	1.45	3	4.35		
Prices were falling	4	5.80	2	2.90	3	4.35	9	13.04		
Red chili can increase appetite	5	7.25	14	20.29	3	4.35	22	31.88		
Total	27	39.13	27	39.13	15	21.74	69	100		

3.2. Relationship between motivations for purchase and a total consumption of Red Chili

In the study of consumer behavior, motivation is driven by the need [4]. Consumer motivation determines consumption patterns, the amount and frequency of purchase. If the motivation to buy for necessities, consumers tend to buy in bulk and routine every time. Although pepper prices are high, consumers will continue to buy even in small amounts. The dinner menu is available in the household who frequently Indonesia is a condiment. The respondents were predominantly motivated respondents buy red chili because it can increase appetite and consume as much as 1-2 kg in a week that as many as 14 people (20%), followed by respondents buy red chili reasons affordable prices by consuming red chili <1 kg a week as many as 13 people (19%),

Based on statistical analysis obtained from Chi Square value of 17.743 with a P-Value of 0,023 where the value is smaller than α (0.023> 0.05), in which H_0 is rejected, thus it can be concluded there is a relationship between motivation purchase and the amount of consumption of red chili, with a confidence level of 95%.

3.3. Relationship between number of family members and consumption of Red Chili

According to Berg (1986) in [5], the number of family members affects food and non-food consumption patterns. If income is low, then a great family food and nutrition potentially is less four times greater than small families.

Table 3. Relationship between number of family members by total consumption

Number of Family Members	Total Consumption						Total		χ2	P(5%)
	< 1 Kg		1 - 2 Kg		> 2 Kg					
	f	%	f	%	f	%	f	%		
< 2 People	2	2.90	4	5.80	0	0.00	6	8.70	9.809	0.044
2 - 5 People	25	36.23	23	33.33	13	18.84	61	88.41		
> 5 People	0	0.00	0	0.00	2	2.90	2	2.90		
Total	27	39.13	27	39.13	15	21.74	69	100		

Most of the families have 2-5 members and consuming red chili <1 kg in a week. Based on statistical analysis obtained from Chi Square value of 9.809 with a P-Value of 0.044 where the value is smaller than α(0.044 <0.05), in which H_0 is rejected, thus it can be concluded that the degree of 95% of the data shows that there is a relationship between the number of family members and the amount of consumption of red chili. consisting of 2-5 people pretty much consume more chili. Based on the results of many studies more dominant family members consume lots of chili compared with family members who bit or the number of family members who are more relative.

With regard to characteristics mostly within a family, there are 2-5 people who consumed chili less than 1 kg. Generally, women are able to take into account the needs of red chili in daily to be taken into account when the chili will be depleted to meet the needs. Each family member has a number of different consumption.

Table 4. Relationship between taste and frequency of red chili purchase

Taste	Frequency of Purchase						Total		χ2	P(0,05)
	> 5 Times		3 - 5 Times		< 3 Times					
	f	%	f	%	f	%	f	%		
Like Spicy	51	73.91	2	2.90	8	11.59	61	88.41	22.519	0.000
Less Like it Spicy	0	0.00	1	1.45	3	4.35	4	5.80		
Dislike Spicy	1	1.45	0	0.00	3	4.35	4	5.80		
Total	52	75.36	3	4.35	14	20.29	69	100		

3.4. Relationship between taste and frequency of red chili purchase

A consumer tastes are respondents to the red chili can be connecting with the frequency of purchase of red pepper. According to [3] appetite will encourage consumers to buy more and more often. Respondents housewife tastes can be assumed to be equal to the tastes of the whole family.

The respondents who most often bought the chili is the respondents who liked spicy and bought as many red chili more than 5 times in a week, Chi Square correlation value of 22.519 with a P-Value of 0,000 where the value is smaller than α (0.000 <0.05) H_0 is rejected, thus it can be concluded that there is a relationship between the frequency of purchase and taste of red chili, with a confidence level of 95%.

The above data shows that consumers do not like spicy and less like spicy possessed the same total percentage of 11.6 percent. Consumers who liked spicy chili were affecting the intensity purchase as much as 88.4 percent. Chili demand increased because there was excessive consumption patterns with a high purchase frequency reached more than 5 times a week. This means that there is a relationship between the frequency of purchase and spicy taste of chili. The chili is known to increase appetite and therefore consumers buy more chili to meet the increasing frequency of purchase chili.

3.5. Relationship between Income and Frequency of Red Chili Purchase

Family income is the most decisive factor of the quantity and quality of food consumed. The low income causes the purchasing power of the food to be low and the family food consumption will be reduced [5]

Table 5. Relationship between income and frequency of red chili purchasing

Income	Purchase Frequency						Total		χ2	P(0,05)
	> 5 Times		3 - 5 Times		< 3 Times					
	f	%	f	%	f	%	f	%		
< 3 Million	9	13.04	0	0.00	2	2.90	11	15.94	42.115	0.000
3 - 5 Million	42	60.87	3	4.35	2	2.90	47	68.12		
> 5 Million	1	1.45	0	0.00	10	14.49	11	15.94		
Total	52	75.36	3	4.35	14	20.29	69	100		

The respondents Consumption Patterns of Urban Household: Case in Sarimas Regency, Sukamiskin, Bandung most often bought chili is that family with income of 3-5 million per month and bought red chili more than 5 times in a week. Almost every day these respondents were shopping chili. Based on statistical analysis obtained Chi Square value of 42.115 with a P-Value of 0,000 where the value is smaller than α(0.00 <0.05) H_0 is rejected, thus it can be concluded that the 95% confidence level data indicate that there is a relationship between income and red chili purchase frequency. This means the family income related to the frequency of chili purchase. With sufficient income, the family will be able to meet the needs of food, including chili though expensive. These results are based on the value $\chi 2$ count smaller than $\chi 2$ table and the level of significance acquired greater significance.

4. Conclusion

The average income of red chili consumer was 3-5 million per month, with age of 34-42 years old, high school educational background, skilled cooking, like spicy, and ethnic Sundanese.

Red chili consumption patterns include the amount of consumption of less than 1 kg per week and frekeusi purchase of more than 5 times per week. While the chili price raised, the amount and purchase frequency was not reduced.

There is a significant relationship between family income, purchase motivation, and the number of family members and the red chili consumption patterns. The reason for the purchase of red chili was for seasoning and increase appetite. The price increase does not affect the pattern of consumption of red chili. The consumption pattern includes the number and frequency of purchase. Membership many families red chili tend to consume in large quantities.

IOP Conf. Series: Earth and Environmental Science **334** (2019) 012056 doi:10.1088/1755-1315/334/1/012056

Acknowledgment
This publication is part of the research output of Participatory Plant Breeding Red Chili Unpad. Acknowledgements to DRPM, KemenRistek-Dikti which has provided funding of research through National Strategic Research (STRANAS) Competitive Grant scheme for fiscal year 2017.

References
[1] Badan Pusat Statistik Susenas. 2009. Outlook Komoditas Pertanian (Perkebunan). Pusat data dan informasi pertanian. Departemen Pertanian.
[2] Siegel S 1997 Statistika Non Parametrik: Untuk Ilmu-ilmu Sosial. Jakarta : PT Gramedia
[3] Sugiyono 2011. Metode Penelitian Kuantitatif Kualitatif dan R&D. Bandung : Alfabeta.
[4] Pujoharso C 2013 Aplikasi Teori Konsumsi Keynes Terhadap Pola Konsumsi Makanan Masyarakat Indonesia. *Jurnal Penelitian Fakultas Ekonomi dan Bisnis*. Universitas Brawijaya
[5] Diah I 2010 Pola Konsumsi Pangan dan Status Gizi pada Rumah Tangga Peserta Program Pemberdayaan Masyarakat di Kota dan Kabupaten Bogor. *Jurnal Penelitian Fakultas Ekologi Manusia.*

IOP Conf. Series: Earth and Environmental Science **334** (2019) 012057 doi:10.1088/1755-1315/334/1/012057

Value Chain Financing in Coffee Agribusines (the Case of a Cooperative in Pangalengan, West Java, Indonesia)

T Karyani[1], E Djuwendah[1], K Kusno[1] and A S Hermita[1]

Faculty of Agriculture, Universitas Padjadjaran, Indonesia

E-mail : tuti.karyani@unpad.ac.id

Abstract, Economic globalization encourages coffee farmers to prepare themselves to adapt and compete. In dealing with this issue, Margamulya Coffee Farmer Group has transformed into a cooperative under the name of the Margamulya Coffee Producers Cooperative (MCPC). MCPC sold 80% of its coffee production to PT Taman Delta Indonesia (PT TDI) and was committed to its permanent supplier. In fulfilling its role as coffee suppliers, farmers were faced with a lot of capital needs to produce quality coffee. MCPC needs capital to pay farmers cherry as well as processing and marketing costs. This study aims to describe the coffee value chain in MCPC, to identify the perpetrators and their business processes, and to design financing pattern suitable for farmers and MCPC. The method used was case study at MCPC using value stream mapping technique. The results show that there are two value chains in coffee agribusiness in MCPC while the actors in the chain consist of Farmer-Cooperative-Exporter and Farmers-Cooperatives-Retail (café/others). The business process in farmers remained in coffee producing, while the cooperatives focused on processing and marketing. The result of this study shows that the proposed financing model of value chain financing for its success requires the involvement of stakeholders from outside the supply chain. Value chain financing for structured markets gives more guarantee of business sustainability due to market certainty and price. This encourages farmers to maintain both the quantity and quality of their coffee products.

1. Introduction

In Pangalengan sub District, there is a cooperative whose members are coffee farmers named Coffee Producer Coperative Margamulya (MCPC). This cooperative holds and distributes 80% coffee beans to PT. Taman Delta Indonesia and the rest to other buyers such as Indocom, Sarimakmur and café around Bandung. The receipt of MCPC coffee by the structured market was due to assistance from PT Taman Delta in obtaining UTZ certificates essentially in the management of its business on the environment. With regard to the environment and social aspects can be obtained a high price [1]. In addition, in some countries, cooperatives can increase coffee farmers' access to wider markets [2].

The demand for 70 tons of green beans per year means that 350 tons of chery is required, and if the average price of cherry is Rp 8.500 per kg, which means the needs of funds to pay farmers around Rp. 3 billion. In addition, processing from chery to green bean certainly requires time and the availability of large funds to pay coffee farmers and for the cost of logistics and processing. This is certainly very difficult for cooperatives who are only 3 years old. This condition is because the payment system from PT Taman Delta Indonesia (TDI) has a grace period of about 1-2 weeks, while farmers want to be paid in cash because they need money to meet their daily needs. Meanwhile, if not immediately paid, then farmers would sell coffee to middlemen who roam the Pangalengan sub District by offering in cash payment but lower prices. In other words, the abundant coffee harvests from farmers were not currently accompanied by the readiness of the downstream subsystem (agroindustry) owned by MCPC as a result of the weak capital of the cooperative.

From several previous studies, financial problems in a chain involve actors whose needs are different, especially during pre-harvest and post-harvest [3]. In the case of the MCPC, fee is required to pay the raw material of cherry coffee from farmers and the cost for processing cherries into green beans. Therefore, this study aims to examine how exactly business processes in the value chain of coffee and how appropriate financing patterns can be developed to meet the needs of actors involved in the Java Preanger coffee value chain in the MCPC.

2. Materials and Methods

Research object is coffee value chain financing with structure market orientation. Meanwhile, research subject was covered all coffee agri-business actors along the supply chain of every agri-business sub system from up-stream and down-stream who were involved in the chain.

Qualitative research was used in this research, due to the necessity of describing research study on agriculture institution roles and comprehension on research object phenomena as envisaged in the reality [4]. Case study was applied as research technique that is used to give in-depth and intensive study of an element or a target group on research subject [5].

The needs of value chain financing identified by value chain mapping that described the flow of information, goods, and money. Business process and needs of financing of some particular actors in the chain will create financing scheme with each actor capacity in the chain as its consideration.

3. Results and Discussion

3.1 Value Chain Mapping

MCPC has various marketing channels. The diverse marketing channels are determined by the various forms of coffee products produced by MCPC. Each product has a different target market. Each type of products produced has a different value, and there are different actors and activities in different chains. The various value chains contained in the MCPC are simply explained by the actors and the product forms in Figure 1.

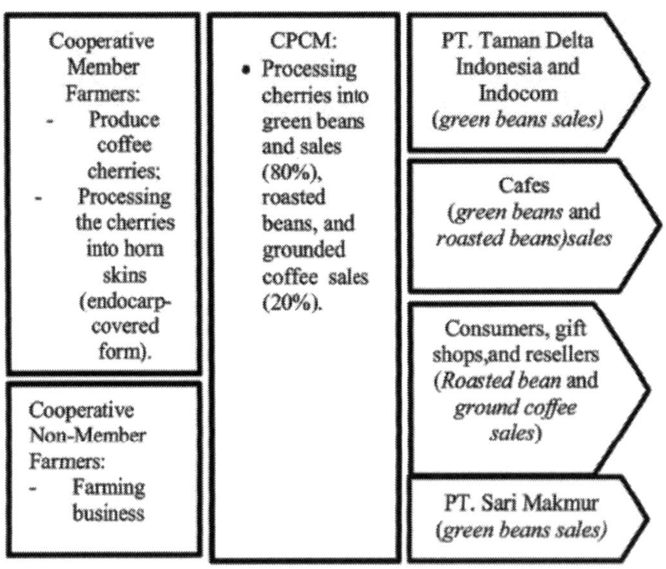

Figure 1. Value Chain Mapping on Coffee Agribusiness at MCPC

However, the value chain mapping through the cooperative actually is divided into 2 VCs, which are structured market oriented (PT TDI) and unstructured markets (cafe, PT Sarimakmur, Indocom and end consumers). In the structured market, the quantity and quality of coffee demand is continuous, so there is a market guarantee

3.2 Actors in Coffee Agribusiness Value Chain and Business Process

There are several actors along the value chain that involve in coffee agribusiness. These actors are farmers member and nonmember farmers, MCPC and Buyers who have different kind of role and activities in order to support of coffee agribusiness.

Table 1. Value chain actors of coffee agribusiness

Actors	Business Process
Farmers	Member farmers conduct cultivation of UTZ-certified Arabica coffee crops of *Sigarar Utang* variety and harvesting. The harvesting technique used by the member farmers of the cooperative was selective harvesting. Non-member farmers were producers of raw materials for the processing of green bean coffee of careless quality conducted by MCPC
Cooperative (MCPC)	The activities conducted by MCPC were the purchase of cherries, coffee processing, packaging, coffee marketing, and shipment. MCPC accommodated Arabica coffee farmers, established partnerships with various parties, conducted coffee processing management, and had a role as a distributor of aids to coffee farmers.
Buyers/exporter, PT Indocom, PT Sarimakmur, Cafes	Exporters are actors who need commodities based on contracts. Buyers purchase the harvested coffee through farmer group according to a contract that has been agreed in regards to the quantity, quality, continuity, prices, and payment method. PT. TDI finances UTZ certification for the coffee processing of MCPC as well as all farmers who were currently registered as MCPC members. The local coffee companies included PT. Indocom, who bought green beans of grade II and III specialties, and PT. Sari Makmur, who bought the green beans of careless quality. MCPC sold specialty coffee in the form of green beans of Grade I, and roasted beans to cafes and some individual end consumers

3.3 Pattern of Financing on Coffee Value Chain Actors

Value chain actors have partial financing depending on each one to the financial institution around them. But not all can access financial institutions, especially farmers. Farmers' lack of access to formal financial institutions due to ignorance and inability to reach them.

Coffee farmers largely finance their coffee budgets from their own capital, because for seeds and fertilizers, at the beginning of planting, are provided free assistance from the government through cooperatives. Some farmers (40%) obtained loans from cooperatives amounting to Rp. 500,000 up to Rp. 5,000,000. Returned when the harvest arrives by cut from receipt of the sale of coffee to the cooperative.

Cooperatives in managing their business, first were assisted by the government in the form of land for processing plants, offices, coffee processing equipment and machines, buildings for factories and offices. For working capital, among others, to buy cery from farmers who should sell, but because farmers need cash money then to buy cery, cooperatives use funds from loans from BNI and other banks. Cooperatives can not rely on their own capital from member savings because there are still many members who do not understand cooperative. The need of this fund is very big because if to meet the needs of PT TDI buyer which is 70 tons of green bean per year, it means the cooperative requires 350 tons of cery worth Rp. 3 billion a year. Cooperative until now can only get credit up to Rp. 500 million in PKBL BNI and BRI approximately Rp.900 million used for working capital and business development, so when viewed from the needs are still not met. Therefore then there are some investors invest funds in cooperatives including the board by applying the concept of profit sharing.

Formal financial institution in Pangalengan could only provide funds or payments in the conventional way (old paradigm approach) that was only providing credit according to government programs such as KUR with interest rate of 9% per year, the amount given in Office Unit maximum Rp. 200 million, in case more than Rp. 200 million must go to the branch office. Currently with President Jokowi's policy on social forestry, farmers get legal certainty in managing their land, so the possibility of accessing credit for KUR becomes easier.

Social forestry is the government policy of providing access for communities in and around forests to manage forest areas sustainably. This step is expected to create jobs and improve the welfare of the people, while taking care of the environment of forests with sustainable. The program targets 12.7 million hectares of community management access through five schemes, namely Community Forest, Village Forest, Community Plantation Forest, Partnership, and Indigenous Forest.

With the existence of social forestry is open for farmers to utilize forest land with plants that still maintain environmental sustainability, among others, by planting coffee. Even farmers who at the same time become a coffee processor will get a higher profit than just as a farmer. Therefore, in addition to increasing the productivity of coffee, farmers can apply for credit or financing for processing machines.

As explained before, that 80% of coffee farmers who are members of MCPC already have a structured market that is to exporter (PT TDI) means there is a certainty of market both volume and price so that it is called as a value chain structured market orientation. Therefore, along with the Value Chain (VC), the financing offered is the financing of the value chain not the partial per perpetrator.

Currently in some countries there has been a growing value chain approach of finance in rural areas. The relevance of this rural financing approach arises from the observation that an integrated activity between the financial sector to facilitate the flow of goods and services from producers to consumers. As stated by [6] that "Value chain finance - financial products and services flowing to and/or through VC to address the needs of those involved in that chain, need for finance, a need to secure sales, procure products, reduce risk and / or improve efficiency within the chain". Thus the financing of this value chain will complement the rural and agricultural finance system approaches. In other words, this value chain is another way to improve farmers' access to rural finance [7]. Therefore, referring to this VC financing model, then for the case of coffee business in MCPC, it is suggested that the financing pattern as depicted in Figure 2

Figure 2 shows that the stakeholders involved directly are farmers and cooperatives, while as a supporting institution is a bank financial institution and the government and universities are as facilitators. This model is a triple helix implementation, all of which have important role to succeed the proposed design. For more details as follows:

1. There are 2 types of financing required by farmers, i.e. working and investment capital. Working capital includes agro-input such as fertilizer and labor costs, while investment capital refers to a simple processing machine to obtain higher value added than selling only in the form of cherry. The required loan is around Rp. 7 million up to Rp. 10 million. The mechanism of farmers through cooperatives proposes financing to banks for farmers' need to obtain agroinput credit. Agroinput company shows the agroinput value invoice purchased by farmers through cooperative, then bank pay to agroinput agent. Furthermore, farmers will repay the loan from the fruit of coffee sold through cooperatives.
2. For cooperatives required is temporary loan for cherry payment to farmer before receiving payment from buyer. This need is because farmers can not delay payments from cooperatives to meet their life needs. Thus a suitable financing model is factoring, the cooperative shows an exported invoice and paid by the bank about 70-80% of the green bean value sold to the exporter. Once the exporter pays the entire green bean value of the cooperative through the bank, then the bank pays the rest. .

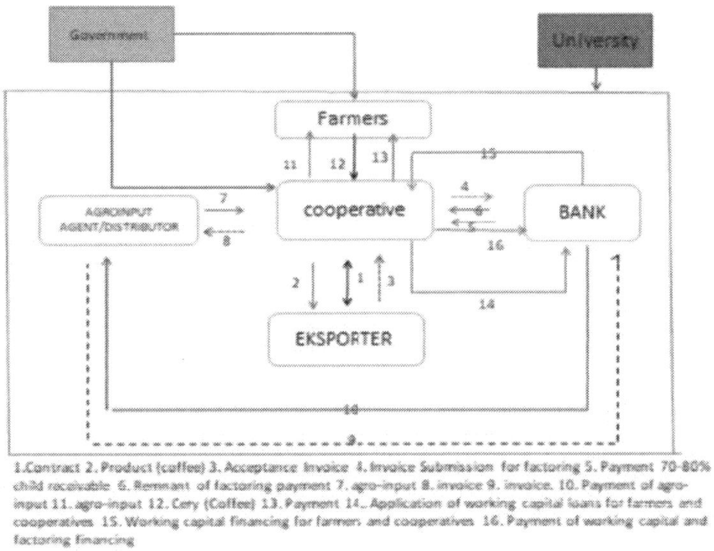

Figure 2. Design of Financing Pattern for Coffee Agribusiness at MCPC.

The role of higher education institution is to provide facilitation and provide assistance for the smooth financing of this VCF. The government through the Department of plantation and Department of Cooperatives and SMEs to provide technical guidance for coffee farmers in both quantity and quality, and to facilitate the form of coffee development program through cooperatives such as seedlings, engagement MCPC in events organized by the government both at the district, province and national level.

4. Conclusion

The result of value chain mapping indicates that there are 2 market orientation of coffee produced by MCPC, i.e. export orientation (80 percent) and retail orientation such as cafe and individual (20 percent). The channel is very short of the Farmers-Cooperatives - Exporters (PT Taman Delta Indonesia) and Farmers-Cooperatives-Retail (Indocom, cafes, and consumers). The business process of farmers is an on-farm activity, only about 15% have done simple processing that causes 'horn skin' or grain coffee. MCPC's business processes include providing production facilities for member farmers, providing guidance to members and facilitation of various coaching or mentoring from third parties to farmers. In addition, the MCPC is processing cherries into green beans, roasted beans and grounded coffee.

The financing model designed for MCPC conditions with the farmers as members is Value Chain Financial where farmers can be given financing for their farming from banks through cooperatives. The cooperative requires bailout funds for farmers chery payment with factoring.

Need to strengthen institutional capacity of cooperative both human resources and capital, hence, the cooperative in the future can be a direct exporter. Implement VCF patterns for actors in the coffee value chain for a sustainable value chain.

Acknowledgments
This research is supported by Unpad Internal Grant in RKDU scheme 2017//2018, therefore, through this opportunity, the research team would like to thank all the support of DRPM Unpad. Likewise, we would like to thank the management of MCPC for their cooperation.

IOP Conf. Series: Earth and Environmental Science **334** (2019) 012057 doi:10.1088/1755-1315/334/1/012057

References

[1] Gonzalez-Perez and Santiago Gutierrez 2012 Cooperation in coffee markets: the case of Vietnam and Colombia. *Journal of Agribusiness in Developing and Emerging Economies* **2(1):** 57-73.

[2] BeucheltTina D and Manfred Zelle, 2012 The role of cooperative business models for the success of smallholder coffee certification in Nicaragua: A comparison of conventional, organic and Organic-Fairtrade certified cooperatives. *Renewable Agriculture and Food Systems* **28(3):** *195–211*

[3] KaryaniT, Agriani S, Hesti N U, Elly R 2006 *Mango Agricultural Supply Chain: Actors, Business Process, And Financing Scheme. International Journal of Applied Business Economics Research*. **14:** 117751-7764

[4] Idrus M 2007 *Metode Penelitian Ilmu-Ilmu Sosial* (Pendekatan Kualitatif dan Kuantitatif). UII Press. Jogjakarta

[5] Moehar D 2002 *Metode Penelitian Sosial Ekonomi*. Bumi Aksara. Jakarta

[6] Miller, Calvin and Jones, Linda. 2010. *Agricultural Value Chain Finance Tools and Lessons.* Food and Agriculture Organization of the United Nations and Practical Action Publishing

[7] Chalmers, Geoffrey 2005 “A Fresh Look at Rural and Agricultural Finance,” Washington, D.C.: AMAP/USAID, Rural and Agricultural Finance Note No. 1 (www.microlinks.org).

IOP Conf. Series: Earth and Environmental Science **334** (2019) 012058 doi:10.1088/1755-1315/334/1/012058

The Analysis of Poverty Trap on Smallholder Dairy Farms in Pangalengan Area

M Sulistyati[1], Hermawan[1], and A Firman[1]

[1]Animal Husbandry Faculty, Padjadjaran University, Sumedang 45363, Indonesia

E-mail: marina.sulistyati@unpad.ac.id

Abstract. Dairy farming is dominated by smallholder producers in Pangalengan area. The characteristics of poverty trap are lack of capital, low productivity, low income, low savings and investment. Therefore, the purposes of this research were: 1) to find out the ability of smallholder farmers to increase the family income; 2) to analyze which factors influence on the existence of keeping dairy cows in small scale; and 3) to know the efforts of smallholder producers release from poverty trap. The research method used a survey method with quantitative and qualitative approaches. Number of respondents that chosen in study location were 100 respondents. The ordinary least square (OLS) is used to analyze the factors that affect the keeping dairy cows in small scale. The results showed that: 1) small scale dairy farms (one to three heads) have not been able to meet the needs of family life; 2) factors that influenced on the ownership of dairy cows in small scale those are cows asset, dairy farming income, cow productivity, and non-formal education; 3) the efforts of smallholder farmers to quit from poverty trap must increase number of cows, productivity, earnings, and improve non-formal education in the form of training and guidance.

1. Introduction

Dairy farmers are still exist due to the dairy business is to be a livelihood for most farmers. In general, farmers in Pangalengan area are members of cooperative Koperasi Peternak Sapi Bandung Selatan (KPBS)). The cooperative has established five Milk Collection Point (MCP) program in five Cooperative Service Centers area, such as Los Cimaung, Cipanas, Warnasari, Citere and Mekarmulya. The MCP assists farmers to get the individual of price of fresh milk using KPBS standard where the justification of milk price was used to Cooperative Service Centre price. This price standard could decrease farmers income who has a good milk quality.

The phenomenon of emerging poor farmers is due to their inability to compete to increase production and compete with the outside world. In according to [1], various efforts have been made by small-scale dairy farmers, such as adding dairy cows, but due to limited land so that they still place them on a small scale of ownership, so that opportunities to increase income are still difficult to reach.

Poverty in small-scale dairy farmers is a social phenomenon that cannot be assessed as something that must be eradicated from the surface of the earth. They are trapped in a circle of poverty and difficult to escape from this condition. Trap poverty or known as the poverty trap is a mechanism that makes it very difficult to get out of poverty. The trap of poverty is created when the economic system requires a large amount of various forms of capital to generate enough money to get out of poverty. When individual farmers lack capital, they may also find it difficult to gain access to financial institutions to get loans.

This theory has actually been stated by Ragnar Nurkse in [2], which states that poor countries remain poor because of underdevelopment, market imperfections and lack of capital causes low productivity. Low productivity levels can cause low income. The derivative has an impact on low savings and investment. Low investment can have an impact on low employment and have implications for the emergence of unemployment and so on [3].

The problems expressed by Nurkse in [2] are closely related to what happens in the dairy cow business. The dominance of small-scale livestock businesses raises more questions, what really happens with small-scale businesses. Do they themselves not want to leave the business community that only maintains 1-3 lactation cows? In addition, the low productivity of dairy cows further makes them in a vicious circle of poverty.

Some interesting things to be studied further in this study, namely:

1. Domination of small-scale animal husbandry businesses in the South Bandung Cattle Breeders Cooperative (KPBS).
2. Small businesses have supported farmer households
3. Low productivity of dairy cattle business has an impact on income
4. Ability to maintain small-scale dairy businesses
5. Are there households that can release from small-scale ownership circles

The above issues are the first step to detect the cycle of poverty that occurs in small-scale cattle farming. Poverty is not only assessed from an economic perspective but can also be assessed from a social perspective that leads to the limitations of individuals or groups in accessing networks and social structures that support opportunities to increase productivity. These inhibiting factors generally include internal factors, including individual farmers' characteristics such as age, formal education, experience, livestock ownership and cultural barriers. While external factors come from outside one's ability, such as bureaucracy or official regulations that prevent someone from getting resources. Therefore, to overcome this problem, a sharp analysis and comprehensive and sustainable handling by all parties is needed.

The objectives of this research are: 1) to find out the ability of smallholder farmers to increase the family income; 2) to analyze which factors influence on the existence of keeping dairy cows in small scale; and 3) to know the efforts of smallholder producers release from poverty trap.

2. Materials and methods

The location of the study was selected purposively at 5 MCPs, namely Los Cimaung, Cipanas, Warnasari, Citere and Mekarmulya from 23 group farmer in the KPBS Pangalengan area. Primary data was obtained based on the results of interviews with households of dairy farmers who have dairy cattle lactation of 1-3 heads. Interviews were also conducted to managers and cooperative leaders to obtain comprehensive data. For the purposes of calculation or analysis, secondary data support from various agencies is indispensable, such as from KPBS, the Bandung Regency Livestock and Fisheries Agency, the Bandung Regency Statistical Center (BPS), and other institutions related to this research. The research location was selected purposively at 5 Cooperative Service Centers namely: Los Cimaung, Cipanas, Warnasari, Citere and Mekarmulya from 23 Cooperative Service Centers in the KPBS Pangalengan area. Primary data were obtained based on the results of interviews with dairy farmer households that have 1-3 lactating dairy cows. Interviews were also conducted to managers and cooperative leaders to obtain comprehensive data. For the purposes of calculation or analysis, secondary data support from various agencies is indispensable, such as from KPBS, the Bandung Regency Livestock and Fisheries Agency, the Bandung Regency Statistical Center (BPS), and other institutions related to this research.

Respondent determination technique used in this research is Stratified Random Sampling, according to Paturochman random sampling stratification is a sampling technique that is carried out in layers with the characteristics of members who are heterogeneous or not uniform. Respondents taken from this study were members of breeders who kept dairy cows from five Cooperative Service Centers with total population of 506 members of the cooperative, taking samples of 100 people representing five Cooperative Service Centers.

Dairy business income is used to analyze the ability of small-scale business farmers to sustain their lives. The formula for finding income contribution from dairy cattle is as follows:

$$\pi = \sum \mathbf{TR} + \sum \mathbf{TCi}$$

TR= Total Revenue   Π = Revenue [4]
TC= Total Cost

To find out the factors that influence small-scale dairy cattle business ownership, the Ordinary Least Square (OLS) method is used.

The analytical method used in this study is quantitative approach, such as calculating the contribution of income, besides the frequency distribution is calculated. Qualitative approaches are used with in-depth interviews (independent interviews).

3. Results and discussion

The characteristics of respondents could be seen at Table 1, such as age, formal education, and livestock experiences. This is considered important because it is a factor that exists in farmers and is considered as the driver of its business. The first factor of age will influence the knowledge and mindset of respondents. The higher a person's age, the more knowledge or knowledge he has because the respondent's knowledge comes from his own experiences and experiences obtained from others. Younger age, farmers have a higher enthusiasm to know things they don't know yet.

Table 1. The characteristics of respondents in five

No	Description	**Cooperative Service Centers** *(person)*				
		Cipanas	Citere	Los	Mekar Mulya	Warna Sari
1	Age (years)					
	< 50	18	11	16	6	16
	≥ 50	11	6	2	6	8
2	Formal Education					
	Elementary school	16	12	8	10	18
	Junior high school	11	4	8	2	2
	Senior high school	2	1	1	0	4
3	Experience (years)					
	< 10	4	5	5	1	5
	≥ 10	25	12	13	11	19
4	Length of membership (years)					
	< 10	9	5	6	1	10
	≥ 10	20	12	12	11	14
5	Ownership (heads)	2,2	2,2	2,9	2,2	2,3
	Number of respondents	**29**	**17**	**18**	**12**	**24**

3.1. Dairy farming is a livelihood for most farmers

Dairy cattle business is the main business for most farmers involved in dairy farming. At present, small-scale dairy farming business still dominates the milk business. Based on the survey results in five KPBS Pangalengan Cooperative Service Centers shows that the ability of small-scale farmers to support their households with dairy cattle business can be seen in Figure 1. This shows the income and costs of milk production from 100 farmers who were respondents. Based on the picture it is clear that if farmers rely solely on income from milk income, the dairy cattle business is not profitable because generally, milk income is not enough for production costs. This shows that the ability of dairy cattle businesses with only 1-3 heads cannot meet the livelihood needs of livestock families if they only rely on their dairy farming business. This is in accordance with the opinion of [3] which shows that the income of small-scale livestock households cannot meet the needs of a decent life.

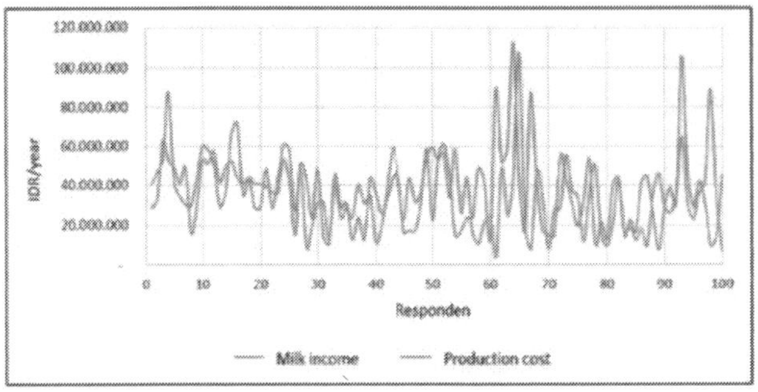

Figure 1. Income of Milk and Production Costs of Small-Scale Dairy Cattle Businesses

Figure 1 shows that the combination of income with other income obtained by the families of small-scale farmers, such as the sale of calves, heifers, bulls, or rejected cattle. This means that there is another income from the dairy farming business that can sustain their lives. The results of the combination of revenue from dairy cattle business activities that generate income compared to the cost of dairy cattle production can be seen in Figure 2.

Furthermore, if the income from the dairy cattle business is compared with the poverty line in each respondent's family, see Figure 2. Determination of the poverty line refers to the poverty line of Bandung Regency in 2016, the poverty line is IDR 275,562 /capita/month or equivalent to IDR 3,306,744 /capita/year. The poverty line value is then calculated by the number of family members of each respondent, the results are seen in the Figure. In general, farmers who only rely on the sale of milk to the cooperative and there is no sale a calf, virgin, or male parent, will be below the poverty line. Therefore, poverty vulnerability occurs in these small-scale farmers if there is no income from other sources. Thus, one effort that needs to be attempted so that small-scale farmers are able to earn income other than dairy cattle businesses can help them get out of poverty through male calf sales.

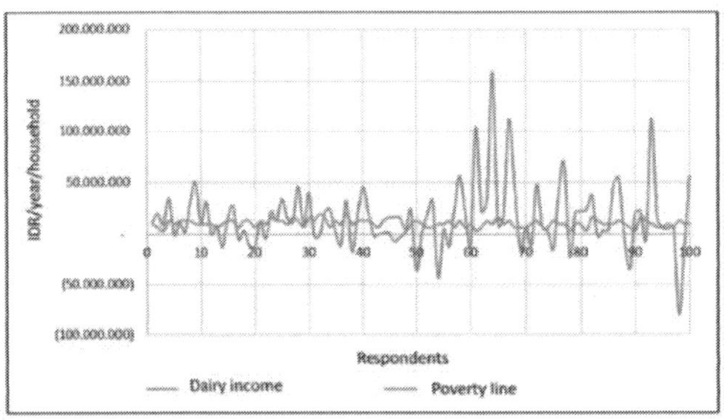

Figure 2. Income of Dairy Farmers

Respondents would continue to do business if the income received from livestock products is better than other businesses. However, if the income from livestock is insufficient or classified as low income then it will seek additional jobs in other fields other than livestock breeding. In accordance with the opinion [5] which states that if a community's income or income obtained is relatively insufficient for the primary needs, the community will seek additional income. This is supported by the opinion [6] stating that income generated from livestock businesses, and outside the non-livestock sector, will play a key role in determining to keep raising or migrating to the non-livestock sector. The results [7] research also support that states that income results have a significant effect on regeneration. Based on the results of the above study it is known that the cognitive aspects possessed by respondents regarding the desire for regeneration of dairy cattle business are good enough because some respondents are quite understanding of the desire to regenerate dairy cattle business.

3.2. Networks and social structures that support productivity improvement

Social networks are relationships that are created between farmers in a group or between groups with other groups. The relationships that occur can be in the form of formal or informal forms. Social relations is a reflection of cooperation and coordination among citizens based on active and reciprocal social ties [1]. The cosmopolitan level is one indicator of social networks. Farmers' activities in dealing with other farmers or other parties. The level of cosmopolitan is also interpreted as an orientation outside the social system with broader interpersonal relationships. The cosmopolitan in this study was seen based on the activities of respondents out of the village, visiting and/or visited by other breeders in one month to get information. Farmers who like to join people outside the social system are generally more innovative.

Most respondents only do social networks with farmers in one group, close ones or communication occurs when depositing milk. Communication with farmers outside the group is only carried out at certain times such as meetings at KPBS. The Chairperson of the group has a wider social network than its members because the attachment of the chairman to the KPBS in the context of monthly meetings causes the chair to have a higher frequency for interpersonal and external communication. Farmers' social structure is partly formed because of the social networks that have been formed. The Social structure formed formally because as a group member. Position in groups also determines interpersonal and interpersonal communication networks.

3.3. The efforts of smallholder farmers out of poverty trap

The poverty trap is a circle that the family could not out of the trap if there is no effort to release from the cycle even that the situation will be transferred generation to generation. The vicious cycle of poverty can actually be solved through education or training or with other factors. The results of this study indicate that the cycle of poverty in small-scale dairy cattle business can be solved through 4 factors, namely:

1. Increased Dairy Cattle Assets
2. Increasing Farmer Income
3. Increasing Productivity of Dairy Cows
4. Improving Non-Formal Education

These results have shown that the cycle of the vicious cycle of poverty in dairy cattle businesses in the KPBS Pangalengan work area can be solved by increasing the number of dairy cows, especially the parent. Only this increase in dairy cattle must be accompanied by an increase in the area of the cage that must be provided by the breeder (see Figure 6). This is in accordance with the regression table above, even though land assets are not significant, but an increase in land area will increase the scale of the dairy cattle business.

Increasing cattle business income is an absolute thing to do so that the scale of dairy cattle business increases. Increased income can be done by increasing the price of milk at the farmer level, the ability to sell non-productive dairy cows, or generating other income from outside the dairy cattle business. This is in line with the research [3] which shows that the participation of the wife of a farmer can increase the income of a dairy cow family business in Bangladesh. The wife can break the circle by providing value-added income to the family of dairy farmers, for example working in the non-livestock sector or in line with the dairy cattle business that can generate additional income for the family. The productivity of dairy cows is another factor that can solve the vicious cycle of poverty. If livestock that is kept have high productivity, they just need to maintain a few livestock because by only raising livestock with high productivity, farmers can generate quite good income. Non-formal education will also increase the dairy business. If all farmers get training, they will be interested in maintaining large quantities. However, the survey results show that most farmers have not received direct training from instructors, either from cooperatives or from other extension officers. This is in line with what was suggested by the World Bank [8] that education can solve the cycle of poverty that occurs in rural communities.

4. Conclusion

Small-scale dairy cattle businesses are able to support farmers' households. Supporting networks and social structures to get opportunities to increase the productivity of small-scale dairy farmers in the work area of KPBS Pangalengan. Efforts of farmer households to get out of the small-scale business circle or *poverty trap* in the work area of KPBS Pangalengan can be done through an increase in livestock assets, increase in dairy products, increase income, and improvement in non-formal education in the form of training and counselling.

Acknowledgement
The research is funded by HIU Program organized by DRPM Padjadjaran University. We would like to thank for the effort of DRPM and staffs to facilitates this study.

References

[1] Damsar 2002 Sosiologi Ekonomi (edisi revisi) *Rajawali Pers* Jakarta.
[2] Kuncoro M 2000 Economic Development: Theory and policy issues *YKPN* Yogyakarta.
[3] Yasmin S, Ikemoto Y 2015 Women's participation in small scale dairy farming for poverty reduction in Bangladesh *American International Journal of Social Science* **4(5)**.
[4] Paturochman M 2012 Determination of Amounts and Sampling Techniques *Unpad Press* Bandung.
[5] Subarno A 2012 Relationship between the level of social economy and development at PPI Pantai Sadeng GunungKidul *Thesis Magister Science UNS* Surakarta.
[6] Perloff J 1991 The impact of wage differential on choosing to work in Agriculture *American Journal of Economic* **73(3)**: 671-680.
[7] Hennessy T 2002 Modelling Succession on Irish Dairy Farms *Pape prepared for presentation at the Xth EAAE Congress Exploring Diversity in the European Agri-Food System* Zaragoza-Spain.
[8] World Bank 2006 Equality and development report *World Bank* Washington DC.

IOP Conf. Series: Earth and Environmental Science 334 (2019) 012059 doi:10.1088/1755-1315/334/1/012059

Financial access of farmers and factors associated with the access: empirical evidence of banana farms in West Java, Indonesia

E Wulandari[1], Ernah[1] and D Supyandi[1]

[1]Faculty of Agriculture, Universitas Padjadjaran,
Jl. Raya Bandung – Sumedang KM. 21, Jatinangor 45363, West Java, Indonesia

E-mail: eliana.wulandari@unpad.ac.id

Abstract. Access to finance has an important role in managing agriculture. Farmers in Indonesia can access finance from different sources of finance such as from banks, micro finance institutions, government through farmers' associations, buyer/traders, agricultural input kiosks, and from other sources of finance including family and friends. The objective of this research was to analyse the financial access of farmers and factors associated with the access. A survey was conducted to collect data especially from one hundred banana farmers in the centre of banana production in West Java, i.e. Cianjur and Ciamis District. The data were analysed using descriptive statistics and logistic regression. The results show that the majority of farmers had access to finance from government distributed through farmers' association, and from family and friends. Furthermore, the results show that factors associated with the financial access of farmers differ among the sources of finance. Age and farming experience have statistically significant association with the access to finance from bank. Moreover, education and farm size significantly associate with the access to finance from buyer/trader.

1. Introduction

Agricultural sector in Indonesia has faced problems in the provision of business capital, particularly in small and medium business scale farmers. Small farmers in terms of farm size have relied on the part of profit that they earned from previous season.

In general, the provision of finance for agriculture can be accessed from formal and informal financial sources [1, 2, 3]. Formal financing institutions provide finance in the form of credit from bank, while informal financing can be provided by agricultural input seller, agricultural product buyer, money lenders, friends, relatives and government aid in the form of agricultural inputs such as fertilizers and seeds [3]. The study of Wulandari et al [4] has shown that farmers also have access to finance from agricultural input kiosk in term of the flexibility payment of inputs.

The main challenges in accessing finance from formal sector such as from banks relate to administrative procedure [5] and the lack of collateral [6]. Most of farmers in rural areas have access to finance from non-formal financial sources such as from middleman, though a relatively higher interest but they provide easier access to finance [7].

Banana is one of agricultural commodities that have been developed in Indonesia. In Indonesia, the production of banana has a positive growth during the last five years, i.e. from 2013 to 2017 as much as 14.07%. On average, the banana production has reached 6922181 tonnes during the years. West Java is one of the centres of banana production in Indonesia, which has a positive growth as much as 3.04%.

Farmers in Indonesia have not implemented high technology yet in their banana farming [8], hence may affect to banana production. This may because of limited access of the farmers to financial sources. Lack of farming capital, causing limited costs to fulfil agricultural inputs required in farming such as fertilizers, pesticides and other needs [9]. This study aims to analyse the financial access of farmers and factors associated with the access.

2. Materials and methods

The focus of this study is on banana farms since banana has been identified by Indonesian Ministry of Agriculture as one of the important crops for horticultural development in Indonesia [10]. The methods used in this study was survey, which involves one hundred banana farmers in the centre of banana production in West Java, i.e. Cianjur and Ciamis District. According to the data of Indonesia Statistic that is presented in Figure 1, on average, Cianjur and Ciamis District had the highest production of banana from 2012 to 2016 in West Java.

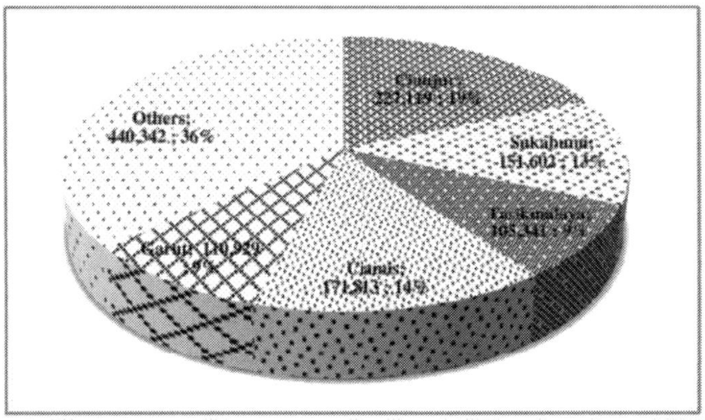

Figure 1. The average of banana production in West Java from 2012 to 2016

This study was conducted to collect data from May to September 2017 using a structure questionnaire. The data that were used in the study were the experience of farmers in obtaining access from various financial sources and socio-economic variables. Farmers' access to finance was analysed in terms of access to commercial credit from bank and micro finance institution (MFI), in-kind finance (non-cash finance) from trader and farmers' association, flexible payment of inputs from agricultural input kiosk and finance from other financial sources including from relatives, neighbours and friends. The socio-economic variables include gender, age, education level, farming experience and farm size.

The data were analysed using descriptive statistics and logistic regression. Logistic regression was used in this study to determine factors related to farmers' access to finance from various financial sources. The form of logistic regression used in this study is presented below.

$$y_i^* = \beta_1 + \beta_2 gender + \beta_3 age + \beta_4 education + \beta_5 farming\ experience + \beta_6 farm\ size + \varepsilon_i \quad (1)$$

$$i = 1,2, \ldots\ldots\ldots, 100\ (farmers)$$

Where the dependent variable y_i^* denotes farmers' access to finance from various financial sources. The independent variables (x_i) are the socio-economic factors, which were expected to relate to farmers' access to finance. The socio-economic variables are gender, age, education level, farming experience and farm size.

All the variables, except for dummy variables, were standardised prior to the regression. Moreover, the problem of homoscedasticity was checked using the Breusch-Pagan test. Following the method of VIF of Rook et al. [11], a test of multicollinearity was performed by calculating the Variance Inflation Factors (VIF) for each variable.

3. Results and discussion

3.1. Farmers' access to finance from various sources of finance

The descriptive statistics of the respondents in 2017 is presented in Table 5. As shown in Table 5, the majority of the respondent was male (93 per cent). The Table 1 also shows that, the average age of the farmers was 51 years and most of the farmers obtained primary school as the highest level of education. On average, the farmers had 23 years in farming experience and had less than one hectare of farm size. The accessibility of farmers to different financial providers is presented in Figure 2.

Table 1. The descriptive statistics of the respondents

Variables		Value
Gender (portion)	Male	0.93
	Female	0.07
Age (years)		51
Education (portion)	No formal education	0.02
	Primary school	0.67
	Junior high school	0.11
	Senior high school	0.17
	University	0.03
Farming experience (years)		23
Farm size (hectares)		0.58

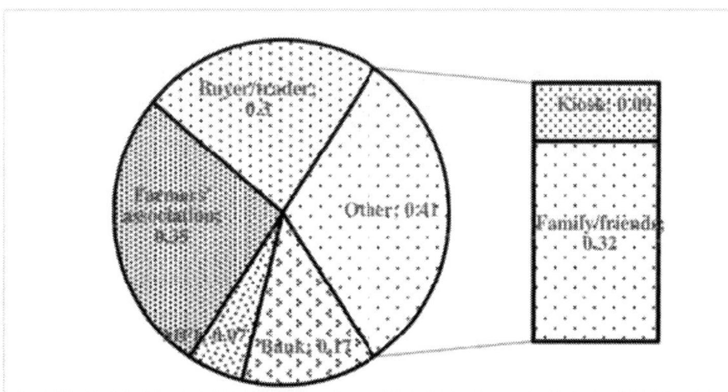

Figure 2. The accessibility of farmers to different financial providers

As shown in Figure 2, finance that could be accessed by the majority of farmers was finance from family and friends (32 per cent). Furthermore, Figure 2 shows that finance provided by formal institutions such as bank and MFI could not be accessed by many farmers since not more than 20 per cent of the farmers who could obtained finance from those finance providers. Farmers in Indonesia has faced difficulty in accessing finance from formal institution [6]. Informal finance has been a major financial source for households [12]. With regard to access to in-kind finance, the Figure 1 shows that this type of finance could be accessed by farmers from government through farmers' association (35 per cent). Government has provided agricultural finance subsidies or other kinds of incentive for stimulating farmers in improving their production [13]. Another study on finance has shown that finance provided by government had helped farmers in increasing their production [14]. Figure 2 also shows that farmers had access to in-kind finance from buyer/trader (30 per cent). Improving access to agricultural inputs can be attempted by having a contract with trader [15]. Of the financial access from agricultural input kiosk, the result shows that only few farmers (9 per cent) had access to finance from the kiosk in form of the flexibility of input payment.

3.2. Factors related to farmers' access to finance from various sources of finance

The results of logistic regression model in determining factors related to access to finance from different financial sources are presented in Table 2.

Table 2. Factors determining access to finance from different financial sources

Variable	Bank	MFI	Farmers' association	Trader	Agricultural input kiosks	Others
Constanta	-3.655 (1.665)	-3.367 (2.347)	-1.266 (1.427)	-1.022 (1.780)	-1.664 (2.105)	-1.990 (1.692)
Gender	-	-1.657 (1.382)	-0.010 (0.916)	0.350 (1.180)	-0.960 (1.251)	0.918 (1.151)
Age	0.058 [b] (0.031)	0.069 (0.045)	0.004 (0.024)	0.018 (0.030)	-0.002 (0.040)	0.013 (0.025)
Education level	-0.136 (0.285)	-0.341 (0.467)	0.150 (0.220)	-0.860 [a] (0.345)	0.105 (0.335)	-0.328 (0.254)
Farming experience	-0.037 [b] (0.022)	-0.023 (0.030)	0.025 (0.018)	-0.016 (0.020)	0.025 (0.029)	-0.010 (0.018)
Farm size	0.299 (0.647)	-0.780 (1.185)	-0.700 (0.520)	0.902 [b] (0.542)	-0.980 (0.987)	0.714 (0.488)

[a] Significant at 5 per cent and [b] significant at 10 per cent level.
Standard error in parentheses

The outcomes of the logistic regression model show that socio-economics factors have different relation to access to finance from different financial sources. The results presented in Table 2 show that some socio-economics factors have relation to access to finance from bank and buyer or trader, while the relation has not been found to access to finance from the other financial sources. For instance, age and farming experience have statistically significant association with the access to finance from bank. Age positively relates to access to finance from bank implying that older farmers have higher opportunity to have access to finance from bank. The positive relation may associate with the fact that older farmers focus in improving efficiency [16], which may lead to a higher opportunity to have access to finance from bank. However, farming experience has a negative association with the access to finance from bank indicating that farmers who have less experience in farming experience have more access to finance from bank.

With regard to access to finance from buyer or trader, education and farm size significantly associates with the access to finance from this type of financial source. Education has a negative association with the access to finance from buyer or trader implying that more educated-farmers have less access to finance from buyer/trader. The negative association might be because farmers who have higher education level tend to have off-farm jobs [17], in which farming may not become their main activity and thus less effort to look for finance from financial sources including from buyer. Conversely, farm size positively relates to access to finance from buyer/trader implying that farmers who have larger farm size have higher opportunity to have access to finance from buyer/trader. A possible explanation for this finding is that farmers with larger farm size have more knowledge of the important requirements to access finance to banks [18], hence may have higher probability to obtain finance from bank.

4. Conclusion

This study aims to analyse the financial access of farmers from various sources of finance and factors associated with the access. The study used a survey method involving one hundred banana farmers in the centre of banana production in West Java, i.e. Cianjur and Ciamis District. Descriptive statistics and logistic regression were performed for the data analysis. The results show that the majority of farmers had access to finance from government distributed through farmers' association, and from family and friends. Furthermore, the results show that factors associated with the financial access of farmers differ among the sources of finance. Age and farming experience have statistically significant association with the access to finance from bank. Age positively relates to access to finance from bank, while farming experience has a negative association with the access to finance from this financial source. Moreover, education and farm size significantly associate with the access to finance from buyer/trader. Education has a negative association with the access to finance from buyer/trader, on the other hand, farm size positively relates to access to finance from buyer/trader. In the light of the findings, this study suggests the need to enhance farmers' access to finance especially from bank and buyer/trader. For instance, the enhancement may be performed through a higher opportunity for younger farmer to be able to access to finance from formal financial institution such as bank.

Acknowledgments

The authors would like to gratefully acknowledge the farmers for the participation in this study. The grateful also goes to the directorate of research, community service and innovation of Universitas Padjadjaran for the financial support of this study.

References

[1] Nurmanaf AR 2007 Micro finance informal institution is closer with farmers *Analisis Kebijakan Pertanian* **5(2)** 99-109.
[2] Hamka AA and Danarti T 2012 The existence of Thithil Bank in traditional market activity: a case study of Kota Batu market *Journal of Indonesian Applied Economics* **4(1)**.
[3] Mulyaqin T and Astuti Y 2013 Availability and utilization of rice paddy farming financing sources in Pandeglang Regency, Banten Province *Bulletin Ikatan BPTP Banten* **3(1)**.
[4] Wulandari E, Meuwissen M, Karmana MH and Oude Lansink A 2017 Performance and access to finance in Indonesian horticulture *British Food Journal* **119(3)** 625-638.
[5] Hastuti EL and Supadi 2005 Community's accessibility on agricultural financing institutions in rural areas *SOCA (Socio-Economic of Agriculture and Agribusiness)* **5(2)**.
[6] Hartono R, Hadi S, Juanda B, and Rusastra I W. 2013. Development of Alternative Business Model of Farming Credit Institutions in Rural Areas. Informatika Pertanian, 22 (2): 121 – 135
[7] Yuliana RRRD 2017 Optimization of economics marine resources: fisherman funding system *Jurnal Ekonomi dan Pembangunan 22*(**2**)195-196.
[8] Sayuti R 2016 Prospect of production and marketing of bananas in the province of DI Aceh *Forum Penelitian Agro Ekonomi* **12(1)** 38-48.
[9] Mirza M, Amanah S and Sadono D 2017 The dynamic level of women farmer group in supporting the sustainability of family business in medicinal plants in Bogor Regency, West Java *Jurnal Penyuluhan* **13(2)** 181-193.
[10] Agricultural Ministry of Indonesia 2011 Technical guidelines for operationalization of horticultural development 2012 Directorate General of Horticulture Agricultural Ministry of Indonesia Jakarta.
[11] Rook AJ, Dhanoa MS and Gill M 1990 Prediction of the voluntary intake of grass silages by beef cattle 2. Principal component and ridge regression analyses *Animal Production* **50** 439-454.
[12] Pham T, Lensink R 2007 Lending policies of informal, formal and semiformal lenders: evidence from Vietnam *Economics of Transition* **15(2)**181–209.

[13] Elizabeth R 2017 The restructuritation of implementation and effectivity of financial program to the improvemnent of rice capacity and productivity *UNES Journal of Agricultural Scienties* **1(1)** 88-104

[14] Wulandari E, Ernah and Supyandi D 2018 Agricultural production and access to in-kind finance from government: a study of banana farms in Cianjur District, West Java. *IOP Conference Series: Earth and Environmental Science* **166(1)** 012042. IOP Publishing.

[15] Schipmann C and Qaim M 2011 Supply chain differentiation, contract agriculture, and farmers' marketing preferences: the case of sweet pepper in Thailand *Food Policy* **36(5)** 667-677.

[16] Zhengfei G and Oude Lansink A 2006 The scurce of productivity growth in Dutch agriculture: a perspective from finance *American Journal of Agricultural Economics* **88** 644-656.

[17] Reardon T, Berdegue J and Escobar G 2001 Rural nonfarm employment and incomes in Latin America: overview and policy implications *World Development* **29(3)** 395-409.

[18] Wulandari E, Meuwissen MPM, Karmana MH and Oude Lansink AGJM 2017 Access to finance from different finance provider types: Farmer knowledge of the requirements *PloS ONE* **12(9)** e0179285 available at https://doi.org/10.1371/journal.pone.0179285.

IOP Conf. Series: Earth and Environmental Science **334** (2019) 012060 doi:10.1088/1755-1315/334/1/012060

Food Security Level of Farmer Households in Rice Field Agroecosystem

I Rosada[1], Nurliani[1] and F A Gobel[2]

[1]Study Program of Agribusiness, Faculty of Agriculture Universitas Muslim Indonesia
[2]Study Program of Public Health, Faculty of Public Health, Universitas Muslim Indonesia

E-mail : idarosada@yahoo.com

Abstract. Food security at the household level essentially shows the ability of households to meet the food adequacy. This ability is influenced by complex factors but generally related to socio-economic aspects, food production behavior, consumption and allocation of household resources. The objectives of the research are 1) to identify the social and economic characteristics, 2) to identify the food consumption pattern, and 3) to analyze the level of household food security based on the quality of food consumption. This research was conducted in Lanrisang Sub-district, Pinrang District, South Sulawesi Province. This research is a depth survey research in 50 farm households in rice field agroecosystem that were chosen through *simple random sampling*. The analysis method was descriptive analysis, farm activities analysis and food consumption quality. The results showed that the description of socio-economic conditions of farmer household is 60% low educated, 58% had < 7 years of working experience on farm, 52% have ≥ 3 family members, 74% with the income of ≥ IDR 1,800,000, 86% of household expenditure on food, and only 9 people (18%) have alternative job. The level of household food security of farmers in rice field agroecosystem is categorized as "food insecure."

1. Introduction

Food security for a country is paramount, especially for a country with a large population like Indonesia. The population of Indonesia is projected to be 270 million by 2025. [1] stated that the history of Indonesia's development shows that food security issues are closely linked to economic stability (particularly inflation), living costs, and national political stability. Therefore, food security becomes an absolute requirement for national development [1].

Poverty and food security are two interrelated phenomena. In this case, the unsafe food security condition becomes the source of poverty; on the contrary, because of poverty, oneself does not have food security [2]. Indirect poverty is an indication of the weakness of the fulfillment of food needs at the household level either because of the low income of the community or the uneven distribution of food, thus making them as food-prone communities. Poverty has become the fundamental problem so that households are unable to access food well. Besides, poverty has also caused households unable to meet the needs of adequate food, both in quantity and quality.

The local government of South Sulawesi Province has succeeded in increasing the production of various types of food crops. Data during four years (2011-2014) showed that the production of almost all types of food crops has increased considerably [3]. However, the income of the community, especially the farmers' households, is still low, thus affecting the consumption pattern and household food security. This is very ironic considering the principal livelihood of farmers as food producers [4].

The increased food security is directed at self-reliance of resource-based local communities/farmers through improving food production programs, maintaining food adequate, safe and halal food availability in every area at all times, and also preparing anticipation to avoid food insecurity.
As a basic need and one of the human rights, food has a meaning and a significant role for the life of a nation. The availability of food that is smaller than its needs can create economic instability. Various social and political upheavals can also occur if food security is disrupted. This critical food condition can even endanger economic stability and national stability. For Indonesia, food is often identified with rice for it is as the primary staple food. Food security disturbances such as skyrocketing rice prices can lead to multi-dimensional crises. This condition occurred in 1997/1998 and could trigger social vulnerabilities that endanger economic stability and national stability. Indonesia was predicted to experience a food crisis in 2017 when it sees the imbalance between the population and the increasingly unbalanced availability of food today. The high number of percentage of households that have not been prosperous, the consumption rate which decreases each year and the high nutritional status of children under five years will enable the occurrence of food insecurity at the household level [5].

Samaulue Village is one of the villages located in Pinrang District that area belongs to the type of agro-ecosystem of rice field. This village is located in Lanrisang Sub-district. There are 432 households that work as farmers. The social life of the community in rice field agro-ecosystem is usually low educated; productivity is very dependent on seasons, limited business capital, lack of supporting facilities, poor market mechanism and long transfer of technology and communication [6]. These things resulted in low income of farmers. Therefore, this research aims: (1) to identify the socio-economic characteristics of farm households, (2) to identify the pattern of household food consumption of farmers, (3) to analyze the level of household food security of farmers based on the quality of food consumption in rice field agroecosystem.

2. Research Method

2.1. Population and Sample

Population in this research was all farmer households who live in Samaulue Village, Lanrisang Sub-district, Pinrang District, South Sulawesi Province, as many as 502 households. Then, the sample was determined by simple random sampling by taking 10% so that the total sample is 50 households.

2.2. Data Analysis

2.2.1. Descriptive Analysis. Descriptive analysis is used to describe socio-economic characteristics of respondents consisting of education, knowledge of respondents about nutrition and food, experience as farmers, number of family dependent, side job, household income and household expenditure.

2.2.2. Farm Activities Analysis.

Analyzing the household income [7], applied the following formula:

$$P_{RT} = P_{UT} + P_{LUT} + P_{ART} \quad (1)$$

Where:
P_{RT}: a total of household income (Rp/month)
P_{UT}: income from farming activities
P_{LUT}: income from other activities (alternative income)
P_{ART}: the income of a household member

In analyzing household expenditure [7], applied

$$C_{RT} = \sum_{i=1}^{n} C_P + \sum_{j=1}^{m} C_{NP} \qquad (2)$$

Where:
C_{RT} :a total of household consumption (Rp/month)
C_P :Expenditure for food (Rp/month)
C_{NP} :Expenditure for non-food (Rp/month)
$i_{1..n}$:Consumption for some types of food
$j_{1..m}$:Consumption for some types of non-food

2.2.3. Analysis of Household Food Security. Measuring the level of household food security of farmers on the type of agro-ecosystem rice was determined based on the quality of food consumption using the method of Food Diversification Score (SDP) [8]. The scoring of actual household consumption on the amount of food needed per unit of the consumer for adult men (UK) in each food group (staple food, side dishes, vegetables, fruits and milk).

Table 1. Measurement of Household Food Security Based on Quality of Food Consumption

Type of Food	Total of needed food per unit consumption of one man (UK)²⁾	Score¹⁾
Rice, cereal, sweet potatoes³⁾	500 g	0 1 2
Animal & Vegetable Side Dishes	200 g	0 1 2
Vegetable	150 g	0 1 2
Fruits	200 g	0 1 2
Milk	25 g	0 1 2
Total maximum score		10

In which:
1) 0 = if the portion of factual consumption: < 0,5 UK
1 = if the portion of factual consumption: 0,5< UK < 1
2 = if the portion of factual consumption: > 1
2)Medium activities
3) 500 g = if the portion of sweet potatoes < 20%, corn < 10%
600 g = if the portion of sweet potatoes 20 – 50 %
700 g = if the portion of sweet potatoes > 50 %
Criteria for assessment of household food security:

- If Score of Food Diversification (SDP) ≥ five so it is included as criteria for food-secure households, and
- If Score of Food Diversification (SDP) < 5 so it is included as criteria for food-insecure households.

3. Result and Discussion

3.1. Socio-economics Characteristics

The socio-economics characteristics of the respondents can be seen in Table 2 below.

Table 2. Respondents' Socio-economics Characteristics in Samaulue Village, Lanrisang Sub-district, Pinrang District, South Sulawesi Province, 2018

No.	Characteristics	Number of Respondents	Percentage (%)
1.	Education		
	• Lower level	30	60
	• Higher level	20	40
2.	Farming experience		
	• < 17 years	29	58
	• ≥ 17 years	21	42
3.	Family members		
	• < 3 people	24	48
	• ≥ 3 people	26	52
4.	Household income		
	• < IDR 1,800,000	13	26
	• ≥ IDR 1,800,000	37	74
5.	Household expenditure		
	• on food		86
	• on non-food		14
6.	Alternative job		
	• Yes	9	18
	• No	41	82

Table above shows that 60% of the total respondents has lower educational level; 58% have worked as farmers for < 7 years; 52 % have ≥ 3 family members; 74 % gain ≥ IDR1.800.000 of income; household expenditure on food as much as 86 %; and only 9 people have alternative job beside farming.

3.2. Food Consumption

The purpose of food consumption is to obtain the nutrients the body needs [8]. In general, the type of food consumed by the farmer households was less varied, i.e., only two basic foods (rice and corn), other main food consumption such as cassava and sweet potato generally consumed by respondents only as a distraction. Types of side dishes consumed by respondents and their families were animal side dishes (fish and eggs) and vegetable side dishes (tofu and tempe). Furthermore, the types of vegetables consumed by household respondents were spinach, kangkung, beans, and eggplant. Types of fruits consumed by respondents were only two types, i.e. banana and papaya. All respondents did not consume milk as a complement to four healthy dishes five (milk) to make it perfect. The average of staple food consumption is 255.37 grams/person/day, while the standard food requirement according to [9] is 500 grams/person/day. It indicates that the food consumption of staple foods have not met the standard of staple food needs yet. Average of side dishes consumption was229.74 grams/ person/day, while the standard needs of side dishes according to [9] is 200 grams/person/day. The results showed that the consumption of side dishes had met the standard side dishes needs. The average of vegetable consumption is 129.03 grams/ person/day, while the standard needs according to [9] is 150 grams/ capita/day,so the result did not meet the standard of vegetable needs. The average of fruits consumption is 55.54 grams/person/day, while the standard needs of fruits according to [9] is 200 grams/ person/day. These results indicate that the consumption of fruits in the household respondents didnot meet the standard needs of fruits. Consumption of milk in Samaulue Village did not meet the milk standard needs either, because no respondents consumed milk in the menu of household consumption. Food consumption can be seen in the following table 2.

Table 3. Average of Food Consumption in Respondent's Household in Samaulue Village, Lanrisang District, Pinrang Regency, 2018

No	Description	Average of Food Consumption (gram/day)	Average of Food Consumption per person (gram/person/day)*	Standard of Food Needs(gr/person/day)
1.	Staple	766.10	255.37	500
2.	Side Dishes	684.25	229.74	200
3.	Vegetables	387.09	129.03	150
4.	Fruit	166.59	55.54	200
5	Milk	0	0	25

Source: Primary data, 2018

In which : *Average of households member : 3 people

3.3. Status of Food Security

Household food security status was analyzed using Food Diversification Score (SDP) by [9]. SDP was calculated based on the Quality of Food Consumption (MKP) by using the actual household consumption score on the amount of food needed per unit of consumption (UK) in each food group. The criteria for assessment of household food security are assessed based on the value of Food Diversification Score.

Based on Table 3 it showed that the status of household food security in Samaulue village is mostly in the category of "food insecure", which is 34 households with a percentage (68%), and households that are categorized as "food secure" are 16 households with percentage 32%). This condition becomes a specific finding considering Pinrang Regency is one of the areas of food barn especially rice in South Sulawesi. This finding is in line with the study of [10] stated that although at the regional level food security status is assured, it is not enough to ensure food security at the household level. Food security at the household level essentially shows the ability of households to meet their food needs. This ability is influenced by complex factors but generally related to changes in the aspects of food production behavior, consumption and allocation of resources in the family.

Table 4. Distribution of Farmers Household Based on Food Security Status in Samaulue Village, Lanrisang Sub-district, Pinrang District, 2018

No	Status of Food Security	Total of Households	Percentage (%)
1	Food Secure households	16	32.00
2	Food Insecure households	34	68.00
	Total	50	100.00

Source: Primary data, 2018

4. Conclusion

From the results of research on farm households on the rice fields agroecosystem in Samaulue Village, Lanrisang Sub-district, Pinrang District, concluded as follows:

1. Description of social condition of farmer household was 60% low educated; long experience as a farmer in average of 17 years; household member was 52% in high category (≥ 3 people); 74% have low income which is IDR 1,800,000; household expenditure allocation was 86% on food, and only 18% of the respondents had an alternative job.
2. Food security level of farmer household in the rice field agro-ecosystem in Samaulue Village, Lanrisang Sub-district, Pinrang District included as "food insecure".

Acknowledgment
The authors would like to acknowlegde the General Director of Research and Technology& Higher Education (RISTEK-DIKTI) for the research fund through *Penelitian Dasar Unggulan Perguruan Tinggi (PDUPT)* scheme for the fiscal year 2018/2019.

References

[1] Rita , 2010 *Pengantar Ekonomi Pertanian* Penerbit Andi Jogjakarta
[2] Hasan 2005 *Ketahanan Pangan Rumahtangga*, http://repository.ipb.ac.id. Accessed on 22 December 2015
[3] Dinas Pertanian Tanaman Pangan dan Hortikultura, Sulawesi Selatan, 2014 Peningkatan Produktivitas Padi Terpadu di Sulawesi Selatan. Makassar.
[4] Lakitan B 2012 Kesejahteraan Petani dan Kedaulatan pangan. Media Indonesia online. 24 November 2017.
[5] Yulian P, wan Abbas Zakaria, Rabiatul Adawiah 2013 Ketahanan Pangan Rumahtangga Nelayan di Kecamatan Teluk Betung Selatan, Kota Bandar lampung *Jurnal JIIA* **Vol.1** No. 2
[6] Bulkis 2012 *Ketahanan Pangan Rumahtangga Pedesaan. Arus Timur*, Universitas Hasanuddin. Makassar.
[7] Rahim dan Hastuti 2007 *Pengantar, Teori dan Kasus Ekonomi Pertanian.* Penebar Swadaya, Jakarta.
[8] Hardinsyah 2012 Kecukupan Energi, Protein, Lemak dan Karbohidrat *Jurnal Gizi dan Pangan* Volume **7 (1)** 27
[9] Hardinsyah, 1996. *Ketahanan Pangan Rumahtangga*, http://repository.ipb.ac.id. Accessed on 22 December 2015.
[10] Purwantini TB, Rachman HPS dan Marisa Y 2005. Analisis Ketahanan Pangan Regional dan Tingkat Rumahtangga (Studi Kasus di Provinsi Sulawesi Utara) dalam Penguatan Ketahanan Pangan Rumahtangga dan Wilayah Sebagai Basis Ketahanan Pangan Nasional. Monograph Series No. 26. Pusat Analisis Sosial Ekonomi dan Kebijakan Pertanian. Bogor.

IOP Conf. Series: Earth and Environmental Science **334** (2019) 012061 doi:10.1088/1755-1315/334/1/012061

Study of Empowerment Smallholders Cocoa Based on Agro-Industry Approach

L Trimo[1], S Hidayat[2] and Ernah[1]

1.Department of Socio-Economics Agriculture, Faculty of Agriculture, Universitas Padjadjaran
2.Department of Plant Pest and Diseases, Faculty of Agriculture, Universitas Padjadjaran

E-mail : lucy.trimo@gmail.com; lucyana.trimo@unpad.ac.id

Abstract. Empowering smallholders' cocoa based on agro-industry requires readiness of the cocoa commodity itself from on-farm to off-farm. The research technique conducted in this study is a descriptive survey approach. The research site is Luwu Raya in South Sulawesi Province. This study uses literature and interviews as a data collection technique. Interviews were conducted with related parties such as government agencies, cooperatives, manufacturers, cocoa farmers associations, and farmer groups which were selected purposively. Collecting secondary data is done by collecting documents in the relevant agencies. Data and information obtained were analyzed descriptively correlatively. The results showed that empowerment of community cocoa into an independent community through agro-industry became a necessity, considering the potential of the community's cocoa in producing cocoa beans was quite large and only a small portion was processed into finished products. To that end, in order to reach an independent community on cocoa commodities, various things are needed such as transferring technology based on local potential, facilitating the government in providing bailouts and processing machines, and supporting regulations that favor farmers.

1. Introduction

Cocoa is one of the export commodities that contribute to increasing Indonesia's foreign exchange. Cocoa commodities ranked third in the plantation sector after rubber and Crude Palm Oil (CPO) as foreign exchange earners. Indonesia is the third largest supplier in the world after Ivory Coast and Ghana with a market share of 13.6% of total world production until 2011. Based on this, the government's efforts to increase the competitiveness of Indonesian cocoa on the world market have been carried out by launching *Gerakan Peningkatan Produksi dan Mutu Kakao Nasional* (GERNAS), especially in cocoa commodity centres in South Sulawesi Province, which is the province with the highest cocoa production growth reaching 8.6% followed by Central Sulawesi, Southeast Sulawesi and West Sulawesi. In South Sulawesi province cocoa production centres are located in Luwu Raya where cocoa production has contributed 63 percent of the national cocoa production with a total cocoa production of 63,259.21 tons and a total area of 133,469.70 hectares consisting of three districts, namely districts: Luwu, North Luwu and East Luwu and Palopo City [1].

Unfortunately, some problems still confront the development of the people's cocoa industry, namely low industrial utilization (<40%); export in the form of cocoa beans (70%); competitiveness; standardization of flavour and food safety; competition with imported and diversified products [2]. This is caused by many factors, one of which is that there are still obstacles to the transfer of technology that take place from Research Centres and Universities to cocoa farmers. In the process of technology transfer, it cannot depend entirely on the dynamics of the transfer of technology (Research Centres and Universities), but can also come from within the farmer itself. Farmer's independence in adopting technology is needed. Therefore, the empowerment of cocoa farmers by the relevant agencies so that the concerned are more independent is an urgent matter to do. This is because the future government programs are more directed towards cocoa industrialization.

2. Materials and Methods

The research technique used in this study is a survey descriptive study approach. The chosen research site is Luwu Raya in South Sulawesi Province, which is the centre of cocoa. Luwu Raya is one of the centres of cocoa in South Sulawesi Province and still tends to produce cocoa beans (not in processed form), therefore, to make the community cocoa more independent, empowerment must be directed to the agro-industry, and it becomes an urgent matter. To empower community cocoa into an independent community based on cocoa agro-industry, it must be supported by its own cocoa commodities from on-farm to off-farm which are interrelated with each other, and are integrative and sustainable. Luwu Raya consists of Luwu District, North Luwu District, East Luwu District, and Palopo City. Data collection techniques are carried out by means of: literature study and interviews with related parties, namely officials in government institutions/agencies, cooperatives, manufacturers, cocoa farmers associations selected purposively, and farmer groups. Secondary data collection was also carried out to supplement primary data, namely by collecting and studying written data in the form of documents or transcripts, newspapers, journals, bulletins, and opening access through the internet looking for websites related to this research. Data and information obtained are then analyzed descriptively correlatively.

3. Results and Discussion

3.1. Potential of Smallholders' Cocoa in the Research Area

In general, cocoa beans produced in Luwu Raya (except East Luwu District and Palopo City) are still not fermented. This condition will affect the quality and selling value of the products. The results of the study showed the reason farmers did not ferment because of the costs that had to be incurred, besides that they (smallholders) also stated that by not doing fermentation their cocoa was sold even at low prices. The quick desire to get money from their cocoa harvest is because of the additional impetus the need to fulfil their daily lives. East Luwu Districts produces seeds that have been fermented and are generally processed into intermediate products, and a small portion in the form of end-products. This area can be the exit gate of the final product from cocoa due to Luwu Timur Regency has a pioneer airport (Sorowako Airport) and an international port (Sorowako Port), thus facilitating the flow of goods to enter the surrounding area; in other words, the existence of international ports and airports in East Luwu Districts strongly supports the entry and exit of goods (trade) [3] Furthermore, the results of research conducted (in 2011-2012) both the Directorate General of PPI Ministry of Industry and the Indonesian Coffee and Cocoa Office (2011), it can be identified that the Luwu Raya region is divided into three types of industry categories, namely: upstream industry, intermediate industry and downstream industry. The purpose of the grouping is to accelerate and equalize the development of the cocoa industry in Luwu Raya.

Based on these groupings, the transfer of information in a system can be easier to do. The grouping of the Luwu Raya region by industry are :

1) Luwu Districts is included in the upstream industry criteria
2) North Luwu Districts falls into the industry criteria
3) Luwu Timur Districts is included in the upstream industry criteria
4) Palopo City falls into the criteria of the downstream industry

The all regions will run if the guidance to farmers in the area is carried out systemically as well as regional coordination with one another supported by Universities, Research Centres and the Central and Regional Governments.

3.2. Transfer of Industrial Technology in Empowering Cocoa Farmers

Technology transfer is the transfer of a technology that is produced and used somewhere to be applied to other places as an innovation technology [4] In order for technology transfer to work well or be acceptable to the target, in this case the community cocoa farmers in Luwu Raya must interact and be integrated based on the potential of each district / city in the region. The potential referred to is not only from the human side (independence), but also from the side: policy support, the existence of supporting stakeholders, technology information system for cocoa processing industry, and other supporting facilities (road facilities and infrastructure, communication, equipment/ processing machinery). In addition, the system must be interrelated with each other. The system referred to here is the nature of giving and receiving technology that is already owned by each region based on their respective competencies independently, while adopting the technology that is continuously provided by the Research Institute and Higher Education.

Coordination and integration are needed to strengthen the system above and facilitate the link between cocoa farmers in an area with cocoa farmers in other regions. Therefore, in empowering cocoa farmer groups in Luwu Raya merged into the Cocoa Industry SME group to facilitate access to all facilities provided by the government and accept technology transfer or seek information as widely as possible from its stakeholders. In order for information transfer to run quickly and smoothly, then determining the right person (member of the social system) can be trusted to run the technology transfer well to other members of the social system. Therefore, to accelerate the process of technology transfer in District / City and out of Regency / City, it takes a type of farmer who has a pioneer character and early application. This type must be in the Cocoa Industry SME group, and is more likely to be selected in technology transfer activities at the District / City level.

Through this Cocoa Industry SME group, technology transfer from existing Universities and Research Centres will make innovative cocoa industry SME groups have high "bargaining power" in carrying out partnerships with larger companies. In order for the dynamics of the fictionalization of Research Institutions, Universities and Government to run well, the coaching model that can be applied can be through several models, namely :

1) Mentoring Model: guiding intensively (industries that have been running).
2) Incubation Model: maturing the beginner's business until it can run (IKM cocoa).
3) Cluster Model: establish partnerships with SME industry and large industries.
4) Mediation Model : open market access

The four models mentioned above can be done in sequence, starting from the model (1) continuing to the model (4). Palopo city which is the basis of downstream industries can be used as an incubator for other regions that want to increase production (quantity and quality of products); Likewise for North Luwu District which is an industry-based intermediate region, it can be a forum for regional development based on upstream industry (Luwu District and East Luwu District).

Industry characteristics that are different in the four regions in the Luwu Raya will lead to coaching that will be carried out by the government (related agencies) will refer to the concept of "*accepting and giving each other in technology transfer*". Community cocoa which has a higher technology level industry can be used as a place of internship for the community cocoa industry which is still lagging behind its technological level. Furthermore, Universities and Research Centres can help, namely, acting as tutors and the central government (the Directorate General of Industrial Regional Development of the Ministry of Industry) and the regional government to function as facilitators.

3.3 Empowerment of SMEs as Independent Farmers

The readiness of cocoa farmers in facing the globalization era is related to the quality of farmers' behaviour, especially in adopting technology transfer from related institutions (research institutions and universities). Farmers' readiness will determine the ability of farmers to be able to be independent. [5] states that there are external and internal factors that can affect the independence of farmers; and external factors of farmers show a real role even more interesting than the internal factors of farmers in the development of farmers' independence. Based on the research of [5], there are several things that influence the independence of farmers, one of which is: the affordability of farmers to agricultural support facilities shows the most sensitive, positive, and real influence on modern behavior and efficient farmers. In addition, there are other things that also influence, namely: a) farmers 'accessibility to markets for agricultural products, b) modern behavior of farmers, c) farmer efficiency behaviour, d) affordability of farmers for farming support facilities, e) farmers' perception of physical feasibility farming, f) farmers 'accessibility to information resources, g) pressure on the development of the non-agricultural sector, h) quality of counselling, i) penetration of other products into farmer's household needs, and j) farmers' perceptions of agricultural development policies. Based on the description, to develop farmers in order to become independent human beings is needed an integrated, comprehensive and continuous collaboration between the community, government, universities and other stakeholders to jointly change the development mentality based on these cultural values. In the other side, [6] also states that for development to succeed, the people of Indonesian must have the mentality of self-confidence, discipline, and responsibility for what is done. Indonesian people who have the mentality as mentioned above will eventually be able to be independent and fully engaged in national development.

4. Conclusions

Empowerment of smallholders cocoa becomes an independent smallholders cocoa through the community cocoa-based agro-industry in Luwu Raya, South Sulawesi Province. Based on the results of the study, empowerment of smallholders cocoa into an independent smallholders through agro-industry becomes a necessity, considering the potential of smallholders cocoa in producing cocoa beans is quite large and only a small portion is processed become finished products. Furthermore, in order for the community cocoa-based agro industry to create independence and sustainability, it must: (a) material information technology must be adapted to the needs of farmers, (b) material information technology must be easy and inexpensive so that it can be affordable by the capital owned by farmers, (c) material information technology must be adapted to local social and cultural conditions, (d) create coordination, integration, synchronization and simplification among cocoa industry stakeholders (stakeholders), (e) create synergic cooperation, (f) regulatory support from the government in favour of farmers, (g) the existence of bailouts to buy cocoa beans that are not fragmented, and (h) the help of fermentation and processing machines that are locally specific.

Acknowledgements

We would like thanks to:

1. Directorate General of PPI, Ministry of Industry, Republic of Indonesia for their trust in the preparation of the Cocoa-Based Industrial Innovation Master Plan in Luwu Raya, South Sulawesi.
2. BPIHP South Sulawesi for good cooperation and support during conducting field surveys and final reports.
3. Head of Service and Staff in related Districts: East Luwu, North Luwu, Luwu and Polopo City.
4. All parties to their participation and support in research

References

[1] Direktorat Jenderal Perkebunan. 2008. Gerakan Umum Nasional Peningkatan Produksi dan Mutu Kakao
[2] Balai Penelitian Kopi dan Kakao Indonesia, 2011. Pengembangan Inovasi Industri Berbasis Kakao.
[3] Dinas Koperindag Kota Palopo. 2010. Rencana Pengembangan Industri Pengolahan Kakao Kota Palopo.
[4] United National Iidustrial Development Organization (UNIDO). 1996. The Role of Transfer of Technology in The Development Process. Manual on Technology Transfer Negotiations.
[5] Soemardjo. 1999. Alih Model Penyuluhan Pertanian Dimensi Partisipatif penyuluh dan Petani. Disertasi. Program Pasca Sarjana IPB; Bogor.
[6] Koentjaraningrat.1983. Kebudayaan, Mentalitas dan Pembangunan. PT. Gramedia; Jakarta.

IOP Conf. Series: Earth and Environmental Science **334** (2019) 012062 doi:10.1088/1755-1315/334/1/012062

Development Model of Community Forest in Bulukumba Regency, South Sulawesi, Indonesia

Nuraeni[1], R Rasyid[1], I Rosada[1], M Ilsan[1] and Saida[2]

[1]Study Program of Agribusiness, Faculty of Agriculture, Universitas Muslim Indonesia, Jl. Urip Sumoharjo Km 5, Makassar, Sulawesi Selatan, 90231 Indonesia
[2]Study Program of Agrotechnology, Faculty of Agriculture, Universitas Muslim Indonesia, Jl. Urip Sumoharjo Km 5, Makassar, Sulawesi Selatan, 90231 Indonesia

E-mail: neni_basri@yahoo.com

Abstract. Forests as a model of national development have significant benefits for the living and livelihood of society. It gives balance and dynamic benefits in either ecological, socio-cultural or economic sides. The study aims to develop a development model of community forest in Bulukumba Regency, South Sulawesi, Indonesia. The selection of respondents was carried out purposively, namely selecting respondents who were involved in community forest management. Methods of data collection in this study were conducted through interviews, discussions, questionnaires and field surveys. The respondents came from various experts and stakeholders related to community forest development activities. The secondary data collection was obtained from literature sources and documents from related institutions with this research. The analytical method used is a dynamic system analysis method for community forest development in Bulukumba Regency consisting of needs analysis, problem formulation, system identification and model simulation. The research result showed that the dynamic model of community forest management was built with three (3) sub-models, namely; community forest area, community forest production, and economics of the community forest. The scenarios built in this community forest development model are a combination of four (4) aspects, namely the policy of community forest development, the effectiveness of the nursery, the management system and the existence of counselling. The simulation result of the community forest development model in Kajang District was the optimistic scenario..

1. Introduction

Forests as a national development model have real benefits for the living and livelihood of the Indonesian society. It happens either ecological, socio-cultural or economic benefits are balanced and dynamic. For these reasons, the forest must be sustainably managed, protected and utilized for the welfare of Indonesian society, for both present and future generations.

Community forests are generally recognized as a successful program in rehabilitating forest condition and contributing to the development of local communities. A research was done by Chapagain, and Banjade [1] indicated that community forest is significant in increasing the income of communities around the forest. It needs collaborative action between government, community, and stakeholders in reducing the poverty of community living around the community forest.

Cronkleton, at all, stated that reducing emissions from deforestation and degradation will depend on rural communities to manage forest resources. Thus, forest management by the local community can be an option by recognizing the factors that influence the institution in community forest management [2].

Community forest management in Indonesia has developed since many years ago. It is cultivated by Indonesian people; even the management is still done traditionally. Community forest management conducted by the community is still self-supporting with pure self-service without any government intervention, either done in monoculture pattern or with mixed cropping pattern with agroforestry system [3]. Besides producing wood, the community forest in Indonesia also producing nontimber or wood which has economic value and can support food security. The areas of community forests in South Sulawesi Province are about 295,000 hectares scattered in all regencies/cities [4]. Moreover, the vast forest in Bulukumba District ± 22,148,04 Ha spread in ten districts.

One of the patterns of forest vegetation rehabilitation is by developing community forests. Therefore, community forest management by involving the community should be supported by a proper understanding of the function and role of forests for the living and life itself. A research done by Nuraeni et al. [5] showed that the variables of community knowledge about the benefits of forests have a direct and significant effect on people's behavior in managing community forests. It was found that the higher the knowledge of farmers about the benefits of forests, the better the behavior of farmers in managing community forests.

The communities managing forests are profoundly influenced by their traditions and cultures, where the kinship system is integrated with local knowledge and wisdom that matches the way of life and community needs. The concept of conservation is instilled in youth through various activities, and leaders act as good examples of the way of living their life [6].

Race & Millar stated that stakeholder analysis is being used in identifying communities or group of society most affected by the impact of development activities [7]. This analysis also gives massive benefit in determining priorities about the community or community groups needed in the implementation of activities and beneficial to their development. An activity can benefit some people, but it can also harm other people. It is inline that the stakeholder analysis is usually associated with several elements, such as; the existence of community groups, the impacts, and consequences that arise from the implementation of development programs. Thus, this study aims to develop a model of community forest area development in Bulukumba Regency, South Sulawesi, Indonesia.

2. Materials and Methods

This research was carried out in the community forest area of Kajang District as one of the districts that have the most extensive community forest in Bulukumba Regency, South Sulawesi, Indonesia. The respondents are community forest communities and stakeholders involved in community forest management in the region.

The selection of respondents was carried out purposively, namely selecting respondents who were involved in community forest management. Methods of data collection in this study were conducted through interviews, discussions, questionnaires and field surveys. The respondents came from various experts and stakeholders related to community forest development activities. The secondary data collection was obtained from literature sources and documents from related institutions with this research. The analytical method used is a dynamic system analysis method for community forest development in Bulukumba Regency consisting of needs analysis, problem formulation, system identification and model simulation. The formulation of alternative design or scenario of community forest development model was constructed from three sub-models, namely; community forest area, community forest production and economic sub-model of the community forest.

3. Results and Discussion

3.1. Design and Model Structure

3.1.1. Sub Model of Community Forest Area.

The sub model of community forest area includes components that cause changes in community forest area, namely; conversion of plantation land to community forest which causes the increase of community forest area. Component of land utilization of nursery and conversion of forest land to other function is an aspect which caused the decreasing of community forest area. This sub-model is heavily influenced by community forest development policies and nursery effectiveness. The width of the community forest affects production through the number of trees planted. It was based on available land capacity and varies by type of timber and commodities.

3.1.2. Sub Model of Community Forest Production.

The community forest production sub-model described the dynamics of the production of various timber and other commodities. This sub-model includes aspects of growth, as factors that increase production (inflow) and the proportion of natural dies and harvest as an aspect that causes the decrease in production (flow). This sub-model is strongly influenced by the management system due to the management system itself and the existence of extension agencies. It is as a result of the management system affects the proportion of land use per type of timber and commodities whereas counseling can affect the annual production. The harvest component is part or fraction of production appeared in the form of a volume of processed wood and fruit or rubber latex. The harvest rate is directly influenced by the economic sub-model because it determines the amount of income.

3.1.3. Sub Model of the Economics of Community Forest.

The sub-model of the economics of community forest shows the dynamics of revenue generation based on the difference between revenue and expenditure. In this sub-model, the harvest receipt component and expenditure component are included which consisting of nursery cost, tax, and maintenance cost. Besides explaining the change in income and expenditure, in this sub model, it is also calculated per capita income from the amount of income and the number of farmers managing the community forest.

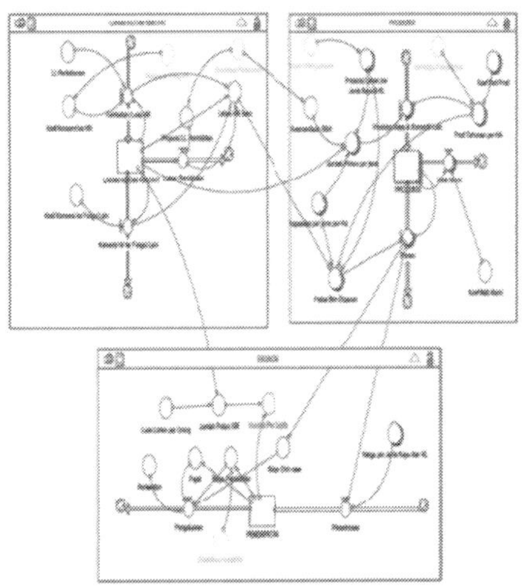

Figure 1. Structure of Dynamic Model of Community Forest Development in Kajang District, Bulukumba Regency

3.2. Scenario and Model Formulation

The scenarios built in this community forest development model are a combination of four (4) aspects, namely; community forest development policy, nursery effectiveness, management system and counseling presence. The policy of community forest development consists of two (2) categories, namely: (1) Without the widening of community forest area and (2) Widening community forest area for about 5-10%. The effectiveness of the nursery consists of three (3) levels, namely; (1) To supply about 25% of the land capacity, (2) To support about 50% of land and (3) To support 100% of the land. The management system consists of three (3) levels, namely; (1) Monoculture (bitti), (2) Polyculture (mixture), i.e., bitti, local teak, white teak and mahogany, sengon; and (3) Agroforestry, i.e., bitti, local teak, white teak, mahogany, sengon, cocoa, langsat, and rubber. The counseling consists of; (1)

Ineffective counseling and (2) Effective extension. The combination of four (4) factors that are scenarios will be simulated with a 35-year time rime with an interval (time state) each year.

3.3. Scenario Simulation of Model Dynamic

Based on the dynamic model simulation results, the estimation result within 35 years in four (4) aspects, namely; community forest management policy, nursery effectiveness, management system and extension existence. The combination of these aspects resulted in four scenarios for the development of community forests:

1. The pessimistic scenario, it was done without widening the community forest area, supplying about 25% of the land capacity for nursery effectiveness, monoculture, and ineffective counseling.
2. The optimistic scenario, it was done with widening the community forest area, supporting 100% of nursery land, agroforestry, and effective counseling.
3. The Moderate scenario 1, it was done without widening the community forest area, supporting about 50% of nursery land, polyculture, and effective counseling.
4. The Moderate scenario 2, it was done by widening the community forest area, supporting about 50% of nursery land, polyculture, and ineffective counseling.

3.3.1. Pessimistic Scenario.

The simulation results of pessimistic scenario 1 indicated that the policy without widening the land area, then the forest area will decrease. The effectiveness of bitti seed breeding from year to year will also decrease due to ineffective counseling. Bitti plants are produced in the 12th year and will decline until the 35th year so that will affect the farmer's income which will decrease as well.

3.3.2. Optimistic Scenario.

Scenario simulation results showed that the widening of the community forest area about 25% will increase income and income per capita, which is the 16th year per capita income becomes 103,810,626 with the widening of the community forest area up to 259.39 Ha. The number of timber trees, i.e., local teak, white teak, bitti, mahogany, and sengon increased up to the 25th year as well as other commodities such as cocoa, rubber, and langsat. The income per capita and community forest farmers' income increase as well. The number of community forest farmers increased until the 25th year. Similarly, incomes increased until the 34th year, and income per capita increased until the 16th year (103,810,626.60). The Bitti plants were produced in the 12th year of 3,430 which experienced peak increases compared to other plants. The local teak is harvested in the 15th year with a production of 1,542.16. Sengon is produced in the 9th year (220.82), white teak is produced in the 7th year (1,379.87) and mahogany in the 10th year (1,265.85).

In other commodities, such as; cocoa, rubber, and leachate show that the highest commodity is rubber, compared to cocoa and langsat commodities. It appears where the rubber commodity produced 545,275.97 in the 6th year followed by cocoa harvested 41,166.12 in the 5th year and langsat 3,066,920.92 in the 10th year. Harvest for the white teak, local teak, bitti, mahogany, sengon, rubber described the bitti harvest 3,235.22 in the 12th year; the local teak harvested 1,581,66 in the 16th year. It was followed by white teak harvested 1,323.91 in the 8th year, sengon is harvested 206.12 in the 10th year, and mahogany was harvested 1,241.36 in the 11th year. For rubber commodities harvested 543,348.35 in the 6th year, cocoa is harvested 41,166.12 in the 5th year and langsat 2,454,489.55 in the 10th year. This indicates that the rapidly harvested white teak is in the 8th year, but the massive harvest of bitti is harvested in the 12th year. For the commodity of Multipurpose Trees Species (MPTS) that is cacao which is quickly harvested in 5th year but rubber with the more significant amount of harvest that is in 6th year.

3.3.3. The Scenario of Moderate 1.

The simulation result of scenario 3 is moderate 1 indicates that without widening the community forest area and the number of community forest farmers will decrease from year to year. Moreover, the peak increases in income per capita and income in the 16th year. It shows that the number of trees per timber decreases every year. This is because without widening the land area of the community forest.

The number of community forest peasants declined from year to year. The increase in per capita income and income increased in the 16th year and there was a moderate decline. The production of white teak, local teak, bitti, mahogany and sengon shows that the highest production can be seen on sengon up to 190.3 in the ninth year, bitti production up to 2,711.16 in the 12th year, white teak up to 1,240.04 in year the 7th, local teak up to 1,123.56 in the 15th year and mahogany up to 1,061.00 in the 10th year.

3.3.4. The scenario of Moderate 2.

The simulation result of the scenario of moderate 4 shows that the widening of community forest area about 5-10% per year is along with the increasing number of community forest farmers. The number of white teak, local teak, bitti, mahagony and sengon plants increased from year to year. Income and income per capita increase in the 16th year while the number of farmers from year to year increase.

The production of white teak increased in the 7th year, local teak at a the15th year, mahogany at 10th year, bitti at the 12th year and sengon at the 9th year. While harvest for a mahogany increase up to 979.31 in the 11th year, bitti up to 2565.31 in the 12th year, Sengon up to 164.48 in the 10th year, local teak up to 1,274.72 in the 16th year and white teak 1,064.49 in the 8th year.

3.4. The Ratio of Scenario Simulation

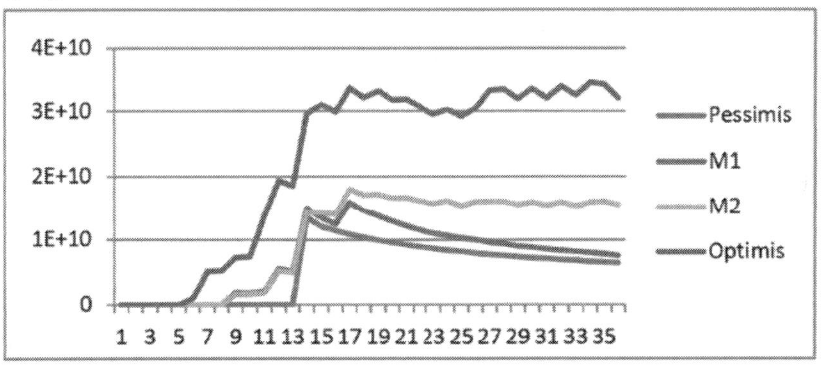

Figure 2. Comparison of Pessimistic, Moderate 1, Moderate 2 and Optimistic Scenario Simulation

Based on Table, there is a simulation comparison between pessimistic scenario, moderate 1, moderate 2 and optimistic. The optimistic model is increasing year by year and the highest in the 16th year. The effectiveness of 4 times from pessimistic simulation is 4,083564214; the increase happened because of the addition of variable of community forest area that is 296,34 Ha; income was in 32,003,793743,875.36, and income per capita was in 86,398,222,05 and amount of community forest farmers were equal to 370,42. As for moderate 1, the effectiveness numbered in 3.056895486 and for moderate 2 the effectiveness numbered in 2,133104684.

Table 1. Comparison Table of Community Forest Revenue Forecast over the next 35 years with Simulation in Pessimistic, Moderate 1, Moderate 2 and Optimistic Scenarios.

Year	Pessimistic	Moderate 1	Moderate 2	Optimistic
0	0	0	0	0
1	0	0	0	0
2	0	0	0	0
3	0	0	0	0
4	0	0	0	0
5	0	0	0	939,846,789.53
6	0	0	0	5,065,357,046.10.
7	0	0	0	5,180,755,112.04
8	0	1,917,717,820.31	1,644,577,944.82	7,205,349,178.34
9	0	1,793,398,511.56	1,567,651,773.18	7,290,476,747.96
10	0	2,118,161,470.30	1,864,683,842.21	13,906,574,440.40
11	0	5,494,245,459.88	5,126,230,523.75	19,278,222,065.25
12	0	5,085,953,788.46	4,913,655,576.99	18,427,093,965.52
13	13,514,144,358.57	14,995,030,564.64	14,363,324,746.52	29,572,451,787.31
14	12,110,257,208.15	13,500,362,862.41	14,315,256,365.17	30,969,352,150.58
15	11,475,064,019.66	12,504,765,821.87	13,964,402,340.92	29,847,874,097.43
16	10,915,760,542.14	15,872,868,592.29	17,971,234,560.19	33,659,406,417.77
17	10,398,601,894.80	14,665,430,754.30	17,087,910,331.52	32,101,049,418.96
18	9,939,683,578.88	13,791,402,745.43	17,110,867,732.80	33,209,748,801.12
19	9,563,137,350.72	12,898,975,805.61	16,637,951,155.58	31,797,652,479.79
20	9,214,721,010.72	12,208,561,038.57	16,566,444,605.65	31,877,043,140.29
21	8,917,718,651.00	11,548,014,882.01	16,162,503,120.41	30,642,390,562.54
22	8,632,147,083.01	11,048,840,229.75	15,752,485,655.70	29,540,956,560.31
23	8,369,830,095.23	10,654,779,314.08	16,118,983,726.30	30,225,630,689.69
24	8,132,244,744.29	10,279,088,622.57	15,389,492,502.27	29,245,019,915.46
25	7,899,200,518.38	9,992,896,387.04	16,042,111,721.12	30,472,894,681.01
26	7,686,996,261.48	9,701,380,328.28	16,021,298,878.51	33,235,422,168.85
27	7,495,611,312.07	9,381,769,708.22	16,000,489,736.53	33,488,625,789.63
28	7,334,866,945.00	9,016,396,186.30	15,545,919,908.99	31,906,857,375.65
29	7,164,266,111.31	8,785,214,456.33	15,944,056,398.60	33,579,368,885.12
30	7,020,039,853.25	8,531,268,462.81	15,481,648,340.40	32,066,325,198.28
31	6,866,990,201.77	8,341,673,655.09	15,951,690,609.99	33,898,316,336.21
32	6,722,035,218.24	8,182,456,317.82	15,406,835,985.59	32,478,139,054.85
33	6,580,700,794.25	7,994,474,072.79	15,954,910,562.94	34,470,177,384.77
34	6,467,680,842.89	7,743,637,242.33	16,040,177,302.72	34,157,800.337,45
35	6,360,512,024.89	7,495,126,140.82	15,597,079,176,13	32,003,743,875.36
TOTAL	1.98783E+11	2.65544E+11	3,80544E+11	8.1174E+11
			Effectiveness 4 times (O)	4.083564214
			M1	3.056895486
			M2	2.133104684

Source: The Analysis Result of Program Stella, 2017

4. Conclusion

A dynamic model of community forest management is built with three (3) submodels, namely; community forest area, community forest production and the economy of the community forest. *The best scenario of the simulation results of the community forest development model in Kajang district is the optimistic scenario.* This is supported by the pattern of agroforestry management, availability of seeds, effective counseling and community forest development policy through the widening land area. Development of community forest in Kajang District, Bulukumba Regency can be developed if the government issues regional development policies and streamlines counseling and encourages farmers to create independent nurseries and agroforestry management systems.

References

[1] Chapagain N and Banjade M R 2009 Community Forestry and Local Development: Experiences from the Koshi Hills of Nepal. *Journal of Forest and Livelihood 8 (2)*

[2] Cronkleton P, Bray D B, and Medina G 2011 Community Forest Management and the Emergence of Multi-Scale Governance Institutions: Lessons for REDD Development from Mexico, Brazil, and Bolivia. *Journal Forests 2*

[3] Usman M 2001 Memposisikan Hutan Rakyat Sebagai Aktualisasi Ekonomi Kerakyatan. *Makalah Hutan Rakyat* (Indonesia-Riau: Seminar MPI Reformasi)

[4] Paembonan S 2016 *Pengelolaan Hutan Rakyat Berbasis Klaster Provinsi Sulawesi Selatan* (Makassar: Temu Usaha Kemitraan Hutan Rakyat dan Perhutanan Sosial)

[5] Nuraeni, Rasyid R, Boceng A, Ilsan M and Amran FD 2018 Model of Community Behavior in the Management of the Community Forest in Bulukumba Regency, South Sulawesi, Indonesia. *Journal of Advanced Agricultural Technologies. Vol 5 (3)*

[6] Kumsap K and Indanon R 2016 Integration of Community Forest Management and Development Activities: Lessons Learned from Ubon Ratchathani Province. *Kasetsart Journal of Social Sciences. 37*

[7] Race D and Millar J 2006 Training Manual: Social and Community Dimensions of ACIAR Projects. *Australian Center for International Agricultural Research-Institute for Land, Water, and Society of Charles Sturt University.* (Australia)

IOP Conf. Series: Earth and Environmental Science **334** (2019) 012063 doi:10.1088/1755-1315/334/1/012063

Effects Of Fermented Milk And Soybean With Probiotic On Blood Lipid In Rats

L Adriani[1], D Latipudin[1] and R Lesmana[2]

[1]Faculty of Animal Husbandry, Universitas Padjadjaran, Sumedang-West Java, Indonesia
[2]Faculty of Medicine, Universitas Padjadjaran, Sumedang-West Java, Indonesia

E-mail : lovita_yoghurt@yahoo.co.id

Abstract. Probiotics have several positive benefits such as reducing pathogenic bacteria population to help food digestion and lowering blood lipid levels. Therefore, a study was conducted to evaluate the effect of fermented cow milk, soybean milk, and combination of both with probiotic bacteria on HDL, LDL, and blood cholesterol level of rats. A total of 30 rats,8 weeks old were used in this study in a Complete Randomized Design. The rats were randomly assigned into 5 treatment and replicated 5 times of 1 rat each. The rats in first group (P0) were given basal feed without probiotic, P1 = basal feed + fermented cow milk ; P2= basal feed +75% fermented milk + 25% soybean milk ; P3 = basal feed + 50% fermented cow milk + 50 % fermented soybean milk P4= basal feed + fermented cow's milk with different bacteria. There was a non-significant reduction in LDL, and cholesterol level of rats, also non-significant addition in HDL due to probiotic supplementation and the highest reduction is in group fed P4. Addition of probiotic in diet of rats gave the same respond, but have beneficial effect in improving blood cholesterol although non-significant different

1. Introduction

Cholesterol plays a major role in human heart health. Cholesterol can be both good and bad. High-density lipoprotein (HDL) is good cholesterol and low-density lipoprotein (LDL) is bad cholesterol. High cholesterol in serum is a leading risk factor for human cardiovascular disease such as coronary heart disease and stroke America's number one killer [1]. Excess cholesterol in the bloodstream can form plaque (a thick, hard deposit) in artery walls. The cholesterol or plaque build-up causes arteries to become thicker, harder and less flexible, slowing down and sometimes blocking blood flow to the heart. When blood flow is restricted, angina (chest pain) can result. A heart attack will result when blood flow to the heart is severely impaired and a clot stops blood flow completely. When there is too much LDL cholesterol in the blood, it is deposited inside the blood vessels, where it can build up to hard deposits and cause atherosclerosis, the disease process that underlies heart attacks.

Cholesterol is formed by fatty acid oxidation (β-oxidation) in the mitochondria liver cells, into Acetyl CoA. The cholesterol is also synthesized by the liver (endogenous), and from food (exogenous). The triglycerides and cholesterol esters, which are derived from food, will be absorbed by the intestinal mucosal cells. Then, it will be re-synthesized and wrapped with the amount of protein and secreted in the kilo micron form to the lymph system and in the end, it circulated throughout the body [2]. The high synthesis of triglycerides and cholesterol will elevate the levels of triglycerides and cholesterol in the blood., the normal blood cholesterol levels have ranged from 52-148 mg/dL [3].

There are various ways to regulate and decrease the triglycerides and cholesterol blood levels. One of the ways is the use of probiotic. Probiotic is a live microbial as a dietary supplement that has beneficial effects for health by improving the balance of microbes in the digestive tract [4]. Probiotic bacteria such as lipolytic bacteria produces Bile Salt Hydrolase (BSH) enzyme. Lactic acid bacteria (LAB) grown in fermented milk as probiotic can also produce BSH enzymes. The BSH enzymes hydrolyze or break the C-24 N-Acyl amide bond formed between bile acids and amino acids in conjugated bile salts to decrease the blood cholesterol levels [5].

The ability of probiotics to decrease the blood triglyceride levels is associated with the activity of acetyl CoA-carboxylase as an enzyme which plays a role in the rate of fatty acid synthesis [6]. [7]stated that the acetyl CoA-carboxylase produces malonyl CoA as the key of metabolism to regulate the synthesis of fatty acids and oxidation, but it can be affected by changes in diet and intestinal activity. Moreover, probiotics can produce statins. Statin is 3-hydroxy-3-methyl-glutaryl-CoA reductase (HMG-CoA reductase) inhibitors, which are cholesterol biosynthesis regulators, Low Density Lowering (LDL)-lowering, Very Low Density Lipoprotein (VLDL), and blood triglyceride levels [8]. The mechanism of decreasing triglyceride levels by statins begins when the inhibitor reduces cholesterol concentrations in hepatocytes and improves the performance of LDL-receptors [9]. The receptors are also related to the VLDL components, which can reduce triglyceride level. Furthermore, [10] indicated that decreasing the synthesis of fatty acid in the liver is a primary factor to decrease the triglyceride synthesis in the liver, which resulting in low triglyceride concentration in the blood plasma.

A number of studies have been carried out to investigate the role of probiotics in lipid metabolism [10, 11]. For example, [12] have studied the combination of chocolate and probiotic yoghurt that can improve blood lipid, and reduce LDL level in the blood of rat. [11] have investigated the role of probiotics on lipid profiles and reported the effect of probiotics on total cholesterol and LDL. However, there is a little study on the role of probiotics in reducing blood lipid concentration particularly improving HDL blood in broilers.

The product that contains probiotic is fermented milk. Fermented milk can be made from various types of milk such as soybean milk and cow milk. These two products contain compounds or substances that can control the blood cholesterol, HDL and LDL levels.

The combination of fermented soybean milk and cow milk in decreasing blood triglyceride and cholesterol levels in broilers have done, however, there is no study on the combination of fermented soybean milk and cow milk in decreasing others blood lipid levels in rats, so the research is very important done

The purpose of this study is therefore to determine the effect of fermented cow milk and soybean milk, and the combination of both with probiotic bacteria on blood cholesterol, HDL and LDL of blood of rats.

2. Materials And Methods

A total of 30 female rat , age 2 month were used in the experiment. The experiment was conducted for 30 days. The fermented milk as a probiotic was given by force feeding. The composition is based on the variant fermented milk of each treatment. The dose of fermented milk is 1.25% of the body weight [13]. Cow milk and soybean milk were used as the basic ingredient for fermentation milk. The fermentation process used bacteria, namely *Streptococcus thermophillus, Lactobacillus bulgaricus, Lactobacillus acidophillus* and *Bifidobacteria sp.* The balance of probiotic bacteria used is 1:1 [14] Rat's standard food (pellet) in this research, and 5 treatment was used. Probiotic yoghurt is taken from product of Lovita yoghurt . The concentration of probiotic yoghurt was given to the rats was calculated following the research of [14] The experiment design used Completely Randomized Design with 5 treatments and 6 replications. The experiment units were 30, and each unit contained 1 female rats.

The treatments consisted of: T0 = Basal ration (without fermented milk), T1 = Basal ration + fermented cow milk , T2 = Basal ration + probiotic 50% fermented cow milk : 50% fermented soybean milk, T3 = Basal ration + probiotic 75% fermented cow milk : 25% fermented soybean milk
T4 = Basal ration mixed with combination of fermented cow milk with different probiotic, *Bifidobacteria sp* and *Lactobacillus acidophilus.*

Cholesterol, HDL and LDL were analyzed using Biolabo products. Cholesterol was analyzed by CHOD-PAP (*Cholesterol Oxidase Phenylperoxidase Amino Phenozonphenol*) methods.

Data measurements obtained using spectrophotometers (Shandong, China) were analyzed statistically using the One way Analyzed of Variance (ANOVA) and followed by Tukey-test (Honestly Significant Difference / HSD) to test the average value between treatments. The hypothesis applied in this study is the level of giving probiotic with ratio 50% cow milk: 50% soybean milk with dose 1.25% of body weight (T4) can give the most significant influence. In this study, three parameters were measured: blood cholesterol , HDL and LDL levels.

3. Results And Discussion

3.1. Blood Cholesterol Level

Table 1. showed that the average blood cholesterol levels ranging from 80.27-113.03 mg/dl. According to [4] total normal blood cholesterol levels in broilers ranged from 52-148 mg/dL Hence, most cholesterol levels in each treatment were at normal levels, except in T1 and T5 can decrease the blood cholesterol until 27-29%. compare to control Nevertheless, the result of the statistical analysis showed that the fermented milk as probiotic supplementation have non significant effect (P >0.05) in decreasing blood cholesterol level.

Table 1. Effect of Fermented Milk and Soybean using Probiotic on Blood Cholesterol, High Density Lipoprotein (HDL), and Low Density Lipoprotein (LD) in Rats

	P0	P1	P2	P3	P4
Cholesterol (mg/dl)	113.03 ± 50.03	80.27 ± 43.18	93.02 ± 21.49	96.9 ±17.28	81.52 ± 9.62
HDL	45.27 ± 22.52	41.90 ± 18.67	45.33 ± 23.42	49.46 ± 13.83	41.54 ± 14.84
LDL	36.17 ± 29.70	23.97 ± 51.57	23.65 ± 8.07	30.53 ± 31.56	29.27 ± 15.61

Carbohydrate type in soybean milk consists of oligosaccharide group, it cannot be used by LAB as energy sources or carbon sources. This is supplied by cow milk as it contains lactose that can be used as a food source for LAB. The lactose helped LAB to maintain them to keep alive and to be able to remodel and ferment carbohydrates, which are contained in fermented cow milk and also the combination of fermented cow milk and soybean milk.

Biliary salt deconjugation and cholesterol assimilation process in T1 and T5 have better performances compared to other treatments. However, the LAB in T5 has better performances than others. The LAB produced BSH enzymes that hydrolyze or break the C-24 N-Acyl amide bond formed between bile acids and amino acids in conjugated bile salts. The bile salts are deconjugated by separating the glycine or taurine from steroids to produce free or conjugated bile salts in the less absorbed free cholesterol acids formed by the small intestine [15, 16]. Unconjugated bile salts have a lower solubility rate at physiological pH when produced. If the production makes it more hydrophobic, it will be less ionic. It will have an impact on low intestinal absorption rates. As a result, secondary bile acids will be settled in the intestine and released through the feces [17]. The release of bile acids causes a decrease in the number of bile acids absorbed in the enterohepatic cycle. In that cycle, bacteria released bile salt hydrolase enzymes, to break the glycine or taurine from steroids, and resulting in free or deconjugated bile salts [18]. Thus, it has an effect in increasing de novo bile synthesis [19]. The formation of bile acids requires cholesterol as a precursor, so then biliary salt deconjugated indirectly. It can decrease blood cholesterol levels because the existing cholesterol is used to form new bile acids. In the mechanism of cholesterol assimilation, LAB will absorb the micelle cholesterol in the lumen intestine and will enter on the cellular membrane of bacteria and converted into carotenol, which is a sterol that cannot be absorbed by the intestine; so it will not be absorbed and tends to be released with faces [20, 21].

The content of isoflavones in soybeans also plays a role in reducing blood cholesterol levels. It can be seen in T2 and T3. The fermentation process can activate isoflavone in the soybean milk and changed into aglycone by a β-glucosidase enzyme. The aglycone acts to increase soybean milk activity in order to lower blood cholesterol level [22]. Aglycone in genistein and daidzein soybean can increase bile acid excretion and regulate LDL receptor activity [23]. Moreover, it was also reported that the flavonoids in the soybeans were able to inhibit the activity of 3-hydroxy-3-methyl-glutaryl CoA (HMG CoA) enzyme, which inhibits synthesis of cholesterol and ACAT (acyl-CoA). The compound will compete with HMG CoA to bind the HMG CoA reductase enzyme, then to inhibit the cholesterol synthesis [24]. Cholesterol acyltransferase acts in lowering cholesterol esterification in the intestine and liver [25]. Hence, the blood cholesterol levels will decrease.

The soybean also contains a phytosterol, which acts to bind and inhibit the absorption of cholesterol from food in the intestine. It means that the Niemann-Pick C1 Like 1 (NPC1L1) enzyme transporter cannot promote cholesterol absorption into the lymph system. The cholesterol-binding phytosterol enters the intestinal lumen by the ABCG5 enzyme transporter. Then, it will leave through the feces. However, another study indicated that the succeeding cholesterol moves to the endoplasmic reticulum by Acyl-CoA Cholesterol Acetyl Transferase 2 (ACAT 2) enzyme, which is further esterified into Kilo micron. That action will inhibit the activity of HMG CoA reductase to stimulate the formation of cholesterol, which will reduce cholesterol synthesis, and thereby can lower blood cholesterol [26].

3.2. Low Density Lipoprotein (LDL)

The Duncan test results in Table 1. Showed that LDL levels between treatments were not significantly different. All treatments showed a positive effect of lowering LDL levels although statistically not significantly different ($p > 0.05$) compare to control. Adding probiotic bacteria can decrease up to 33.7%. Reduced levels of LDL in rats on fermented milk, fermented soy milk and the combination also occur due to the activity of lactic acid bacteria producing bile salt hydrolase enzymes that hydrolyze or break the C-24 N-Acyl amide bond formed between bile acids and acids amino in conjugated bile salts. BAL activity. Bile salt hydrolase (BSH) enzyme deconjugs bile salts by separating glycine or taurine from steroids, produce free bile salts or deconjugate in the form of free cholic acid less absorbed by the small intestine [5].

Aside from lactic acid bacteria the decrease in blood LDL levels is also caused by the presence of isoflavones contained in soybeans. The fermentation process in soybeans will hydrolyze isoflavones into free isoflavone compounds called aglycons. Isoflavones of flavonoids that are polyphenolic compounds. Isoflavones can liberate one hydrogen atom from one reducing group associated with one free radical, synthesizing the 3-hydroxy-3-methylglutaril-CoA (HMG-CoA) synthesizer becomes inhibited which acts as a cholesterol-forming precursor and then the oxidation reaction of LDL cholesterol is inhibited.

Isoflavones in the treatment can dilate blood vessels and lower LDL oxidation levels, also inhibit the activity of HMG-CoA enzyme that plays a role in inhibition of cholesterol synthesis and acylCoA enzyme: Cholesteryl acyltransferase which plays a role in decreasing esterification of cholesterol in the intestine and liver, and able to bind LDL lipophilic flavonoid properties. The role of isoflavones here as antioxidants is a compound capable of inhibiting the oxidation of other molecules, which demonstrate the ability to prevent oxidation of Low Density Lipoprotein (LDL) by capturing free radicals.

The next factor, the HDL present in the blood plasma that binds to cholesterol or its ester and carries it along the bloodstream from the peripheral cells to the liver cells. The cholesterol that is bound to undergo the process of reshuffle produce liver cholesterol reserves, cholesterol reserves required for the synthesis of VLDL and biosynthesis of other compounds. VLDL binds LDL by receptor molecules on the peripheral tissue membranes inhibition by HDL, so that high levels of HDL will prevent LDL accumulation in blood vessel walls.

Based on the results of this study showed that LDL level of blood rats was lower than control of either fermented milk treatment or the combination of fermented milk and fermented soy milk .

HDL levels have an inverse relationship with each other with various diseases, so the ratio of LDL / HDL cholesterol is an important predictive parameter [27]. Low blood levels of LDL are good for health, where the risk of constricts of blood vessels will be low because cholesterol is transported throughout the body slightly.

A product can know the level of safety and health of the ratio of LDL / HDL. This research can increase the ratio of LDL / HDL of blood that is on control, is 1: 1,2 become respectively 1: 1.69 ; 1: 1.75; 1: 2.8; 1: 1.63, and 1; 1.42 / Based on these results, in this study obtained the ratio of LDL / HDL above the ideal number, the ratio of LDL/HDL of monogastric blood is 1: 1.6, therefore obtained security and health is higher than control .

4. Conclusion

Supplementation basal rations with fermented cow's milk (T1) and the combination of fermented cow milk and soybean milk with *Lactobacillus acidophilus* and *Bifidobacteria* microbes (T5) showed the highest decreases in blood cholesterol compare to others. High Density Lipoprotein and Low Density in all treatments gave the same results. However the balance of LDL/HDL have improved. Using fermented cow milk, fermented soybeans or the combination of both with *Lactobacillus acidophilus, Lactobacillus bulgaricus*, and *Streptococcus thermophilus*, therefore it is recommended compare to without adding probiotic.

Acknowledgements

We gratefully would like to thank a group of students in conducting the study of the utilization of probiotic. Furthermore, we would also like to thank the PUPT Project in 2016 of The Ministry of Research, Technology and Higher Education and also Padjadjaran University.

References

[1] Tabas I 2002 Cholesterol in Health and Disease *J. Clin. Invest.* **110(5)**: 83-90. http://www.sciencepub.org.

[2] Lehninger A L 1982 Dasar-Dasar Biokimia Jilid 1 Diterjemahkan oleh Maggy Thenawidjaja *Penerbit Erlangga* Jakarta.

[3] Basmacioglu H, Mkahel E 2005 Research on the Factor Affecting Cholesterol and Some Other Characteristics of Eggs in Laying Hens *J. Vet. Anim. Sci.* **29**:157-164.

[4] Manoppo M R A, Rahmi S, Tatang S A, Yanti D 2007 Pengaruh Pemberian Crude Chlorella terhadap Kadar Total Kolesterol Darah Ayam Broiler *Fakultas Kedokteran Hewan* Universitas Airlangga.

[5] Surono I S 2004 Probiotik Susu Fermentasi dan Kesehatan. Yayasan Pengusaha Makanan dan Minuman Seluruh Indonesia (YAPMMI) Jakarta.

[6] Santoso U, Tanaka K 1995 Effect of Dried Bacillus subtilis Culture on Growth, Body Composition and Hepatic Lipogenic Enzyme Activity in Female Broiler Chicken *British Journal of Nutrition* **74**: 523-529.

[7] Abu E L, Jayakumar A, Baldini A, Chirala S S, Wakil S J 1995 Human acetyl-CoA carboxylase: Molecular cloning, characterization, chromosomal mapping, and evidence for two isoforms *BioMed Central* **15**: 1-8

[8] Cavallini D C U, Bedani R, Bomdespacho L Q, Vendramini R C, Rossi E A 2009 Effects of Probiotic Bacteria, Isoflavones and Simvastatin on Lipid Profile and Atherosclerosis in Cholesterol-Fed Rabbits: A Randomized Double-Blind Study *BioMed Central* **8**: 1-8.

[9] Grundy S M 1988 HMG-CoA Reductase Inhibitors for Treatment of Hypercholesterolaemia N. Engl. *J.Med.* **319**: 24 – 32.

[10] Scorve J, Al-Shurbaji A, Denil A, Ivone B, Leint B, and Robert K B 1993 On the Mechanism of the Hypolipidemic Effect of Sulfur-Substituted Hexadecanedionic Acid (3-Thiadicarboxylic Acid) in Normolipidemic Rats *J.Lipid Res.* **34**: 1117-1185.

[11] Guo Z, Liu X M, Zhang Q X, Shen Z, Tian F W, Zhang H, Sun Z H, Zhang H P, Chen W 2011 Influence of Consumption of Probiotics on the Plasma Lipid Profile: a Meta-Analysis of Randomised Controlled Trials *Nut.Metab.Cardiovasc. Dis.* **21**:844–850.

[12] Adriani, Nenden I, Sjafril 2012 The Effect of Cocoa and Fermented Milk Mixture on Triglyceride and Blood Glucose Contents of Hyperlipidemia Rats CMU *J.Nat.Sci. Special Issue on Agricultural & Natural Resources 11*(**1**)**a**.

[13] Lovita, Nenden I 2015 The Addition Of Cocoa (Titeobroma Cacao) Powder In Milk Fermented To Reduce The Uric Acid Level On Hyperlipidemi Rats.

[14] Lovita A, Nenden I, Sjafril D 2012 The Effect of Cocoa and Fermented Milk Mixture on Triglyceride and Blood Glucose Contents of Hyperlipidemia Rats CMU *J.Nat.Sci. Special Issue on Agricultural & Natural Resources* **11(1)**: 383- 386.

[15] Fuller R 1992 History and Development of Probiotics. Edited by Fuller *Probiotic: The Scientifis Basis* London: Chapman And Hall.

[16] McGowan M W 1983 A Peroxidase-Coupled method for The Calorimetric Determination of Serum Triglycerida *Clin.Chem* pp. 529-538.

[17] Lengkey H A W, Adriani L 2013 The Effect of Lactobacillus bulgaricus and Streptococcus thermophilus as Ration Supplement on Broiler Carcass Weight, Carcass Fat Content and The Serum Cholesterol Carcass Content *International Prosiding of Biotechnology*.

[18] Baroutkoub A, Roushan Z M, Razmik B, Julayi H, Sohrabi Z, Mazloomi S.M, Eskandari M H 2010 Effect of Probiotic Yoghurt Consumption on the Serum Cholesterol Levels in Hypercholestromic Cases in Shiraz, Southern Iran *Scientific Research and Essays* **5(16)**: 2206-2209.

[19] Noh D O, Son H K, Shen G 1997 Incorporation of Cholesterol into The CellulerMembrane of Lactobacillus acidophilus ATCC 4321 *J. Dairy Sci.*

[20] St-Onge P M, Farnworth E R, Jones P J H 2000 Consumption of Fermented and Nonfermented Diary Products : Effects of Cholesterol Concentrations and Metabolism *J.Clin.Nutr.* **71**: 674-681.

[21] Puryana I G P S 2011 Populasi Lactobacillus rhamnosus SKG34 dalam saluran Pencernaan dan Pengaruhnya terhadap Kadar Kolesterol Tikus Putih (Rattus norvegicus) *Tesis* Universitas Udayana Bali.

[22] Fadhilah A N, Hafsan, Nur F 2015 Penurunan Kadar Kolesterol oleh Bakteri Asam Laktat Asal Dangke Secara In Vitro *Prosiding Seminar Nasional Mikrobiologi Kesehatan dan Lingkungan* Makasar : dakultas Sains dan Teknologi UIN ALAUDDIN pp. 174-180.

[23] Winarsih H 2005 Isoflavon, Berbagai Sumber, Sifat dan Manfaatnya pada Penyakit Degeneratif *Gadjah Mada University Press* Yogyakarta.

[24] Borradaile N M, Dreu L E, Wilcox L J, Edwards J Y, Huff M W 2002 Soya phytoestrogens, Genistein and Daidzein, Reduce Apoliprotein B Secretion from HepG2 cells Through MultipleMecanisms *Biochem. J.* **366**: 531-539.

[25] Hardaningsih R, Nurhidayat N 2006 Pengaruh Pemberian Pakan Hiperkolsterolemia terhadap Bobot Badan Tikus Putih Wistar yang diberi Bakteri Asam Laktat *Bioverditas*.

[26] Fuhrman B, Aviram M 2001 Flavonoids Protect LDL from Oxidation and Attenuate Atherosclerosis *Curr. Opin. Lipidol* **12**: 41-8.

[27] Murray R K, Granner D K, Rodwell V W, Pendit B U, Wulandari N 2009 Harper's Biochemistry 27th ed. *Jakarta : EGC*.

Paddy Ratoon (*Oryza Sativa*) Nutrients Quality as Feed Forages at Various Cutting Height and Cutting Ages

R Z Islami[1], I Susilawati[1], H K Mustapa[1], Y Budiawati[2], S Nurjanah[1] and A Rochana[1]

[1]Faculty of Animal Husbandry, University Padjadjaran, Bandung Indonesia
[2]Faculty of Agriculture, University Sultan Ageng Tirtayasa, Banten Indonesia

E-mail : romi.zamhir@gmail.com

Abstract. This study was aimed to determine the nutrient quality of paddy ratoon (*Oryza sativa*) at various cutting ages. The design used was completely randomize design with factorial design. The first factor is the cutting height with a cutting height of 10 cm, 20 cm. The second factor is cutting ages which is 14 days cutting ages, 21 days cutting ages and 28 days cutting ages and 35 days cutting ages. The result showed that cutting age of paddy ratoon as forages that produce optimal nutrient quality is at the age 14 days after cutting, with 10.34% crude protein and 28.45% crude fiber and 1.15% crude fat. Height cutting of paddy ratoon is not significantly affected nutrient quality

1. Introduction

Paddy plants are a major component in the national food security system. Paddy is used as a staple food but the byproduct of paddy have enormous potential for animal husbandry. Efforts to increase the productivity of livestock continue to be carried out, one of which is by utilizing paddy straw and other products, namely paddy ratoon is a clump of paddy plants that grow back after being harvested. This paddy ratoon advantage is practical, and has nutritional value that is good enough to increase livestock productivity. The potential of crude protein forage of paddy ratoon reached 8%. This shows that the quality of the forage is almost equivalent to field grass / nature. The production of a second rice crop in one growing season is known as ratooning. Ratoon plants develop by seedling rice regeneration from nodal buds from straw left behind after the first seasonal rice harvest. A study showed that cultivars with a long duration of vegetative growth, high weight per tillers can produce high dry matter that produces in the first and second plants. The increase in ratoon grain yield due to harvesting of main crops at a lower than normal height increases the ratoon yield. Lowering the height of the harvest has shifted the panicle from the origin and delayed appearance of panicles. The yield advantage of the first 20-cm high plant straw can be attributed to panicles that are heavier than tillers coming from basal nodes. Genetics, hay health, availability of soil and water nutrients, and / or weather during the process of growth of the weeds precedes the harvesting factor as the most limiting factor for producing potential. The producer must evaluate the potential yield before applying it lower than the traditional harvest height. The crossing point of the origin can differ between cultivars, by using a standard high 40-cm cutting, it is reported that 'Jasmine 85' ratoon seedlings appear basipetally (starting from the node in the top of the straw and developing down towards the base) and that 60% of tillers originate from the node. The length of the planting season allows for the production of the ratoon plant, which is the second crop that develops from the main straw plant. Rice is traditionally harvested with a cutting height of 40 cm above ground level. However, reducing crop cutting height has the potential to increase crossing yield.

2. Materials and Methods

The research material was paddy plant (*Oryza sativa*) Ciherang that had been planted for six month in area of 320 m^2. The research location was area in Jatinangor West Java with altitude of about 2874 m above sea level, the temperature was 23-27° C and average rainfall of 2963 mm per year. Experimental plots were made with the size of 4x5 m. Paddy planted land was weeded and cleaned from weeds, which were scattered around the plant.

The paddy ratoon were crude protein, crude fiber and crude fat analyzed by proximate method [1], and lignin analyzed by the method [2]. The design used was completely randomize design with factorial design. The first factor is the cutting height with a cutting height of 10 cm, 20 cm. The second factor is cutting ages which is 14 days cutting ages, 21 days cutting ages and 28 days cutting ages and 35 days cutting ages. Duncan's multiple range test was conducted to see the difference between treatments.

3. Results and Discussions

3.1. Crude Protein Content of Paddy Ratoon at Various Cutting Ages and Cutting Height

Crude protein is an essential nutrient in ruminant feed, It is required by ruminants for growing and maintenance the cells and tissues of the body. The average of crude protein content of paddy ratoon at various cutting and cutting height can be seen in Table 1.

Table 1. Average of Crude Protein of Paddy ratoon at Various Cutting ages and Cutting Height

Cutting Height (cm)	Cutting Age (day)				
	14	21	28	35	
10	11.18^{e}	7.98^{c}	6.91^{b}	5.06^{a}	7.78
20	9.51^{d}	8.52^{cd}	6.76^{b}	5.83^{ab}	7.65
	10.34	8.25	6.83	5.45	

The data in Table 1 show that the increased the cutting ages makes the percentage of crude protein content decreased. Result showed that the ages of cutting significantly effect on crude protein content of the paddy ratoon. Further test with Duncan's multiple range test was conducted to determine the effect of differences between treatments. The data in table 1 also showed that the various cutting height not significantly affected on crude protein content of paddy ratoon. The data in table 1 indicate that the crude protein content of each treatment show significant differences. Plant at a young age has a higher protein than the old plant. Cutting age of 14 days containing the highest crude protein (10.34%) and the lowest is in the cutting ages of 35 days (5.45%). This is because in the cell wall older plant largery composed from cellulose and hemicelluloses. The value of nutrient including protein is decreasing. This is in accordance [3] that the plant at a young ages has better quality because has a lower crude fiber and higher crude protein content. Low level of protein in older plants can also be caused by the percentage of stems is higher than leaves.

3.2. Crude Fiber Content of Paddy Ratoon at Various Cutting Ages and Cutting Height

Forages fiber fraction varies depending on the growth, chemical composition and structure of these plants. The average of crude fiber of paddy ratoon at various cutting ages and cutting height can be seen in Table 2.

Table 2. Average of Crude Fiber of Paddy ratoon at Various Cutting ages and Cutting Height

Cutting Height (cm)	Cutting Age (day)				
	14	21	28	35	
10	28.67^{a}	32.95^{c}	30.93^{bc}	31.53^{bc}	31.02
20	29.80^{ab}	31.52^{bc}	31.59^{bc}	31.85^{bc}	31.19
	29.23	32.23	31.26	31.69	

The data in table 2 show that the crude fiber content of the paddy ratoon increase with cutting ages. The lowest average of the crude fiber content obtained in the treatment or 14 days cutting ages (28.45%). While the highest average of the crude fiber content obtained in the treatment of cutting ages 35 days (31.69%). The various cutting height was not significantly increasing the crude fiber percentages. Result of variance analyzed showed that the cutting age significantly affected on crude fiber content of paddy ratoon ($p<0.05$). Further test with Duncan's multiple range test was conducted to determine the effect of differences between treatments.

The data in Table 2 indicates that the crude fiber content of each treatment show significant difference. In the older age of the cell wall formation process still continues to rise. The cell walls of plants composed of cellulose, hemicellulose and lignin. Which are component of crude fiber. In Addition, the increase production of crude fiber was also caused by the process of lignification are increasing as the cutting ages increase. Research result [4], showed that the production of crude fiber gliricidia increases along with the cutting age. The study [5] showed that the content of the fiber fraction on mulberry plants continues to increasing along with the cutting ages.

3.3. Crude Fat Content of Paddy Ratoon at Various Cutting Ages and Cutting Height

The average of crude fiber of paddy ratoon at various cutting ages and cutting height can be seen in Table 3.

Table 3. Average of Crude Fat of Paddy ratoon at Various Cutting ages and Cutting Height

Cutting height (cm)	Cutting age (cm)				
	14	21	28	35	
10	1.23^{d}	1.01^{c}	0.97^{bc}	0.79^{a}	1.00
20	1.06^{c}	0.84^{ab}	1.01^{c}	0.95^{bc}	0.96
	1.15	0.93	0.99	0.87	

Crude fat content of the paddy ratoon increased with cutting ages and cutting height (Table 3). The lowest average of the crude fiber content obtained in the treatment 35 days cutting ages (0,86%). While the highest average of the crude fat content obtained in the treatment of cutting ages 14 days (1.15%). All the above are due to the fact that leaves have a higher content of total fat, compared to the stems [6] and that the content of the plants in total fat decreases as the plants mature [7]

4. Conclusion
Cutting age of paddy ratoon as forages that produce optimal nutrient quality is at the age 14 days after cutting, with 10.34% crude protein and 28.45% crude fiber and 1.15% crude fat. Height cutting of paddy ratoon is not significantly affected nutrient quality.

Acknowledgement
This work was supported by Academic Leadership Grant from University Padjadjaran, Indonesia

Reference
[1] AOAC 2005 Official Method ofAnalysis 13th Ed *Assiciation of the Official Analytical Chemist* Washington DC.
[2] Bahar F A, De Datta S K 1977 Prospects of Increasing Tropical Rice Production *J Agron.* **69** : 536-540.
[3] Dewi U, Tarmidi A R, Hernaman I 2011 Local Lambs Performans Fed Cassava Waste Silage *J. Nutr. Feed Sci.* pp. 136-140.
[4] Djuned H, Mansyur, Wijayanti H B 2005 The influence of cutting age on fiber fraction of mulberry (*Morussindica* L. Var Kanva-2) Proc. National of livestock technology and veteriner Puslitbangnak Bogor pp. 859-864.
[5] Savitri M V, Sudarwati H, Hermanto 2014 Effect of cutting age on Gamal (*Grilricidia Sepium*) Productivity *J livestock Sci.* **23(2)** : 25-35.
[6] Cook C W 1972 Comparative nutritive values of forbs, grasses and shrubs In: McKell, Blaisedl, Goodin (Eds) Wildland shrubs, their biology and utilization pp. 303-310.
[7] SCA 1990 Feeding Standards for Australian Livestock. Ruminants *Standing Committee on Agriculture and CSIRO* Melbourne pp. 266.

Peanut Productivity Under The Albizia Stand In Agroforestry System

A Susanto[1], Mujiyo[2], D Purnomo[2] and MTh S Budiastuti[2]

[1]Graduate School of Sebelas Maret University, Jl. Ir. Sutami 36A, Surakarta, Central of Java, Indonesia
[2]Departemen of Agricultural Science, of Sebelas Maret University, Jl. Ir. Sutami 36A, Surakarta, Central of Java,nIndonesia 57126

E-mail: Asmadiun@yahoo.com

Abstract. An effort to meet the national food need is by utilizing productive forest areas under albizia stands. Food stuff such as peanuts can grow under albizia forest stands. The purpose of this research was to know success of agroforestry system implemented to tree after initial release of contract. This study was conducted under albizia stand of state forest Blitar in East Java. Research design was used in this study was randomized block design. 4- year- old albizia stages with 0%, 25%, 50 % trimming intensity with plant spacing of 2 m x 3 m, 3 m x 3 m, 3 m x 4 m. Albizia trees that used for this comparison as many as 120 albizia trees. Results showed that soil lies under 4-year-old albizia trees still produced peanut with highest weights 129 g/m^2 and lowest weight 117 g/m^2 while highest biomass about 115.10 g and lowest biomass about 98.23 g. Application of agroforestry system under 4-year-old albizia stand is still good and intensive, creating work, improving social welfare, local community opinion change into positive perceptions for forestry development, forest protection, forest fire prevention, reducing rapid forest degradation and environmental quality conservation of forest areas.

1. Introduction

Food availability is very important for the security and stability of a nation. Increasing demand for food along with increasing population, will affect availability of food produced by community. An effort to meet national food demand by utilizing productive forest areas under albizia stands. Peanuts are a commodity with high economic value and can grow under albizia forest stand. Indonesia produces peanuts around 83.73% while 16.27% remaining must be imported from abroad [1]. Maintaining soil fertility can be done with usage of beans varieties as green manures are widely practiced in modern farming systems due to its ability to inhibit nitrogen, fall of leaves and stems will maintain physical properties of soil [2] .

Utilization of dry land in state forest area is still not optimal, type of woody plant such albizia proved to be favorite choice by many peoples, especially Blitar district of East Java Province. In addition, albizia sale value in wood industry is quite high, albizia wood type plants proved to have been able to grow well on state forests and communities. Efforts to increase land productivity and fulfill short-term needs for community can be overcome with agroforestry technology that combines timber crops with short-cycle crops. Though, there is problem in agroforestry technology, such as interactions between plants compilers are sometimes competitive in nature with each other, over factors growth (sunlight, water and nutrients). This happens because both plants are adjacent equally require existing resources either in soil (water and nutrients) or above the ground (sunlight). Therefore, it is necessary to arrange good wood plants spacing and canopy pruning as silvicultural actions of agroforestry. The purpose of this study is to determine productivity of peanut (*Arachis hypogeae* L) and growth of albizia stands on agroforestry systems in dry land state forests.

2. Materials And Methods

2.1. Time and Location
The study was conducted on below 4 year-old albizia stand in the state forest area which administratively belongs to Blitar state forest East Java region for 4 months from November 2017 to March 2018.

2.2. Materials and Research Tools
Materials needed in this research were albizia stands, peanut seeds, NPK and Urea fertilizers. Tools needed in this study are cameras, thermohydrometer, lux meters and writing tools.

2.3. Research Procedures
The research design used in the study was randomized block design. Albizia stand is 4 years old with 0%, 25%, 50% pruning intensity and a spacing of 2 m x 3 m, 3 m x 3 m, 3 m x 4 m. A number of trees were used in data preparation, which was plot tress for 30 trees to deduct edge effect where they have adequate sunlight. Plant design was been experiment trail such as agroforestry albizia. Underside stand plantation has done when plant season in first season a year, that was peanut plantation.

To know fertility of plant status was done with soil analyzes. Soil sample has taken in depth of 0-20 cm and 20-40 cm every plot. Repetition has taken from a few points then composted by one sample. Soil chemical analyze such as C degree of land organic was measured by Walkey and Black method, N totally was defined by semiautomatic Kjedhal Digestion (AOAC) method, Phosphor was measured by P. Bray and the pH of land Electrode glass method.

2.4. Data Analysis
Data collection in this study was biophysical data (physical and chemical properties of soil, temperature, humidity, and light intensity), growth of albizia (diameter and height) and growth of peanuts (peanut plant weight and dry weight of peanut pods yields). Plant growth and productivity data were analyzed quantitatively and descriptively. Data obtained by statistic analyzing shows influence from main plant growth progress and productivity. If analysis value shows positive result, it means $\alpha = 5\%$ according Duncan, this calculation is done with software SPSS ver.16 software and Microsoft excel.

Table 1. Criteria for assessing soil chemical properties according to Soil Research

Nature of the soil	very low	low	medium	high	very high
N	<0.10	0.10-0.20	0.21-0.50	0.51-0.75	>0.75
P_2O_5HCl (me/100gr)	<10	10-20	21-40	41-60	>60
K_2OHCl 25%(me/100gr)	<10	10-20	21-40	41-60	>60
Ca(me/100gr)	<2	2-5	6-10	11-20	>20
Mg(me/100gr)	<0.4	0.4-1.0	1.1-2.0	2.1-8.0	>8.0

Source: Survey of land capability LPPT Bogor

3. Results And Discussion

3.1. Biophysical Conditions of Research Plots
Forest area used as characteristics of soil research can be seen from level description of soil fertility so that growth factors of albizia and peanut agroforestry can be identified. Based on soil samples analysis

before agroforestry planting and after agroforestry planting, it can be known characteristics of physical and chemical properties of soil at research plot location (Table 2).

Table 2. N,P,K , C Organic, soil pH under tree Agroforestry

No.	Variabel	Method	P.2	P.3	P.4	P.5
1	N total (%)	Kjelhdal	0.28	0.28	0.22	0.20
2	P available (ppm)	Bray II	10.63	13.63	7.94	10.86
3	K confused (me%)	Eks. Amn. Act 1N	0.30	0.25	0.16	0.17
4	C. organic (%)	Walkley& Black	1.08	1.14	0.85	0.98
5	pH	Electrode glass	5.17	5.24	5.70	5.30

Source: Soil chemistry physics lab UNS agriculture faculty

Table 2 shows that soil chemical fertility on research site is categorized as low-medium. Nitrogen content and organic carbon are medium category, while phosphorus content is very low and low category. Soil pH on research location is categorized as acid soil. Nitrogen (N) and organic carbon are both in moderate condition. This is because organic matter is main source of N elements. Decomposition speed from that organic higher nutrient content and nitrogen content matter will affect N element cycle [3]. P element in acid soil will generally be bound by acid-causing elements like Al, Fe and Mn, so that elements become unavailable for plants. Albizia root growth can be inhibited due to low availability of phosphorus (P) element, especially in acid soil. Another factor that affects soil fertility is rainfall factor indicator that plants grow normally [4]. Number of rainy days in 2015-2016 was 182 with rainfall value 40,081 mm. This shows that rainfall at research site is very high. Tropical regions using quality seeds because have high temperature characteristics and high rainfall [5]. This condition causes vulnerability to soil quality because on open land will experience erosion and loss of nutrients easier due to rainwater flow. Perennial plants existence in tropical region is needed as a source of raw organic material seems to have met the standards [6].

As well as protect soil surface from rainfall. Litter Existence from perennial plants is an important source of nitrogen for non-fertilized agricultural plant mentioned that organic presence derived from plant biomass gives benefits to agricultural systems without fertilization such as: a) providing a large portion of nitrogen (N) and carbon (C) also half of phosphorus (P) to be absorbed by plants, b) Increase exchange value of soil (CEC), c) Vegetation has a large role in influencing the movement of water [7]. Assisting soil aggregate so improves soil physical properties and reduces susceptibility toward leaching on sandy soil, d) Improving ability to bind water especially to sandy soil, e) Binding micro nutrients so it prevents washing those nutrients. Interactions between species in it can produce various kinds of benefits [8].

Albizia is a type of legume that has ability to bind N in air. N accumulation in albizia biomass then provide benefits fertility to surrounding soil through decay of upper biomass and dead plant roots. Soil fertility Increase can in the addition of organic matter to the soil and reduces the occurrence of soil damage [9]. Litter fall from albizia plants and annual plant litter biomass from albizia tree as well as waste from annual plant should be maintained its existence. Land conservation causes a decrease in run off [10].

Litter functions as a source of organic matter and also it can protect soil from mechanical damage due to rainfall and Agroforestry will reduce surface flow by 9.2% [11]. Litter existence will improve soil physical and chemical fertility. Litter loss upon harvesting will impact on increasing soil temperature by 2.5 °C and will reduce soil moisture consequently, evaporation will increase and soil will be easily drying and losing water in soil human intervention in the ecosystem [12].

Litter is main source of nutrients after experiencing decomposition process in identifying morphological characteristics [13]. Stated that albizia leaves contain 35 g/kg N, whereas albizia root contains13 g/kg N, moreover albizia litter can also be used as effort to overcome problem on soil with acidic ph. Albizia rooting has root nodules as a result of symbiosis with rhizobium bacteria, it is benefit to surrounding soil because it helps supply nitrogen in soil spur the decomposition and nutrient cycle processes [14].

3.2. Biomass Conditions of Research

Fertilizers treatment and their combinations did not significantly affect growth of wet pods. Based on Table 3 above, wet pods weight is larger in N1P1K1 shade with 50 g of 115.10 g. Wet pods weight of or so-called pod weights grow quite well. It has same effect on plant height. This means growth of pod weight is along with height growth of peanut crops. Food supplies in pods will also affect other growth parameters.

Fertilizers have no significant effect on wet pods weight growth statistically. However, when viewed from a farmer scale, 50 g of fertilizer can increase wet pods weight by 16.87 g to control. Main product of peanut plants is the pod, with an increase of 16.87 g for one pod, which will have a huge impact on farmers' income if they harvest peanuts per ha table 3.

Table 3. Peanut biomass value against fertilizer treatment

Treatment	Biomass average (g)
P0K0	98.23
P1K1	100.21
P2K2	106.49
P3K3	105.94
N0P0K0	99.93
N1P1K1	115.10
N2P2K2	101.82
N3P3K3	103.54

The condition of albizia stand which already has fertile soil conditions because albizia leaf litter provides a lot of organic matter to soil under its stand and albizia trees. It is included in Leguminosae tribe containing root nodules. [15] explained that rooting system of albizia contains many root nodules as a result of symbiosis with rhizobium bacteria. It is advantageous to roots and surroundings. Root nodules existence can help soil porosity and nitrogen provision in soil, thus albizia tree can make surrounding soil more fertile. Such soil conditions can be planted with nuts crops so as to increase income of share croppers.

3.3. Age and Results of Peanuts

Age of albizia used in research about 4 years and highest pruning intensity about 50%, make light intensity easy to penetrate soil influence peanut growth and production. With 50% pruning intensity, it is expected increase light intensity that can be tolerated by peanut plants for its production process. Peanuts need an open place for optimal growth. Open space without shade, is a place that optimal for peanut growth and peanut production results. Research Results in [16] showed that peanuts can grow well under albizia stands. Planting peanuts under *Acacia mangium* and *Eucaliptusdeglupta* stand did not work well due to shade and roots of main plants [17]. Teak and peanut agroforestry show that in general, results obtained from agroforestry systems are lower than monocultures, both for actual and potential production [18].

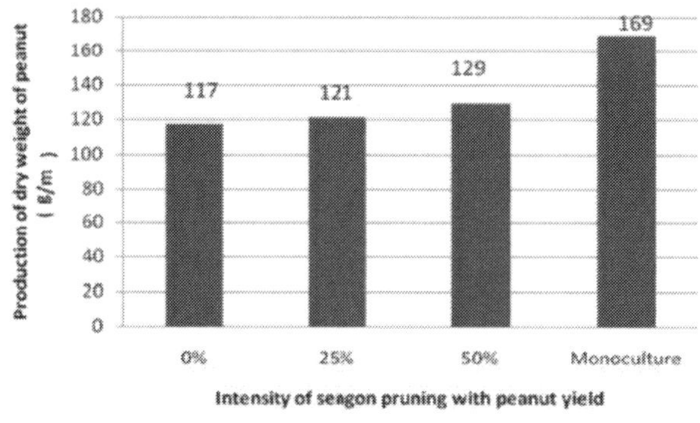

Figure 1. Peanut yield in various pruning albizia

This was due to competition in obtaining growth factors (water, nutrients and sunlight). Nonetheless, data presented in figure 1 and figure 2 shows that wider spacing result in better growth and production of peanuts in relation to age at which peanuts growth is getting older. This shows that at 4 years old albizia spacing has not given a difference effect in obtaining growth factors. Differences in growth factors are more caused by treatment of pruning intensity. This is in accordance with [19] which states that tree spacing does not have a significant effect on production of peanut annual crops.

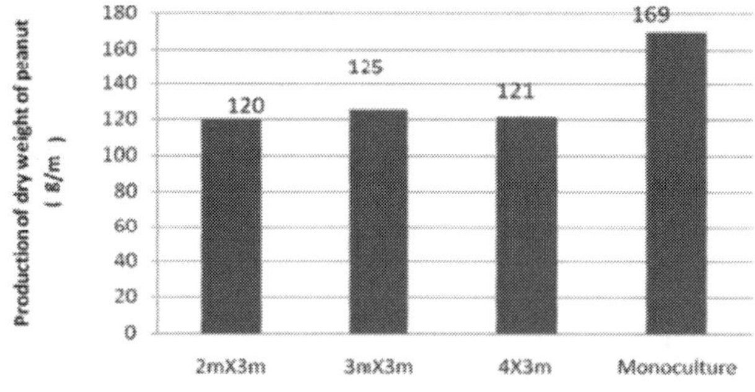

Figure 2. Dry weight of peanuts of various distances sengon

3.4. Peanuts Growth in Agroforestry Planting Patterns

According to [20] plants that suitable for inclusion in intercropping patterns are short-type plants, small leaf crowns, not many branches, mature age and annual, resistant to pest and disease attacks, high yield and not sensitive to sunlight duration. Peanut plants have a lower canopy that can absorb more sunlight and its plants are faster to produce. Peanuts are legume plants that can be symbiotic with Rhizobium so they are able to bind Nitrogen free in air and form root nodules can fertilize soil. According to [21] that intercropping between plants legume and non-legume is perfect because Legume plants can tie N free from air through rhizobium on its root nodules, 30% from N fixation was contributed to other plants in intercropping system. Albizia growth in agroforestry is relatively faster than pattern of albizia planting on monoculture shown by delta (higher diameter growth of lower planting). Albizia and peanuts in agroforestry pattern do not affect competition in obtaining nutrients and water from soil. This is due to different root characteristics. Existence of peanuts in agroforestry planting patterns helps adding nutrients process, especially N this because albizia growth in cropping patterns agroforestry is relatively faster than monoculture. That growth of teak diameter is better in system intercropping peanuts [22].

4. Conclusion

Albizia spacing 2 m x 3 m, 3 m x 3 m, 3 m x 4 m has not given a difference effect in obtaining growth factors. Provision of fertilizer in peanut plants has no significant effect on wet pods weight growth statistically. Albizia and peanuts in agroforestry pattern do not affect competition in obtaining nutrients and water from soil. This was due to different root characteristics

Acknowledgments

The authors would like to thank to Perum Perhutani and Indonesian Directorate General of High Education for research funding granted.

References

[1] National Food Security Agency 2008 Planning Seed Collection Indonesia Tree Seed Source *Develovement Project Ministry of Forestry* **17**: 85-100.

[2] Lakitan, B 1993 Basics of plant physiology Jakarta: *PT Rajagrafindo Persada*

[3] Nishimua T B, Suzuki E, Kohyama T, Tzuyuzaki S 2007 Moratlity and growth of trees in peat swamp and heath forest in Central Kalimantran after severe drought *Plant Ecol.* **188:** 165-177.

[4] Barnett J R, Jeronimidi G 2003 Wood Quality and Its Biological Basis Victoria : *Blackwell Publishing CRC Press.*

[5] Roshetko J M, Pramono A A, Rohadi D, Widyani N, Manurung G S, Fauzi A, Sumardamto P 2012 Smallholder Teak Systems on Java, Indonesia, Income for Families, Timber for Industry. *Conference Proceedings. IUFRO 3.08.00 Small-Scale Forestry Conference 2012: Science for Solutions* 24-27

[6] Rayburn E B 2009 Nutrient Requirments for Beef Cattle *Forage management* West Virginia University : wvu.edu

[7] Bruijnzeel L A 2009 Tropical Reforestation and Streamflow The Need for a Balanced Account Amsterdam : *Vrije Universiteit.*

[8] Jose S 2009 Agroforestry for ecosystem services and environmental benefits: an overview *Agroforest Sys* **76:** 1-10.

[9] Ketema H, F Yimer 2014 Soil property variation under agroforestry based conservation tillage and maize based conventional tillage in Southern Ethiopia *Soil and Tillage Research* **141**: 25-31.

[10] Dou L, Huang M, Hong Y 2008 Statistical Assessment of the Impact of Conservation Measures on Streamflow Responses in a Watershed of the Loess Plateau, China *Water Resour Manage* **6:** 25-29

[11] Wang Y, Zhang B, Ling L, Zepp H 2011 Agroforestry System Reduces Subsurface Lateral Flow and Nitrate Loss in Jiangxi Province China *Agriculture, Ecosystem and Environment* **140:** 441-453

[12] Hossain M K, M L Rahman, Hoque A T M R 2004 Comparative regeneration status in a natural forest and enrichment plantations of Chittagong (south) forest division Bangladesh *Journal of Forestry Research* **15:** 255-260
[13] Fuhuda W M G, Guevara C L, Kawuki R, Ferguson M 2010 Selected morphological and agronomic descriptors of cassava IITA *Ibadan. Nigeria* **142:** 76-79
[14] Zimmer M 2002 Is decomposition of woodland leaf litter influenced by its species richness *Soil Biology & Biochemistry* **34:** 277 – 284.
[15] Yulipriyanto H 2010 Land Biology and Management Strategy *Graha Ilmu.First Edition* Yogyakarta **9:** 45-62
[16] Lakitan B 1993 Basics of plant physiology Jakarta : *PT Rajagrafindo Persada*
[17] Haryanto Y, Dwiriyanto H 1988 Trial of Agroforestry Food Crop Development *BTR Benakat* Palembang **64:** 89-90
[18] Hermansyah P A, Taufik M 1997 Development of Biological Tethering N s. *Faculty Agriculture University of Bengkulu* Bengkulu 54-67
[19] Sindhoesarojo 1995 Development of Biological Tethering N *Proceedings/Workshop Minutes, Research Biological Nitrogen tethering in nuts* 68 – 120
[20] Imran S 2009 Relationship of Temperature and Plant Growth Yogyakarta : *Graha Ilmu.*
[21] Elkan G H 1987 Symbiotic Nitrogen Fixation Technology Carolina : *CRC Press.*
[22] Sabarnurdin M S 1992 The effect of seasonal plants on teak growth (*Tectona grandis*) as well as soil fertility in intercropping system in Wanagama I *Buletin FKT UGM Yogyakarta* **21:** 35-51.

IOP Conf. Series: Earth and Environmental Science **334** (2019) 012066 doi:10.1088/1755-1315/334/1/012066

Genetic Parameter Contributing to Lodging Resistance of F_2 Population in Red Rice

E D Mustikarini[1], G I Prayoga[1*], R Santi[1], Z Nurqirani[1] and H Saragi[1]

[1]Department of Agrotechnology, University of Bangka Belitung. Jl. Raya Balunijuk, Gedung Semangat, Kampus Terpadu UBB, Balunijuk, 33172, Indonesia

E-mail : gigihibnuprayoga@gmail.com

Abstract. The effectiveness and successful of selection depend on the value of genetic variability, heritability and genetic advance estimation. The purpose of this research is to measure the value of genetic parameters (variability, heritability and genetic advance) of F_2 lines derived from crossing of mutant red rice, local accession of red rice, and commercial upland rice varieties. This research used single plant design. The rice materials used in this research was the F_2 seeds derived from hibridization between MR1512 X Inpago 8, MR1512 X Banyuasin, Inpago 8 X Balok, Balok X Banyuasin and Balok X Inpago 8. Selection intensity used is 10%. The filled grains and grain weight have a wide phenotypic variability, whereas the plant height and the productive tiller numbers have wide genotypic variability. The high heritability value was found in the plant height, flowering time, and harvest time. The high genetic advance was found in plant height, productive tillers number, flowering time, and harvest time. In the present work, it has been shown that the plant height, productive tiller numbers, flowering time, and harvest time are the important traits for use in rice lodging tolerance breeding program.

1. Introduction

Local upland rice is one of important ecotypes of rice beside lowland rice because the upland rice contains the essential genetic resources for plant breeding program such as water saving, drought tolerance, and quality of grain. However, the local upland rice also has some weaknesses including low productivity and lack of lodging resistance [1]. Therefore, plant breeding program to mend this weakness is necessary. One of the varieties is red rice which has good quality of grain and high demand. However, the constraint to increase the production of this variety is the high level of lodging. The decrease of rice yield due to lodging is up to 50% according to several research [2-3]. One of the solutions is introgression of lodging resistance into such variety through hybridization.

Hybridization is one of the methods to extend genetic diversity, and combine desirable characters from the parents to obtain new populations and produce the new superior variety [4]. The value of variability and heritability are the important components to measure before determining the method and time to perform the further selection. The effectiveness and success of the selection depend on the value of genetic variability, heritability and genetic advance estimation [5]. The wide variability, high heritability, and high genetic advance in population will support the successful of selection [6].

The important parameter traits must be measured in relation to the purpose of the hybridization and the characters associated with the major traits in breeding program. Recent progress in rice breeding, the genotypes that exhibit precocity, less dense panicles, low sterility and greater 100-grain weight should be prioritized for drought tolerance [7], meanwhile leaf area index and Na-Ca selectivity could be used as selection criterion for salt tolerance [8]. Analysis of genetic parameters is performed on the traits that affect lodging resistance and yield potency. Plant height, stem length, and stem diameter are the most correlated traits with lodging resistance [9-11]. In other study, plant height and bending moment were positively correlated with lodging tolerance index [12]. The identified characteristics can be used in the process of selecting resistant red rice lines.

This research aims to determine the value of variability, heritability and genetic advance of F_2 red rice lines. The F_2 lines used in this study were derived from crosses among the commercial rice varieties (Inpago 8 and Banyuasin (upland rice)), mutant red rice, and Bangka local red rice. Inpago 8 and Banyuasin have been chosen as donor for lodging resistant trait [13]. The mutant red rice (MR1512) was chosen because this rice have a high yield and early maturity in ultisol soil [14]. Balok variety is Bangka local red rice with high yielding which does not have lodging resistance.

2. Materials and Methods

This research was conducted in the Experimental and Research Garden, at University of Bangka Belitung. The materials used in this study are F_2 rice seeds from the hybrid between the varieties of MR1512 X Inpago 8, MR1512 X Banyuasin, Inpago 8 X Balok, Inpago 8 X Banyuasin, Inpago 8 X MR1512, Balok X Banyuasin, Balok X Inpago 8, Banyuasin X Balok, Banyuasin X MR1512, Banyuasin X Inpago 8, and the parental varieties i.e. Banyuasin, Inpago 8, MR1512, and Balok as check.

The method used was single plant design without replication. Each genotype consists of 90 rice seeds. The parameters observed are plant height, productive tillers numbers, flowering time, harvest time, filled grains, and grain weight. Plant height was measured from the plant base to the tip of the highest panicle. Productive tiller numbers were determined at 30 days after flowering for each plants. Flowering time was determined at 80% of the plants are heading. Harvest time was determined on which 80% of the grains on the panicles are fully ripened. Filled grains were separated from unfilled grains and measured with a seed counter meter. Grain weight was determined from filled grains in each plant.

The phenotypic coefficient of variation (PCV) and genotypic coefficient of variation (GCV) were calculated following the procedures suggested by [15], thus

$$\%PCV = \left[(\sigma^2_p)^{1/2}(\bar{x})^{-1}\right] \times 100\% \text{ where } \sigma^2_p = \frac{1}{N}\left[\sum_{i=1}^{N}(xi-\mu)^2\right] \quad (1)$$

$$\%GCV = \left[(\sigma^2_g)^{1/2}(\bar{x})^{-1}\right] \times 100\% \text{ where } \sigma^2_g = \sigma^2_p - \sigma^2_e \text{ and } \sigma^2_e = \frac{1}{n}\left(\sigma^2_{P1}+\sigma^2_{P2}+\ldots+\sigma^2_{Pn}\right) \quad (2)$$

where, $\bar{x}$ = grand mean of the character, xi = value of plant-i, μ = mean of population, N = total of plant observed, σ^2 = variance, σ^2_p = phenotypic variance, σ^2_g = genotypic variance, σ^2_e = environmental variance, Pn= check plant-n, n = total of check plant.

The analysis of the heritability (board sense) following the procedures suggested by [16], and expected genetic advance by [17].

$$\text{Broad Sense Heritability (H)} = (\sigma^2_g)(\sigma^2_p)^{-1} \quad (3)$$

Where, σ^2_p = phenotypic variance, σ^2_g = genotypic variance

The criteria of heritability value are [18] :

$0 \le H < 0.20$, Low; $0.20 \le H \le 0.50$, Moderate; $0.50 < H \le 1.00$, High

$$\text{Expected Genetic Advance (EGA)} = \frac{GA}{\mu} \text{ x } 100\% \text{ where } GA = i.\ H.\sigma_p \quad (4)$$

Where, GA = genetic advance, i = selection intensity (the value is 1.76 at 10%), H = broad sense heritability, σ_p = phenotypic standard deviation, , μ = mean of population.

The criteria of EGA value are :

0 < EGA < 3.3%	= low
3.3% < EGA < 6.6%	= quite low
6.6 % < EGA < 10%	= quite high
EGA > 10%	= high

3. Results and Discussion

According to the results of the phenotypic and genetic variability parameters, the coefficient variation was range from 5.9 to 91.6% for PCV and from 0 to 26.4% for GCV as displayed in Table 1. The phenotypic and genetic coefficient of variation (PCV and GCV) were the basis of variability criteria. Based on the relative coefficient of variation, variability criteria were divided into four criterion [19]. The criterion of phenotypic variability is 0.0-23.0% (narrow), 23.0-46.0% (quite narrow), 46.0-69.0% (quite wide), and 69.0-92.0% (wide). The criterion of genotypic variability is 0.0-6.75% (narrow), 6.75-13.5% (quite narrow), 13.5-20.25% (quite wide), and 20.25-27.0% (wide). According to this value, the phenotypic variability for the flowering and harvest time is narrow, plant height is quite narrow, the productive tiller numbers is quite wide, and the filled grains as well as rain weight have a wide phenotypic variability. The genotypic variability for the filled grains and harvest time is narrow, flowering time and grain weight is quite narrow, plant height and productive tillers number have a wide genotypic variability.

Table 1. Estimates of mean, phenotypic and genetic variability parameters.

Parameters	$\bar{x}$	σ^2_f	σ^2_g	PCV (%)	Criteria	GCV (%)	Criteria
Plant height	108.37	1188.25	647.27	31.81	Quite narrow	23.48	Wide
Productive tiller numbers	10.84	41.74	8.17	59.58	Quite wide	26.36	Wide
Flowering time	82.96	167.67	86.11	15.61	Narrow	11.18	Quite narrow
Harvest time	105.34	38.12	37.49	5.861	Narrow	5.81	Narrow
Filled grains	417.17	137437.02	-10674.51	88.87	Wide	0	Narrow
Grain weight	11.19	105.14	1.62	91.63	Wide	11.36	Quite narrow

$\bar{x}$ = grand mean of population, σ^2_p= phenotypic variance, σ^2_g= genotypic variance, PCV = phenotypic coefficient of variation, GCV = genotypic coefficient of variation

The broad variation in genotypic and phenotypic variability was probably caused by the F_2 population which has the highest level of segregation. Plant height and productive tiller numbers could be used as parameters selection rather than the other parameters with quite narrow and narrow genetic variability. The wide variability indicated that the selection on the parameters could be performed and would be effective [20]. Narrow genetic variability indicated that the individuals were in a relatively homogenous population so that the selection which is based on the character will be less effective [21].

According to the results of the heritability (broad sense) parameters, the heritability value was range from -0.1 to 0.98 as could be seen in Table 2. The negative heritability value was considered as 0 [22]. Based on this value, the heritability criteria for the filled grains and grain weight is low, the productive tillers numbers has a moderate heritability, and the plant height, flowering time, and harvest time has a high heritability.

In the current results, the plant height, productive tiller numbers, flowering time, and harvest time were the important traits in lodging traits selection and could be used in early selection rather than low. The selection of properties with high heritability values can be performed in the early generation, meanwhile if the heritability value is low, the selection can be performed in the final stage of selection program [23]. According to [24], heritability determines the progress of selection, the greater the heritability value, the greater the progress of selection, on the contrary, the lower the heritability value makes the progress of selection slower. High heritability value means that the selection can be applied efficiently due to less effect of environment [25]. In contrast to the high heritability, the parameters with low heritability values indicates that it will be less effective as the next selection criteria [16]. This is due to the possibility of the characters change when planted in different environments, because the influence of environmental factors is quite large on the characters.

Table 2. Estimates ofheritability parameters

Parameters	Broad Sense Heritability (H)	Criteria
Plant height	0.54	High
Productive tiller number	0.2	Moderate
Flowering time	0.51	High
Harvest time	0.98	High
Filled grains	-0.1	Low
Grain weight	0.02	Low

Criteria: Low = $0 \leq H < 0.20$, Moderate = $0.20 \leq H \leq 0.50$, High = $0.50 < H \leq 1.00$

Besides using wide variability and high heritability values, the high genetic advance should also be considered in selecting a population. The results of expected genetic advance (EGA) calculation in Table 3, showed that the EGA value was range from 0 to 30.5%. The parameters which have a high EGA were plant height, productive tillers number, flowering time, and harvest time. The plant height has the highest EGA value which is 30.5%. The parameters which has a low EGA were the filled grains and grain weight.

However, in the present study, the productive tiller number showed moderate heritability and high EGA. It may occur because the value of genetic advance was calculated using the value of phenotypic standard deviation [26]. Productive tillers number parameters can be either applied or not for the next selection, because even though the genetic advance is high, the heritability is moderate and poor for selection [27]. The characters with high genetic advance value that are aligned with high heritability indicate that subsequent selection of these characters will be more effective. The selection will be effective if the population has wide genetic diversity, high heritability, and high genetic progress [28]. High heritability value accompanied by expectations of genetic advance indicates the effect of the additive gene that controls the trait [29]. The low genetic advance value is not necessarily caused by low heritability but may also be influenced by non-additive genes. Low genetic advance can be caused by the control of traits from non-additive genes [30], because non-additive gene is a temporary and inherited gene.

Table 3. Estimates of expected genetic advance (EGA) parameters

Parameters	Mean	GA	EGA (%)	Criteria
Plant height	108.37	33.05	30.5	High
Productive tiller number	10.84	2.23	20.53	High
Flowering time	82.96	11.7	14.11	High
Harvest time	105.34	10.69	10.15	High
Number of pithy grains	417.17	0	0	Low
Weight of pithy grains	11.19	0.28	2.48	Low

Criteria : Low = 0 < EGA < 3.3%, quite low = 3.3% < EGA < 6.6%, quite high = 6.6 % < EGA < 10%, High = EGA > 10%

IOP Conf. Series: Earth and Environmental Science **334** (2019) 012066 doi:10.1088/1755-1315/334/1/012066

Table 4. The average of promising lines selected from plant height, productive tiller numbers, flowering time, and harvest time parameters.

No	Lines	Plant Height (cm)	Productive tiller numbers	Flowering time (dap)	Harvest time (dap)
1	F2-I8xBl-21B-45	71.5	20	71	97
2	F2-I8xBl-21B-34	72.2	21	71	97
3	F2-I8xBl-21B-26	75.5	11	72	97
4	F2-I8xBl-21B-38	76.5	24	60	97
5	F2-I8xBl-21B-28	80.2	15	60	97
6	F2-I8xBl-21B-56	81.4	18	60	97
7	F2-BlxI8-23F-34	84	20	74	103
8	F2-MxI8-19J-59	84.4	30	83	106
9	F2-I8xBl-21B-24	85	14	70	97
10	F2-I8xBl-21B-33	86	22	71	97
11	F2-I8xBl-21B-52	86.7	18	60	97
12	F2-I8xBl-21B-35	92.8	17	60	97
13	F2-MxI8-23E-54	100.5	16	74	106
14	F2-BlxI8-23D-26	102	17	79	110
15	F2-BlxI8-23D-13	107	19	79	110
16	F2-BlxI8-23D-24	107	15	81	110
17	F2-BlxI8-23D-53	113	11	58	110
18	F2-MxI8-23E-52	113.5	14	70	106
19	F2-I8xBl-21C-47	114.5	27	77	108
20	F2-BlxI8-23F-44	114.8	27	70	103
21	F2-MxI8-9D-5	115.7	17	75	104
22	F2-BlxBa-23A-54	117	17	65	106
23	F2-BlxI8-23D-59	117.2	19	58	110
24	F2-BlxI8-23F-40	117.5	16	76	103
25	F2-MxI8-23E-46	118	16	76	106
26	F2-MxI8-23E-51	121.2	15	77	106
27	F2-MxBa-15C-48	123.5	19	77	115
28	F2-BlxI8-23F-47	124	31	65	103
29	F2-BlxI8-23D-36	126.2	26	58	110
30	F2-BlxI8-23D-46	126.8	20	58	110
31	F2-I8xBl-25A-28	128.2	15	74	104
32	F2-BlxI8-23F-38	128.2	19	70	103
33	F2-BlxI8-23D-38	128.8	18	58	110
34	F2-MxBa-13T-53	129	17	64	114
35	F2-MxI8-9D-34	133.3	16	83	103
36	F2-MxBa-15C-25	133.7	21	81	113
37	F2-BlxI8-23D-17	137	29	79	112
38	F2-MxI8-11A-33	140	14	86	113
39	F2-MxI8-9R-39	140.3	17	74	103
40	F2-MxI8-21B-45	140.5	20	71	107

Based on genetic parameters value, the selection was performed with 10% selection intensity. Thus, 40 out of 392 lines were selected as promising lines (Table 4). The selection was conducted based on plant height as a primary selection criteria, followed by productive tillers number, flowering time, and harvest time parameter. Plant height used as primary selection because it has wide GCV, high heritability, and high genetic advance value. In addition, plant height was considered to be the main trait for improving lodging resistance [2] [31-34]. From the selection process, 40 promising lines were obtained, considering that it has a height <141 cm, with the best-supporting characters (productive tillers number, flowering time, and harvest time) by rank method. Mostly selected lines were the F_2 lines derived of a cross between Inpago 8 and Balok (I8xBl). The I8xBl lines mostly selected because the plant height is short (<100 cm) and have many productive tillers. Supposedly, this is derived from the female parent of Inpago 8 which has short height and many tillers. According to [35], plant height is negatively correlated to the productive tillers number. More short plants probably mean more number of productive tiller.

The value of the genetic parameters influenced by parental genotype and growing environment [36]. According to [37], there is the role of parents in the effort to inherit the superior traits to the offspring. The role of the parents in the hybridization will also affect the genetic of offspring in addition to the influence of other factors such as the environment. Occurrence of lodging is common after strong wind accompanied by heavy rains during the grain filling period [2].

About 60 dwarf genes (*d1-d60*) genes and 7 semi-dwarf genes (*sd1-sd7*) have been found [38]. In our previous research, through molecular marker assessment showed that Inpago 8 varieties has an *sd-1* gene that controls plant height [39]. However, the *semi-dwarf1* (*sd-1*), a recessive, semi-dwarfing gene, one of the most important genes used in rice lodging resistance breeding program [40]. The *sd-1* gene has been reported to reduce plant height by 25% through approximately proportional reductions in lengths of the top five internodes; with practically no effect on panicle length [41]. The dwarf phenotype of *sd-1* which resulted from a deficiency in gibberellin (GA) plant growth hormones [42]. The presence of this gene resulted in a shortened culm with high lodging resistance. The adoption of semi-dwarfing rice varieties has significantly increased grain yield potential of rice [4]. MONOCULM 1 (MOC1), a gene that is important in the control of rice tillering. MOC1 may act as a master regulator in the control of rice tillering [43]. The *sd-1* and MOC1 genes could make a significant contribution to improve the lodging resistance and high yielding in rice breeding program.

4. Conclusion

The purpose of this research was to measure the value of genetic parameters of F_2 red rice lines. The filled grains and grain weight have a wide phenotypic variability, while the plant height and the productive tiller numbers have wide genotypic variability. The high heritability value was found in the plant height, flowering time, and harvest time. The high genetic advance was found in plant height, productive tillers number, flowering time, and harvest time. In the current study, it shows that the plant height, productive tiller numbers, flowering time, and harvest time are the important traits for use in rice lodging tolerance breeding program. From the selection process, 40 promising lines were obtained based on the approach of genetic parameters analysis.

Acknowledgements

This research was funded by Penelitian Strategi Nasional Grant 2018 from Ministry of Research, Technology, and Higher Education of the Republic of Indonesia.

References

[1] Riyanto A, Suwarto, Haryanto T A D 2011 Yield and Yield Component of 14 Upland Rice Genotypes in Banjarnegara Regency *Agronomika* **11:** 111–121.

[2] Zhu G, Li G, Wang D, Yuan S, Wang F 2016 Changes in The Lodging Related Traits Along with Rice Genetic Improvment in China *PloS one* **11** e0160104.

[3] Lang Y Z, Yang X D, Wang M E, Zhu Q S 2012 Effects of Lodging at Defferent Filling Stages

on Rice Yield and Grain Quality *Rice Science* **19:** 315-319.

[4] Biswas M K, Mondal M A A, Hossain M, Islam R 2008 Utilization of Genetic Diversity and its Association with Heterosis for Progeny Selection in Potato Breeding Programs *American-Eurasian J. Agric. & Environ. Sci* **3:** 882–887.

[5] Djuariah D 2007 Genetic Variability, Heritability, and Phenotypic Performance of 50 *Ipomea reptans* Poir in Medium Land *J. Agrijati* **5:** 48–53.

[6] Syukur M, Sujiprihati S, Yunianti R, Kusumah A 2011 Estimation of Genetic Variance and Heritability for Yield Component Characters in Chili Pepper Genotypes *J. Agrivigor* **10:** 148–156.

[7] Guimarães C M, Castro A P D, Stone L F, Oliveira J P D 2016 Drought Tolerance in Upland Rice: Identification of Genotypes and Agronomic Characteristics *Acta Scientiarum. Agronomy* **38:** 201-206.

[8] Zeng L, Poss J A, Wilson C, Draz A S E, Gregorio G B, Grieve C M 2003 Evaluation of Salt Tolerance in Rice Genotypes by Physiological Characters *Euphytica* **129:** 281-292. 10.1023/A:1022248522536.

[9] Kelbert A J, Spaner D, Briggs K G, King J R 2004 The Association of Culm Anatomy with Lodging Susceptibility in Modern Spring Wheat Genotypes *Euphytica* **136:** 211-221.

[10] Kelbert A J, Spaner D, Briggs K G, King J R 2004 Screening for Lodging Resistance in Spring Wheat Breeding Programmes *Plant Breeding* **123:** 349-354.

[11] Kong E, Liu D, Guo X, Yang W, Sun J, Li X, Zhan K, Cui D, Lin J, Zhang A 2013 Anatomical and Chemical Characteristics Associated with Lodging Resistance in Wheat *The Crop J.* **1:** 43-49.

[12] Yang H, Yang R, Li Y, Jiang Z, Zheng J 2000 The Relationship Between Culm Traits and Lodging Resistance of Rice Cultivars *Fujian Journal of Agricultural Sciences* **15:** 1-7.

[13] Setyono A, Indrasari S D, Lesmana O S, Sembiring H 2009 Description of Rice Varieties. Subang: *Indonesian Center for Rice Research.*

[14] Mustikarini E D 2016 Adaptation of Red Rice Rice Mutants and Upland Rice Varieties in the Three Sub-Optimal Lands in Indonesia *J. Lahan Suboptimal* **5:** 17–25.

[15] Nur A, Iriany N R, Takdir A M 2013 Genetic Variability and Heritability of Agronomic Characters of Maize Inbred Line with Tester 14 *J. Agroteknos* **3:** 34–40.

[16] Jameela H, Sugiharto A N, Soegianto A 2014 Genetic Variability and Heritability of Yield Component Characters in F_2 Population of Common Bean (*Phaseolus vulgaris* L.) Derived from a Cross Between Introduced and Local Variety *J. Produksi Tanaman* **2:** 324–329.

[17] Widyawati Z, Yulianah I, Respatijarti 2014 Heritability and Genetic Gains of F_2 Population in Chilli (*Capsicum annuum* L.) *J. Produksi Tanaman* **2:** 247–252.

[18] Aryana I G P M Test of Uniformity, Heritability and Genetic Gain of Red Rice Obtained from Back Cross Selection in a Dryland Environment 2010 *J. Crop Agro* **3:** 13–20.

[19] Rahmannisa S L, Waluyo B, Karuniawan A 2011 Performance of Genetic Parameters of Local Sweet Potato Varieties from Cilembu, West Java *Proceeding of Seminar Hasil Penelitian Tanaman Aneka Kacang dan Umbi* p. 675–684.

[20] Yenny R F 2010 Genetic Variability of Seed Sizes of F_2 Soybean (*Glycine max* L.) *J. Agroekotek* **2:** 49–52.

[21] Waqar-Ul-Haq, Malik M F, Rashid M, Munir M Q, Akram Z 2008 Evaluation and Estimation of Heritability and Genetic Advancement for Yield Related Attributes in Wheat Lines *Pakistan Journal of Botany* **20:** 1699–1702.

[22] Wolf D P, Peternelli L A, Hallauer A R 2000 Estimates of Genetic Variance in an F_2 Maize Population *Journal of Heredity* **91:** 384–391.

[23] Zen S 2012 Estimation of Selection Criteria Soybean Genotypes *J. Penelitian Pertanian Terapan* **12:** 75–80.

[24] Lubis K, Sutjahjo S H, Syukur M, Trikoesoemaningtyas 2014 Estimation of Genetic Parameters and Morphophysiological Characteristics Selection of Introducing Corn Lines in Acid Soil

Environments *J. Penelitian Pertananian Tanaman Pangan* **33:** 122–128.

[25] Kristamtini, Sutarno, Wiranti E W, Setyorini W 2016 Genetic Advance and Heritability of Agronomic Characters of Black Rice in F_2 Population *J. Penelitian Pertananian Tanaman Pangan* **35:** 119–124.

[26] Arifiana N B, Sjamsijah N 2017 F_3 Plant Response Selection of Soybean (*Glycine max* (L.) Merrill) Genotypes *Agriprima, J Applied Agricultura Science* **1:** 50–58.

[27] Hartati R S, Setiawan A, Heliyanto B, Sudarsono 2012 Genetic Variability, Heritability, and Correlation Among Characters of 10 Selected Genotypes of Physic Nut (*Jatropha curcas* L.) *J. Littri* **18:** 74–80.

[28] Syukur M, Sujiprihati S, Yunianti R, Kusumah A 2011 Estimation of Genetic Variance and Heritability for Yield Component Characters in Chili Pepper Genotypes *J. Agrivigor* **10:** 148–156.

[29] Hapsari R T, Adie M M 2010 Estimation of Genetic Parameters and Yield Component Relationship of Soybean *J. Penelitian Pertanian Tanaman Pangan* **29:** 18–23.

[30] Suprapto, Kairudin N M 2007 Genetic Variation, Heritability, Gene Action and Genetic Advance of Soybean on Ultisol *J. Ilmu-Ilmu Pertanian Indonesia* **9:** 183–190.

[31] Guirful L, Harming X. Jian Y, Jun Z 2006 Genetic Analysis on Tiller Number and Plant Height per Plant in Rice (*Oryza sativa* L) *Journal of Zhejiang Univ. (Agriculture and Life Science)* **32:** 527-534.

[32] Zhang F, Jin Z, Ma G, Shang W, Liu H, Xu M, Liu Y 2010 Dynamics Between Lodging Resistance and Chamical Contencts in Japonica Rice During Grain Filling *Rice science* **17**.

[33] Chen W Y, Liu Z M, Deng G B, Pan Z F, Liang J J, Zeng X Q, Tashi N M, Long H, Yu M Q 2014 Genetic Relationship between Lodging and Lodging Components in Barley (*Hordeum vulgare*) Based on Unconditional and Conditional Quantative Traits Locus Analyses *Genetic and Molecular Research* **13:** 1909-1925.

[34] Matusinsky P, Suobodova I, Misa P 2015 Spring Barley Stand Structure as An Indicator of Lodging Risk *Zemdirbyste-Agriculture* **102:** 273-280.

[35] Yahumri, Damri A, Yartiwi, Afrizon 2016 Growth and Yield Performance of Three New Paddy Rice Superior Varieties in Seluna Regency, Bengkulu *Proceeding of National Conference: Biodiversifikasi Indonesia* **1**: 1217-1221.

[36] Sarker Z I, Shamsuddin A K, Ara R 2007 Gene Actions of Traits Contributing to Lodging Resistance in Wheat (*Triticum aestivum* L) *Bangladesh Journal of Plant Breeding and Genetics* **20:** 23-30

[37] Ashari A H, Haryadi N T, Poerwoko M S 2011 The Maternal Effect of 5 Parents and 12 Crossbreeding Results of Soybean Crops on Armyworm *Spodoptera litura* (Fab.) Attacks *Berkala Ilmiah Pertanian* **1:** 1–5.

[38] Kinoshita C T 1995 Report of Committee on Gene Symbolization, Nomenclature and Linkage Groups *Rice Genet. Newsl.* **12**: 9.

[39] Mustikarini E D, Prayoga G I, Kartika 2017 Development of Superior Red Rice Varieties through the Introduction of Lodging Resistant Genes to Maintain High Production of Red Rice Mutants *Report of Penelitian Produk Terapan*: University of Bangka Belitung Unpublished.

[40] Mani D 2008 Characterization and Genetic Analysis of a Very High Tillering and Dwarf Rice (*Oryza sativa* L.) Mutant *Doctoral Dissertation*: Texas A & M University.

[41] Rutger J 1984 Induced Semidwarf Mutants *Rice Genetics Newsletter* **1:** 92-93.

[42] Spielmeyer W, Ellis M H, Chandler P M 2002 Semidwarf *sd-1*, "Green Revolution" Rice, Contains a Defective Gibberellin 20-Oxidase Gene *Proceedings of the National Academy of Sciences of the United States of America* **99**: 9043-9048.

[43] Li X, Qian Q, Fu Z, Wang Y, Xiong G, Zeng D, Wang X, Liu X, Teng S, Hiroshi F, Yuan M 2003 Control of Tillering in Rice *Nature* **422:** 618-621.

IOP Conf. Series: Earth and Environmental Science **334** (2019) 012067 doi:10.1088/1755-1315/334/1/012067

Study of The Potential Expansion of New Rice Fields in Central Maluku District to Support Food Security in Maluku Province

N Rusmin[1], S Edy [2], A Robby [3] and K R Dwi[2]

[1]Student Master Program of Agro-Industrial Technology, Faculty of Agricultural Industrial Technology, Padjadjaran University, Jatinangor, Indonesia
[2]Soil and Water Engineering, Department of Agriculture Engineering Biosystem, Faculty of Agricultural Industrial Technology, Padjadjaran University, Jatinangor, Indonesia
[3]Food Processing Technology Laboratory, Department of Food Technology, Faculty of Agricultural Industrial Technology, Padjadjaran University, Jatinangor, Indonesia

E-mail: rusminnuryadin30@gmail.com

Abstract. Central Maluku district is one of the priorities region in the development of paddy fields in the province of Maluku. Maluku province was only able to meet the needs of 58% of the rice. Fulfillment needs rice is one of the Government's efforts in food self-sufficiency. Increased agricultural productivity through the extension of new rice fields became one alternative settlement in fulfilment of rice. The increase in rice production through the expansion of rice fields is still possible. The success of the process of the expansion of paddy fields depend on the expansion of the activities in the rice production. The process of the expansion of rice fields is preceded by feasibility investigation location. Feasibility study of the site was conducted to find out the feasibility of potential land with the observations in the field which is then processed and spatial dianalis in using ArcGIS software. Results of a survey investigating the Central Maluku with the total area of 594.29 ha based on the suitability of the land for the potential expansion of new acres of rice paddies 587.35 ha from 2 (two) subdistricts include North Eastern Seram Subdistrict Kobi m2 170.87 ha and North Seram Subdistrict covering 416.82 ha.

1. Introduction

Maluku province is a region of islands which consists of 559 Islands, and from a number of the island, there are a few islands that belong to the big island. The Mainland province of Maluku is inseparable from the mountain and Lake clusters contained almost throughout the district/city, which numbered four (4) and 11 (eleven) of the Lake.

Central Maluku district is one of the priority regions in the development of paddy fields in the province of Maluku. Administratively, the Central Maluku has 17 (seventeen) subdistrict consists of 172 countries, and 6 (six) wards, with the flagship sector or commodity there is the same and there are different scattered in 17 districts. Geographically, Central Maluku district after expansion is located between 2 ° 30 '-7 ° 30 ' LS and 250 ° – 132 ° 30 ' BT, and is the area of the Islands with a population of as many as 53 Island fruit, where inhabited by as many as 17 and never inhabited as much as 36 pieces. The shape of the Central Maluku district area is grouped based on the physiography approach (macro-relief), Plains, beaches, hills and mountains with a varied slope. Recorded as much as 2 3 mountains, Plains, 2 lakes and rivers are in fruit 144 regions of Central Maluku Regency. According to the BPS Maluku province in the year 2015, Government Maluku province was only able to meet the needs of 58% of the rice. Fulfillment needs rice is one of the Government's efforts in food self-sufficiency, it is because of the availability of food is one of the primary needs [1].

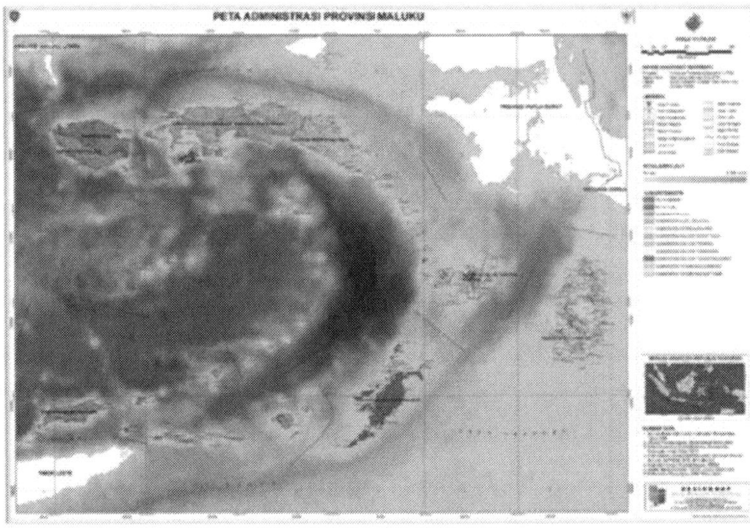

Figure 1. Administration Of Maluku Province Map
(Source: Thematic Map Of Indonesia)

Increased agricultural productivity through the extension of new rice fields became one alternative settlement in fulfilment of rice. The increase in rice production through the expansion of rice fields is still possible, because of the potential land is suitable for the expansion of rice fields was still quite spacious. The success of the process of the expansion of paddy fields depend on the expansion of the activities of the mechanism of the rice is done. The process of the expansion of rice fields is preceded by business feasibility investigation location [2].

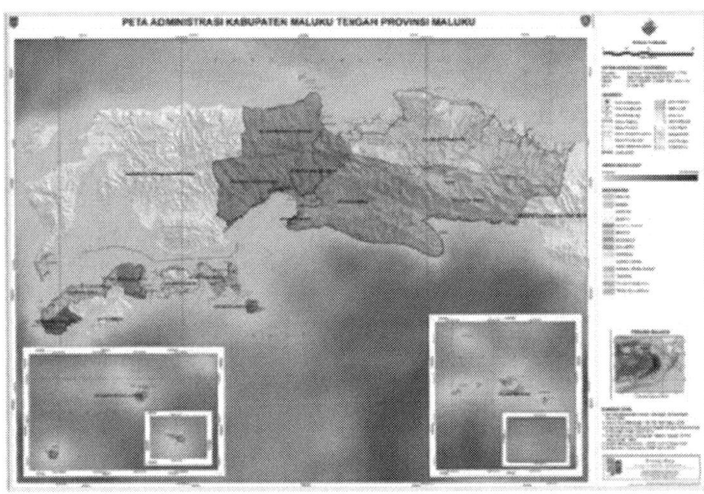

Figure 2. Adiministration Of The Central Maluku District
(Source: Thematic Map Of Indonesia)

2. Materials and Methods

Feasibility study of the site was conducted to find out the feasibility of potential land with the observations in the field which is then processed and spatial analysis in using ArcGIS software. Activities survey investigate prospective location and candidate farmers (prospective location and candidate farmers) expansion of paddy fields using some method that collaborates be integrated activities either in descriptive, analytic or survey. a holistic approach is used in order to obtain optimum results. stages of work in the activities of si-prospective location and candidate farmers includes preparations consisting of data acquisition and mobilization, survey the field include a spatial location of land measurement and measurement and identification of characteristics of land and water resources. Socio-economic surveys were done to obtain information the willingness of farmers in the rice field expansion program. Preparation phase carried out the preparation of the list of prospective locations. The list of prospective locations for expansion of paddy fields in the survey and investigation. The location of the candidate list was signed by the head of Department of agriculture Provinces or appointed by the KPA [3]

Field survey includes measurement of the activity of land are the spatial candidate. Prospective land that had already been planned then measured in terrestrial by directly surrounding it. Characteristics of prospective land observed in field intensity observation ground one observation represent 25 ha. Ground observations made through the soil profile or drilling as deep as 1.2 meters or more shallow when there are rock solid for mineral soils, whereas in peat up to a depth of 1.5 meters if the thickness of the peat is less than 1 meter or substratum (mineral lands) if the thickness of the peat is more or equal to 1 meter. Evaluation methods of land use refer to the Technical Evaluation of land for agricultural commodities (BBSDLP, 2012) [4,5] with modifications to suit local conditions. All the data obtained are then analyzed to produce classroom suitability of land for the extension of new rice fields. Land suitability class refers to the framework of FAO [6] as follows:

Class S1 : Highly suitable : The land does not have a limiting factor or real meaning against the use of sustainable, or limiting factor are minor and will not affect the productivity of the land.

Class S2 : Moderately suitable : The land has a limiting factor, and this will be a limiting factor to productivity, requiring additional input (input). The delimiter can usually be overcome by farmers themselves.

Class S3 : Marginally suitable : The limiting factor will be very effected on productivity, requiring additional input more than land that belongs to the S2. To overcome the limiting factor on S3 requires high capital, so the need for assistance or intervention (intervention) the Government or private parties.

Class N : Not suitable : The land has a limiting factor that is very heavy and/or difficult to overcome [7].

3. Results and discussion

3.1. Analysis of The Suitability Land

At first the plan land use potential to be developed into paddy fields proposed by the Department of Agriculture of Central Maluku district covering an area of 500 ha, the targets in the northern Seram of Huahulu village Trans Kilolima 350 m2/ha, Seram Subdistrict The North East of the village of Waetonipa and Seti Marasaua covering an area of 100 ha (each covering an area of 50 ha village) and North Eastern Seram Subdistrict Kobi Kobi Village Conscious area of 50 ha. However, further developments, the district changed his proposal i.e. the redirect target from the villages of Waetonipa and Marasaua to the village of Sariputih and the village of Leawai. Based on the results of the investigation area in 2 (two) subdistricts retrieved extents of 594.29 ha, thus there is excess area of 94.29 ha which are contained in Table 1.

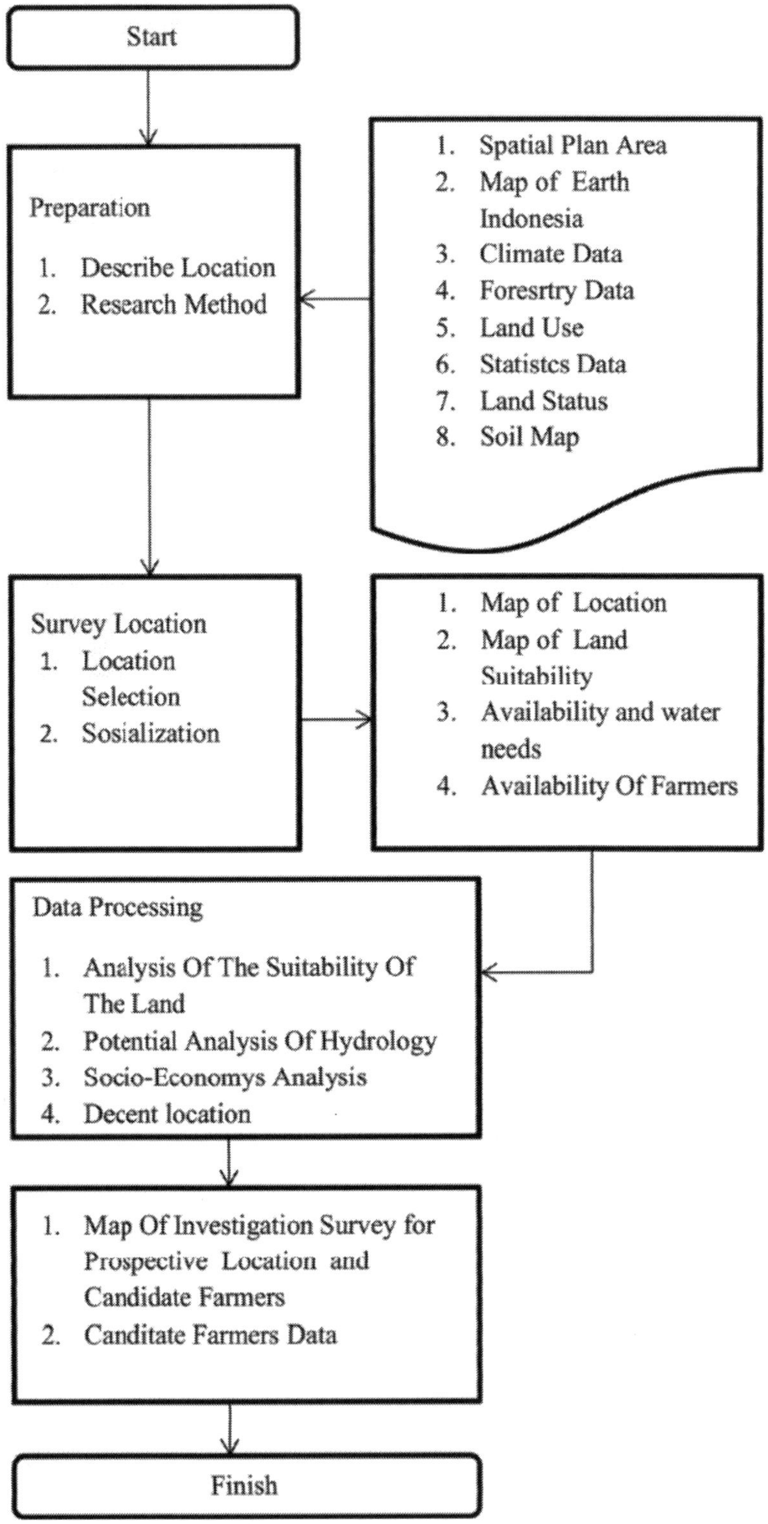

Figure 3. Flowchart Diagram of The Survey Investigation New Rice Field

Results of a survey investigating the Central Maluku district region with a total area of 594.29 achieved ha based on the suitability of the land for the potential expansion of rice paddies covering 587.35 ha from 2 (two) subdistricts include North Eastern Seram Subdistrict Kobi 170.87 ha and North Seram Subdistrict covering 416.82 ha. Seram subdistrict of North East and North Seram Kobi is a subdistrict of land with the potential for the expansion of rice farming is big enough and good enough because it is customary owned land and resettlement with physical characteristics which are quite supportive of plus the community's socially very enthusiastic to participate the expansion of rice fields

Table 1. Area measurement results Suvey Investigation-Prospective Location And Candidate Farmers Distribution Subdistrict and Village in Central Maluku district

No	Subdistrict	Village	Realization of Measurement Prospective Location and Candidate Farmers (Ha)
1	Seram Utara Timu Kobi		171,84
		Sari Putih	128,09
		Leawai	43,75
2	Seram Utara		422.45
		UPT Kilo V	422.45
	Total		594,26

(Source : Survey Location, 2016)

3.2. North Seram Subdistrict
Survey of North Seram subdistrict investigation carried out in the village of Seram UPT Kilo V with extents reach 422.45 ha. Based on the results of the analysis of the eligibility of land for potential printing rice apparently retrieved extents of 416.48 ha.

Table 2. Area of Potential expansion of rice fields on the North Seram Subdistrict

No.	Village	Realization of Measurement (ha)	The Potential Expansion of Rice Fields (ha)
1	Huahulu	422,45	416,48
	Total	422,45	416,48

(Source : Survey Location, 2016)

3.3. Socio-Economic Analysis Of Agricultural
The fulfillment of the basic necessities of farmers. Assuming the production of rice, then ownership of land with an area of Candidates Farmers ha 1.0 average, obtained results 4.0 ton – 4.5 ton/year (assuming 1 harvest time/year). Regional food procurement If estimated opening of the expansion of the wetland area of 3,647 ha (tentative) with production results paddy 4 ton/ha obtained food as much willingness 14,588 tons/year assuming only one harvest time/year. This amount is enough means to meet the food needs of regional development are in place in particular the expansion of rice fields. Land use sleep unproductive land mostly belonging to Prospective farmers in Maluku province is recommended as the expansion of paddy fields have the condition as a land with grass vegetation and trees that are not productive. Therefore, the existence of a plan the expansion of rice fields can make the land became productive which gives the benefit of economic, social and regional food security programmer in particular.

IOP Conf. Series: Earth and Environmental Science **334** (2019) 012067 doi:10.1088/1755-1315/334/1/012067

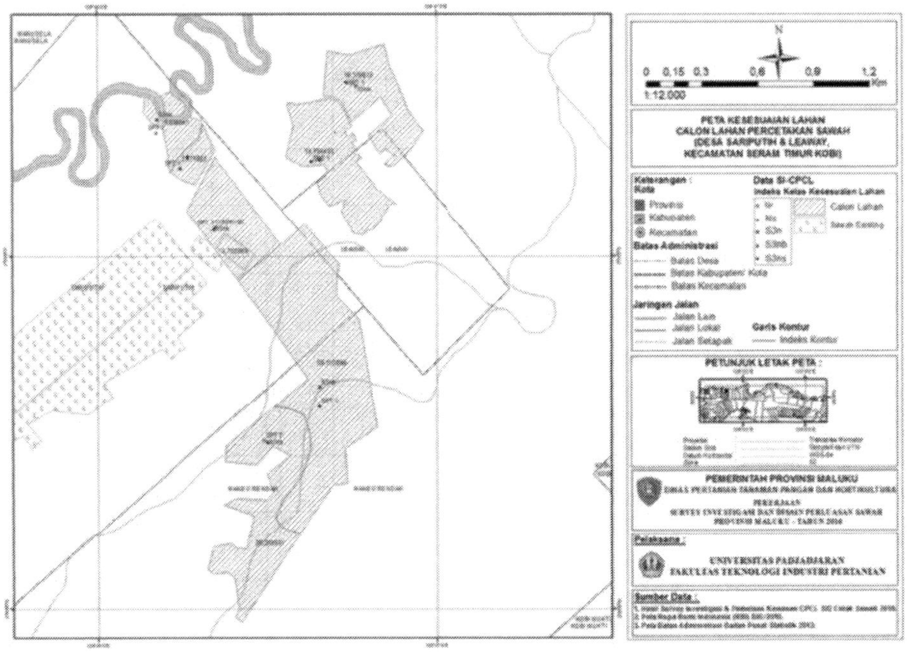

Figure 4. Map Overview Of Prospective Land Seram Timur Kobi Subdistrict, Central Maluku District

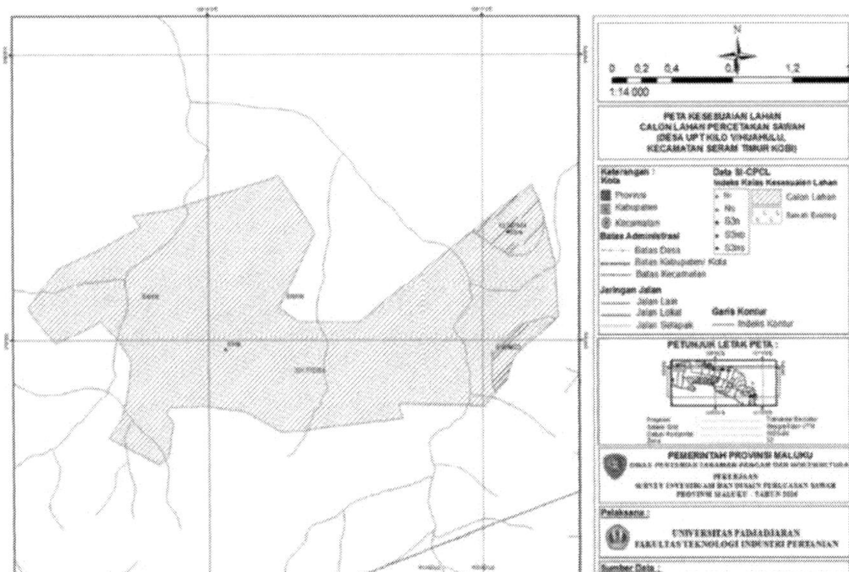

Figure 5. Map Overview Of Prospective Land North Seram Subdistrict, Central Maluku District

Table 3. Prospective Location in Central Maluku District

No	Subdistrict	Village	Measurements (ha)	Land Suitability (ha)		Prospective Location		
				S3nb	S3ns	Physical Suitability/Feasibility Of Land	Hydrology	Social-Economy
1	Seram Utara Timur Kobi	Sari Putih	128,09	101,83	26,26	Yes	Yes	Yes
		Leawai	43,75	43,75		Yes	Yes	Yes
2	Seram Utara		422,45	332,7	89,75	Yes	Yes	Yes

(Source : Survey Location, 2016)

4. Conclusion

A survey investigation of Prospective Location and Candidate Farmers Expansion/Printing New rice fields in Central Maluku Regency of Maluku province has been completed, with the close area of 591.29 ha. From the land that measured over 591.29 ha, after analysis of the suitability of land for land acquired land that was worthy of the S3nb of 468.28 and S3ns of 116.01 ha for socio-economic development is beneficial for the fulfilment of basic necessities farmers, regional food procurement, and land use, which is not productive. Central Maluku district included into the cluster of islands of agricultural development in the province of Maluku, thus the policy and direction of the development of any regions should be focused in the field of agriculture. After a review and evaluation of land with some limiting factor for growing rice plant terms that exist in Central Maluku Regency, Central Maluku District have the potential of land for expansion of new acres of rice paddies do 591.29 ha.

Acknowledgments
The authors thank the Maluku Province Department of Agriculture has funded the research through a cooperation scheme of local government with the Faculty of the technology of the agriculture industry, as well as allowing to publish the results of this research.

References

[1] BAPPEDA Provinsi Maluku 2012 RPJMD Provinsi Maluku Tahun 2014-2019 Pemerintah Provinsi Maluku Ambon.
[2] Hayatuliman M 2017 Analisis Kesesuaian Lahan Untuk Tanaman Padi Sawah di Kabupaten Subang Bagian Tengah Institut Pertanian Bogor Bogor Provinsi Kalimantan Selatan *Pangan* **23(3)**: 232-243.
[3] FAO 2002 FAO Guidline for Crop Water Reuquirements and Irrigation Scheduling Module 4 Harare: FAO. [internet] [2016, 3 Mei]. Retrieved from: ftp://ftp.fao.org/docrep/fao/010/ai593e/ai593e00.pdf
[4] Balai Besar Litbang Sumberdaya Lahan Pertanian 2015 Kriteria Kesesuaian Lahan Tanaman Padi Bogor Balai Besar Litbang Sumberdaya Lahan Pertanian. [internet] [2016, 2 March]. Retrieved from: http://bbsdlp.litbang.pertanian.go.id/kriteria/padi.pdf
[5] Balai Besar Penelitian dan Pengembangan Sumberdaya Lahan Pertanian 2010 Juknis Kesesuaian Lahan Bogor: Balai Besar Penelitian dan Pengembangan Sumberdaya Lahan Pertanian [internet] [2016, 10 July]. Retrieved from: http://bbsdlp.litbang.pertanian.go.id/evaluasi_lahan.php
[6] FAO 1976 A Framework for Land Evaluation. Soil Resources Management and Conservation Service Land and Water Development Division *FAO Soil Bulletin* **32** FAO-UNO Rome.

IOP Conf. Series: Earth and Environmental Science **334** (2019) 012067 doi:10.1088/1755-1315/334/1/012067

[7] FAO 1981 A Framework for Land Evaluation Roma: FAO [internet] [2016, 3 Mei 2016].Retrieved from: https://www.mpl.ird.fr/crea/taller-colombia/FAO/AGLL/pdfdocs/framele.pdf

Effect of Sawdust Biochar and Cow Manure Application on Soil Fertility at Peanut (*Arachis Hypogaea* L.) Land

A Fahreza[1], Zaitun[1], A Marliah[1], Sabaruddin[1], Sufardi[1], T F Karmil[2], Darusman[1], and Samadi[1]

[1] Agriculture Faculty, Syiah Kuala University, Banda Aceh, Indonesia
[2] Veterinary Faculty, Syiah Kuala University, Banda Aceh, Indonesia

E-mail: zaitundara@unsyiah.ac.id

Abstract. The low available nutrient and C-organic cause dryland has a low fertility rate. Biochar and manure application in agricultural cultivation must be reviewed further in order to obtain an effective and applicable way to increase production yields and to improve soil fertility. The purpose of this study was to determine the effect of sawdust biochar and cow manure on soil fertility in peanut land. This includes two types of soil amendment made from locally available raw materials (sawdust biochar and cow manure). Data analysis was conducted by using Randomized Block Design with factorial 3 x 3 and 3 replications. First factor was sawdust biochar (0 t/ha; 2.5 t/ha; and 5 t/ha). Second factor was cow manure (0 t/ha; 2.5 t/ha; and 5 t/ha). Soil samples were analysis on 45 days after planting and the observed variables were : (1) C Organic; (2) Total N; (3) C/N ratio; and (4) soil pH. The sawdust biochar application has a significant effect on C-Organic and soil pH. The best biochar application was 2.5 t/ha. Cow manure application has not significant effect on C-Organic, Total N, C/N ratio, and soil pH.

1. Introduction

Agricultural production on dryland is considerably lower when compared to agricultural production in wetland systems. This happens because wetlands generally have a relatively higher soil fertility rate when compared to dryland. In addition, other constraints are acid soil reactions, low organic C and N content, cation exchange capacity and low base saturation, high phosphate fixation, erosion problems and water availability [1], especially during the dry season, so the index of cropping in dryland is still relatively low. However, when viewed from characteristics of dryland, it is still necessary some action to overcome the limiting factors that become constraints in its development.

Land extensification for food crops is currently experiencing constraints due to limited productive land. Efforts to maintain the land extensification is by extension to dryland. Dryland farming system is agricultural cultivation practice which has high dependency on climatic factors, so it is easily degraded if its management is not appropriate. The low available nutrient and C-organic cause dryland has a low fertility rate. Improvement of soil fertility on dryland can be done with the addition of organic materials, either with manure, commercial compost and plant residues [2] . Biochar and manure application in agricultural cultivation must be reviewed further in order to obtain an effective and applicable way to increase production yields and to improve soil fertility.

Peanut production has decreased by 2015. Peanut production in Indonesia in 2015 was 605,449 tons, while in 2013 and 2014 reached 701,680 and 638,896 tons. The decline in production is also accompanied by a decrease in the area of peanut harvest in Indonesia. In 2013, the peanut harvest area decreased from 519,056 ha to 499,338 ha and decreased by 19,718 ha in 2014. In 2015 the decrease reaches 67,707 ha, so the harvest area in 2015 is only 454,349 ha [3]. This decrease in production shows that the practice of peanut cultivation has problems.

The purpose of this studied to determine the effect of sawdust biochar and cow manure on soil fertility in peanut land.

2. Materials and Methods

This research conducted at Campus experimental site, Syiah Kuala University on May to September 2017. The research was conducted by experimental method in field and laboratory. Soil amendment formulation was evaluated for improving soil and crops productivity. This includes two types of soil amendment made from locally available raw materials (sawdust biochar and cow manure). It was expected that the kind of soil amendment (sawdust biochar and cow manure) can be evaluated to know which the more effective as soil amendment to increase soil fertility at peanut cultivation land.

Data analysis was conducted by using Randomized Block Design with factorial 3 x 3 and 3 replications. Total 27 plots. First factor was sawdust biochar (0 t/ha or 0 kg/plot; 2.5 t/ha or 0.9 kg/plot; and 5 t/ha or 1.8 kg/plot). Second factor was cow manure (0 t/ha or 0 kg/plot; 2.5 t/ha or 0.9 kg/plot; and 5 t/ha or 1.8 kg/plot).

Biochar and cow manure application was 2 weeks before planting according to treatments. Urea fertilizer as base fertilizer was given 25% recommended (50 kg/ha) and applicated at planting time. Plot size was 1.2 m x 3 m and planting size 30 cm x 30 cm. Peanut, Bima variety, harvested 100 days after planting. Soil samples were analysis on 45 days after planting and the observed variables were : (1) C Organic; (2) Total N; (3) C/N ratio; and (4) soil pH.

3. Results and Discussion

3.1. Pretreatment Soil Analysis

The soil type in the Campus Experimental Site The ACIAR Project is Entisol with sandy loam texture with coarse class. Table 1 describes the soil chemical properties of Entisol of the Campus Experimental Site, The ACIAR Project, Syiah Kuala University.

3.2. Effect of Sawdust Biochar on Organic C, Total N, C/N Ratio, and pH on Entisol

Table 2 shows that C-Organic in peanut land was higher in application of sawdust biochar 2.5 t/ha and significantly different application of sawdust biochar 5 t/ha, but not significantly different from application of sawdust biochar 0 t/ha. Furthermore, the soil pH was higher in the application of sawdust biochar 5 t/ha and significantly different from the application of sawdust biochar 0 t/ha, but not significantly different with the application of sawdust biochar 2.5 t/ha. N-Total in peanut land tends to be higher in sawdust biochar 2.5 t/ha, although was not significantly different with N-Total due to sawdust biochar 0 t/ha and 5 t/ha. The C/N ratio in peanut land was higher in application of sawdust biochar 5 t/ha, although not statistically different with C/N ratio on application of sawdust biochar 0 t/ha and 2.5 t/ha.

Sawdust biochar application has significant effect on C-Organic and soil pH. This was presumably due to the provision of biochar can play a role as soil amendment. The biochar function for soil was as soil ameliorant because it has a relatively high pH and cation exchange capacity (CEC) [4]. In addition, carbon in sawdust biochar can increase the content of C in soil (Figure 1). The lower organic content of C in the treatment of 5 t/ha compared to other treatments (Table 2) showed that the carbon was more utilized by soil micororganisms as a source of energy. The amount of pore space on sawdust biochar can provide suitable habitats for the growth of soil microorganisms (Figure 2), so that soil microorganisms can work well.

The results of [5] showed that 1 t/ha biochar was able to increase soil pH. The higher biochar given to the soil causes the pH to increase. [6] also showed that giving pine sawdust biochar increases soil pH and the higher biochar dosage was also soil pH higher. The increasing soil pH and being neutral would be good for the crop, in accordance with [7] statement that the soil pH around neutral nutrients would be widely available to plants.

Sawdust biochar application was not significant effect on N-Total and C/N ratio. This was presumably because sawdust biochar has not functioned properly. This was indicated by a moderate C/N value. Nitrogen competition between plants and microorganisms occurs when the C/N ratio is high. This

IOP Conf. Series: Earth and Environmental Science **334** (2019) 012068 doi:10.1088/1755-1315/334/1/012068

is because more C than N, so there will be many sources of energy and microorganisms using existing nitrogen for its development [3].

Tabel 1. Soil chemical properties at Campus Experimental Site Syiah Kuala University [9]

Parameters		Criteria	Value
pH (H_2O)		Neutral-Slighty Alkaline	7.2-8.6
C organic (%)		Very low to Low	0.74-1.54
Total N (%)		Very low to Low	0.05-0.11
Total P (mg 100 g^{-1})		High to Very High	49-122
Total K (mg 100 g^{-1})		High to Very High	49-42
P available (ppm)		High to Very High	28-85
	Ca	Low to High	5.33-15.97
Exch	Mg	High	3.51-6.27
(cmol kg^{-1})	K	Low to Medium	0.12-0.39
	Na	Very low to High	0.08-1.60
Σ Cations (cmol kg^{-1})		Very High	10.77-19.75
CEC (cmol kg^{-1})		Low	9.42-14.57
Exch Al		Very low	0
Exch H		Very low	0.02-0.04
Base Saturation Basa (%)		Very High	>100
Soil fertility Status		Medium	

Tabel 2. Effect of sawdust biochar on C Organic, Total N, C/N ratio, and pH at peanut land.

Sawdust Biochar (t/ha)	C Organic (%)	Total N (%)	C/N	pH
0	1.06 ^{ab}	0.08	13.68	6.43 ^{a}
2.5	1.17 ^{b}	0.09	13.67	6.64 ^{ab}
5	0.99 ^{a}	0.08	14.52	6.71 ^{b}
HSD 5%	0.12	-	-	0.24

Letters in superscript indicate differences at 5% level

3.3. Effect of Cow Manure on Organic C, Total N, C/N Ratio, and pH on Entisol

The cow manure application was not significant effect on C-Organic, N-Total, C/N ratio and pH. This is assumed because the cow manure has not been decomposed perfectly so it has not given effect on some parameters of soil chemical analysis. This was supported by C-Organic and N-Total were also at a low dose and can be measured from C/N levels was classified as moderate. If the C/N ratio was relatively high, there will be nitrogen competition between plants and microorganisms, there will be a lot of energy sources and microorganisms using existing nitrogen for their formation and development and causing competition with plants [7].

IOP Conf. Series: Earth and Environmental Science **334** (2019) 012068 doi:10.1088/1755-1315/334/1/012068

Figure 1. Sawdust biochar pore with Scanning Electron Microscopy (SEM)

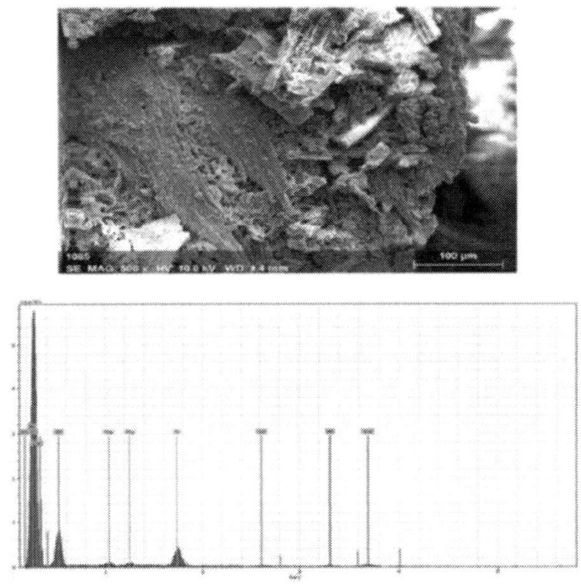

Spectrum: BSG CNSiK

El	AN	Series	unn. C [wt.%]	norm. C [wt.%]	Atom. C [at.%]	Error (1 Sigma) [wt.%]	K fact.	Z corr.	A corr.	F corr.
C	6	K-series	70.85	70.85	78.66	8.89	0.940	0.754	1.000	1.000
O	8	K-series	21.55	21.55	17.96	3.47	0.177	1.216	1.000	1.000
Si	14	K-series	4.36	4.36	2.07	0.23	0.018	2.413	1.000	1.005
Ca	20	K-series	1.62	1.62	0.54	0.13	0.005	3.046	1.000	1.030
Na	11	K-series	0.49	0.49	0.28	0.07	0.002	2.267	1.000	1.002
K	19	K-series	0.47	0.47	0.16	0.07	0.001	3.432	1.000	1.032
Mg	12	K-series	0.42	0.42	0.23	0.06	0.002	2.596	1.000	1.003
Cl	17	K-series	0.23	0.23	0.09	0.05	0.001	3.370	1.000	1.016
		Total:	100.00	100.00	100.00					

Page 3 / 4

Figure 2. Sawdust biochar analysis with Energy Dispersive Analysis (EDS)

Tabel 3. Effect of cow manure on C Organic, Total N, C/N, and pH at peanut land

Cow Manure (t/ha)	C-Organic (%)	Total N (%)	C/N	pH
0	1.02	0.08	13.22	6.68
2.5	1.07	0.08	13.92	6.50
5	1.12	0.08	14.72	6.59

4. Conclusion

The sawdust biochar application has a significant effect on C-Organic and soil pH. The best sawdust biochar application was 2.5 t/ha. Cow manure application has not significant effect on C-Organic, Total N, C/N ratio, and soil pH.

Acknowledgment

This research was funded by the Australian Centre for International Agricultural Research (ACIAR) - SMCN/2012/103 "Improving Soil and Water Management and Crop Productivity of Dryland Agriculture Systems of Aceh and New South Wales".

References

[1] Sanchez P A 2004 Properties and Management of soils in the Tropics *John Wiley and Sons* New York.

[2] Handayanto E, Ariesoesilaningsih, E 2002 Diversity of flora as a healthy farming capital: case studies on degraded critical land of upstream brine basin. in healthy farming roots *Agriculture Faculty Brawijaya University* Malang

[3] Badan Pusat Statistik 2016 Peanut harvested area by province (ha), 1993-2015 *https://www.bps.go.id/linkTableDinamis/view/id/873* Acces on April 4, 2017

[4] Alianti, Zubaidah Y S, Saraswati D 2016 Response of tomato (*Lycopersicum esculentum* Mill.) on biochar application and biofertiliser on peat soil *Journal Agri Peat* **17(2):** 115-125

[5] Endriani 2013 Utilization of oil palm shells biochar as soil amendment ultisol Bahar river Jambi *Research Journal Jambi University* **15(1):** 39 – 46

[6] Sika M P 2012 Effect of biochar on chemistry, nutrient uptake an fertilizer mobility in sandy soil *Thesis* Univeristy of Stellenbosch

[7] Taufiq A 2002 Status of P and K of dryland of Alfisol soil of Java and Madura and optimization of fertilizer for peanut plant *Proceedings of the National Seminar and Annual Meeting of the Regional Commission of Indonesia Soil Science Association* **16-17**: 94-103

[8] Arifah S M 2013 Application of the kind and dosage of manure on potato crops *J. Gama* **8(2):** 80-85

[9] The ACIAR Project Soil Survey 2016 Soil Survey Report at Aceh Province Australian Centre for International Agricultural Research (ACIAR) *SMCN/ 2012/103 "Improving Soil and Water Management and Crop Productivity of Dryland Agriculture Systems of Aceh and New South Wales".*

Measurement Model for Supply Chain Performance of Shallot

D E Pratiwi, B Setiawan, C Puspo, Nugroho[1] and A E Hardana[1]

[1]Department of Socio-economics of Agriculture, Brawijaya University, Jalan Veteran Malang, 65145, Indonesia

E-mail : pratiwi.fp@ub.ac.id

Abstract. Supply chain is an important key for any businesses' existence and development, since this concept can be a solution for them to hold efficient operational practices, minimizing cost, and on time delivery. Therefore, appropriate management practices on supply chain of agricultural commodity could be a substantial matter since it has been facing problems in efficiency, supply, and minimizing cost. The higher the economic value of a commodity, the more important the role of management on its supply chain. One of Indonesia's strategic and high valued commodity is shallot due to its high consumption and great contribution on regional economic development of production areas. Performance of shallot supply chain may impact on the parties involved, as farmers, traders, and suppliers. Thus, in order to formulate an accurate strategy to improve shallot supply chain performance, a measurement model is needed. This research employed Supply Chain Operations Reference (SCOR) method to identify the model, while analytical hierarchy process (AHP) was used to determine the weight of each indicator in the model. Findings of this research indicate that production is the key aspect on shallot supply chain. Furthermore, quality and order-fulfilment the most significant parameter in defining its performance is quality. Cost and order-fulfilment are proven as the highest weight-scores attribute and indicator, respectively, in measuring the performance of shallot supply chain.

1. Introduction

Supply chain refers an alignment among material procurement, production, up to product delivery in which some activities involving suppliers and distributors. As in any network, supply chain in agribusiness includes the strategic management of a network that is aimed at delivering to its demands or customers, in achieving efficiency and competitive advantage. Several studies have shown that supply chain is linked up with product competitiveness, which can be improved through increasing understanding on capacity, capability, even limitation of any entities involved in a supply chain [1-10] The Indonesian government has made financial and policy support to increase the development of agricultural sector through its *Nawacita* Plan (which refers to nine priority program). One of such program is revitalization on food sovereignty, in which shallot is pointed out as one of strategic commodity, due to its economic value and high demand. Fluctuation on its price often is a leading indicator of national inflation since its price is able to give a quick response to any both economic and non-economic "shock" on general condition of national economy, such as aggregate demand shock or natural disasters that might obstruct the commodity's distribution [11] In the other hand, cultivating this commodity requires high cost to provide, as both investment and operational spending, but it might not promising high revenue for farmers at the same time, caused by gap on its demand and supply. Even its consumption is always high, which reaches 2,55 kg/capita/year [12], but it is only can cultivated in certain areas which are concentrated in Java island, especially in Nganjuk (East Java), Brebes (Central Java), and Cirebon (West Java). Regarding to basic characters of agricultural commodity, including shallot, which are: perishable, seasonal, inconsistent on quality, and bulky, then it is reasonable that a good practice on supply chain management holds an important role to solve those congenital issues in order to prevent any price fluctuation and disequilibrium on shallot market.

Moreover, researches on SCM performance modelling were also conducted and developed by employing varied methods, such as balanced scorecard (Bhagwat and Sharma, 2007), multi-criteria analysis [13], Data Envelopment Analysis/DEA [14,15], activity-based costing ([16], and Supply Chain Operations Reference/SCOR [17,18,19]. In measuring the performance of supply chain, most of studies have applied SCOR method since it provides a standard description of supply chain processes, performance metrics, best practice, and enabling technologies [20] The basic structure of the SCOR model including five processes: Plan, Source, Make, Deliver and Return.

This paper aims to design a measurement model for shallot supply chain performance by developing any criteria and alternatives to increase its performance, as well as identifying shallot supply chain. The data was collected from 91 shallot farmers and 36 agents involved in the supply chain in East Java Province for empirical analysis. The agents included input suppliers, *penebas* (marketing agents), traders at city-level and province-level (merchants), port-traders, to the retailers.

2. Overview of Shallot Supply Chain in Indonesia

Shallot has been cultivated intensively by Indonesian farmers since its demands for this commodity tends to increase in line with the increase of the population number, due to there is no substitution commodity for it, both of as spice or herb. Besides domestic demand, demand for shallot in international market is also high. During 2009-2013, Indonesia was the fourth biggest exporter country for shallot, after New Zealand, France, and Netherland [21] Shallot production in Indonesia is mainly in Central Java (Brebes) and East Java Province (Nganjuk and Probolinggo). In particular, East Java Province covers for 23,16% of total production in 2015.

The structure of shallot supply chain in Indonesia involves some agents, from farmer, marketing agent, trader, kiosk/barn, central market, retailer, to consumer (Figure 1). This structure made the farmers received small income caused by too long chain. Farmers cannot maximize their margin share from added value process since too many agents involved and take their own shares. At the same time, the price of inputs tends to increase, especially for seed, fertilizer, and pesticide. In the long run, this condition will be a driving factor that caused farmer is not being able to cultivate shallot anymore since they cannot perform a capital accumulation for next production.

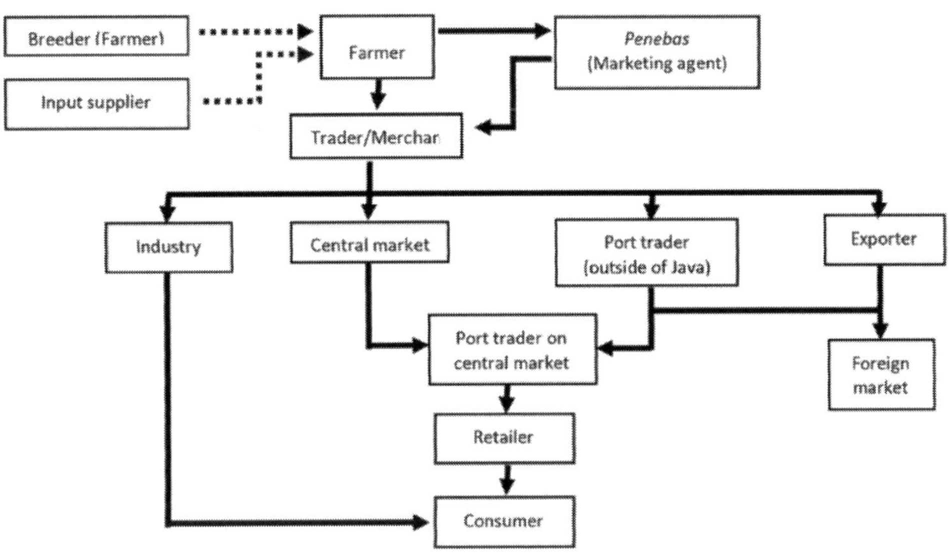

Figure. 1. Commodity flow in shallot supply chain.

These production and marketing constrains is reasonable enough for emerging the need for a good practice on shallot supply chain in order to maintain the continuity of this chain. Most of activities on shallot supply chain are held in central market, at which farmers, traders, even retailers meet, share any information, and make transaction. Roles and activities conducted by each agency involved in shallot supply chain will be discussed as follow :

- Input supplier
  Inputs used include seed, fertilizer, pesticide, and machineries, which are supplied by farm shops/kiosks.Farmers used to get seed from other farmers (breeder farmers), who live at the same town or at the other town. Some farmer groups even provide all inputs for their members. All transaction is made in cash, but some farmer groups usually allow their member to buy input by credit.
- Farmer
  Most o farmers sell their dried crops directly from their fields to *penebas* (marketing agents) or to trader. Price of shallot received by farmers is determined by the quality of the bulbs that could be measured from their size and color. Farmers usually get their revenue paid off on the day the buyer (trader) loads the shallot to his barn.
- *Penebas* (marketing agent)
  Penebas usually defines as local traders, who buy dried shallot only from farmers in their district, then sell them directly to a trader/merchant or send them to the nearest central market, so they are available for bidding.
- Trader/Merchant
  This kind of trader collects shallot from some local traders (but sometimes they buy from farmers directly), since they have more capital. After did some handling activities, such as: cleaning, sorting, and packaging, they send the shallot to other merchants at the other town, or even at the other island (especially Kalimantan, West Papua, and Sumatra) that called as port-trader. Some merchant sell small amount of their shallot to local retailers in the traditional market, and some also sell to an exporter for foreign market, such as Thailand, Taiwan, Vietnam, Singapore, and Malaysia.
- Port-trader
  Port-traders act like importers, who buy shallot from other town (or island) then load them to local central market for auction. Sometimes they sell their shallot directly to the local merchants and retailers.
- Industry
  There are two kinds of industry involved in shallot supply chain, they are: home-industry of fried onion and food manufacturing. Home industry of fried onion usually runs by local small scale enterprise (SME), who sells their product, to traditional restaurant nearby. In the other hand, food manufacturing refers to big factory which has shallot as component for their production.
- Retailer
  Retailers are the last chain of shallot supply chain, at which consumer buy shallot for consumption. Kiosks, grocery shops, vegetable stalls, and even supermarkets can be categorized as retailers since they sell shallot directly to the consumer (household).

3. Data

The data used in this study was collected from all agents involved in shallot supply chain in East Java Province who were determined by multistage sampling procedure, from October to December 2016. First stage, Nganjuk Region and Probolinggo Region were purposively selected based on their contribution on shallot national production. Second stage, respondents at each region were identified used snowball sampling, based on information from key informants (farmers) to identify all agents involved in shallot supply chain, from input suppliers, *penebas* (marketing agents), traders at county-level and province-level (merchants), port-traders, to the retailers. Ninety one farmers and thirty six agents was counted as respondents of this study.

Table 1. The Characteristics and structures of Respondents.

Variables	Farmers (%)	Agents
Age: under 50 years old	62.63 (57)	80.56 (29)
Age: more than 50 years old	37.37 (34)	19.44 (7)
Education: Elementary school	80.22 (73)	27.78 (10)
Education: High school	19.78 (18)	72.22 (26)
Experience: ≤ 20 years	74.73 (68)	58.33 (21)
Experience: > 20 years	25.27 (23)	41.67 (15)

Table 1 show three main characteristic of samples used in this study. It can be observed from the table that most farmers and agents involved in shallot supply chain are aged less than 50 years old, showing that the majority of actors in shallot supply chain are at productive age. Table 1 also indicates that educational background of farmers is contrary to the agents, in which most farmers finished only their elementary school, while most agents' education background are high-schooled (junior high, senior high, and undergraduate). Agents have larger possibility to adopt any technology and innovation, which represents more access to marketing and production information. Even though, they have higher educated rate, but the data on Table 1 presents most agents was less experienced than farmers.

4. Model Specification

Performance measurement system is needed in order to optimize the supply chain network of shallot. Performance measurement is aimed to support goals designing, performance evaluating, and formulating future mission at operational, tactical, and strategic level [22]. In this study, designing model for measuring supply chain performance was divided into two stages, those were: (1) designing measurement model by adopting Supply Chain Operations Reference (SCOR) method; and (2) calculating matrix weights used Analytical Hierarchy Process (AHP) technique. The steps to employ AHP are: (1) comparing scores of each parameter; (2) developing comparison matrix; (3) solving the Eigen values; (4) multiplying among matrixes; and (5) determining Consistency Ratio (CR).

5. Supply Chain Operations Reference (SCOR)

SCOR defines as reference model to describe the operation process of a supply chain [23] This model is usually used to measure the performance of company's supply chain, to develop its performance, and to communicate it to each party involved. It combines some elements (e.g., business techniques, benchmarking, best business practices) and applies them on supply chain system in the form of comprehensive framework as a reference to increase the performance of supply chain management. This model differentiates supply chain process into five main activities as follow.

- Plan
  Plan is defined as processes that balance demand and supply of shallot to develop the best practice to meet the procurement, production, and delivery needs. This process includes distribution need forecasting, capacity planning, stock planning and controlling.
- Source
  Main objective of source process is to meet planned or actual market demand, by providing input and services needed to cultivate shallot, includes shipping scheduling from input supplier, receiving, purchasing, choosing a supplier and evaluating its performance.
- Make
  Concept of "make" process in this study is defined as cultivation process, includes scheduling production, quality control, pest management, and equipment maintaining.
- Deliver

Delivery process is aimed at delivering the final product to meet market demand of shallot by managing order, transportation, and storage, in order to maintain shallot quality until it is in the hands of consumer, in the local and even foreign market.

- Return/Process
  Product return activities include identifying product quality and post-delivery customer support.

6. Matrix design for performance measurement

Hierarchy structure for measurement matrix of shallot supply chain used in this study consists of four levels, which were: (1) supply chain process; (2) performance parameters; (3) performance attributes; and (4) performance matrix. Level 1, supply chain processes, was adopted from SCOR method which included: Plan, Source, Make, Deliver, and Return. Parameters used to measure the performance for each process above were: Added value, Quality, and Risk. Furthermore, this study employed Reliability, Flexibility, Cost, Responsively, and Asset as performance attributes, to analyze, to evaluate, and to compare one supply chain with other chain. Whereas, five variables were used to formulate the performance matrix, they were: Shipping Performance (SP), Orders Fulfilments (OF), Lead-time Order Fulfillment (LTOF), Cash to Cash Cycle Time (CCCT), and Daily Stock (DS). These variables were chosen by considering the real condition of shallot supply chain at the study sites. Figure 2 show hierarchy structure of each level on measuring shallot supply chain .

7. Results and Discussion

7.1. Level 1: Supply chain process

The supply chain processes of shallot consist of: (1) planning process (Plan); (2) procurement process (Resource); (3) process of cultivation (Make); (4) shipping process (deliver); and (5) post-delivery process (return). The process of cultivation (Make), is proved as the first priority in improving the performance of shallot supply chains in East Java Province, since its weight was the highest (0.362) (see Table 2). The reason is that, in this process, members of the shallot supply chain are required to always keep the quantity of harvested commodities (products) able to meet market demand amidst predictable natural conditions (especially weather) and significant price fluctuations.

Table 2. Weight and priority process in shallot supply chain.

Process	Weight	Priority
Plan	0.095	4
Source	0.220	3
Make	0.362	1
Delivery	0.231	2
Return	0.092	5

The shipping process (Delivery) becomes the second most important priority in the supply chain, since most of the shallot production from both production area (Nganjuk Regency and Probolinggo Regency) is sent to other regions (other towns, other islands, and even exported to other countries). This needs to be pursued because the stakeholders (e.g., trading partners, port-traders) want such kind of certainty and continuity of supply to their region to maintain price stability.

7.2. Level 2: Parameters of supply chain performance

The most considered parameter in the shallot chain performance assessment is the quality dimension, with weight of 0.511 (presented in Table 3). Quality dimensions include shallot quality from on-farm crops and shallot which is sent to other regions. The quality of freshly harvested shallot will affect the preferences of the merchant or other marketing agents in determining the purchase price and quantity of product purchased. Assuring quality from farm to fork is one of the key elements of efficient agricultural supply chain management system [24]

While the quality of shallot at the delivery location will affect the loyalty of traders/merchant/port trader that are local merchant trading partners. Therefore, quality parameters become the most important of shallot supply chain performance. Since shallot supply chain tend to take long time to distribute, especially when it has to be sent to other island, it needs an effective supply chain practices which will help ensure the freshness of agro product by helping to reduce the time taken in transportation and by using favourable storage condition and better processes and systems.

Table 3. Weight and priority parameter in shallot supply chain.

Parameter	Weight	Priority
Quality	0.511	1
Risk	0.307	2
Added value	0.182	3

The empiric results on the table also show that supply chain of shallot is a risky business, since the value of risk variable is the second highest weighted parameter of shallot supply chain performance. Two important component of farming risk are market/price and production risks [25]. Market risk is related to uncertainty about the prices of outputs (and sometimes inputs, too) at the time production decisions were taken. Therefore, production risk is determined by weather condition and animal or plant diseases.

7.3. Level 3: Attributes of supply chain performance

Performance attributes consist of supply chain reliability, supply chain flexibility, supply chain responsiveness, supply chain costs, and management of supply chain asset. Cost attribute affects the size of profits to be gained, for both of production and transportation costs. From production side, seed and fertilizer are inputs that highly price-sensitive. Moreover, any changes in government support for fertilizer prices in form of subsidy would affect its demand and supply [24]. Furthermore, another kind of cost that should be considered more is transportation cost. Geographically speaking, the majority of shallot production activities take place in the Java Island. Often, hundreds tons of shallot from Java is moved by ships to other island of the country, which is a quite challenge. Therefore, cost becomes the most priority attribute with the weight of 0.286 in making the decision to improve the performance of the shallot supply chain.

Table 4. Weight and priority attribute in shallot supply chain.

Attribute	Weight	Priority
Reliability	0.283	2
Flexibility	0.114	5
Cost	0.286	1
Responsiveness	0.131	4
Asset	0.186	3

The second priority is reliability, with a weight of 0.283. Reliability of the supply chain is strongly connected to the concept of risk [26]). This attribute has a great influence in maintaining a supply chain, particularly in shallot supply chain, since it helps emerging a long-term relationship among supply chain members, especially between traders and consumers. Reliability of farmers and traders in ensuring the continuity of shallot supply in the market indicates the performance of it supply chain. In contrast, the results showed that performance of flexibility on shallot supply chain was weak since its weight was only 0.114. This kind of performance mainly occurs at farmer level, due to lack of grading and opportunity on trading in small quantity.

7.4. Level 4: Indicators of supply chain performance

The results proved that order fulfilment was the main indicator in assessment of supply chain performance. The number of shallot production in East Java Province tends to fluctuates, due to weather constraint that is not suitable to cultivation process. A delayed cultivation will lead to smaller volume of yields and delivery, causing traders and processors to be unable to fulfil the quantities contracted with other traders (e.g., merchants, traders, or port traders). So, it is important to maintain the performance of order fulfilment process because it helps delivering qualified product at the right time and right place, and also handling uncertainties from environmental factors [27]

Table 5. Weight and priority indicator in shallot supply chain.

Indicator	Weight	Priority
Shipping performance (SP)	0.231	2
Order fulfilment (OF)	0.284	1
Lead-time order fulfilment (LTOF)	0.170	4
Cash to cash cycle time (CCCT)	0.213	3
Daily stock (DS)	0.102	5

Shipping performance becomes the second most important indicator of shallot supply chain performance, since most of shallot is produced in certain production area in Java Island, which is going to deliver to vary consumer places. On time delivery, accuracy of the type of transportation used, as well as the accuracy of the storage method used during the shipping process, determines the performance of this indicator. These factors are important because they affect the income of traders who are being partnered with traders from other region. Lack on shipping performance would increase logistic loss on agricultural product, so it is difficult to gain benefit along a supply chain [28]

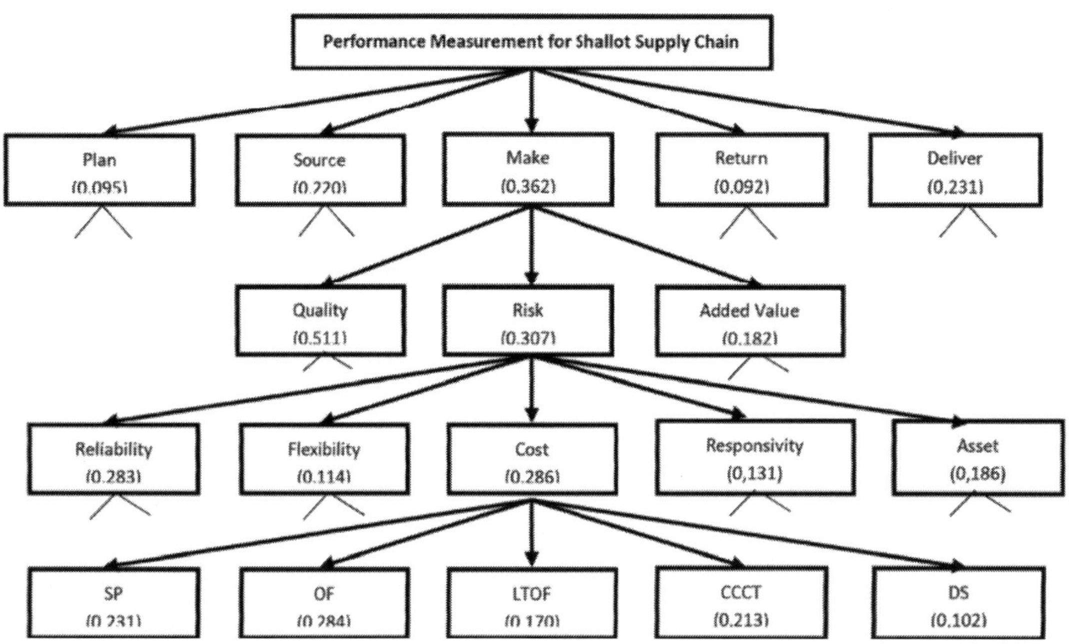

Figure. 2. The Hierarchy structure of each level for performance measurement of shallot supply chain.

Table 5 also presents that Cash to cash cycle time (CCCT) is the next top priority in the shallot supply chain assessment, with a weight of 0.213. CCTC is based upon the strategic elements of inventory management, the determination of terms in trading contracts, and the role of cash in the relationship among all parties involved in shallot supply chain, which depends on a payment cycle that will be used for next transaction. The last indicators influenced the shallot supply chain performance are Lead-time Order Fulfilment (LTOF) and Daily Stock (DS), which have weights of 0.170 and 0.102 respectively.

8. Conclusion

This study identified supply chain network of shallot in East Java Province as the second biggest production area in Indonesia, as well as formulated a measurement model for its performance. This study collected data from a randomly selected sample of 91 farmers, and selected sample of 36 marketing agents involved in this supply chain, included traders/merchants, port-traders, industries, and input suppliers, which were determined by snowball sampling procedure. SCOR approach was chosen to build up model and matrix since it already provided parameters and indicators for measuring the supply chain performance. The results made up a four-levels-hierarchy structure of performance measurement model for shallot supply chain. The results reveal that cultivation process of shallot (Make) was the most important process in this supply chain, in which variable of Quality as the main parameter. Therefore, the government should improve farmers' access to information of technology and innovation on shallot farming to increase the crop quality. Furthermore, Cost was proved as attribute that need to be considered the most in order to maintain the supply chain performance. This finding shows that efficiency on production and marketing cost will enhance the performance of supply chain. The government should continue supporting stabilization of input prices (especially for seed, pesticides, and fertilizers), and other facilities (e.g., transportation and credit access) to maximize the efficiency along the supply chain. Then, Order Fulfilment (OF) became the main indicator in assessing supply chain performance of shallot. So that, each parties involved in this chain should be cooperative in balancing demand and supply of shallot, rather than carry out any market consolidation that leads to collusive behaviour. The implications from this result could be related to synchronization of processes along the chain, cost reduction, transparency and responsibility among parties involved in shallot supply chain in order to improve its performance.

References

[1]Barratt M and A Oliveira 2001 Exploring the experiences of collaborative planning initiatives *International Journal of Physical Distribution and Logistics Management* **31(4)**, pp. 22.

[2] Boyer KK, G K Leong, P T Ward and L J Krajewski 1997 Unlocking the potential of advanced manufacturing technologies. *Journal of Operations Management*pp. 331-347.

[3] Das A and R Handfield 1997 Just-in-time and logistics in global sourcing: An empirical study *International Journal of Physical Distribution and Logistics Management* **3(27)**, pp. 244-259.

[4] Dewett T and G R Jones 2001 The role of information technology in the organization: a review, model, and assessment *Journal of Management* pp. 313-346.

[5] Frohlich M T and R Westbrook 2001 Arcs of integration: An International study of supply chain strategies *Journal of Operations Management*, pp. 185-200.

[6] Hill C A and G D Scudder 2002 The use of electronic data interchange for supply chain coordination in the food industry *Journal of Operations Management*, pp. 375-387.

[7] Lejeune M A and N Yakova 2005 On characterizing the 4C's in supply chain management *Journal of Operations Management*, pp. 81-100.

[8] Mentzer J T, J H Foggin and S L Golicic 2000 Collaboration: The enablers, impediments and benefits. Supply Chain Management Review (September-Oktober), pp. 52-58.

[9] Shah R, S Meyer-Goldstein and P T Ward 2002 Aligning supply chain management characteristics and inter-organizational information system types: An exploratory study. *IEEE Transactions on Engineering Management* **49(3)**, pp. 282-292.

[10] TangK and J Tang 2002.Time-based pricing and leadtime policies for a build-to- order manufacturer. *Production and Operations Management* **11(3)**, pp. 374-392.

[11] Rizaldy, Dicky Zunifar 2017 The impact of commodity prices on inflation in Malang: Case study on shallot and red chilli. UMM Repository (in Bahasa), Malang.
[12] Centre for Horticulture 2013 Study on food crop production. Bogor : IPB Repository.
[13] Romero C dan Rehman T 2003 Multiple criteria analysis for agricultural decisions. Elsevier, Amsterdam.
[14] Zhu J 2003 Quantitative models for performance evaluation and benchmarking: data envelopment analysis with spreadsheets and DEA Excel solver. International Series in Operations Research and Management Science.
[15] Wong WP, dan Wong KY 2007 Supply chain performance measurement system using DEA modelling. *Journal of Industrial Management and Data Systems* **(107),** pp. 361-381.
[16] LapideL 2000 What about measuring supply chain performance? ASCET 2.

[18] Lai KH, Ngai EWT, dan Cheng TCE 2002 Measures for evaluating supply chain performance in transport logistics. Logistics and Transportation Review (38), pp. 439-456.
[19] Supply-Chain Council 2004 SCOR. Available: [http://www.supply-chain.org/index.ww].
[20] Delipinar G E and Kocaoglu B 2016 Using SCOR model to gain competitive advantage: A literature review. Procedia-Social and Behavioral Sciences (229), pp. 398-406.
[21] Ministry of Agriculture 2016 Outlook on Shallot. Jakarta: Data Centre and Information System for Agriculture (in Bahasa)[
[22] Van der Vorst JGAJ 2006 Performance Measurement in Agrifood Supply Chain Networks: An overview. on: C., A J. Wijnands, R. Huirne, O. van Kooten (ed.).Quantifying the agri-food supply chain. Springer Science Business Media, Netherland.Wang, N. 2003. Measuring transaction costs: an incomplete survey. Ronald Coase Institute, Chicago. Ronald Coase Institute Working Papers no. 2. http://coase.org/workingpapers/wp-2.pdf. [4 Juli 2008].
[23] Marimin and Magfiroh 2010 Application for decision making for supply chain management. Bogor: IPB Press (in bahasa).
[24] Chandrasekaran N and G Raghuram 2014 Agribusiness supply chain management. CRC Press, Boca Raton.
[26] Nowakowska M and A Tubis 2015 Reliability of the cut flowers; supply chain. Safety and Reliabitlity of Complex Engineered System, pp. 1755-1762.
[27] Goldman S L, Nagel R N, and Preiss K 1995 Agile competitors and virtual organizations: Strategy=ies for enriching the customer. Van Nostrad Reinhold, New York.
[28] Ouyang Huanhuan 2102 Supply chain performance measurement: The integrated project of Shenda market chain and Lijin Agricultural base. Thesis. HAMK University of Applied Science. Forssa.

IOP Conf. Series: Earth and Environmental Science **334** (2019) 012070 doi:10.1088/1755-1315/334/1/012070

Nutrients and Anti-nutrients Content Analysis of Bogor Taro Mutant Clone (*Colocasia esculenta*)

F Nurilmala[1] and D Mardiana[1]

[1] Nusa Bangsa University, Jl. KH Sholeh Iskandar Km 4 Cimanggu, Bogor 16166, Indonesia

E-mail : febi@unb.ac.id

Abstract. Bogor Taro is an important local staple food, a potential substitute to rice as carbohydrate source in Indonesia. However, its production in the last decade tends to decline due to unsuccessful breeding program. Taro contains both nutrients (carbohydrate) and anti-nutrients, such as calcium oxalate that cause itching sensations when consumed. At present, through somaclonal variation technology by gamma radiation, six mutant-clones have been produced. We investigated technology to improve carbohydrate content and to depress calcium oxalate on the six mutant-clones available. We used iodometric titration to examine carbohydrate levels, permanganometric titration to measure calcium oxalate levels, argentometric titration to determine cyanide acid levels, and extraction method-gravimetric for saponins. We found, there were three mutant clones which have higher carbohydrate and lesser calcium oxalate content compared to the parent, namely B521 (5 Gy gamma radiation), B1023 (10 Gy gamma radiation) and B1511 (15 Gy gamma radiation) clones which will be proliferate to be a superior seeds.

1. Introduction

The primary source of carbohydrate for the majority of Indonesians is rice, even though there are alternatives besides rice as the primary source of carbohydrate such as taro. Of the many tubers, taro is a good rice substitute as a food source due to has a high starch content reaching 70-80 % [1]. There exist other nutrients inside taro such as carbohydrate as well as anti-nutrients that could impair the nutrients. Anti-nutrients are chemical matters that have evolved on plants to defend biological functions between plants. Anti-nutrients diminish the utilization of nutrients (proteins, vitamin and minerals in particular) thus effectively preventing optimal utilization of nutrients and decreasing the nutritional value [2]. Cultivation is expected to disable anti-nutrients factors and increase the availability of bioactive compound. However, the health risk for consumers in a high amount of anti-nutrients residue must not set aside.

One of the anti-nutrients found in taro is an oxalate compounds. This compound could cause itches. Food consumption with a high amount of oxalate can be detrimental to health due to the formation of oxalate stones also known as kidney stones, resulting in renal disorder [3]. Besides oxalate compounds, there is also cyanide acid found in taro. Cyanide is a poison to every living thing, could obstruct breathing due to the hindering of ferricytochrome oxidase enzyme process of oxygen intake for breathing, and cause imperfect cell growth [4]. Side effects caused by HCN poisoning also include headaches, vomiting, and dizziness. The human body could tolerate HCN when consumed within the reasonable amount. A standard was assigned by FAO that specifies tubers with the HCN amount of 50 mg/kg below is still safe to be consumed. Another compound which included as anti-nutrients on taro is saponin. Generally tastes bitter and foams when shaken with water. It is also poisonous to some cold-blooded animals [2].

The development of genetic variation on Bogor Taro aims to a higher carbohydrate and lower calcium oxalate compared to the parent has been done through the somaclonal variation induction by gamma radiation treatment at a dose of 5, 10, and 15 Gray to the culture of Bogor Taro apical bud and has acclimatized 5 mutant clones, namely B012, B521, B1022, B1023, and B1511 [5]. The mutant clones were then analyzed to find out the content of nutrition (carbohydrate) and anti-nutrition (calcium oxalate, cyanide, and saponin).

2. Materials and Methods

Plant materials obtained from 4-6 months old acclimated mutant clones and the parent from Nusa Bangsa University's collection. Sample preparation began with washing the taro's corms, peeled, and washed again in running water, then cut into match-sized parts and soaked in water for one night. The sample is blanched for 5 minutes and dried in the oven at 70 °C for 24 hours. Dried sample are then grinding to 100 meshes.

Determining carbohydrate level according to iodometric titration method that 1 gram sample put into 500 ml Erlenmeyer flask, added 25 ml of 1.25% H_2SO_4, and then boiled for 3 hours using upright cooler. Chill and neutralize with 3.25% NaOH using red litmus and move to the 100 ml volumetric flask and squeeze until line marks with distillation water and then strain. Using 25 ml pipette, move the distillate into 500 ml Erlenmeyer flask, add 25 ml of Luff solution, add some boiling stones, and 10 ml of distilled water. The mixture is then heated again in upright cooler, and the mixture should be boiled within 3 minutes, boil for 10 minutes straight, then immediately chill inside an ice-filled tub. After chilled, add ten ml of 30% KI solution and 25 ml of 25% H_2SO_4, then slowly titrate with 0.1 N Tio solution with 0.5% starch indicator.

Calcium oxalate determined using permanganometric titration method. Two grams of the sample are macerated at room temperature with 20 ml of 0.15% citric acid for 6 hours. Strain the solution, and the filtrate is then acidized and precipitated with 5 ml 10% tungstophosphoric acid and then centrifuged for 5 minutes at 1500 rpm. Supernatant and precipitation dissolved with 2 M H_2SO_4, and then heated until 70-80°C and then titrated with 0.01 M $KMNO_4$.

Saponin level decisive according to [6]. Ten grams sample soaked in 100 ml of 20% Ethanol and heated in the water bath at 55°C for 4 hours while being stirred with a stirrer. The mixture then strained, and the residue is extracted again with 100 ml 20% Ethanol. Extraction result is then joined and heated with a water bath at 90 °C until there is 20 ml extraction left. The concentrate then moved into separate funnels (250 ml and ten ml). Diethyl ether is then added and shaken. The liquid layer then moved, and purification process is repeated using 30 ml n-Butanol. The n-Butanol result is washed twice with five ml 5% NaCl. The final mixture then heated with a water bath. After evaporation, the sample is dried with an oven at 105 °C until a constant weight achieved. Saponins level then calculated in percentage.

Determining cyanogenic glycosides as said by [6]. Five grams sample is moved into 800 ml Kjeldahl flask, the sample then macerated with 50 ml water for 2 hours at temperature room. Then added 50 ml water, 10 ml of 0.02 N $AgNO_3$ and 0.5 ml of liquid HNO_3 homogenized and distilled at the temperature of 95 °C until container volume is 75 ml. Fe Alum ($NH_4Fe(SO_4)_2 \cdot 12H_2O$) indicator is then added to the distillate until the color of the mixture turns yellow, then titrated with 0.02 N KCN until brick-red sediments formed.

3. Results and Discussion

We found that the color of the mutant clone flesh corms has changed from its parent. The parent was white, and the mutant clones were yellow and yellowish (Figure 1). The color changed indicated the presence of secondary metabolites such as carotenoid and flavonoid and are essential traits of different cultivars. Compared to the common whitish like the parent of this studies, aspect of the yellow/yellowish corm flesh are favored more by consumers [7]. This result gives hope to the mutant clones to be well accepted by consumers as a new type of variety.

The content of carbohydrate on various taro mutant clones show that there were four mutant clones have higher carbohydrate compared to the parent (Figure 2). Average carbohydrate spread is around 33.95 g/100 g ingredient. The most substantial rise is on the B1511 clones with an increase of 46.93 g/100 g ingredient or about 211% expansion. We also found the decrease of 50% compared to the parent from B1022 clones. Higher carbohydrate content from its parent can be considered to be an alternative food source so that the society no longer has to be reliant upon rice to fulfill their main food needs. There is no specific amount of carbohydrate needed by the human body per day, although [8] and [9] advises that 40 - 80% total energy consumption be from complex carbohydrate and a maximum amount of 10% from sugar.

Figure 1. The color of taro mutant clones corm compares to the parent.

The contents of calcium oxalate on various taro mutant clones show that there were four mutant clones which have lower contents of calcium oxalate and only one mutant clone show a higher calcium oxalate content compared to the parent (Figure 2). Calcium oxalate content obtained is in the range of 3.20 – 41.60 mg calcium oxalate/100g ingredient. Average calcium oxalate decrease is 16.64 mg/100 g ingredient and the most substantial calcium oxalate decrease to 90% found on B521 clone. The increase of calcium oxalate content on B1023 and B012 clones was between 20% – 30%. Calcium oxalate caused acridity and itching sensation when consumed. It caused by needle-like oxalate crystals, raphides, that can penetrate the soft skin and cause discomfort in the tissue. Many studies reduce the effect of oxalate by cooking or boiling the taro corms [10-14]. Our study shows that Gamma irradiation can reduce the oxalate content of Bogor taro corms. We expected that some clone like B521 and B1511 could become new varieties with low oxalate content without going through the cooking process.

The contents of saponin on various taro mutant clones show that there were three clones which have lower contents of saponin compared to their parents, four clones which have an equal amount of saponin, and one that has the higher content of saponin to the parent (Figure 2). The range of saponin content is between 17.89 – 20.64 g/100 g ingredient. The most substantial reduction found in B521 clone with a decrease of 7.54%. Saponin is an active compound which foams when shaken in water and at low concentration could cause red blood cell hemolysis [15]. However, saponin can also be used in the detergents and cosmetics industries as they have excellent foaming properties, moderate detergency and anti-dermatophytic activity [16-18]

The contents of cyanide show that there were four clones have lower cyanide, and only one clone has higher cyanide compared to the parent. The most substantial decrease of cyanide found in B1511 clone, and there is an increase of 34% in cyanide content found in B521 clone. According to [19] the lethal dose of cyanide in a human body is 250mg. So far, all the clones include the parent are lower than the lethal dose of cyanide.

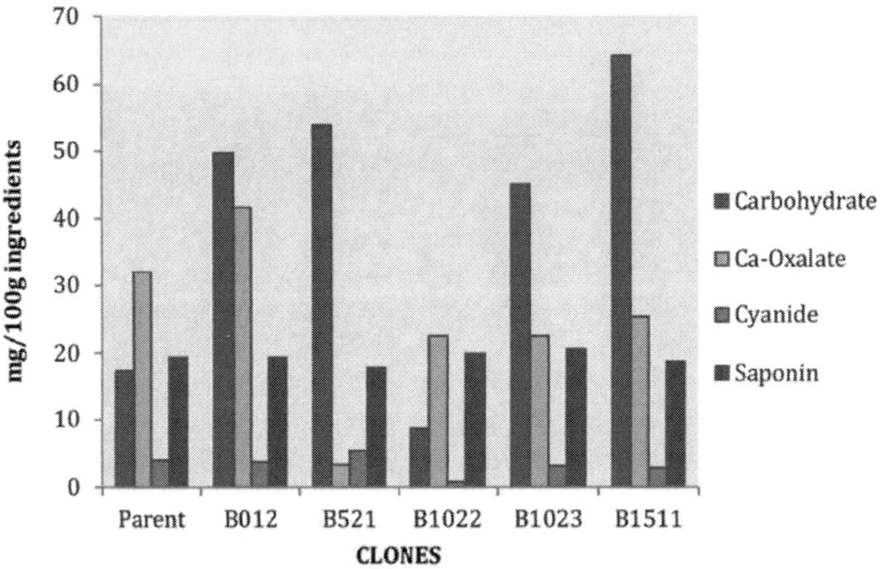

Figure 2. Nutrient (Carbohydrate) and Anti-nutrients content of taro mutant clones compares to the parent.

4. Conclusion

There were three mutant clones of Bogor Taro which have a higher carbohydrate content and lower content of calcium oxalate compared to the parent, namely B521, B1023 and B1511 clones which will then be multiplicated to meet the needs of superior seeds.

Acknowledgements

Acknowledgements delivered to Ministry of Research, Technology and Higher Education that has funded this research through Applied Product Research Contract No. 1598/K4/KM/2017.

References

[1] Kaushal P, Kumar V, Sharma H K 2015 Utilization of taro (*Colocasia esculenta*): a review *J Food Sci Technol* **52(1):** 27–40.

[2] Alcantara R M 2013 The Nutritional Value and Phytochemical Components of Taro [*Colocasia esculenta* (L.) Schott] Powder and its Selected Processed Foods *J. Nutr. Food Sci.* **3(3):** 1-7.

[3] Kumoro A C, Putri R D A, Budiyati C S, Retnowati D S, Ratnawati 2014 Kinetics of Calcium Oxalate Reduction in Taro (*Colocasia Esculenta*) Corm Chips during Treatments Using Baking Soda Solution *Procedia Chem.* **9**: 102–112.

[4] Lechtenberg M 2011 Cyanogenesis in Higher Plants and Animals *eLS* pp. 1–11.

[5] Nurilmala F, Hutagaol R P, Widhyastini I M, Widyastuti U, Suharsono 2017 Somaclonal variation induction of Bogor taro (*Colocasia esculenta*) by gamma irradiation *Biodiversitas J. Biol. Divers.* **18(1):** 28–33.

[6] Ubwa S T, Anhwange B A, Chia J T 2011 Chemical Analysis of Tacca *American Journal of Food Technology* **6(10):** 932-938.
[7] Champagne A, Legendre L, Lebot V 2013 Biofortification of taro (*Colocasia esculenta*) through breeding for increased contents in carotenoids and anthocyanins *Euphytica* **194(1):** 125-136 .
[8] FAO/WHO 1998 Carbohydrates in human nutrition Report of a Joint FAO/WHO Expert Consultation *FAO Food Nutr Pap* **66** 1–140.
[9] Englyst K N, Liu S, Englyst H N 2007 Nutritional characterization and measurement of dietary carbohydrates *Eur. J. Clin. Nutr.* **61(1):** S19–S39.
[10] Catherwood D J, Savage G P, Mason S M, Scheffer J J C, Douglas J A 2007 Oxalate content of cormels of Japanese taro (*Colocasia esculenta* (L.) Schott) and the effect of cooking *J. Food Compos. Anal.* **20(3–4):** 147–151.
[11] Oscarsson K V, Savage G P 2007 Composition and availability of soluble and insoluble oxalates in raw and cooked taro (*Colocasia esculenta* var. Schott) leaves *Food Chem.* **101(2):** 559–562.
[12] Azene H 2017 Nutritional Composition and Effects of Cultural Processing on Anti-nutritional Factors and Mineral Bioavailability of Colocasia *J. Food Nutr. Sci.* **5(4):** 147-154.
[13] Kumoro A C, Budiyati C S, Retnowati D S 2014 Calcium oxalate reduction during soaking of giant taro (*Alocasia macrorrhiza* (L.) Schott) corm chips in sodium bicarbonate solution *Int. Food Res. J.* **21(4):** 1583–1588.
[14] Ayele E, Urga K, Chandravanshi B S 2015 Effect of cooking temperature on mineral content and anti-nutritional factors of yam and taro grown in southern Ethiopia *Int. J. Food Eng.***11(3):** 371–382.
[15] Bissinger R, Modicano P, Alzoubi K, Honisch S, Faggio C, Abed M, Lang F 2014 Effect of saponin on erythrocytes *Int. J. Hematol.***100(1):** 51–59.
[16] Moses T, Papadopoulou K K, Osbourn A 2014 Metabolic and functional diversity of saponins, biosynthetic intermediates and semi-synthetic derivatives *Crit. Rev. Biochem. Mol. Biol.* **49(6)**: 439–462.
[17] Chen Y F, Yang C H, Chang M S, Ciou Y P, Huang Y C 2010 Foam properties and detergent abilities of the saponins from Camellia oleifera *Int. J. Mol. Sci.* **11(11):** 4417–4425.
[18] Tamura Y, Miyakoshi M, Yamamoto M 2012 Application of Saponin-Containing Plants in Foods and Cosmetics Creative Commons Attribution License (http://creativecommons.org/licenses/by/3.0).
[19] Jethava D, Gupta P, Kothari S, Rijhwani P, Kumar A 2014 Acute cyanide Intoxication: A rare case of survival *Indian J. Anaesth.* **58(3)**: 312–314.

IOP Conf. Series: Earth and Environmental Science **334** (2019) 012071 doi:10.1088/1755-1315/334/1/012071

The Effect of Organomineral on pH, Nitrogen Content, Organic-C Content and Yield of Upland Rice (*Oryza sativa L.*) on Inceptisols, West Java Indonesia

O Mulyani[1], R Sudirja[1], B Joy[1], and R A Hanandiva[1]

[1]Department of Soil Science and Land Resources Management Faculty of Agriculture, Universitas Padjadjaran

E-mail: oviyanti.mulyani@unpad.ac.id

Abstract. Generally, Inceptisols has low to moderate fertility characteristics. This condition can make a problem on the development of upland rice production caused by poor soil quality. Upland rice is one of the cultivated plants that is cultivated on dry land. This research was aimed to find the perfect combination of organomineral including humic acid, zeolite, natural phosphate, and dolomite on soil and yield of upland rice in Inceptisols. This research was conducted at Ciparanje, Faculty of Agriculture, Universitas Padjadjaran from July 2017 to February 2018. This experiment applied a Randomize Block Design (RBD) consisted of 10 treatments and three repetitions with the combination of humic acid (6; 8; 10 kg/ha); zeolite (150; 200; 250 kg/ha); dolomite (100; 150; 200 kg/ha) and natural phosphate (250; 300; 350 kg/ha). The result of this experiment showed that there was a significant effect of the organomineral on pH, Organic-C and yield of upland rice but no for total N. However the combination of humic acid 10 kg / ha + zeolite 150 kg / ha + dolomite 100 kg / ha + raw phosphate 250 kg / ha is the best doses compared to other combinations.

1. Introduction

Paddy (*Oryza sativa* L.) is a very important cultivating plant for human, because more than half of the world's population depends on this plant as a main food. Furthermore, the rice production in Indonesia is expected to increase as population increases. In 2015, the demand of rice reached 30.92 million ton with assuming an average per capita consumption is 132.98 kg/capita/year. To meet the amount of production, various methods need to be done. One of them is to expand the area of rice cultivation to dry land. Moreover, this land has a big potential to develop, because around 5.1 million ha spread in various provinces in Indonesia [1]. On the other hand, the main problems in this land is low water content and nutrients. This conditions can lead to low productivity of rice and requires paddy rice plants that can survive in this condition, one of them is upland rice. This land in West Java reached 110.424 ha with average yield 2.85 t/ha and the highest yield is 3.1 t/ha. In 2006, this production can contribute on food security in West Java until 315.082 t/ha. Even so, the yield of upland rice is still lower than paddy rice in West Java [2].

One of the factor that can increasing upland rice production is application of material that can change characteristics of poor soil, such as ameliorant like zeolite. The application of this material in industry and agriculture is a lot, because zeolite has characteristics like high cation exchange capacity, high water holding capacity, and high adsorption capacity [3]. The other materials is dolomite, it is a mineral which contains nutrients of Magnesium (Mg) and Calcium (Ca) with the chemical formula $CaMg(CO_3)_2$. This type is commonly used for agricultural land, plantation land, industrial and fishery needs and applied to stabilize soil pH. Last material is natural phosphate, this fertilizer is important for plant growth as a source of P nutrients and has ability to slow release of P nutrient. Furthermore, to complete this role, we need to add organic fertilizer. This fertilizer can increase plant growth and production, because they are able to play a role in improving soil structure, increasing CEC, increasing water retention, increasing soil biological activity, and as a source of plant nutrition. Regarding [4,5], organic fertilizer is material made by composting process from waste (animals or plants) that contain 67-82 kg CO_2 from one ton food waste and garden waste. Furthermore, the active form from organic matter is humic substances that has chemically active compounds, with high CEC contains [6,7]. This material has benefits in improving soil properties because rich of carbon (41% - 47%) and nitrogen [8]. Moreover, humic acid has an important role in supporting the life of microorganisms that live in the soil, increasing membrane permeability, increase production of chlorophyll and photosynthesis. Based on the explanation above, it is expected that through application of the combination of organomineral can improve the efficiency of the use of inorganic fertilizers, it is also expected to increase the characteristics of soil on dry land.

2. Materials and Methods

The experiments were conducted at Experimental field, Soil Fertility and Soil Nutrition Laboratory Department Soil Science and Resource Management, Faculty of Agriculture, Universitas Padjadjaran. The organ0mineral combination consisting of humic acid, zeolite, dolomite, and natural phosphate from Yogyakarta, insecticide (sihalotrin 106 g/L, tiametoksam 141 g/L, profenofos 500 g/L, carbofuran) and fungicide also Situ Bagendit rice seed variety. The soil that used is Inceptisols from Ciparanje region.

The experiment was arranged in Randomize Block Design (RBD), consisted of 10 combination treatments with 3 replications, there are humic acid (6; 8; 10 kg/ha); zeolite (150; 200; 250 kg/ha); dolomite (100; 150; 200 kg/ha); and natural phosphate (250; 300; 350 kg/ha). Data analysis was performed with SPSS 16.0 application. If F test shows significant result (sig <0,05), then do advanced test by using Duncan test 5%. The observations consisted of Percentage of organic C using the Walkey and Black method ; pH-content using electrode glass (pH meter); N contain using Kjeldahl method; and yield of upland rice

3. Results and Discussion

3.1 Organic C

Based on the data in Table.1, the organomineral treatments had significant effect on soil organic C content, for all treatments.

Table 1. Effect of Organomineral Combination (Humic Acid, Dolomite, Natural Phosphate and Zeolite) on Organic C

Treatments		Organic C (%)
A = Control		1.37 a
B = Combination of HA 6 kg/ha +	Z 150 kg/ha + D 100 kg/ha + NF 250 kg/ha	1.48 ab
C = Combination of HA 6 kg/ha +	Z 200 kg/ha + D 150 kg/ha + NF 300 kg/ha	1.49 ab
D = Combination of HA 6kg/ha +	Z 250 kg/ha + D 200 kg/ha + NF 350 kg/ha	1.61 bc
E = Combination of HA 8 kg/ha +	Z 150 kg/ha+ D 100 kg/ha + NF 250 kg/ha	1.70 bc
F = Combination of HA 8 kg/ha +	Z 200 kg/ha + D 150 kg/ha + NF 300 kg/ha	1.86 c
G = Combination of HA 8 kg/ha +	Z 250 kg/ha + D 200 kg/ha + NF 350 kg/ha	1.66 bc
H = Combination of HA 10 kg/ha+	Z 150 kg/ha + D 100 kg/ha + NF 250 kg/ha	0.52
I = Combination of HA10 kg/ha +	Z 200 kg/ha + D 150 kg/ha + NF 300 kg/ha	0.52
J = Combination of HA 10 kg/ha +	Z 250 kg/ha + D 200 kg/ha + NF 350 kg/ha	0.55

Note : HA : Humic Acids ; Z : Zeolite ; D : Dolomite ; NF : Natural Fosfat

For general, we can see that all the treatments gives high organic C contain than control. In table.1 above, the dose of humic acid treatment 10 kg/ha + dolomite 200 kg/ha + natural phosphate 350 kg/ha + zeolite 250 kg/ha give the best result (7%-40%) compared to control and other treatments. This condition can happen because organic material such as humic acid, contain high C organic and other nutrients for plants. Although the percentage of organic matter is no more than 5% in every type of mineral soil, but humic acid is usually rich in carbon (more than 40%), high oxygen content, nitrogen and low hydrogen content [8]. Organic materials can also improve soil structure, aeration and the soil ability to bind water. Furthermore, this is happen because all material that we add in treatments can increase the organic C contain.

3.2 Total N

Based on the data in Table.2, the organomineral treatments had no significant effect on total N content, for all treatments. When we see in Table. 2 above, the N contain for all the treatments gives high nitrogen contain than control. On the other hand, this is not gives significant effect on statistics value. This is caused by sorption of nitrogen as nitrate ions (NO_3^-) and ammonium ions (NH_4^+) by plant at vegetative stage is used for increasing the height and number of tillers of plant. The absorption of N by plants is influenced by several factors, such as physiological conditions of the plant, type of plant, and needs of plants in certain nutrients. Moreover, external or environmental factors that affect the absorption of N by plants are light, air, water, and soil pH [9]. According to [8], humic acid also give a source of nitrogen from 2%-5% but on the other side, zeolite application can reduce N contain this is due to the coating by zeolite or activated charcoal that can make a nitrogen not dissolved immediately.

Table 2. Effect of Organomineral Combination (Humic Acid, Dolomite, Natural Phosphate and Zeolite) on Total N

Treatments		Total N (%)
A = Control		0.41
B = Combination of HA 6 kg/ha +	Z 150 kg/ha + D 100 kg/ha + NF 250 kg/ha	0.46
C = Combination of HA 6 kg/ha +	Z 200 kg/ha + D 150 kg/ha + NF 300 kg/ha	0.47
D = Combination of HA 6kg/ha +	Z 250 kg/ha + D 200 kg/ha + NF 350 kg/ha	0.45
E = Combination of HA 8 kg/ha +	Z 150 kg/ha+ D 100 kg/ha + NF 250 kg/ha	0.49
F = Combination of HA 8 kg/ha +	Z 200 kg/ha + D 150 kg/ha + NF 300 kg/ha	0.46
G = Combination of HA 8 kg/ha +	Z 250 kg/ha + D 200 kg/ha + NF 350 kg/ha	0.52
H = Combination of HA 10 kg/ha+	Z 150 kg/ha + D 100 kg/ha + NF 250 kg/ha	0.52
I = Combination of HA10 kg/ha +	Z 200 kg/ha + D 150 kg/ha + NF 300 kg/ha	0.52
J = Combination of HA 10 kg/ha +	Z 250 kg/ha + D 200 kg/ha + NF 350 kg/ha	0.55

Note : HA : Humic Acids ; Z : Zeolite ; D : Dolomite ; NF : Natural Fosfat

3.3 Soil Acidity (pH)

Based on the data in Table.3, the organomineral treatments had significant effect on pH content, for some treatments. Regarding Table.3 above, we can see that the pH value is increasing for all treatments than control. The pH value of this soil is obtained from active (actual) acidity and potential acidity. Furthermore, actual acidity is caused by the presence of H^+ ions in the soil solution. This ions can be produced when cations reactions, humification process of organic matter or root activities.

Table.3. Effect of Organomineral Combination (Humic Acid, Dolomite, Natural Phosphate and Zeolite) on pH

Treatments		pH
A = Control		1.37 a
B = Combination of HA 6 kg/ha +	Z 150 kg/ha + D 100 kg/ha + NF 250 kg/ha	1.48 ab
C = Combination of HA 6 kg/ha +	Z 200 kg/ha + D 150 kg/ha + NF 300 kg/ha	1.49 ab
D = Combination of HA 6kg/ha +	Z 250 kg/ha + D 200 kg/ha + NF 350 kg/ha	1.61 bc
E = Combination of HA 8 kg/ha +	Z 150 kg/ha+ D 100 kg/ha + NF 250 kg/ha	1.70 bc
F = Combination of HA 8 kg/ha +	Z 200 kg/ha + D 150 kg/ha + NF 300 kg/ha	1.86 c
G = Combination of HA 8 kg/ha +	Z 250 kg/ha + D 200 kg/ha + NF 350 kg/ha	1.66 bc
H = Combination of HA 10 kg/ha+	Z 150 kg/ha + D 100 kg/ha + NF 250 kg/ha	0.52
I = Combination of HA10 kg/ha +	Z 200 kg/ha + D 150 kg/ha + NF 300 kg/ha	0.52
J = Combination of HA 10 kg/ha +	Z 250 kg/ha + D 200 kg/ha + NF 350 kg/ha	0.55

Note : HA : Humic Acids ; Z : Zeolite ; D : Dolomite ; NF : Natural Fosfat

The treatment of Humic Acid 10 kg/ha + Dolomite 100 kg/ha + Natural Phosphate 250 kg/ha + Zeolite 150 kg/ha give the high pH value compared to control and other treatments. It can be seen that the application of ameliorant with a maximum dose of humic acid can increase the value of soil pH, because humic acid can reduce the solubility of Al and Fe soils by make complex compounds that are difficult to dissolve. Also application of dolomite with contains elements of Ca and Mg, can shift the position of H^{+} in reaction and increase soil acidity. Furthermore, dolomite is provides H supply to the soil which reacts with H^{+} to water and causes H^{+} levels is decrease and soil pH is increases [10].

3.4 Yield of Upland Rice

Based on the data in Table 4 that the organomineral treatments had significant effect for some treatments than control. As general, yield of upland rice parameter gives increasing pattern than control.

Table 4. Effect of Organomineral Combination (Humic Acid, Dolomite, Natural Phosphate and Zeolite) on Yield

Treatments		Yield of Upland rice (g)
A = Control		1.37 a
B = Combination of HA 6 kg/ha +	Z 150 kg/ha + D 100 kg/ha + NF 250 kg/ha	1.48 ab
C = Combination of HA 6 kg/ha +	Z 200 kg/ha + D 150 kg/ha + NF 300 kg/ha	1.49 ab
D = Combination of HA 6kg/ha +	Z 250 kg/ha + D 200 kg/ha + NF 350 kg/ha	1.61 bc
E = Combination of HA 8 kg/ha +	Z 150 kg/ha+ D 100 kg/ha + NF 250 kg/ha	1.70 bc
F = Combination of HA 8 kg/ha +	Z 200 kg/ha + D 150 kg/ha + NF 300 kg/ha	1.86 c
G = Combination of HA 8 kg/ha +	Z 250 kg/ha + D 200 kg/ha + NF 350 kg/ha	1.66 bc
H = Combination of HA 10 kg/ha +	Z 150 kg/ha + D 100 kg/ha + NF 250 kg/ha	0.52
I = Combination of HA10 kg/ha +	Z 200 kg/ha + D 150 kg/ha + NF 300 kg/ha	0.52
J = Combination of HA 10 kg/ha +	Z 250 kg/ha + D 200 kg/ha + NF 350 kg/ha	0.55

Note : HA : Humic Acids ; Z : Zeolite ; D : Dolomite ; NF : Natural Fosfat

The dosage of Humic Acid formulation Humic Acid 10 kg/ha + Dolomite 100 kg/ha + Natural Phosphate 250 kg/ha + Zeolite 150 kg/ha give the highest yield of upland rice (23.05 g). According to [9], supply of N in the generative phase is very important in postpone leaf aging process and maintaining photosynthesis during the grain filling phase and increasing protein. This adequacy during the generative phase is very important to achieve high rice yield. Furthermore, increased yield of dry milled grain also caused by the application of humic acid because this material has growth hormones such as auxins, cytokines, and gibberellins. This hormones can serves to stimulate the process of seed germination and push ahead the process of roots formation, stimulate cell, increase flowering, percentage of flowers and fruit, and reduce the loss of flowers and fruit [12].

4. Conclusion

The combination of organomineral (humic acid, zeolite, dolomite, and natural phosphate) give significantly affect on organic C, soil pH and yield of upland rice plants, but not significantly affect on the percentage of N-total content. The combination dose of humic acid 10 kg / ha + zeolite 150 kg / ha + dolomite 100 kg /ha + natural phosphate 250 kg / ha is the best dose for this experiment.

Acknowledgements

This research was supported by Academic Leadership Grand (ALG). We thank our colleagues from this team who provided insight and expertise that greatly assisted the research.

References

[1] Hidayat A, M Soekardi and BH Prasetyo 1997.Availability of Land Resources and Direction for Utilization of Several Commodities. Proceedings of the Land and Agro-climate Research. Land and Agro-climate Research Center, Bogor. Pp. 1-20.

[2] Central Statistics Agency 2007 West Java in Figures. Available: http://www.bps.go.id/awabarat. Retrieved on November 22, 2016.

[3] Mumpton FA 1999 La roca magica: uses of natural previous zeolites in agriculture and industry. *Proc. natl. Acad. Sci. USA* **96** 3463–3470.

[4] Boldrin A, Andersen JK, Møller J, Christensen TH, Favoino E. Composting and compost utilization: accounting of greenhouse gases and global warming contributions. 2009. *Waste Manage Res* **27**:800–12.

[5] Marzouk HA, Kassem HA 2010 Improving fruit quality, nutritional value and yield of Zaghloul dates by the application of organic and/or mineral fertilizers. 2010. *Sci Hortic* **127**:249–54.

[6] Stevenson F J 1994 Humus Chemistry: Genesis, Composition, Reactions, 2nd ed. New York: Wiley.

[7] Koopal LK, T Saito, JP Pinheiro and WH van Riemsdijk 2005 Ion binding to natural organic matter: General considerations and the NICA- Donnan model. Colloids and Surfaces A: *Physicochemical and Engineer*- ing Aspects **265**: 40–54.

[8] Tan KH 1993 Environmental Soil Science. Marcel Dekker. Inc. New York

[9] Chaturvedi I 2005 Effect of Nitrogen Fertilizer on Growth, Yield, and Quality of Hybrid Rice (*Oryza sativa* L.). *J Eur Agric* **6(4):** 611-618.

[10] Maftu'ah E, A Mas, A Syukur and BH Purwanto 2013 Effectiveness of Amelioran on Degraded Peatlands to Increase Growth and Uptake of NPK Sweet Corn (*Zea mays* L.) *Plants. Agron Indonesia Journal* **41 (1):** 16-23

[11] Soplanit R and Nukuhaly R 2012 The Effect of Nutrients Management (NPK) on N Availability and Yield of Rice Plant (*Oryza Sativa* L.) In Waelo Village, Buru District. *Agrogolia* **1(1)**

[12] Suwardi and H. Wijaya. 2013. Increased Food Crop Production with Active Material of Humic Acid with Zeolite as Carrier. *Jurnal Ilmu Pertanian Indonesia (JIPI)* **18 (2):** 79–84.

Combined application of native mycorrhizal and cellulolytic fungi to manage drought effects on maize

F Fikrinda[1,2], S Syafruddin[2], S Sufardi[2] and R Sriwati[2]

[1]Doctoral Program in Agricultural Science, Syiah Kuala University, Banda Aceh 23111, Indonesia
[2]Faculty of Agriculture, Syiah Kuala University, Banda Aceh 23111, Indonesia

E-mail : fikrinda@unsyiah.ac.id

Abstract. Drought become the foremost abiotic stress limiting plant growth and leading to crop-yield loss. This study was aimed to evaluate the effect of native arbuscular mycorrhizal (AM) and cellulolytic fungi (CF) on the growth, yield, mycorrhizal colonization, and the phosphorous (P) and proline contents of maize under drought. This study was conducted using a 3 x 3 factorial arranged in randomized complete block design with three replications The treatments were inoculation of native AM fungi (no AM fungi, *Acaulospora tuberculata*, and *Gigaspora* cf. *gigantea*) and native celluloltic fungi (no CF, *Talaromyces pinophilus* strain MR107 and *Talaromyces pinophilus* isolate OK3SP103P) evaluated under 50% field capacity (FC). Our results clearly indicated that increasing plant height, plant diameter, mycorrhizal colonization, and P content and decreasing proline content were affected either by native AM or cellulolytic fungi, but dry weight of 100-grain was only increased by native cellulolytic fungi. Combined inoculation of native AM and cellulolytic fungi improved P content and mycorrhizal colonization. *Gi.* cf. *gigantea* and *T. pinophilus* strain MR107 were more potential native inoculants to eliminate negative effect of water stress on maize.

1. Introduction

Increasing the global demand of food to feed 9.1 billion people by 2050 [1] may create a food crisis in the future if do not followed by increasing crop production. One of the major constraints to the global crop production is the availability of water [2]. Drought is considered as the main environmental factor limiting plant growth and leading to reduce more than 50% of the average crop-yield [3,4,5] and it becomes increasingly severe in some regions due to the changes in the global climate [6]. This water deficit affects a vast range of plant characteristics including plant morphology, physiology and biochemistry [7] and reduces growth and productivity [8,9,10,11]

Maize (*Zea mays* L.), an important cereal crop grown all over the world [12], is sensitive to drought stress. Limited water availability negatively affects the performance of maize at almost all growth stages [13] and productivity of maize [13,14]. According to [15], its average annual yield loss due to drought is around 15 % of its potentzal yield. Study of 14 tropical-adapted, registered maize hybrid conducted by [16] showed that grain yield of water-deficit plants were also approximately 62% lower than in well-watered plants.

There may be opportunities to improve plant productivities under water stress. Application of beneficial microorganisms could be among sustainable mitigation practices but there is a lack of information on their soil microorganism functionalities in relation to water deficit. Amongst these microorganisms, arbuscular mycorrhizal (AM) fungi are received increasing attention because of their numerous benefits for their host plants. These fungi are key components of soil microbiota forming symbiotic relationships with the roots most terrestrial plants, including maize [17,18].

Mycorrhizal colonization of the roots has been shown to increase drought tolerance in some plants [19-22]. Recently published reports have indicated that AM fungi ameliorate plant tolerance to drought stresses through a variety of mechanisms such as enlargement of root areas and stomatal regulation [23-25], improvement of nutritional, physiological, and genetical plant [26,27], and changes in soil structure in terms of the quantity and quality of aggregate stability [28]. Their effect of mycorrhization on plant adaptation to water stress depends on the intensity of the stress and the type of the fungal isolate [29,30,31]. Besides, it is well known that drought adapted and tolerant microbial ecotypes native to a particular soil or environment are the most efficient [32-34] Even, [35] showed that agroforestry trees can serve as a source of AM inoculants to intercropped annuals such maize.

Some soil microorganisms are known to enhance the AMF symbiosis with vascular plants [36-38] but studies about the effect AM and cellulolytic fungi is still rarely reported especially under water stress. Fungi such as *Neurospora crassa, Talaromyces cellulolyticus, T. pinophillus, Trichoderma asperellum, T. virens,* and *T. harzianum* had cellulolytic ability [39-43]. The experiment conducted by [43] indicated that dual treatment of native AMF and *T. harzianum* was better for biomass production of *Ricinus communis* L. than dual treatment of AMF and *Asergillus niger* or mycorrhiza alone. Contrary to previous results, [44] found that co-inoculation of plants with AMF and *T. harzianum* did not result in an additive effect on plant growth and nutritional status. Even, [45] exhibited negative effect of this CF on the development of AM fungi. Because of these unconcistency results, this study was aimed to test the hypothesis that both native AM and cellulolytic fungi act synergistically to alleviate drought stress on maize by increasing the growth, yield, mycorrhizal colonization, and the phosphorous (P) and reducing proline contents of maize under drought.

2. Materials and methods

2.1. Experimental Design

This experiment was conducted on the basis of a factorial experiment in a randomized complete block design (RCBD) with three replications. The first factor was application of native AM fungi including without AM fungal inoculation, *Acaulospora tuberculata*, and *Gigaspora* cf. *gigantea*. The second factor was cellulolytic fungi (CF) consisting of without CF, *Talaromyces pinophilus* strain MR107 and *T. pinophilus* isolate OK3SP103P.

2.2. Soil collection and preparation

Soils (0–20 cm) were collected from Aceh Besar (05°34'25"N, 095°32'1,2"E, 47 m a.s.l). The soils were classified as Typic Dystrudepts according to USDA soil classification (Soil Survey Staff, 2010). The main soil characteristics were pH 5.77 (1:1 soil:water), 1.81% organc carbon (Walkley and Black method), 0.15% total nitrogen (Kjeldhal method), 4.68 ppm (Bray 1 method), 1.92 cmol (+) kg^{-1} exch. K, 11.94 cmol (+) kg^{-1} exchangeable (exch.) Ca, and 4.75 cmol (+) kg^{-1} exch. Mg.

Soils were air-dried after collection, crushed and passed through a 5 mm sieve to remove coarse organic matter and stones. Then, 10 kg soil samples were placed in each plastic pots, 26 cm high and 35 cm in diameter. At each pot, 50 g sterilized siam weed compost and the soil were mixed throughoutly and incubated during seven days. Pots were covered by black plastic and the water status of growth medium was maintained at 100% of field capacity (FC) before water stress treatment. The pots were placed on a screenhouse with the diurnal temperature range between 26°C and 34°C (night and day, respectively).

2.3. Preparation and multiplication of native mycorrhizal and cellulolytic fungi inoculum

These AM and cellulolytic fungi, obtained from rhizosphere of some plant at which soil sampling area, were selected inoculants from previous studies. These native spores of AM fungi (*A. tuberculata* and *Gi.* cf *gigantea*) were multiplicated in an open pot culture of *Z. mays* as host and sterile zeolite medium. After two months of plant growth the shoots were eliminated and the cultures were dried gradually for a month. The under-grown part (mycorrhizal roots, fungal spores and mycelium) and the medium was used as a stock culture.

The native cellulolytic fungi isolates (*T. pinophilus* strain MR107 and *T. pinophilus* isolate OK3SP103P used were first grown on Mandel medium in petri plate and were further multiplicated in 250-mL flasks containing 50 mL of Mandel medium in shake culture for 48 h at 28 °C. The Mandel medium contained $(NH_4)_2SO_4$ 1.4; urea 0.3; KH_2PO_4 2.0; $CaCl_2$ 0.3; $MgCl_2.H_2O$ 0.3; $FeSO_4.7H_2O$ 0.005; $MnSO_4.H_2O$ 0.016; $ZnCl_2.2H_2O$ 0.014; $CoCl_2.2H_2O$ 0.002; Tween 80 0.2 % v/v. This was prepared following standard procedure (Azcón, 1993).

2.4. Inoculation trials and growing conditions

Ten gram of AM inoculum (having an average of 100 spores and 70% of AM colonization of root) from the stock culture was inoculated at the appropriate pots immediately at sowing time just below the maize seeds whereas the cellulolytic inoculum was applied with 10 mL of the suitable fungal culture (10^8) at the soil surface of each pot around the seeds. Three seeds of maize (variety Hybride Bisi 2) previously surface sterilized were sown 2 cm deep into each pots which was covered by the black plastic with a small growing hole at the middle of the pot to reduce evaporation. The plants were thinned to one per pot after seven days.

Plants were maintained under natural light and temperature conditions, and were watered daily to keep soil moisture at 100% of field apacity (FC) until water stress treatment. At the sowing time, amounts of basal nutrient were added i.e. 250, 75 and 100 $kg{\cdot}ha^{-1}$ as urea, TSP and KCl to keep the water condition.

2.5. Parameters

2.5.1. Plant growth and yield

Plant growth parameter measured consisted of plant height and stem diameter. Plant samples were collected respectively. Half doses of urea and full doses of TSP and KCl were applied during the early stage of planting before the initiation of the drought stress while another part of urea was given 14 days after planting.

2.5.2. Procedure for water stress application

The crops were watered at 100 % of field capacity (FC) until the water stress treatments (30 days after sowing). The methodology for the application of water stress is described by [46, 47] To determine 100% FC, we weighed pots containing 10 kg of dry soil (P1 = 10 kg dry soil) and these pots were then watered until soil saturation (flow of excess water under gravity). The watered pots were covered with black plastic bag to avoid the evaporation and they were soaked for 48 hours in screenhouse. Pots were weighed again in order to determine P2, which was considered as saturated soils. The difference between P2 and P1 (P2–P1) provided the amount of water that was required to saturate the soil, which represents 100% of field capacity. In order to determine the watering volumes for 50% FC, we added to the pots containing dry soil, water volumes corresponding to 0.5 X (P2-P1). Pot cultures containing maize seeds were watered regularly using tap water at 50% FC. Lost water was replaced with fresh distilled water each day at 18:00, at 45 days after sowing (DAS). Plant height was also measured from the ground level to the point of the plant from where the tassel emerges using a steel ruler while stem diameter was measured using Vernier calipers. Maize yields (100 grain) was measured at crop maturity (100 DAS) and presented at 14% moisture.

2.5.3. Proline analysis

Proline was analysed by High-Performance Liquid Chromatographic (HPLC) method. This amino acid was analysed at Environmental Biotechnology Laboratory (EBL), Indonesian Center for Biodiversity and Biotechnology (ICBB), Bogor.

2.5.4. Root colonization

Root colonization by AM fungi was measured after 45 DAS. The subsample of roots for determination of AM colonisation was cleared in potassium hydroxide (KOH, 10% W/V) then stained with ink and vinegar using a modification of the method of Vierheilig et al. (1998) and using a modified line-intersection method for 100 intersections (McGonigle et al., 1990).

2.5.5. P concentration analysis

Tissue phosphorous status was analysed afer wet digestion with HNO_3 + H_2O_2 by molybdate blue method.

2.5.6. Data analysis

All data was analysed by analyzed by variance analysis using SPSS statistical software version 22.0. Means were compared by one-way analysis of variance and *Duncan*'s multiple range test at the 5% level of significance.

3. Results and Discussion

3.1. Plant growth and yield

Plant growth as indicated in plant height and diameter under water stress was significantly affected by native AM fungi and maize inoculated with *Gi.* cf *gigantea* presented better plant growth (Table 1). Compared to *A. tuberculata,* this AM fungi improved plant height significantly and plant diameter not significantly. Contrary to the plant growth, there were no significant differences in dry weight of 100 maize grain due to AM fungi inoculation. This study also showed that native cellulolytic fungi significantly improved plant growth and yield under water stress (Table 1). Native cellulolytic fungi (*T. pinophilus* strain MR107 and *T. pinophilus* isolate OK3SP103P) led to a significant increase in plant height, plant diameter, and dry weight of 100 grain in comparison to non- inoculated plant under water stress.

Maize growth under water stress was significantly improved by native AM fungi, indicating this inoculant could improve unfavourable conditions caused by water deficit stress and make the plants more tolerant to this stress. These results were in line with [48,49]. Contrary to expectations, there were no statistical differences detected on maize yield, indicating that the grain yield of maize was similar for either AM fungi inoculated or uninoculated treatments. Similar to this result was also reported by [50].

In addition to the intensity of the stress, the effect of mycorrhization on plant adaptation to water stress also depends on the type of the fungal isolate [51] AM fungi species even within the same genus have different effects on plant response to drought stress [52]. This study indicated that *Gi.* cf *gigantea* was superior to increased plant growth than *A. tuberculata*. This result was in agreement with several previous studies [50,52,34] and confirms the results of [53]. *Gigasporaceae* species having a larger soil mycelium may be better at nutrient uptake than AMF with small mycelia (i.e. *Acaulosporaceae*) [54].

Inoculation with cellulolytic fungi could also increase significantly maize growth and yield. Both cellulolytic inoculants had similar effects on those parameters. Applying cellulase/cellulolytic fungi to crops has shown to promote plant growth performance [55]. Among cellulolytic fungi studied with regarding to plant productivity, *Trichoderma* was the most popular. For examples, [56] reported that seeds among three varieties of maize treated with *T.harzianum* showed a significant higher germination than untreated seeds grown under water stress. A study carried out by [40] suggested that inoculation of *T. asperellum* not only enhanced the growth and yield of rice, but also significantly increased the number of healthy seed and reduced the percentage of dirty panicle infected seed and empty (unfertile, undeveloped) seed. Another study on clover showed that *T. hammatum* enhanced plant height and growth rate [57].

3.2. Proline content

Proline content of maize leaf was significantly affected by inoculation of native AM fungi inoculation. Maize inoculated by *Gi. cf gigantea* accumulated lowest proline content (Table 1) and the level of this amino acid at this plant was 14.59% lower than non-inoculated plant. Otherwise, inoculation of native cellulolytic fungi did not affected significantly the proline content of maize leaf (Table 1). Nevertheless, cellulolytic fungi slightly decreased the proline content by 11% dan 4.52% by *T. pinophilus* isolate OK3SP103P and *T. pinophilus* strain MR107 respectively.

This study showed that both AM fungi inoculants significantly decreased proline content in maize under drought. A number of previous studies also show similar results [58-61]. The amount of proline content inceased with increase in stress [62] but the lower content of this amino acid in inoculated plant may reflect an increased drought tolerance [36]. Accumulation of proline is an inevitable physiological response of plants experiencing severe water stress [14] . Proline not only acts as cytoplasmic osmoticum but also serves as a reservoir of carbon and nitrogen sources for post stress growth [63]. Contrary to AM fungi, the effect of cellulolytic inoculation on proline content was rarely reported. This study indicated that the proline content in maize inoculated by native cellulolytic microorganisms was lower than uninoculated plant exposed to water stress. This result was in contrast to the experiment by [39] on tomato inoculated with *T. Harzianum*.

3.3. Mycorrhizal colonization

Mycorrhizal colonization of maize root was affected significantly by interaction between myorrhizal and cellulolytic fungi. This interactive effect is presented in Figure 1. Root colonization level of maize varied between 37% and 77%. The best AM fungi inoculant for mycorrhizal colonization was *G.* cf *gigantea* (Figure 1a). This AM fungi inoculation treatment gave similar effect on mycorrhizal colonzation to *A. tuberculata* inoculation treatment regardless of cellulolytic fungi inoculation treatment. Mycorrhizal colonization was least in uninoculated plant either by AM fungi or cellulolytic fungi (as control) but the colonization was increased differently after being inoculated by each cellulolytic fungi.

Table 1. Effect AM and cellulolytic fungi inoculation on plant height, plant diameter, dry weight of 100 grain and proline content of maize

Treatments	Plant height (cm)	Plant diameter (cm)	Dry weight of 100 grain (g)	Proline content (ppm)
AM inoculation				
Non- AM	171.35 ab	2.71 a	23.64 a	7.62 b
A. tuberculata	168.40 a	2.81 ab	24.47 a	7.15 ab
G. cf *gigantea*	174.97 b	2.88 b	24.42 a	6.51 a
CF inoculation				
Non- CF	167.99 a	2.70 a	30.28 a	7.52 a
T. pinophilus strain MR107	175.92 b	2.81 b	33.23 b	7.05 a
T. pinophilus isolate OK3SP103P	170.81 ab	2.89 b	33.21 b	6.71 a
Significance				
AM inoculation	*	*	ns	*
CF inoculation	*	**	**	ns
AM x CF	ns	ns	ns	ns

AM, arbuscular mycorrhizal; CF, celluololytic fungi; ns, not significant

The effect of cellulolytic fungi inoculation on mycorrhizal was varied depending on AM fungi species (Figure 1b). Mycorrhizal colonization was highest in the uninoculated plant by cellulolytic fungi but AM fungi treatment. At plant inoculated by *T. pinophilus* strain MR107, its coinoculation with both species of native AM fungi gave better effect on mycorrhizal colonization. The different result was indicated by plant inoculated by *T. pinophilus* isolate OK3SP103P where inoculation of this cellulolytic fungi alone or coinoculation with each AM fungi species gave similar effect to mycorrhizal colonization.

The mycorhizal colonization of maize was successful both for plants inoculated with AM fungi and for those inoculated with cellulolytic fungi and the high rate of root colonization of maize confirmed the compatibility between the plants and the AM fungi. Both AM species had similar effect on mycorrhizal colonization whether the the plants were colonized with cellulolytic fungi or not. Eventhough, Gi. cf *gigantea* more success (77%) than *A. tuberculata* (62.33%) in colonizing maize root in non-cellulolytic treatments and well cooperate with *T. pinophilus* MR107. The tendency of plants to be colonized by mycorrhizal fungi depends on the strain of mycorrhizal fungi [64]. Besides, both cellulolytic microorganisms also enhance effectively AM colonization either combined with *Gi.* cf. *gigantea* or *A. tuberculata*. Therefore, *Gi.* cf *gigantea* could be as the best native inoculant AM fungi to improve mycorrhizal colonization in maize root under water stress alone and/or in combination with *Talaromyces pinophilus* MR107. The synergistically efects on this parameter wer also observed between AM fungi with other beneficial microorganisms such as *Thiobacillus* [65], *Bradyrhiobium* [66], *Paenibacillus mucilaginosus* [67], and yeast [68].

3.4. P content

This study indicated that phosphorus in maize under water stress was significantly affected by interaction between native AM and cellulolytic fungi (Figure 2). As at the mycorrhizal colonization, *Gi.* cf *gigantea* was superior to *A. tuberculata* on P content (Figure 2a). The highest of plant P content was shown at co-inoculation *Gi.* cf *gigantea* and *T. pinophilus* strain MR107 while its'coinoculation with the other cellulolytic fungi has similar effect to the cellulolytic-uninoculated treatment. Maize inoculated with *A. tuberculata* alone had significant higher P content in its tissue than after being coinoculated by each cellulolytic fungi. Inoculation of both cellulolytic fungi at uninoculated plant by AM fungi indicated similar effect on P content.

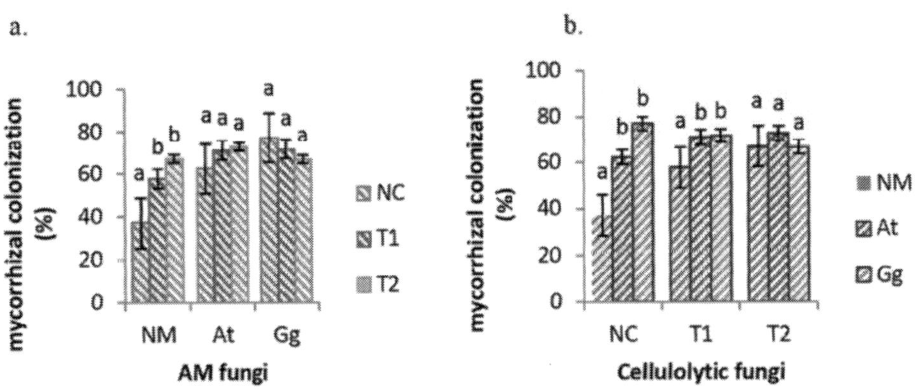

Figure 1. Effect of AM fungi inoculation at different cellulolytic fungi inoculation (a) and effect of cellulolytic fungi at different AM fungi (b) on mycorrhizal colonization of maize root. Means followed by different small letters above the bars (represent standard error) indicate a significant difference at P = 0.05. NM, non mycorrhiza; At, *A. tuberculata*; Gg, *Gi.* cf *gigantea*; NC, non cellulolytic fungi; T1, *T. pinophilus* strain MR107 and T2, *T. pinophilus* isolate OK3SP103P

P content of maize inoculated *T. pinophilus* strain MR107 and *Gi.* cf *gigantea* had highest value and this value was significantly different from other AM fungi inoculation treatment in the celluloytic fungi treatments. *T. pinophilus* isolate OK3SP103P also showed better cooperation with *Gi.* cf *gigantea* to improve P content.while *A. tuberculata* increased significantly P content in cellulolytic-uninoculated plan

AM contribution to plant P uptake has been long studied but there was a limited reports about their interaction with cellulolytic fungi. Many studies revealed improvement leaf P concentration due to the presence of AM fungi [21] and different AM fungi varied in their potential on nutrient acquisition [60]. Improved P uptake and leaf P concentration by AM fungi under water shortage has been postulated as a primary mechanism for the enhanced drought tolerance of the host plants.

This study showed that interaction of native AM and cellulolytic fungi was beneficial ($p \leq 0.05$) for the P status of maize plants. *Gi. cf gigantea* showed its best effect on P content of maize after being coinoculating with cellulolytic inoculant *T. pinophilus* MR107. The positive effect of interaction between these microorganisms was also reported by Medina [51]. The ability of both microorganisms to improve P uptake was related to their potency as phosphate solubilizer. They secrete phosphatases and several organic acids having ability to solubilize insoluble phosphate.

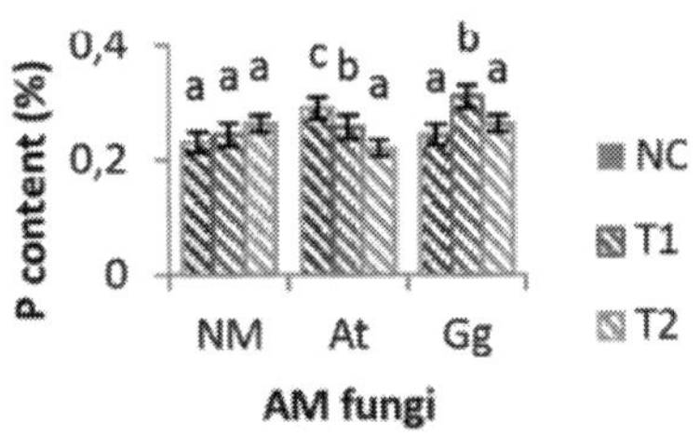

(a)

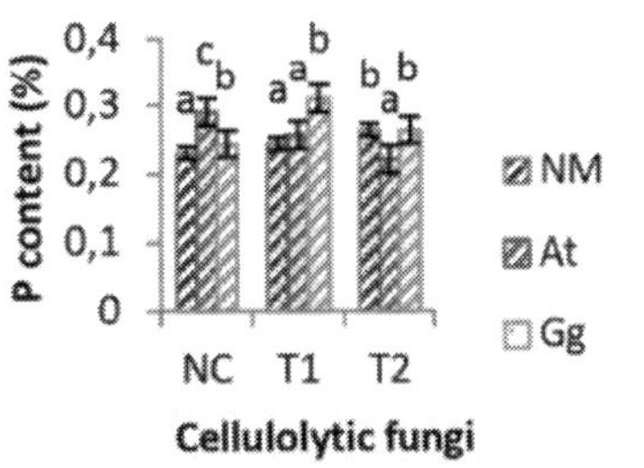

(b)

Figure 2. Effect of AM fungi inoculation at different cellulolytic fungi inoculation (a) and effect of cellulolytic fungi at different AM fungi (b) on P content of maize leaf. Means followed by different small letters above the bars (represent standard error) indicate a significant difference at $P = 0.05$. NM, non mycorrhiza; At, *A. tuberculata*; Gg, *Gi. cf gigantea*; NC, non cellulolytic fungi; T1, *T. pinophilus* strain MR107 and T2, *T. pinophilus* isolate OK3SP103P

4. Conclusions

Coinoculation of native AM and cellulolytic fungi to manage water stress could improve mycorrhizal colonization and P content in maize. Increasing plant height and diameter, mycorrhizal colonization, and P content and decreasing proline content were significantly affected either by native AM or cellulolytic fungi, but significantly increased of dry weight of 100-grain was only by native cellulolytic fungi. *Gi.* cf. *gigantea* and *T. pinophilus* strain MR107 were more potential native inoculants to eliminate negative effect of water stress on maize.

Acknowledgements

This work received financial support from The Ministries of Research, Technology, and Higher Education of Republic Indonesia under scheme of Doctoral Dissertation Research No. 042.061.401516/2018 and from Australian Center for International Agricultural Research (ACIAR) PROJECT SMCN 2012/103.

References

[1] Alexandratos N, Bruinsma J 2012 World Agriculture Towards 2030/2050: The 2012 revision. ESA Working paper No. 12-03. Rome, FAO

[2] Silva H, Sagardia, S, Seguel O, Torres C, Tapia C, Franck N, Cardemil L 2010 Effect of Water Availability on Growth and Water Use Efficiency for Biomass and Gel Production in Aloe Vera (*Aloe barbadensis* M.) *Industrial Crops and Products* **31**, 20–27

[4] Aliche EB, Oortwijn M, Theeuwen TPJM, Bachem CWB, Visser RGF, van der Linden CG 2018 Drought Response in Field Grown Potatoes and the Interactions Between Canopy Growth and Yield. *Agricultural Water Management* **206** 20–30

[5] Korres NE, Norsworthy JK, Burgos NR, Oosterhuis DM 2017 Temperature and Drought Impacts on Rice Production: An Agronomic Perspective Regarding Short- and Long-Term Adaptation Measures *Water resources and rural development* **9** 12-27

[6] Foyer CH, Lam HM, Nguyen HT, Siddique KHM, Varshney RK, Colmer TD, Cowling W, Bramley H, Mori TA, Hodgson JM, Cooper, J.W., Miller, A.J., Kunert, K., Vorster, J, Cullis C, Ozga JA, Wahlqvist ML, Liang Y, Shou H, Shi K, Yu J, Fodor N, Kaiser BN, Wong FL, Valliyodan B, Considine MJ 2016 Neglecting Legumes has Compromised Human Health and Sustainable Food Production *Nat. Plants*, 16112. http://dx.doi.org/10.1038/nplants.2016.112

[7] Shinohara T, Leskovar DI 2014 Effects of ABA, Antitranspirants, Heat and Drought Stress on Plant Growth, Physiology and Water Status of Artichoke Transplants. *Sci Hortic.* **165**:225-234

[8] Eppel A, Rachmilevitch S 2016 Photosynthesis and Photoprotection under Drought in the Annual Desert Plant *Anastatica hierochuntica* *Photosynthetica* **54 (1)** 143-147

[9] Moles TM, Mariotti L, De Pedro LF, Guglielminetti L, Picciarelli P, Scartazza A 2018 Drought Induced Changes of Leaf-to-Root Relationships in Two Tomato Genotypes. *Plant Physiology and Biochemistry* **128** 24–31

[10] Xiong JL, Li J, Wang HC, Zhang CL, Naeem MS 2018 Fullerol Improves Seed Germination, Biomass Accumulation, Photosynthesis and Antioxidant System in *Brassica napus* L. under Water Stress *Plant Physiology and Biochemistry* **129**, 130–140

[11] Zhang C, Zhan DX, Luo HH. Zhang YL, Zhang WF 2016 Photorespiration and Photoinhibition in the Bracts of Cotton under Water Stress *Photosynthetica* **54 (1)** 12-18

[12] Verma NK, Pandey BK, Singh UP, Lodhi MD 2012 Effect of Sowing Dates in relation to Integrated Nitrogen Management on Growth, Yield and Quality of Rabi Maize (*Zea mays* L.) The Journal of Animal & Plant Sciences **22(2)** 324–329

[13] M Etesami, H Alikhani, HA 2018 Improvement of Growth and Yield of Maize under Water Stress by Co-Inoculating an Arbuscular Mycorrhizal Fungus and a Plant Growth Promoting Rhizobacterium together with Phosphate Fertilizers. Agriculture *Ecosystems and Environment* **258,** 59–70

[14] Shaw AK, Bhardwaj PK, Ghosh S, Roy S, Saha S, Sherpa AR, Saha SK, Hossain Z 2016 ß-Aminobutyric Acid Mediated Drought Stress Alleviation in Maize (*Zea mays* L.) *Environ Sci Pollut Res* **23** 2437–2453

[15] Ziyomo C, Bernardo R 2013 Drought Tolerance in Maize: Indirect Selection through Secondary Traits Versus Genomewide Selection. *Crop Sci* **53** 1269–1275

[16] Oyekale KO, Daniel IO, Kamara AY, Akintobi, DCA, Adegbite AE., Ajala, M.O. 2008. Evaluation of Tropical Maize Hybrids under Drought Stress. *Journal of Food, Agriculture & Environment* **6(2)** 260-264

[17] Fikrinda F, Syafruddin, Sufardi dan R Sriwati 2016 Keanekaragaman Fungi Mikoriza Arbuskula di Rizosfer Beberapa Varietas Jagung pada Inseptisol. In: SME Susilowati, karman, Y Vitner, DJ Selanno, F Yacob, Mursito SB, R Koneri, dan Mudatsir (Eds.). Prosiding Seminar Nasional Biotik 2016. Pendidikan Berbasis Bio-Enterpreneurship dan Pelestarian Sumber Daya Alam Hayati dalam Mempersiapkan Sumber Daya Manusia pada Era Masyarakat Ekonomi Asean (MEA). FTK Ar-Raniry Press, Banda Aceh, pp. 259-265

[18] Sangabriel-Conde, W, Maldonado-Mendoza IE, Mancera-López ME, Cordero-Ramírez JD, Trejo-Aguilar D, Negrete-Yankelevich S 2015 Glomeromycota Associated with Mexican Native Maize Landraces in Los Tuxtlas, Mexico. *Applied Soil Ecology* **87** 63–71

[19] Gholamhoseini M, Ghalavand A, Dolatabadian A, Jamshidi E, Khodaei-Joghan, A 2013 Effects of Arbuscular Mycorrhizal Inoculation on Growth, Yield, Nutrient Uptake and Irrigation Water

Productivity of Sunflowers Grown under Drought Stress *Agricultural Water Management* **117** 106-114
[20] Püschel D. Rydlová J, Vosátka M 2014 Can Mycorrhizal Inoculation Stimulate the Growth and Flowering of Peat-Grown Ornamnetal Plants under Standard or Reduced Watering. *Applied Soil Ecology* **80** 93-99
[21] Shi ZY, Mickan B, Feng G, Chen YL 2015 Arbuscular Mycorrhizal Fungi Improved Plant Growth and Nutrient Acquisition of Desert Ephemeral *Plantago minuta* under Variable Soil Water Conditions *J Arid and* **7(3)** 414-420
[22] Mirshad PP, Puthur JT 2017 Drought Tolerance of Bioenergy Grass *Saccharum spontaneum* L. Enhanced by Arbuscular Mycorrhizae. *Rhizosphere* **3** 1-8
[23] Li T, Lin G, Zhang X, Chen Y, Zhang S, Hen B 2014 Relative Importance of an Arbucular Mycorrhizal Fungus *Rhizophagus intraradices* and Root Hairs in Plant Drought Tolerance *Mycorrhiza* DOI 10.1007/s00572014-0578-3
[24] Petipas RH, González JB, Palmer TM, Brody AK 2017 Habitat-Specific AMF Symbioses Enhance Drought Tolerance of a Native Kenyan Grass *Acta Oecologica* **78** 71-78
[25] Zhang F, Zou YN, Wu QS 2018 Quantitative Estimation of Water Uptake by Mycorrhizal Extraradical Hyphae in Citrus under Drought Stress *Scientia Horticulturae* **229** 132–136
[26] Bitterlich M, Sandmann M, Graefe J 2018 Arbuscular Mycorrhiza Alleviates Restrictions to Substrate Water Flow and Delays Transpiration Imitation to Stronger Drought in Tomato *Frontiers in Plant Science* **9** 1-15
[27] Xu L, Li T, Wu Z, Feng H, Yu M, Zhang X, Chen B 2018 Arbuscular Mycorrhiza Enhances Drought Tolerance of Tomato Plants by Regulating the 14-3-3 Genes in the ABA Signaling Pathway *Applied Soil Ecology* **125** 213–221
[28] Martin SL, Mooney SJ, Dickinson MJ, West HM 2012 The Effects of Simultaneous Root Colonisation by Three *Glomus* Species on Soil Pore Characteristics *Soil Biology & Biochemistry 49* 167-173
[29] Bolandnazar S 2009 The Effect of Mycorrhizal Fungi on Onion (*Allium cepa* L.) Growth and Yield under Three Irrigation Intervals at Field Condition. *Journal of Food, Agriculture & Environment* **Vol.7 (2)** 360-362
[30] Mehraban A, Vazan S, Naroui Rad, M R, Ardakany AR 2009 Effect of Vesicular-Arbuscular Mycorrhiza (VAM) on Yield of Sorghum Cultivars *Journal of Food, Agriculture & Environment* **Vol.7 (3&4)** 461-463
[31] Meddich A, Jaiti F, Bourzik W, El Asli A, Hafidi M 2015 Use of Mycorrhizal Fungi as a Strategy for Improving the Drought Tolerance in Date Palm (*Phoenix dactylifera Scientia Horticulturae* **192** 468–474
[32] Alguacil MM, Caravaca F, Azcón R, Pera J, Díaz G, Roldán A 2003 Improvements in Soil Quality and Performance of Mycorrhizal *Cistus albidus* L. Seedlings Resulting from Addition of Microbially Treated Sugar Beet Residue to A Degraded Semiarid Mediterranean Soil *Soil Use Manage* **19** 277–283
[33] Kavamura VN, Santos SN, da Silva JL, Parma MM, Ávila LA, Visconti A, Zucchi TD, Taketani RG, Andreote FD, de Melo IS 2013 Screening of Brazilian Cacti Rhizobacteria for Plant Growth Promotion under Drought *Microbiological Research* **168** 183–191
[34] Séry D, Kouadjo Z, Voko DRR, Zézé A 2016 Selecting Native Arbuscular Mycorrhizal Fungi to Promote Cassava Growth and Increase Yield under Feld Conditions *Front. Microbiol*. 1–13.
[35] Hailemariam M, Birhane E, Asfaw Z, Zewdie S 2013 Arbuscular Mycorrhizal Association of Indigenous Agroforestry Tree Species and Their Infective Potential with Maize in the Rift Valley, Ethiopia *Agroforest Syst* **87** 1261-1272
[36] Armada E, Barea JM, Castillo P, Roldán A, Azcón R 2015 Characterization and Management of Autochthonous Bacterial Strains from Semiarid Soils of Spain and Their Interactions with Fermented Agrowastes to Improve Drought Tolerance in Native Shrub Species *Applied Soil Ecology* 96, 306–318
[37] Arpana J, Bagyaraj DJ, Prakasa Rao EVS, Parameswaran TN, Abdul Rahiman B, 2014 Response of Patchouli to Inoculation with an Arbuscular Mycorrhizal Fungus and Plant Growth Promoting Rhizomicroorganisms *J. Soil. Biol. Ecol.* **34** 70–83.
[38] Bagyaraj DJ 2014 Interaction between Arbuscular Mycorrhizal Fungi and The Soil Organisms and Their Role in Sustainable Agriculture In: Singh, D.P., Singh, H.B. (Eds.) Trends in Soil Microbial Ecology. Studium Press LLC, USA, pp. 257–280

[39] Alwhibi MS, Hashem A, Abd_Allah E F, Alqarawi AA, Soliman D W K, Wirth S, Egamberdieva D 2017 Increased Resistance of Drought by *Trichoderma harzianum* Fungal Treatment Correlates with Increased Secondary Metabolites and Proline Content *Journal of Integrative Agriculture* **16(8)** 1751–1757

[40] Charoenrak P, Chamswarng C 2016 Efficacies of Wettable Pellet and Fresh Culture of *Trichoderma asperellum* Biocontrol Products in Growth Promoting and Reducing Dirty Panicles of Rice *Agriculture and Natural Resources* **50** 243-249

[41] NihiraT, Saito Y, Nishimoto M, Kitaoka M, Igarashi K, Ohtsubo K, Nakai H 2013. Discovery of Cellobionic Acid Phosphorylase in Cellulolytic Bacteria and Fungi *FEBS Letters* **587** 3556–3561

[42] Kishishita S, Fujii T, Ishikawa K 2015 Heterologous Expression of Hyperthermophilic Cellulases of Archaea *Pyrococcus* sp. by Fungus *Talaromyces cellulolyticus J Ind Microbiol Biotechnol* **42** 137–141

[43] Sawant VS, Bansode KG, Bavachkar SN, Bhale UN 2013 Potential of Various Fungi for Biomass Production of Castor *Pak.J.Biol.Sc* **16(21)** 378-1382

[44] Martínez-Medina A, Roldán A, Pascual JA 2011 Interaction between Arbuscular Mycorrhizal Fungi and *Trichoderma harzianum* under Conventional and Low Input Fertilization Feld Condition in Melon Crops: Growth Response and *Fusarium* Wilt Biocontrol *Applied Soil Ecology* **47** 98–105

[45] Green H, Larsen J, Olson PA, Jensen DF, Jakobsen I 1999 Supreession of the Biocontrol agent *Trichoderma harzianum* by Mycellium of Arbuscular Mycorrhizal Fungus *Glomus intraradices* in Rroot-Free soil *Appl. Envir. Microbiol* **65** 1428-1434

[46] Meddich A, Jaiti F, Bourzik W, El Asli A, Hafidi M 2015 Use of Mycorrhizal Fungi as a Strategy for Improving the Drought Tolerance in Date Palm (*Phoenix dactylifera Scientia Horticulturae* **192** 468–474

[47] Zarik L, Meddich A, Hijri M, Hafidi M, Ouhammou A, Ouahmane L, Duponnois R, Boumezzough A 2016 Use of Arbuscular Mycorrhizal Fungi to Improve the Drought Tolerance of *Cupressus atlantica* G. C.R *Biologies* **339(5-6)** 185-196

[48] Alahdadi I, Tajik M, Iran-nejad H, Armandpisheh O 2009 The Effect of Biofertilizer on Soybean Seed Vigor and Field Emergence *Journal of Food, Agriculture & Environment* **7 (3&4)** 420-426

[49] M, Naghavi MR, Aboughadareh AP, Rad HN 2013 Effects of Drought Stress on Yield and Yield Components in Maize Cultivars. *International Journal of Agronomy and Plant Production* **Vol 4(4)** 809-812

[50] Cozzolino V, Meo DM, Piccolo A 2013 Impact of Arbuscular Mycorrhizal Fungi Applications on Maize Production and Soil Phosphorus Availability *Journal of Geochemical Exploration* **129**, 40–44

[51] Medina A, Roldán A, Azcón R 2010 The Effectiveness of Arbuscular-Mycorrhizal Fungi and *Aspergillus niger* or *Phanerochaete chrysosporium* Treated Organic Amendments from Olive Residues upon Plant Growth in a Semi-Arid Degraded Soil *Journal of Environmental Management* **91**, 2547-2553

[52] Olawuyi OJ, Odebode AC, Babalola BJ, Afolayan ET, Onu CP 2014 Potentials of Arbuscular Mycorrhiza Fungus in Tolerating Drought in Maize (*Zea mays* L.). *American Journal of Plant Sciences* **5** 779-786

[53] Anozie H I, Orluchukwu JA 2018 Response of Three Varieties of Maize (*Zea mays*) to Arbuscular Mycorrhizal Fungi (*Gigaspora Gigantea*) in the Humid Tropics. International Journal of Agriculture and Earth Science **4(3)** 30-36

[54] Hart MH and Reader RJ 2002 Taxonomic Basis for Variation in the Colonization Strategy of Arbuscular Mycorrhizal Fungi. *New Phytologist* **153** 335–344

[55] Phitsuwan P, Laohakunjit N, Kerdchoechuen O, Kyu KL, Ratanakhanokhai K 2013 Present and Potential Applications of Cellulases in Agrigulture, Biotechnology, and Bioenergy *Folia Microbiol* **58** 163-176

[56] Chepsergon J, Mwamburi LA, Kiprop KE 2018 Augmented Maize Seed Germination and Seedling Growth under Water Stress Using *Trichoderma harzianum* from Semi-Arid Soils *Plant Pathology & Quarantine* **8(1)** 37–46

[57] Alcántara C, Thornton C R, Pérez-de-Luque A, Le Cocq L, Pedraza V, Murray P J 2016. The Free-Living Rhizosphere Fungus *Trichoderma hamatum* GD12 Enhances Clover Productivity in Clover-Ryegrass Mixtures *Plant Soil* **398** 165–180

[58] Bhosale, K.S., Shinde, B.P. 2011. Influence of Arbuscular Mycorrhizal Fungi on Prline and Chlorophyll Content in *Zingiber officinale* Rose Grown under Water Stress. Indian Journal of Fundamental and Applied Life ScienceVol 1(3), 172-176

[59] Yooyongwech S, Phaukinsang N, Cha-um S, Supaibulwatana K 2014. Arbuscular Mycorrhiza Improved Growth Performance in *Acadamia tetraphylla* L. Grown under Water Deficit Stress Involves Soluble Sugar and Proline Accumulation. *Plant Growth Regul* **69** 285-293

[60] Zardak SG, Dehnavi MM, Salehi A, Gholamhoseini M 2018 Effects of Using Arbuscular Mycorrhizal Fungi to Alleviate Drought Stress on the Physiological Traits and Essential Oil Yield of Fennel *Rhizosphere* **6** 31–38

[61] Zhang Z, Zhang J, Huang Y 2014 Effects of Arbuscular Mycorrhizal Fungi on the Drought Tolerance of *Cyclobalanopsisglauca* Seedlings under Greenhouse Conditions *New Forests* **45** 545-556

[62] Song H, Li Y, Zhou L, Xu Z, Zhou G 2018 Maize Leaf Functional Responses to Drought Episode and Rewatering *Agricult. For. Meteorol* **249** 57–70.

[63] Fukutaku Y, Yamada Y (1984) Source of Proline Nitrogen in Water-Stressed Soybean (*Glycine max*) II. Fate of 15 N-Labelled Protein *Physiol. Plant* **61** 622-628

[64] Rydlová J,S ýkorová Z, Slaviková R 2015 The Importance of Arbuscular Mycorrhiza for *Cyclamen purpurascens* subsp. *immaculatum* Endemic in Slovakia *Mycorrhiza Vol . 25* Issue 8, 599-609

[65] Mohamed AA, Eweda WEE, Heggo AM, Hassan E A 2014 Effect of Dual Inoculation with Arbuscular Mycorrhizal Fungi and Sulphur-Oxidising Bacteria on Onion (*Allium cepa* L.) and Maize (*Zea mays* L.) Grown in Sandy Soil under Green House Conditions *Annals of Agricultural Science* **59(1)** 109–118

[66] Omirou M, Fasoula DA, Ioannides IM 2016 *Bradyrhizobium* Inoculation Alters Indigenous AMF Community Assemblages and Interacts Positively with AMF Inoculum toImprove Cowpea Performance *Applied Soil Ecology* **108** 381–389

[67] Wang P. Wu SH, Wen M-X, Wang Y, Wu Q-S 2016 Effects of Combined Inoculation with *Rhizophagus intraradices* and *Paenibacillus mucilaginosus* on Plant Growth, Root Morphology, and Physiological Status of Trifoliate Orange (*Poncirus) trifoliata* L. Raf.) Seedlings under Different Levels of Phosphorus. *Scientia Horticulturae* **205** 97–105

[68] Sarabia M, Jakobsen I, Grønlund M, Carreon-Abud Y, Larsen J 2018. Rhizosphere Yeasts Improve P Uptake of a Maize Arbuscular Mycorrhizal Association. *Applied Soil Ecology* **125** 18-25

AUTHOR INDEX

AUTHOR INDEX

AUTHOR INDEX